# Laboratory Manual
## for Anatomy
## and Physiology

# Laboratory Manual for Anatomy and Physiology

## *WITH CAT DISSECTIONS*

### FOURTH EDITION

**Patricia J. Donnelly**
**George A. Wistreich**

*East Los Angeles College*

HarperCollinsCollegePublishers

Acquisitions Editor: Bonnie Roesch
Developmental Editor: Meryl R. G. Muskin
Project Coordination and Text Design: York Production Services
Cover Design: Kay Petronio
Cover Photo: © 1989 Howard Sochurek, The Stock Market
Compositor: Modern Graphics Inc.
Printer and Binder: Von Hoffmann Press
Cover Printer: The Lehigh Press, Inc.

An attempt has been made to obtain permission from all suppliers of photographs used in this edition. Some sources have not been located, but permission will be requested from them upon notification to us of their ownership of the material. All material without a credit line is from *Laboratory Manual for Anatomy and Physiology: With Cat Dissections,* Third Edition.
LABORATORY MANUAL FOR ANATOMY AND PHYSIOLOGY: With Cat Dissections, Fourth Edition

Library of Congress Cataloging-in-Publication Data

Donnelly, Patricia J.
    Laboratory manual for anatomy and physiology : with cat dissections / Patricia J. Donnelly, George A. Wistreich.—4th ed.
        p.    cm.
    Includes bibliographical references and index.
    ISBN 0-06-500903-7
    1. Physiology, Experimental—Laboratory manuals.   2. Cats—Physiology—Laboratory manuals.   3. Cats—Anatomy—Laboratory manuals.   4. Dissection—Laboratory manuals. I. Wistreich, George
A.  II.  Title.
QP44.D66   1993
599′.01′078—dc20                                                                92-44384
                                                                                              CIP

93 94 95 96 9 8 7 6 5 4 3 2

# CONTENTS

## SECTION XIV

**The Urinary and Reproductive Systems** 663

### EXERCISE

## APPENDICES

# PREFACE

T he purpose of *Laboratory Manual for Anatomy and Physiology: With Cat Dissections,* fourth edition, is to provide students with functional and effective experiences that will enable them not only to observe but also to understand fundamental principles and concepts of human anatomy and physiology. The exercises in this manual are organized into complete units of instruction so that students study each organ system and observe and recognize fundamental similarities and differences among the systems.

As the manual's subtitle indicates, we have used the cat as the primary dissection animal. Tables in the exercises on the muscular system and certain other systems relate information learned in the dissection procedure on the cat to human anatomy. The choice of the cat as a dissection specimen was made for two reasons: first, human cadavers are difficult to obtain and maintain; second, we feel that a student's own dissection has more instructional value than mere observation of demonstration material. In designing the physiology experiments we kept in mind the need to provide modern and functional approaches to demonstrate physiological concepts and techniques. The experiments and studies use relatively new, inexpensive, and easily obtainable specimens and laboratory setups. It is our feeling that basic principles of physiology can be demonstrated with simple techniques and a minimum of complex equipment.

## ORGANIZATION OF THE MANUAL

The 52 laboratory exercises in this manual include a variety of special features that are designed to stimulate interest in the subject matter, to involve students in the learning process, and to guide them through specifically planned activities.

The manual offers a broad range of exercises from which instructors may choose. It may be used for a one- or two-semester course in anatomy and physiology at the undergraduate level. Some of the exercises presume a knowledge of basic high school biology and chemistry. Although the manual was designed for courses related to allied health curricula, it can be used equally well in courses for general mammalian anatomy and physiology.

The 52 exercises are arranged in 14 sections as follows:

*Section I* introduces the cellular and tissue levels of organization. Exercises are designed to show the features of cellular structure and function and the chemical and physical aspects of cellular permeability. This section begins with a functional exercise emphasizing the proper use and care of the microscope.

*Section II* surveys anatomical orientation and terminology and serves as an introduction to many terms used in subsequent exercises.

*Section III* demonstrates the general features of the skin, nails, and hair. Particular attention is given to fingerprints and the general appearance of blood vessels in response to different temperatures.

*Section IV* familiarizes students with the structure and functions of the skeletal system. Exercises in this section provide many color drawings of the support system in the human body.

*Section V* contains numerous exercises designed to demonstrate the structure and properties of mammalian muscle. The cat is the laboratory animal, and there are black-and-white photographs with accompanying line drawings and color photographs to assist the students in their dissections. Other exercises cover the structure of muscle and the chemistry of muscular contraction.

*Section VI* provides students with a comparative and functional treatment of nervous system coordination and control. The anatomy of the human system is stressed and is used to demonstrate cranial and spinal reflex responses. The sheep and cat brains are dissected.

*Section VII* provides an overall view of the human endocrine system. The interrelationships, locations, and functions of the glands are covered in sufficient detail to emphasize the importance and special properties of the endocrine system.

*Section VIII* contains numerous simple experiments that emphasize the importance and special properties of the senses.

*Section IX* familiarizes students with basic dissection procedures and points out the organ systems that can easily be seen without more extensive dissection.

*Section X* presents the major anatomy and physiology of the cardiovascular system. There are exercises on the heart, lymphatic system, and blood structure. An exercise demonstrating the clinical features of coronary circulation is also included.

*Section XI* presents coverage of the components of blood, the immune system, and various aspects of diagnostic hematology, including factors involved with coagulation, hemoglobin determinations, blood typing, cross-matching, and the immune response.

*Section XII* familiarizes students with the basic anatomical and physiological features of the respiratory system. Procedures for conducting pulmonary function tests are incorporated in this section.

*Section XIII* treats the structure of the digestive system. Exercises dealing with enzymatic action, digestion chemistry, teeth, and cholesterol testing are important parts of this section.

*Section XIV* deals with the anatomy and physiological features of both the urinary and reproductive systems. It includes an examination of the urogenital organs and demonstrates the principles associated with dialysis, fluid and electrolyte balance, and acid–base balance. An exercise dealing with routine diagnostic urinalysis is contained within this portion of the manual. In addition, attention is given to gamete formation, ovulation prediction, and pregnancy testing.

## LEARNING AIDS

A number of learning aids have been incorporated to make this manual an effective tool for students. These include the following:

*Section overviews.* Each major section of the laboratory manual is preceded by a description of the range of topics, exercises, and experiments to be covered. A list of general references is included to provide background information and a basis for further study.

*Illustrations.* Photographs and diagrams have been included to provide a greater understanding of anatomical relationships and physiological principles, techniques, and test results. Photographs showing important structures, reactions, and results are used throughout the manual. Special sections containing color photographs of cat muscles and histology are presented to simplify identification.

*Procedure diagrams.* Exercises that introduce basic techniques contain step-by-step illustrations of the procedure to show the proper handling of equipment, animals for dissection, and related items.

*Laboratory review.* At the end of each exercise there are a number of selected completion, multiple-choice, matching, and other forms of questions to test students' comprehension of the major principles and results of the exercises performed. In addition, this section encourages students to think critically.

*Index.* An index has been included in the manual. Students should consult this index to facilitate location of anatomical parts, tests, and procedures.

## INSTRUCTOR'S MANUAL

An instructor's manual to accompany the laboratory manual is available. It contains specific directions for the preparation of laboratory materials and descriptions of alternative procedures and the use of supplementary aids. There are suggested sources for audiovisual aids, laboratory equipment and supplies, and questions for laboratory examinations and quizzes.

## ACKNOWLEDGMENTS

We thank the following colleagues who read and reviewed the manuscript versions of our work and whose comments we found constructive:

Judi Brown, Olympic College

David Evans, Pennsylvania College of Technology

Mark Shoop, Macon College

Eric Sun, Macon College

The authors also gratefully acknowledge the Schleicher & Schnell Corporation; IsoLab, Inc.; ChemTrak Corporation, and the OvuGen Corporation for providing samples of their products for use in developing this manual.

Special thanks to College Biological Supply Company, Escondido, California, for supplying the dissection specimens: Dr. Joel Schechter of the Department of Anatomy, University of Southern California Medical School, and Leonard D. Dank, President of Medical Illustrators, Inc., for art work; and Dr. Daniel Fertig of East Los Angeles College and Richard Holdredge for photographic assistance.

We are also deeply grateful to the members of the editorial and production staffs of HarperCollins Publishers.

P. J. Donnelly
G. A. Wistreich

# TO THE STUDENT

## LABORATORY OPERATING PROCEDURE

In the laboratory you have an opportunity to study and apply anatomical and physiological concepts and principles at close range and to see for yourself the details of structures, processes, and relationships characteristic of living systems, Your observations should provide the answers to most questions that arise concerning the material under study. Your laboratory instructor will offer helpful suggestions, but will not be a source of easy answers for unprepared students. The following rules should assist you in getting the most out of laboratory sessions.

## LABORATORY PROCEDURES AND REGULATIONS

1. Laboratory work begins promptly at the time designated in the schedule.
2. Laboratory demonstrations, which include general directions designed to make your work easier and more efficient, usually will be given at the beginning of laboratory sessions.
3. You will be assigned a microscope, a laboratory drawer with other necessary equipment, and a laboratory work space for the duration of the class. It is your responsibility to keep all equipment issued to you in good working condition. Report damaged or missing items to your instructor immediately. You are expected to read each laboratory exercise and carry out suggested reading assignments before you come to class. The instructor will designate in advance which parts of each exercise will be undertaken during each class period.
4. Try to acquire self-reliance in performing procedures. Read all directions carefully and follow them closely. You may find it helpful to underline procedural steps. Learn to observe your own work critically so that you can correct and improve your performance as you do each procedure. However, if you have any doubts about any portion of the exercise, ask your instructor for help.
5. Place items such as extra textbooks, coats, and other personal belongings in the area specified by your instructor. Never let such items clutter your operating space.
6. In exercises requiring the use of chemicals, take only the quantity of materials you need.
7. Leave all laboratory facilities and equipment in good order at the close of each period.
8. Conserve electricity, water, and gas. Turn off water faucets and gas burners when they are not needed.

## LABORATORY RECORDS

1. Your lab notebook is a valuable record. Keep it up to date, using procedures outlined by your instructor.
2. As you perform your experiments, record your data. Always perform exercises in the sequence assigned.

3. The illustrations in this manual suggest the way drawings should be made, spaced, and labeled. Try to draw plain, clear-cut outlines of structures and other items studied. Use pencil. Coloring or shading will be specified as needed.

4. The labeled drawings supplied in this manual are important objects of study and should be used as guides in locating anatomical parts and as an aid in recalling concepts and principles.

## GENERAL LABORATORY SAFETY PRECAUTIONS

1. Your instructor will show you the location and demonstrate the use of emergency equipment such as fire extinguishers, fire blankets, and first-aid kits. Know their locations and how to use them.

2. Never eat, store food, or smoke in the laboratory.

3. To prevent damage to street clothes, always wear a laboratory coat or apron during the class period. Remove this protective apparel before leaving the laboratory.

4. Tie back long hair to prevent it from being a laboratory fire hazard.

5. Double-check the label of any chemical you use.

6. When diluting acids, always add acid (AAA) to water.

7. Never heat an inflammable liquid over an open flame.

8. When heating materials in a test tube over an open flame, never point the open end toward anyone. Hold the test tube at an angle, not upright, over the flame, heating the sides and bottom. Keep moving the test tube so it does not overheat. Use only glassware marked Pyrex or Kimex. Other glassware will shatter when heated.

9. Do not inhale the fumes of any substance directly. When required to note the odor of a chemical, first take a deep breath of fresh air and exhale normally. Then with your hand waft the vapors toward your nose and sniff slowly.

10. Do not pipette chemicals by mouth.

11. Do not use your fingers as a stopper for a tube or a bottle when shaking the tube to mix its contents.

12. Never eat, smoke, drink, taste chemicals or lick labels in the laboratory. DO NOT BRING FOOD TO THE LABORATORY.

13. If you get a chemical on your skin or in your eyes, immediately flood with water and call for help.

14. Report any fire or personal injury such as a cut or burn to your instructor immediately.

15. Never remove equipment, solutions, or other materials associated with the exercise from the laboratory.

16. Dispose of chemicals, biological materials, and paper as indicated by your instructor. Do not put paper or dissected materials in the sink.

17. In cleaning dissection instruments, always hold sharp edges away from you.

18. Wash your hands thoroughly with detergent and water before leaving the laboratory. Carry out this procedure even in cases of coffee and health breaks.

19. If you are in doubt about a personal health problem, always ask your instructor first before you attempt to do anything.

## PRECAUTIONS FOR THE HANDLING OF BLOOD OR OTHER BODY FLUIDS

The *Centers for Disease Control* has developed a series of guidelines for health care workers in order to prevent the transmission of human immunodeficiency viruses (HIVs) as well as other blood-related infectious diseases such as hepatitis

B virus infection. The guidelines in the form of laboratory precautions and procedures are clearly specified in the appropriate exercises of this manual. Following the specific steps will enable you to work with body fluids safely and effectively and without any danger to your safety and well being.

## MATERIALS TO BE PROVIDED BY THE STUDENT

1. One clean laboratory coat or other type of protective apparel
2. Textbook and laboratory manual
3. One 4H drawing pencil
4. One eraser (not art gum)
5. Colored pencils—red, blue, yellow, and green
6. One wax marking pencil or a washable-ink felt pen
7. One package of glass slides and cover slips
8. One package of lens paper
9. Two packages of matches
10. One clean, lint-free, soft cloth
11. Slide box (optional)
12. Surgical or rubber gloves for protection against stains and contact with chemicals
13. Safety goggles or glasses (optional)
14. One dissection kit or individual instruments to include:
    a. One pair of coarse scissors
    b. One pair of fine forceps
    c. One scalpel with at least three disposable blades
    d. Two dissecting needles

## PRONUNCIATION GUIDE

In the exercises relating to bones and muscles, you will find *phonetic pronunciations* for some anatomical terms. Taking time to sound out new terms and to say them aloud once or twice will help you master one of the tasks of the anatomy and physiology course—learning its specialized vocabulary. The following key explains the system used for the pronunciations.

1. The strongest accented syllable appears in capital letters (e.g., bī-LAT-er-al, dī-ag-NŌ-sis). A syllable that has a secondary accent is marked by a single quote mark (e.g., kon'-sti-TOO-shun, fiz'-ē-OL-ō-jē).
2. Vowels pronounced with long sounds are indicated by a line above the vowel and are pronounced as in the following common words.
    ā as in *māke*
    ē as in *bē*
    ī as in *īvy*
    ō as in *pōle*
3. Vowels not marked for long sounds are pronounced with the short sound, as in the following words.
    e as in *bet*
    i as in *sip*
    o as in *not*
    u as in *bud*
4. Other phonetic symbols are used to indicate the following sounds.
    a as in *above*
    soo as in *sue*
    kyoo as in *cute*
    oy as in *oil*

Through the use of this laboratory manual you should gain a deeper understanding of human anatomy and physiology. Such knowledge will demand not only a genuine effort but also an organized coordination of text reading and laboratory experimentation. A solid foundation in the basic principles of anatomy and physiology will make the effort well worthwhile.

# Microscopy, Cellular Structure, Organization, and Function

*P*hysiologic systems usually exhibit two characteristic properties: *transduction* and *translocation*. The first of these properties involves the changing of energy from one form to another, while the other property, translocation, is associated with the movement of either material or information from one location to another. Because the cell is considered to be the most basic component of a physiologic system, the properties of such units and the way in which they are organized into higher systems, such as tissues and organs, determine their overall function. Furthermore, the performance of cellular structures, as well as the coordination of their activities, helps to maintain the state of dynamic equilibrium in the body's internal environment; this is known as *homeostasis*.

Among the different tools essential for studying cells and their activities are microscopes. The primary function of a microscope is to enable the observer to distinguish structures and points that are separated by short distances. This function is *resolution* or *resolving power*.

Resolution is more important than magnification: it is not always desirable to obtain the largest image possible, but it is necessary to obtain sharp detail. Resolution is dependent on many factors, including illumination and the nature of the specimens. The resolving power of a microscope is inversely related to the wavelength of light used—that is, the shorter the wavelength, the greater the resolving power.

Two categories of microscopes in general use are the *bright-field instrument*, which contains optical lenses and uses visible light as the source of illumination, and the *electron microscope,* which contains electromagnetic lenses and uses a beam of electrons as the source of illumination. Because the electron beam has a considerably shorter wavelength than ordinary light, the resolving power and magnification obtainable with electron microscopes far exceeds those obtained with other optical systems (Table I-1). Two general types of electron microscope are currently in use: the transmission and the scanning instruments. Another

**Table I-1  Comparison of Microscope Types**

| Type | Magnification Range | Resolution (in nm) | Applications Include |
|---|---|---|---|
| Compound bright field | 10–2,000 | 300 | Used extensively for the examination of various cell types, either stained or unstained |
| Fluorescence | 10–3,000 | 200 | Employs fluorescent dyes for staining; useful in many medical diagnostic procedures for identifying cells, including cancer and microorganisms and/or their parts |
| Phase contrast | 10–1,500 | 200 | Used to examine and study cellular structures; staining of specimens is not required |
| Transmission electron (TEM) | 200–1,000,000 | 1 | Used to examine and study the ultrastructure of cells and viruses and certain biochemicals |
| Scanning electron (SEM) | 200–500,000 and greater | 5.0–10.0 | Used to examine and study surface features of cells and nonbiological materials; images are presented in three-dimensional form |
| Scanning transmission electron (STEM) | Combination of TEM and SEM | Combination of TEM and SEM | Used for applications similar to those listed for TEM and SEM; also used for analysis of chemical elements in specimens; equipped with computer for measurements and calculations |

instrument, the scanning transmission electron microscope, combines many of the properties of both instruments.

Several elaborate techniques are used in electron microscopy in preparing specimens and providing specimen contrast. Such techniques include ultrathin sectioning (slicing) and staining (Figure I-1a), the use of electron stains, and depositing the atoms of heavy metals in a vacuum onto the surfaces of specimens. The microscope used is a deciding factor in the selection of the preparation technique. The scanning electron microscope is a particularly useful instrument because it provides three-dimensional views of specimens without the need for extensive preparation techniques (Figure I-1b). Such views provide a basis for understanding complex relationships within cells, tissues, and organs and enable observers to gain a perspective as to the true locations and organization of cellular structures.

Specimens examined by electron microscopy must be viewed in a vacuum.

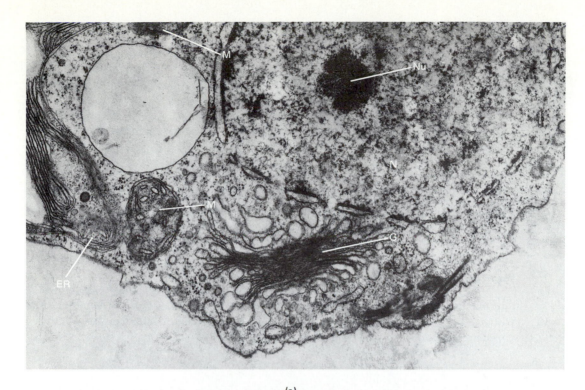

(a)

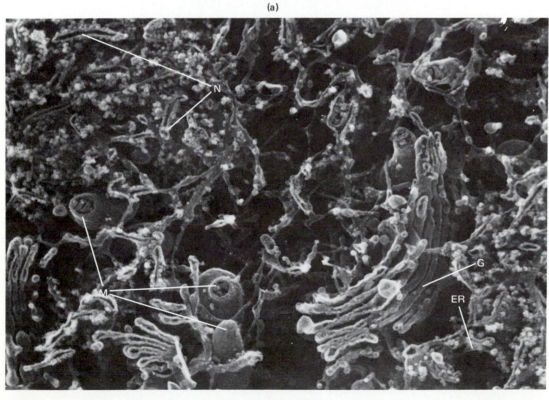

(b)

**Figure I-1(a)** Electron micrographs: (a) a transmission micrograph showing Golgi apparatus **(G)**, mitochondria **(Ms)**, endoplasmic reticulum **(ER)**, nucleus **(N)**, and nucleolus **(Nu)**; (b) a scanning micrograph providing an internal three-dimensional view of an eukaryotic cell—cellular organelles that can be seen include the endoplasmic reticulum **(ER)**, with ribosomes attached; Golgi apparatus **(G)**; nucleus **(N)**; and individual mitochondria **(M).** The original magnification is 26,000 × . (From K. Tanaka and A. Mitsushima, *Journal of Microscopy,* 133: 213–222, 1984.)

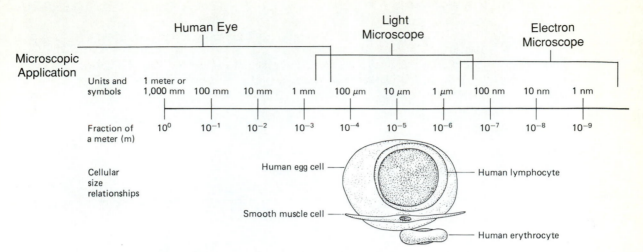

**Figure I-2** Metric units of length commonly used in the biological sciences. 1 meter = 1000 millimeters **(mm)**; 1 mm = 1000 micrometers **(μm)**; 1 μm = 1000 nanometers **(nm)**. The relative sizes of various types of cells and the respective magnification ranges of different microscopes are shown.

Furthermore, specimens viewed or photomicrographs taken of specimens lack color, which is a major means of obtaining contrast in light microscopes. Exercise 1 presents the features of the light microscope, emphasizing that it incorporates optical lenses and visible light as the source of illumination.

Sizes and shapes are important identifying properties of cells. In the biological sciences, several common metric units are used to express the dimensions of cells and their parts. The common metric units of measurement include the millimeter **(mm)**, the micrometer **(μm)**, and the nanometer **(nm)**. The relationships of the units and their application to cells are shown in Figure I-2.

Cells from different forms of life and from different regions of a single multicellular organism differ greatly in various features, including shape, size, rate of metabolic activity, and functions. While all cells share certain basic structural features, some have specialized structures called *organelles*. The exercises in this section not only emphasize numerous properties of cells and their organization but also demonstrate that they are efficient basic units of life.

### *GENERAL REFERENCES*

Burrells, W. *Microscope Techniques: A Comprehensive Handbook for General and Applied Microscopy*. New York: Halsted Press, 1976.

Cargille, J. J. *Immersion Oil and the Microscope*. Technical Reprint 10-51. Cedar Grove, NJ: R. R. Cargil Laboratories, 1975.

de Duve, C. *Blueprint for a Cell: The Nature and Origin of Life*. Burlington, NC: Neil Paterson, 1991.

DeRobertis, E. D. P., and E. M. F. DeRobertis, Jr. *Cell and Molecular Biology*, 8th ed. Philadelphia: Lea & Febiger, 1987.

Gartner, L. P., and J. L. Hiatt. *Atlas of Histology*. Baltimore: Williams & Wilkins, 1987.

Mollring, F. K. *Microscopy from the Very Beginning*. Oberkochem, Federal Republic of Germany: Carl Zeiss.

Tortora, G. J., and S. R. Grabowski. *Principles of Anatomy and Physiology*, 7th ed. New York: HarperCollins, 1993.

## *SPECIFIC REFERENCES*

Dautry-Varsat, A., and H. F. Lodish. "How Receptors Bring Proteins into Cells." *Scientific American,* 250: 52–58, 1984.

Liotta, L. A. "Cancer Cell Invasion and Metastasis." *Scientific American,* 266: 54–63, 1992.

Sandberg, A. A. *The Chromosomes in Human Cancer and Leukemia,* 2nd ed. New York: Elsevier, 1990.

# The Microscope and Specimen Preparation

**After completing this exercise, you should be able to**

1. Identify the parts of the microscope
2. Explain the functions of the parts of the microscope
3. Demonstrate the proper method of focusing, changing objectives, carrying the microscope, and cleaning the microscope
4. Calculate the total magnification of any ocular and objective combination
5. Prepare a temporary wet mount for the examination of a specimen
6. Define and understand the following terms and concepts: *resolving power, parfocal, working distance, depth of field,* and *magnification*
7. Identify several types of human blood cells and their parts

The compound microscope is a precision instrument. It is essentially a system of accurately ground lenses arranged to give sharp, clear, magnified images of minute objects. Magnification that may be obtained with different lens systems ranges from approximately 10 to 2000 times ($\times$) the diameter of the specimen being observed. It should be noted that for each lens system, the maximum magnification usually is obtained by highly trained microscopists using auxiliary equipment under ideal conditions.

The optical lenses of the microscope include the *ocular* or eyepiece, usually three objectives with different magnifying powers, and the substage condenser. All other parts of the microscope are accessory to the main purpose of magnification by these lenses. They are designed to hold the object to be examined steadily in place, to regulate the amount of light necessary for clear viewing, and to aid in the focusing by adjusting the space relationship between the ocular, the objective, and the specimen on the slide.

Microscopes differ greatly from laboratory to laboratory. Your instructor will demonstrate and explain the general construction and manipulation of the particular instrument you will use. There is, however, a set of basic rules that should always be observed for all instruments, to ensure maximum efficiency and proper maintenance.

## USE AND CARE OF THE MICROSCOPE

1. Always carefully remove the microscope from its place of storage by holding it firmly by the arm with one hand and supporting it at the base with the other. Keep the instrument in an upright position.
2. Place the microscope at least 6 inches from the edge of your laboratory table.
3. Do not tamper with or remove any parts. If the microscope does not seem to be functioning properly, call your instructor immediately. Other laboratory sections may use the same instrument, and any part damaged but not reported by someone else might be charged to you.
4. Do not handle the lenses with your fingers. Perspiration contains fatty acids and other substances that can mar the lens glass. Always use the specified type of lens paper for cleaning the optical system. *Never use paper towels or cloths.* These materials will scratch the delicate lens surfaces.
5. Wipe the lenses of the instrument before and after laboratory use.
6. Use a clean, soft, linen cloth to remove dirt, dust, or other materials from parts other than the lenses.

7. Do not allow liquids, particularly acids and alcohol, to come in contact with any part of the microscope.

8. Always use a cover slip when examining objects or organisms mounted in water or other fluids.

9. Remove immersion oil from the microscope with lens paper.

10. Always raise the body tube sufficiently before either placing a slide on or removing a slide from the microscope stage.

11. Before putting the microscope away, always put the low-power objective in working position (i.e., in line with the body tube), open the diaphragm, and raise the condenser to its highest fixed position.

12. When disconnecting the light source, do not pull on the wire; instead, grasp the plug firmly to remove it from the socket.

13. If a cover is provided for the microscope, be certain that it covers the entire instrument.

# PRINCIPLES OF MICROSCOPY

To serve its purpose, a microscope must provide adequate magnifying power so that the finest details of a specimen are separated enough to be visible to the eye. The magnifying power of microscope lenses usually is indicated by markings on objectives and eyepieces (e.g., $45\times$, $100\times$). The total magnification obtainable is the product of the magnifications produced by the lenses individually.

While magnification is an important microscope property, the limiting factor that determines the usefulness of the light microscope is its resolving power. The resolving power of an optical system is its ability to produce separate images of small parts of a specimen that are only a short distance apart—that is, the ability to distinguish fine detail. In practice, the limit of resolution obtainable with a compound light microscope is about 0.2 $\mu$m, or roughly 1000 times the resolving power of the unaided human eye.

The quality of the image obtained or its definition depends on the degree of contrast between the specimen and its surroundings and between parts of a specimen. Such contrast is due in part to differences in light absorption, and various techniques that affect absorption are used to improve contrast. For example, the use of chemical stains results in color differences. Also, reducing the intensity of the light illuminating a specimen makes it easier to distinguish cells that have a refractive index similar to the fluid in which they are suspended.

Direct examination of living specimens can be extremely useful in determining size and shape relationships, motility, and reactions to various chemicals. One of the methods in general use for studies of this type is the wet-mount technique. This method maintains the natural shape of cells and reduces the distorted effects that can occur when specimens are dried and preserved.

In this exercise, the wet-mount technique is used to emphasize particular properties of the light microscope and to introduce features of living cells.

## Identification of Microscope Components

Although compound microscopes from different manufacturers vary in appearance or design, they consist of the same basic components. The parts of a typical microscope and their functions are described in Table 1-1. Use this information as you examine and locate specific parts of the microscope.

### Materials

1. Compound microscope with an appropriate light source (one per student).
2. Large cut-away diagram of a representative microscope for class use.

### Procedure

This procedure is to be performed by students individually.

1. Carefully remove the microscope assigned to you from the cabinet by firmly grasping the microscope's arm with one hand and placing the other hand under the instrument's base for support.
2. Place the microscope approximately 6 inches from the edge of the laboratory table, with the microscope arm facing you.
3. Examine your microscope, and with the aid of Figure 1-1, locate and learn the functions of each component listed in Table 1-1.
4. Answer items pertaining to this portion of this exercise in the laboratory review section.

## Temporary Wet-Mount Specimen Preparation

Specimens for microscopic examination can be prepared in several ways. The temporary wet-mount method described here is sufficient for materials to be examined for a short time period. (*Note:* The specimens prepared in the following two procedures will be used for viewing in a later section of this exercise.)

### Materials

1. The following materials should be provided per four students:

**Table 1-1   Microscope Components and Their Functions**

| Microscope Component | Location or Description | Functions |
|---|---|---|
| Ocular (eyepiece) | Uppermost series of lenses | Magnification |
| Body tube (barrel) | Main cylindrical part | Holds oculars, conducts light rays from specimen |
| Nosepiece | Movable, usually circular plate at bottom of body tube | Holds objectives |
| Low-power objective | Shortest objective, usually magnifies 10× | Magnification |
| High-power (dry) objective | Intermediate-sized objective, usually magnifies 43× to 45× | Magnification |
| Oil immersion objective | Longest objective; marked with either etched, red, or black circles; usually magnifies 100× | Magnification |
| Condenser | Lens system located below central stage opening | Concentrates and directs light beam through specimen |
| Stage | Platform on which specimens for examination are placed | Specimen support |
| Mechanical stage | Control devices that permit movement of slides from left to right and forward and backward on stage | Used to move specimen |
| Iris diaphragm | Located beneath stage in association with condenser unit, controlled with a lever | Regulates brightness or intensity of light passing through lenses |
| Coarse adjustment knob[a] | Generally located below stage | Used for preliminary and coarse focusing by raising or lowering body tube |
| Fine adjustment knob[a] | Generally located below stage | Used for final or fine focusing by raising or lowering body tube |
| Condenser adjustment knob | Control knob located below stage | Used to obtain full illumination by raising or lowering condenser |
| Base | Heavy, bottom portion on which instrument rests | Microscope support |
| Arm | Somewhat curved portion of microscope, used in carrying instrument | Microscope support |
| Mirror[b] | Double-faced mirror for reflecting light through specimen | Reflects light from source through lens systems |

[a]On several instruments, the coarse and fine adjustments are combined in one knob device.
[b]Found on older instruments.

a. One eyedropper
b. One small beaker of water
c. One pair of scissors
d. One dissecting needle
e. One bottle of methylene blue stain
f. Four wooden applicator sticks
g. Container with disinfectant for applicator sticks

2. The following materials should be provided for class use:

a. Microscope slides and cover slips
b. Newspaper
c. Lens paper

## Procedure 1: Letter "e"

This procedure is to be performed by students individually.

**Optical and Mechanical Features of**
# THE MICROSCOPE

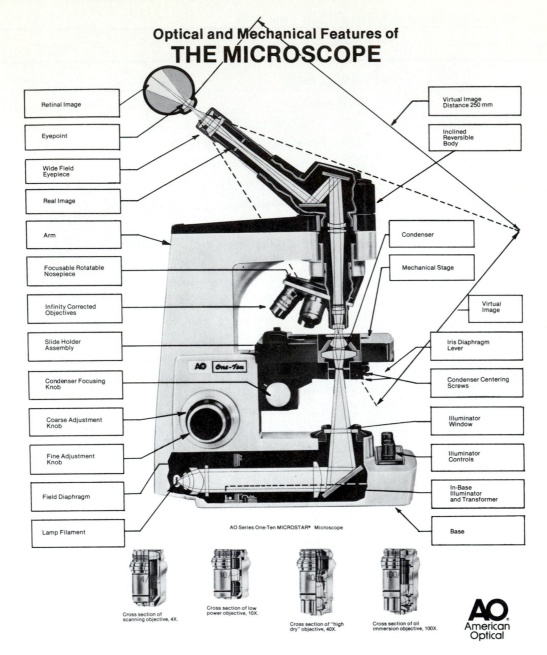

Retinal Image

Eyepoint

Wide Field Eyepiece

Real Image

Arm

Focusable Rotatable Nosepiece

Infinity Corrected Objectives

Slide Holder Assembly

Condenser Focusing Knob

Coarse Adjustment Knob

Fine Adjustment Knob

Field Diaphragm

Lamp Filament

Virtual Image Distance 250 mm

Inclined Reversible Body

Condenser

Mechanical Stage

Virtual Image

Iris Diaphragm Lever

Condenser Centering Screws

Illuminator Window

Illuminator Controls

In-Base Illuminator and Transformer

Base

AO Series One-Ten MICROSTAR® Microscope

AO One-Ten

Cross section of scanning objective, 4X.

Cross section of low power objective, 10X.

Cross section of "high dry" objective, 40X.

Cross section of oil immersion objective, 100X.

**AO** American Optical

**Figure 1-1** A cut-away diagram of a compound monocular microscope showing major components. Many instrument components are combined. Note that the fine-adjustment and coarse-adjustment knobs are combined and that the condenser adjustment is permanently fixed in position. (Courtesy of Reichert Scientific Instruments.)

1. Carefully clean a glass slide and a cover slip, as indicated by the instructor. Place a drop of water on the center of the microscope slide (see Figure 1-2).

2. Cut a row of a few letters from newspaper. Be certain that at least one of the letters is a small "e."

3. Place the string of letters in the drop of water on the slide, and make certain that the liquid covers them.

4. Holding the cover slip at a 45° angle, place one edge on the slide in contact with the drop of water and then slowly lower it into place over the letters (see Figure 1-2).

5. Save this preparation for later viewing.

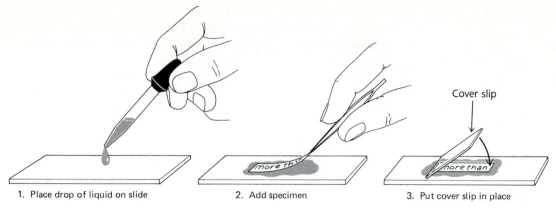

1. Place drop of liquid on slide    2. Add specimen    3. Put cover slip in place

Cover slip

**Figure 1-2** Steps in the preparation of a temporary wet mount.

## Procedure 2: Cheek Cells

This procedure is to be performed by students individually.

The squamous epithelial lining of the human cheek provides a readily available source of cells for microscopic examination. These cells will serve as an introduction to general properties of animal cells.

1. Place 1 drop of water and 1 drop of methylene blue on the center of a clean glass slide.
2. Gently scrape the inside of your cheek with the end of a clean wooden applicator stick. (*Note:* Force is not necessary to dislodge cells.)
3. Gently mix the scraped material with the drops of fluid on the slide.
4. Apply the cover slip as described earlier and shown in Figure 1-2.
5. Save this preparation for viewing in the next part of this exercise.
6. Dispose of the applicator stick, as indicated by your instructor.

## Use of the Microscope

This portion of the exercise is designed to familiarize you with basic features of microscope's parts and operation. The procedures described are for instruments with movable body tubes.

## Materials

1. The following materials should be provided for class use:
   a. Temporary wet mounts of the letter "e" and of squamous epithelial cells prepared in Procedures 1 and 2 of the preceding section of this exercise.
   b. Lens paper

2. Prepared slides of crossed colored silk threads (one per four students).

## Procedure 1: Microscope Operation

This procedure is to be performed by students individually.

1. Wipe all lenses with lens paper.
2. Move the low-power objective into position under the body tube. You will feel a click when it is correctly in place.
3. Connect your light source, and turn it on.
4. Place the letter "e" slide on the mechanical stage, and secure it with the slide holder or other means provided. The finger-control lever for the slide holder is generally located at the rear left-hand corner of the mechanical stage. Forward pressure on the lever will cause the slide holder to open, and the slide may be inserted into the holding area. The bottom of the slide should rest on the microscope stage and not on the mechanical stage apparatus. Release the finger-control lever to allow the slide holder to close and to hold the slide within the mechanical stage.
5. Using the control knobs on the mechanical stage, move the slide until the specimen on the slide is over the center of the condenser opening in the stage. One control knob moves the slide left to right, and the other moves it front to back.
6. Use the coarse-adjustment knob to lower the low-power objective to approximately ¼ inch above the cover slip.
7. Use the condenser adjustment knob to raise the condenser as far as it will go.
8. While looking into the ocular, move the iris diaphragm lever back and forth. Note the change in illumination. The intensity or brightness of illu-

mination is controlled by adjusting the iris diaphragm.

9. Locate your specimen by using the low-power objective and coarse adjustment first. Focus *upward*—that is, move the objectives upward—with the fine-adjustment knob until the specimen comes into clear view. In a microscope with a fixed body tube, the objectives remain in a fixed position. Focus is controlled by moving the stage *downward* until the specimen comes into view. If you have difficulty in locating your specimen, look for a region with a color or intensity different from other portions of the slide.

10. In order to change objectives (switch from low power to high power), do not raise the body tube; simply turn the nosepiece to bring the desired objective into place. The specimen will generally stay in focus with any objective, but a slight turn of the fine-adjustment knob may be necessary. Most microscopes are *parfocal*—that is, once the specimen is in focus, it will remain so when the objective lens is changed.

11. When looking through the microscope, keep both eyes open at all times. This may be difficult at first with a monocular microscope, but it will become easier with practice. Keeping both eyes open prevents eye strain and enables you to observe and draw microscopic views of specimens without moving your head.

12. When the microscope is no longer needed, the following steps should be performed:
    a. Remove any slide from the stage.
    b. Clean the instrument, including the lenses, to remove any fluid from specimens.
    c. Open the diaphragm.
    d. Raise the condenser to its highest position.
    e. Move the low-power objective into position over the stage.
    f. Wrap the electrical cord around the base.
    g. Return the microscope to its proper storage place.

## Procedure 2: Examination of Temporary Wet Mounts

This procedure is to be performed by students individually.

### Letter "e"

1. Examine the letter "e" on the slide, as it appears to the unaided eye in a normal reading position.

Make a sketch of the letter in the results and observations section, following Procedure 3. All that is necessary is an outline of the "e" because we are interested only in the orientation or position of the letter rather than detail. (Refer to Color Plate 1.)

2. Focus and examine the preparation under the low-power objective following the steps listed in Procedure 1.

3. Slowly move the slide, first side to side and then forward (up) and backward (down), with the mechanical-stage-control knobs. Compare this view of the orientation of the "e" with the view with the unaided eye.

4. Sketch a representative microscopic field in the results and observations section.

5. Change to the high-power objective. Adjust the light, and carefully focus with the fine-adjustment knob. A portion of the letter "e" will be off center.

6. Center the portion of the letter while you have it under high power, and then turn to low power, to determine how far off center the "e" is in your particular microscope. This is important and should aid you in all your future microscope work.

7. Practice shifting from low to high power and back again. (Note that only a small amount of manipulation with the fine-adjustment knob is necessary to bring the letter into proper focus whenever the other objective is swung into position.)

8. Sketch the representative high-power microscopic field in the results and observations section.

9. On your sketch of the "e," as observed under low power, draw a circle around the portion of the letter seen under high power.

10. Remove, clean, and return the slide to a tray or other location indicated by your instructor.

11. Answer the laboratory-review questions pertaining to this portion of the exercise.

### Squamous Epithelial Cells

Note that these cells are thin and flat and somewhat irregular in shape. The edges of the cells often become folded over in the process of preparation, and the cells tend to clump together in groups. Some cells may be covered by small, almost dotlike structures. These are usually bacteria from the oral region.

1. Examine your cheek-cell preparation under the low-power objective. Make the necessary adjustments for light intensity.

2. Select a microscopic field for study, in which the cells are separate from each other and exhibit some-

what centrally located and darker-staining nuclei. Refer to Color Plate 2.

3. Using first low power and then high power, draw several squamous cells in the space provided in the results and observations section. Label the following: cytoplasmic (cell) membrane, cytoplasm, and nucleus. If bacteria (small spherical or sticklike forms) are present, indicate them also.

### Procedure 3: Depth of Field

This procedure is to be performed by students individually.

To focus any microscope, the specimen being examined must be brought to the proper distance from the objective lens. This is done as indicated in Procedure 1, by the vertical movement of the body tube. In this portion of the exercise, you will consider the depth of field (focus) or the thickness of a specimen that is in focus at any one time. By use of this technique, determinations can be made as to what structures and location details are in a specimen. In addition, knowledge of the width, length, and depth of a specimen can be used to understand the three-dimensional feature of objects.

1. Place a prepared slide containing three different colored threads on the microscope stage, so that the point where the threads cross is over the center of the light opening in the stage.
2. Using the low-power objective, find the location on the slide where the three threads cross.
3. Use the coarse adjustment to raise the body tube slowly until the place where the threads cross is barely in focus. The colors should be visible, but the image will be fuzzy and blurred.
4. Slowly lower the body tube until the edge of the first thread appears clear. What is the color of the first thread to appear?
5. Use the fine-adjustment knob to focus on the second and third threads. Determine the color of the threads in the middle and on the bottom. Enter your findings in Table 1-2 of the results and observations section.
6. Obtain an unknown thread slide from your instructor. Examine it in the same way you examined the known slide. Determine the color on top, in the middle, and on the bottom, and enter your findings in the results and observations section. Check with the instructor to see whether your observations are correct.

## RESULTS AND OBSERVATIONS

### *Letter "e"*

1. Sketches

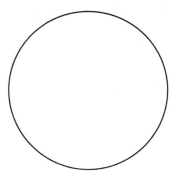

Unaided eye view

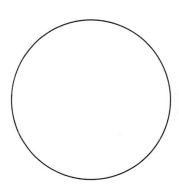

Low power

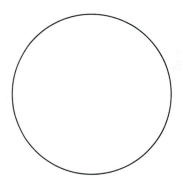

High power

2. List three ways in which the microscope changes the appearance of the letter "e."

   a. _____

   b. _____

   c. _____

## *Squamous Epithelial Cells*

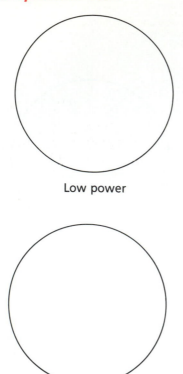

Low power

High power

## *Depth of Focus*

1. Unknown thread slide number: _____
2. Complete Table 1-2.

## Working Distance and the Oil Immersion Lens

The distance between a specimen and an objective lens is known as the *working distance.* Magnification increases as this distance decreases, and more light is needed for better and more accurate viewing of a spec-

imen. Figure 1-3 shows the relationship between the working distances of objectives and the iris-diaphragm adjustments needed to provide adequate illumination. In addition, the more a specimen is magnified, the greater is the chance for the specimen and the lens of the objective to come into contact. Great care should be exercised when the high-power (oil immersion) objective is in position. Improper use of objectives can result in broken slides and damaged objectives.

The objective of highest power on compound light microscopes is usually the oil-immersion objective. The objective must always be immersed in cedar oil or some other substance with about the same index of refraction as glass. The refraction index of a substance determines the angle at which light is bent when it passes through that substance. If a glass slide has air above and below it, as when the high- or low-power objective is used, light is bent upon entering the glass and again when it emerges into air. Because of this refraction, image details become blurred, and some light is lost as it passes by rather than enters the objective (Figure 1-4). With the oil-immersion lens, light passing through the slide proceeds in a straight path through the immersion oil. Immersion oil thus increases the resolving power of the objective so that finer details can be seen and more of the incident light is used.

In this portion of the exercise, a prepared blood smear is used in order to demonstrate the properties of the oil-immersion objective. The general microscopic features of human blood cells are given in Table 1-3, Figure 1-5, and Color Plate 3.

### Materials

The following materials should be provided for class use:

1. Prepared human-blood smears with coverslips
2. Charts and other reference materials showing blood-cell types
3. Immersion oil
4. Lens paper and xylene

**Table 1-2**

| Thread Layer | Color of Thread Layer in Known Slide | Color of Thread Layer in Unknown Slide |
|---|---|---|
| Top | | |
| Middle | | |
| Bottom | | |

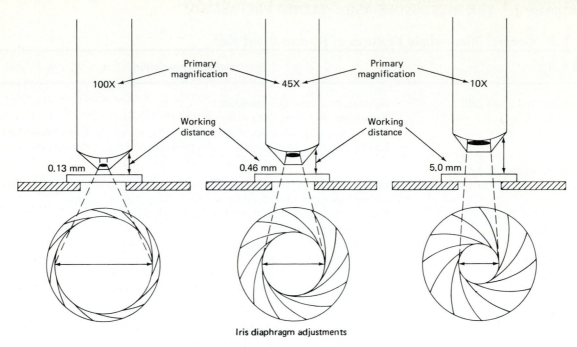

Iris diaphragm adjustments

**Figure 1-3** The relationship between the working distances of the low (10 ×), high (dry) (45 ×), and oil-immersion (100 ×) objectives, and the iris-diaphragm adjustments necessary to obtain satisfactory specimen illumination. The primary magnification and working distance (in millimeters) are given for each objective.

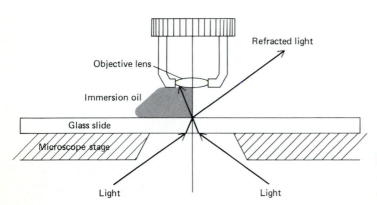

Figure I-4

**Figure 1-4** A diagrammatic comparison of the effects of immersion in oil and in air on the path of light leaving a microscope glass slide.

## Procedure

This procedure is to be performed by students individually.

1. Place a blood-smear slide into position on the microscope stage.
2. Rotate the low-power objective into position.
3. Focus and locate a field showing both red and white blood cells.
4. Swing the low-power objective aside.
5. Put one drop of immersion oil on the slide region over the light opening in the stage.

6. Move the oil-immersion objective into position over the slide. The objective should touch the oil and be just short of touching the slide.
7. Look through the ocular, and move the iris-diaphragm lever to make any necessary light adjustments. If the light is too bright, it will be difficult to distinguish the cells from the background.
8. Focus upward with the fine-adjustment knob until the blood cells come clearly into view. Refer to Color Plates 3, 36, and 37.
9. Examine the slide, and identify all the blood cell types described in Table 1-3.

**Table 1-3  General Microscopic Features of Human Blood Cells**

| Blood Cell | General Description |
|---|---|
| Erythrocyte (red blood cell) | Circular, biconcave disc appearance<br>No nucleus in mature state<br>Most numerous of all types of cells found in the peripheral blood (bloodstream)<br>Usually measures approximately 7 μm in length and 2 μm in thickness |
| Leukocytes (white blood cells)<br>  Granulocyte (granular leukocyte) | Contains distinct cytoplasmic granules, which, upon staining, react with dyes<br>Granulocytes also have irregular and multilobed (lobose) nuclei |
|   Eosinophil | Granules react with dyes to yield characteristic red color |
|   Basophil | Granules stain blue |
|   Neutrophil | Granules stain neutral or orange |
|  Agranulocyte | Does not contain granules<br>Has rounded, rather than lobose, nucleus |
|   Lymphocyte | Has individual nucleus that occupies major portion of cell |
|   Monocyte | Has kidney-shaped nucleus; generally larger than lymphocyte |
| Platelet (thrombocyte) | A circulating, nonnucleated fragment of a large bone-marrow cell (megakaryocyte), which participates in blood clotting |

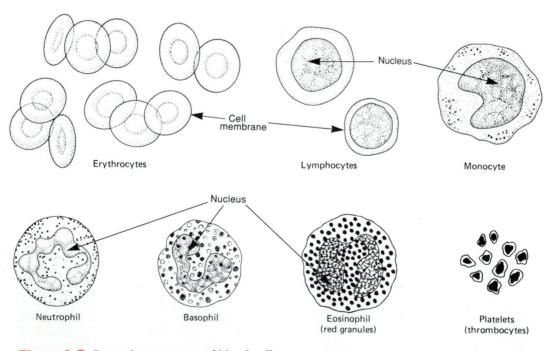

**Figure 1-5** General appearance of blood cell types.

10. Identify the cells indicated in the blood smear in Figure 1-6 in the following results and observations section.
11. Answer the laboratory-review questions pertaining to this portion of this exercise.
12. After completing this procedure, remove the slide, and clean it with the xylene and lens paper. (*Note:* xylene should be used with slides having fixed cover slips.)
13. Wipe the oil from the oil-immersion objective, and swing the low-power objective into position.

## RESULTS AND OBSERVATIONS

1. Identify all labeled cells in Figure 1-6.

   a. _____

   b. _____

   c. _____

   d. _____

   e. _____

   f. _____

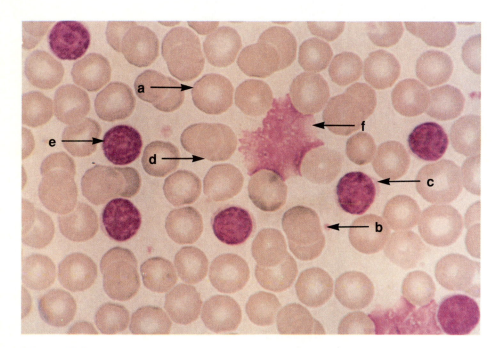

**Figure 1-6** Representative human blood smear preparation.

# LABORATORY REVIEW *1*

NAME _____

LAB SECTION _____ DATE _____

## *THE MICROSCOPE AND SPECIMEN PREPARATION*

### COMPLETION AND FILL-IN

1. Label the microscope parts in Figure Q1-1.

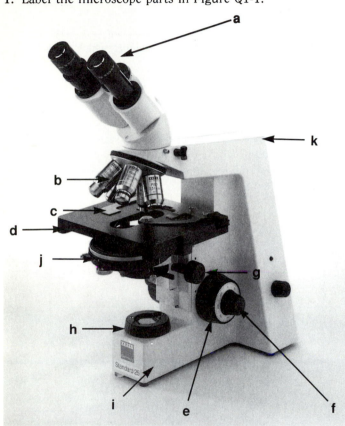

**Figure Q1-1** (Courtesy of Carl Zeiss, Inc.)

2. What is the magnification of the following objectives without an ocular?

    a. Low power _____    b. High power _____    c. Oil immersion _____
3. Which of the objectives comes closest to a specimen and is most likely to break a

    slide if proper precautions are not taken? _____

## ANSWERS

a. _____

b. _____

c. _____

d. _____

e. _____

f. _____

g. _____

h. _____

i. _____

j. _____

k. _____

4. Complete Table Q1-1.

**Table Q1-1**

| Microscope Part | Functions |
| --- | --- |
| Ocular | a. |
| Low power, high power, and oil-immersion objectives | b. |
| c. | Lens, located beneath the stage opening, which controls the light-beam size |
| d. | Controls the intensity of light |
| e. | Platform on which slides or other materials to be examined are studied |
| Coarse-adjustment knob | f. |
| g. | Adjustment device used for final focusing |
| Condenser adjustment knob | h. |
| i. | Used for providing support during transport |

5. What materials should be used for the cleaning of microscope lenses? _____

6. a. Distinguish between resolving power and magnification. _____

   _____

   b. Which of these is the more important property of a microscope? _____

   c. List the factors that contribute to resolving power. _____

   _____

7. Table Q1-2 lists some common problems encountered in microscope usage. Indicate the proper procedure for correcting each of them.

### Table Q1-2

| Problem | Correction Procedure |
|---|---|
| Oil-immersion lens does not enter immersion oil on slide | a. |
| Microscope field is too bright | b. |
| Specimen appears blurred after proper light and focusing adjustments are made | c. |
| Microscope view does not change as slide is moved | d. |
| Only half of the field appears to be illuminated | e. |

8. Indicate which microscope part in Figure Q1-1 is used in the situations described in Table Q1-3.

### Table Q1-3

| Situation | Microscope Part |
|---|---|
| Fine focusing of a specimen | a. |
| Securing a slide for viewing | b. |
| Adjusting the brightness of the light | c. |
| Providing full illumination for viewing | d. |
| Preliminary focusing of a specimen | e. |
| Changing of objectives | f. |

**ANSWERS**

1. _____

2. _____

3. _____

4. _____

5. _____

6. _____

## MULTIPLE CHOICE

Insert the correct answer in the spaces provided.

1. What property of an ordinary laboratory microscope enables a person to change objectives without major focusing adjustments?
   a. resolving power
   b. magnification
   c. parfocal
   d. focusing
   e. resolution

2. How does closing the diaphragm affect image brightness?
   a. decrease
   b. increase
   c. no effect

3. What is the total magnification obtainable using the oil-immersion objective and a $10 \times$ ocular?
   a. 100 $\times$
   b. 450 $\times$
   c. 45 $\times$
   d. 100 $\times$
   e. 1000 $\times$

4. Which of the following objectives has the shortest working distance?
   a. 2 $\times$
   b. 10 $\times$
   c. 45 $\times$
   d. 100 $\times$

5. When changing objectives from low power to high power, it is generally necessary to
   a. lower the condenser
   b. raise the condenser to its maximum
   c. open the diaphragm
   d. close the diaphragm
   e. a and d (only)

6. Changing from the low power to the high power will enable a person to see _____ of a specimen.
   a. less
   b. more
   c. about the same portion

# The Cellular Level of Organization

***After completing this exercise, you should be able to***

1. Distinguish between prokaryotic and eukaryotic forms of cellular organization
2. List the major organelles found in eukaryotic animal cells
3. Give the functions of major eukaryotic organelles
4. Locate and identify the basic parts of animal cells

Living organisms are composed of basic units of structure and function termed *cells*. Furthermore, cells contain subcellular parts, referred to as *organelles* (or little organs), that exhibit distinctive structural and functional properties (see Figure 2-1).

The use of electron microscopes and associated techniques for the preparation of biological specimens has demonstrated clearly the existence of two fundamentally different types of cellular organization—namely, the *prokaryotic* (primitive nucleus) and *eukaryotic* (true nucleus).

The prokaryotic cell, which is the unit of cellular organization for bacteria, is distinguished by its small size and the absence of many of the organelles normally found in eukaryotic cells (Figures 2-1 and 2-2). All living organisms other than bacteria have a eukaryotic type of cellular organization (Table 2-1).

Dyes (coloring agents) termed *biologic stains* frequently are used to distinguish organelles and/or chemical substances found in and between cells. Several categories of biologic stains are recognized; however, tissue stains are the most commonly used in microscopic work. In procedures involving these stains, one, two, and occasionally three dyes are used to differentiate (distinguish) the nucleus from the cytoplasm of cells

and to permit the differentiation among different tissue types. Stains may be used to color both living and preserved tissue. In addition, special techniques employing enzymes or acids can be used to identify specific chemicals in cellular structures. For example, the nucleic acids deoxyribonucleic acid (DNA) and ribonucleic acid (RNA) are identified with the use of the enzymes deoxyribonuclease (DNase) and ribonuclease (RNase) and acids such as trichloroacetic acid and perchloric acid.

In this exercise, major attention is given to the basic parts of eukaryotic cells that are readily visible with the compound bright-field microscope. Specimens for the exercise include stained representatives of animal eukaryotic cells, such as blood and cheek cells, and several of their organelles. Examples of enzyme-treated cells also are considered.

## An Introduction to Cellular Organelles

### Materials

1. Six demonstration microscopes, one for each of the following prepared, stained slides:
   a. Cross-section of frog *(Rana pipiens)* liver
   b. Frog or fish blood smear
   c. Smooth muscle cells (teased preparation or a cross-section of intestinal tissue)
   d. Human blood smear
   e. Squamous epithelial cheek cells
   f. Ciliated epithelial cells
2. Unknown slide preparation, one per student
3. Lens paper

### Procedure

This procedure is to be performed by students individually.

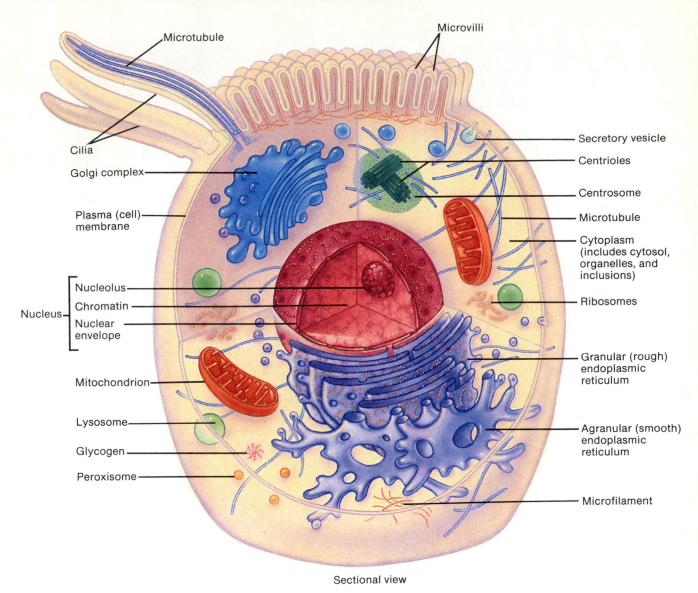

Sectional view

**Figure 2-1** The cellular organization of a eukaryotic cell. (From G. J. Tortora, *Principles of Human Anatomy*, 6th ed., New York: HarperCollins, 1992.)

1. Examine the preparations provided under low and high powers. Refer to Figure 2-3 and Color Plates 2, 3, 4, 5, 6, 7, and 8 as guides.
2. Look for the various organelles associated with eukaryotic cells.
3. Pay particular attention to the size and shape of the various cell types.
4. Complete the table and questions in the following results and observations section.
5. After you have examined the six demonstration slides, obtain an unknown slide from your instructor. Examine the slide, and identify the specific type of cell it contains. The preparation will be similar to one of those studied in the exercise. Enter your findings on the report form in the results and ob-servations section. Check with your instructor to see whether your identification is correct.

## RESULTS AND OBSERVATIONS

### Introduction to Cellular Organelles

1. List four organelles of animal cells.

   a. _____

   b. _____

   c. _____

   d. _____

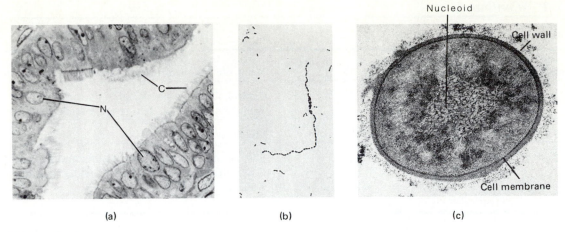

(a)　　　　　　　　(b)　　　　　　　　(c)

**Figure 2-2** Prokaryotic and eukaryotic cells. (a) Eukaryotic cells taken from a human oviduct; note the well-defined nuclei **(N)** and cilia **(C);** such details are obtainable with properly prepared specimens (from I. A. Brosens and G. Vasquez, *Journal of Reproductive Medicine,* 16: 171, 1976). (b) The appearance of the much smaller prokaryotic cells; these bacterial forms are *cocci* (spherical shape). (c) An electron micrograph of a prokaryotic cell clearly showing the absence of membrane-bound internal structures (organelles) and a well-defined nucleus (nucleoid). (Figures (b) and (c) are from C. M. Ward, Jr., and G. W. Claus, *Journal of Bacteriology,* 114: 378–389, 1973.)

**Table 2-1   Eukaryotic Organelles and Their Functions and Activities**

| Organelle | Associated Functions and Activities |
|---|---|
| Cell plasma membrane | Transport of substances into and out of cells<br>Phagocytosis (engulfment of foreign material) in some cells<br>Pinocytosis (uptake of fluid) |
| Cell wall (found only in plants and certain microorganisms) | Imparts shape and strength to the cell<br>Protection against some osmotic imbalances |
| Chloroplast | Photosynthesis |
| Cilium | Motion or movement of substances past the ciliated cell |
| Endoplasmic reticulum | Protein synthesis |
| Flagellum | Propulsion |
| Golgi complex | Storage and sorting structure for cell products<br>Transport of proteins, lipids, and other cell products to the cell surface |
| Microbody or peroxisome | Enzymatic activities |
| Microtubule | Transport of materials within cell<br>Development and maintenance of cell shape<br>Cell division<br>Part of internal structure of cilia and flagella |
| Mitochondrion | Synthesis of the energy-rich compound adenosine triphosphate (ATP) |

**Table 2-1**  *(Continued)*

| Organelle | Associated Functions and Activities |
|---|---|
| Nucleolus | Major site for the formation of ribosomal components |
| Nucleus | Control of cellular physiological process<br>Transfer of hereditary factors to subsequent generations |
| Ribosome | Protein synthesis |
| Vacuole | Storage site for water; for some amino acids, carbohydrates, and proteins; for cell products; or for cell wastes |

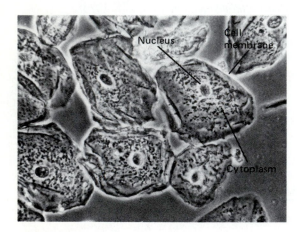

**Figure 2-3** Squamous epithelial animal cells: Note that animal cells show variation in size and shape. (The original magnification is 450×.)

2. Compare the various cell types provided in this exercise, as to the presence of organelles listed in Table 2-2. Indicate the presence of an organelle by a " + " and its absence by a " − ", and briefly describe the shape of the cell under study—for example, round, oval, or spindle.

3. State the type of cell that the unknown slide preparation contains: _____

## Demonstration of Organelles by Special Stains and Treatment

### Materials

1. *Ascaris* egg in mitosis, showing centriole (iron hematoxylin and orange G stain)

**Table 2-2**

| Organelle and Cell Shape | Cell Type | | | | | | |
|---|---|---|---|---|---|---|---|
| | Frog Blood | | Human Blood | | Frog Liver | Smooth Muscle | Squamous Epithelium |
| | *Red* | *White* | *Red* | *White* | | | |
| Cell membrane | | | | | | | |
| Nucleolus | | | | | | | |
| Nucleus | | | | | | | |
| Vacuole | | | | | | | |
| Shape of cell | | | | | | | |

2. *Amphiuma* liver cells, showing mitochondria (iron hematoxylin stain)
3. Golgi apparatus preparation (silver stain)
4. *Necturus* liver cells showing distinctive nucleus and nucleoli (azure B stain)
5. Liver tissue stained with iron hematoxylin and eosin
6. Liver tissue stained for DNA with Feulgen reaction
7. DNase-treated liver tissue and stained with Feulgen reaction
8. RNase-treated liver tissue

## Procedure

This procedure is to be performed by students individually.

1. Examine the preparations provided under the high-power and oil-immersion objectives. Refer to Figure 2-3 and Color Plates 9 and 10 as guides.
2. Look for the various organelles or reactions indicated on each slide. Use the following descriptions to aid you in either finding specific organelles or interpreting reactions.
   a. *Asters*—strands extending from and around a central body
   b. *Mitochondria*—small, individual, threadlike strands, generally blue within the cytoplasm; some are short and sticklike (rod) in appearance
   c. *Golgi complex*—brown or black strands forming a network around a cell's nucleus
   d. *Nucleolus*—small, darkly stained bodies within a cell's nucleus
3. Sketch a representative view of each specially stained organelle, as seen under oil immersion, in the spaces provided in the following results and observations section. Include all organelles observed and the general outline of a cell.
4. Next, examine the liver cells stained by the Feulgen reaction under low and high power. This staining procedure shows the presence of DNA. Refer to

Color Plates 9 and 10. Use these color plates for comparison purposes.
5. Now examine the enzyme-treated preparations. Based on your observations of the stained cells, determine what portion(s) of these cells was (were) affected by the respective enzyme treatments.
6. Complete Table 2-3, and answer the questions in the following results and observations section.

## RESULTS AND OBSERVATIONS

1. Sketch the organelles and their locations.

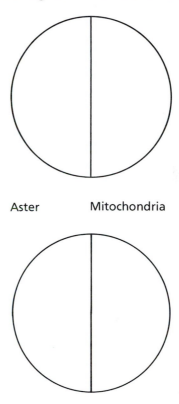

Aster    Mitochondria

Golgi Complex    Centriole

## Table 2-3

| Treatment | Cellular Components | | | |
| | Cell Membrane | Nucleus | Nucleolus | Cytoplasm |
|---|---|---|---|---|
| No treatment | | | | |
| RNase treatment | | | | |
| DNase treatment | | | | |

2. Compare the different enzyme-treated cells, as to the presence of the cellular components listed in Table 2-3. Indicate the presence or absence of a component by a "+" or "−" sign, respectively.

# LABORATORY REVIEW 2

## THE CELLULAR LEVEL OF ORGANIZATION

### COMPLETION AND FILL-IN

1. Identify the cell parts shown in Figure Q2-1.

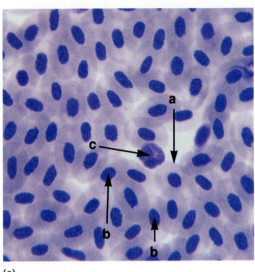

(a)

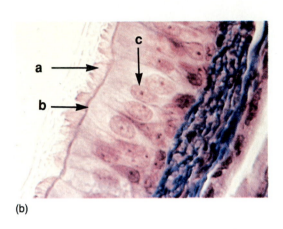

(b)

**Figure Q2-1** (a) Frog blood cells; (b) ciliated epithelial cells. (Courtesy of Carolina Biological Supply Company.)

**(a) FROG BLOOD CELLS**

a. _____

b. _____

c. _____

**(b) CILIATED EPITHELIAL CELLS**

a. _____

b. _____

c. _____

2. Using your textbook or other references, complete Table Q2-1 by giving the functions of the organelles listed.

**Table Q2-1**

| Organelle | Functions |
|---|---|
| Nucleus | a. |
| Nucleolus | b. |
| Endoplasmic reticulum | c. |
| Golgi complex | d. |
| Mitochondrion | e. |
| Centriole | f. |
| Cilia | g. |
| Ribosome | h. |
| Plasma membrane | i. |

3. Is the cell in electron micrograph Figure Q2-2 *prokaryotic* or *eukaryotic?*

_____

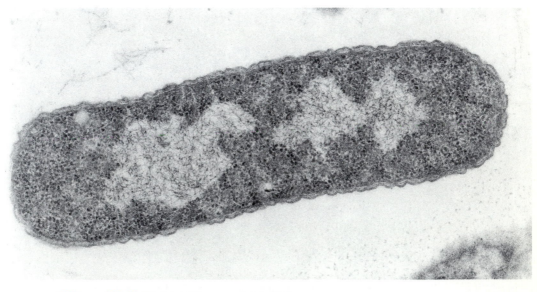

**Figure Q2-2** Prokaryotic or eukaryotic?

4. Identify the organelles in Figure Q2-3.

a. _____

b. _____

c. _____

d. _____

e. _____

f. _____

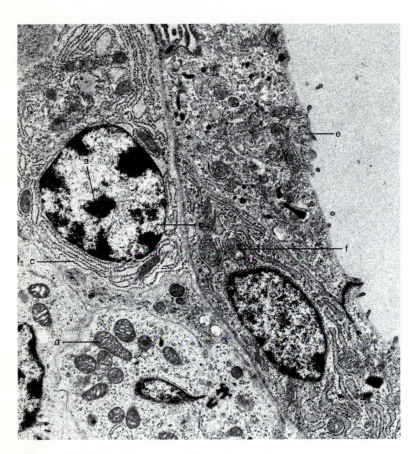

**Figure Q2-3** Eukaryotic cellular organization.

5. State the appropriate metric equivalent units for the following:

a. 11 millimeters = _____ $\mu$m

b. 100 nanometers = _____ mm

c. 1 $\mu$m = _____ mm

d. 17 $\mu$m = _____ mm

e. 10 nm = _____ mm

f. 3 $\mu$m = _____ nm

6. State the function or purpose of the following in the examination of cells:

a. Biologic stains _____

_____

b. DNase _____

_____

c. RNase _____

_____

1. _____
2. _____
3. _____
4. _____
5. _____
6. _____
7. _____
8. _____
9. _____
10. _____
11. _____
12. _____
13. _____
14. _____
15. _____

## MULTIPLE CHOICE

Insert the correct answers in the spaces provided.

1. Subcellular parts that have distinctive and functional properties are called
   a. prokaryotic cells
   b. organelles
   c. small organs
   d. cells
   e. eukaryotic cells

2. Which of the following serves as the external boundary of a cell, through which substances enter and leave?
   a. flagellum
   b. nucleus
   c. endoplasmic reticulum
   d. microfilament
   e. plasma membrane

3. The specific part of the cell that holds or contains hereditary information is known as the
   a. nucleolus
   b. chloroplast
   c. mitochondrion
   d. nucleus
   e. endoplasmic reticulum

4. The cellular structure responsible for or associated with movement is known as a
   a. Golgi complex
   b. cilium
   c. flagellum
   d. nucleus
   e. b and c (only)

5. The cellular sites known for their production of ATP are which of the following?
   a. Golgi complex
   b. endoplasmic reticulum
   c. mitochondria
   d. chromosomes
   e. lysosomes

6. Protein synthesis in a typical cell occurs specifically in or on sites known as the
   a. Golgi complex
   b. cell membrane
   c. centrosome
   d. nucleolus
   e. ribosomes

7. Which of the following organelles direct most if not all activities of the cell?
   a. mitochondria
   b. nucleolus
   c. endoplasmic reticulum
   d. nucleus
   e. cell membrane

8. Short, hairlike organelles that extend from certain cells and may be used to move substances across surfaces are
   a. flagella
   b. microtubules
   c. cell membranes
   d. cilia
   e. cytoskeletons

9. Specific organelles described as being flattened sacks of membranes that collect, package, and distribute molecules within the cell are
   a. chromosomes
   b. lysosomes
   c. centrioles
   d. flagella
   e. the Golgi complex

10. The network of protein microfilaments in a eukaryotic cell's cytoplasm that maintains the shape of the cell, anchors its organelles, and is associated with movements is known as which of the following?
    a. flagella
    b. cytoskeleton
    c. Golgi complex
    d. lysosomes
    e. centrioles

11. _____ are the organelles responsible for protein synthesis.
    a. mitochondria
    b. lysosomes
    c. chromosomes
    d. peroxisomes
    e. ribosomes

12. The term used for membrane-bound structures found in the cytoplasm of eukaryotes and known to perform various functions is
    a. organ
    b. organelle
    c. nucleus
    d. cell membrane
    e. cytoplasm

13. The phospholipid bilayer with embedded proteins that regulates cell permeability is the
    a. nucleus
    b. nuclear envelope
    c. plasma membrane
    d. nucleolus
    e. mitochondrion

14. The powerhouses of a eukaryotic cell, which are prokaryoticlike inclusions, are known as which of the following?
    a. nucleus
    b. ribosomes
    c. Golgi complex
    d. mitochondrion
    e. endoplasmic reticulum

15. The organelle associated with protein synthesis that lacks ribosomes is known as which of the following?
    a. cell membrane
    b. microtubules
    c. smooth endoplasmic reticulum
    d. rough endoplasmic reticulum
    e. mitochondrion

# Cellular Reproduction (Mitosis and Cytokinesis)

### After completing this exercise, you should be able to

1. Describe the sequence of events in the process of mitosis
2. Distinguish between mitosis and cytokinesis
3. Identify specific mitotic stages in prepared slides
4. Describe the general features of a cell cycle
5. Indicate the position of mitosis in the cell cycle

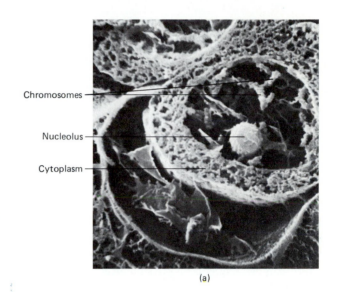

(a)

As indicated in Exercise 2, a eukaryotic cell's internal organization is a complicated assembly of chemical systems designed to perform a number of basic activities and functions to maintain its life on a day-to-day basis. However, cells under various conditions become injured, diseased, or worn and then die. Thus, new cells are needed for purposes of replacement and growth. The nucleus of each cell generally contains a precise set of genetic information built into more or less elongated structures known as *chromosomes* (Figure 3-1). A cell's genetic information is vital to its proper functioning and to determining the properties of the next generation of cells. Nuclear division resulting in the precise distribution of duplicated chromosomes into two newly formed cells is *mitosis. Meiosis,* a specialized type

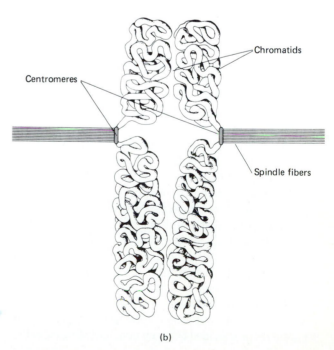

(b)

**Figure 3-1** Cells during division: (a) scanning micrograph showing chromosomes and a nucleolus within a nucleus (reproduced by permission of the National Research Council of Canada from E. D. P. Whelan and others, *Canadian Journal of Botany,* 52:1438–1440, 1974); (b) chromatids attached to spindle fibers at centromeres.

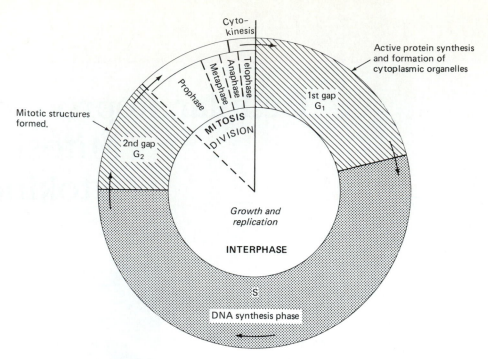

**Figure 3-2** The phases of a cell cycle: The relationship of the four phases ($G_1$, S, $G_2$, and mitosis) of the cycle are shown; events such as cytoplasmic organelle production, DNA synthesis, and cellular and nuclear division are related to cell phases.

of cell division that occurs during the formation of *gametes* (mature sex cells), is described in Section XIV.

During a cell's existence, it passes through a regular, organized sequence of physiological events known as the *cell cycle* (Figure 3-2). The cycle is complex and is divided into four phases: $G_1$ (gap), S (synthesis), $G_2$ (gap), and *mitosis*. As Figure 3-2 shows, $G_1$, S, and $G_2$ compose the interphase. During the $G_1$ phase, active protein synthesis occurs, as well as the formation of new mitochondria, Golgi apparatus, endoplasmic reticulum, and other cytoplasmic materials. Cell size also increases in the $G_1$ phase. In the S phase, DNA synthesis takes place, and chromosomes are duplicated. The DNA in a eukaryotic nucleus, together with a large amount of protein and a small portion of RNA, forms chromatic strands. In the $G_2$ phase, structures directly related to mitosis are produced. Mitosis follows the $G_2$ phase. During mitosis, chromatin strands in the nucleus condense by coiling and folding to form compact chromosomes (Figure 3-1b). Microscopically, each chromosome has a characteristic shape and appears to consist of two chromatin strands termed *chromatids* (see Figure 3-1b). Those chromatids are attached firmly to each other at a specific region termed a *centromere* (Figure 3-1b). Mitosis usually occurs in close association with the division of the cytoplasm *(cytokinesis)*. During this event, the various duplicated cytoplasmic organelles in a parent cell are divided between two newly formed cells.

Mitosis is an extremely efficient process that provides for the even distribution of hereditary material during cell division. In describing the events taking place in mitosis, it is convenient to divide the process into four stages: *prophase, metaphase, anaphase,* and *telophase.* The period between successive stages is *interphase.* It should be noted that mitosis is actually a continuous process and does not occur in discrete steps. It is not always readily apparent when one stage ends and another begins. Table 3-1 summarizes the main events of mitosis.

Normal mitosis during the development of a human from a fertilized egg cell (ovum) produces only cells with the number of chromosomes characteristic of a human. Thus, all cells of the body should contain the same number. There are, however, situations in which abnormalities occur. These include development and mutation.

The stages of mitosis are not limited to normal cells. Abnormal types of cell growth, spoken of as cancers, occur when growth is no longer regulated by the processes that control normal cell growth (Figure 3-3). Any cell in the human body has the potential of becoming a cancer cell. Cancer cells grow and divide more rapidly than normal cells and collectively form a mass termed a *malignant tumor*. Such growths tend to invade surrounding tissues and disrupt the normal structure and function of body organs.

**Table 3-1   Cell Division and Mitosis**

| Stage | General Events | Diagrammatic Representation |
|---|---|---|
| Interphase | 1. Chromosomes are uncoiled.<br>2. Intense cellular metabolic activity includes DNA synthesis (replication). | Centrioles |
| Prophase | 1. Individual chromosome structure becomes visible and recognizable as identical chromatids held together by centromeres.<br>2. Centrioles of cells separate and migrate toward opposite poles of cells.<br>3. Nucleolus and nuclear membrane disappear.<br>4. Spindle appears.<br>5. Asters are present in animal cells only. | Centromere<br>Centrioles<br>Chromosome (2 chromatids) |
| Metaphase | Chromosomes line up by their centromeres at the equator or spindle. | Centrioles<br>Chromosomes |
| Anaphase | 1. Chromatids separate and move to opposite poles of the cell.<br>2. Each chromatid becomes an independent chromosome.<br>3. Cytokinesis begins. | Chromatid<br>Centriole |
| Telophase | 1. Chromosome movement is completed.<br>2. Chromosomes assume the appearance they had in interphase.<br>3. New nucleolus and nucleus appear.<br>4. Cell plate appears only in plants. | Chromosome<br>Centrioles<br>Nuclear membrane |

## Materials

The following materials should be provided for class use:

1. Prepared slides of whitefish (*Coregonus*) blastulas showing mitosis
2. Models or charts showing stages of mitosis

## Procedure

This procedure is to be performed by students individually.

1. Use both low- and high-power objectives to examine the prepared slide showing mitotic stages in the whitefish cells. The slide should contain several

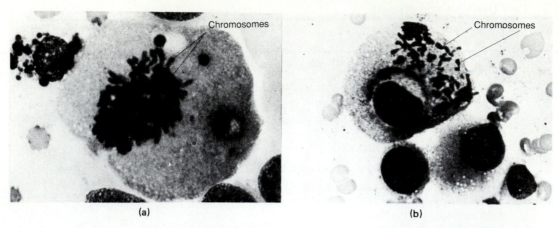

(a)                                                                    (b)

**Figure 3-3** Micrographs of malignant cells dividing: The stages of division are different visually and physiologically from those of normal cells. Note the general appearance of chromosomes. (Reproduced with permission from Freidman and others, "Needle Aspiration of Metastatic Melanoma," Acta *Cytologica,* 24: 715, 1980.)

embryonic cells exhibiting different stages of mitosis. (Refer to Color Plates 7 and 11.)

2. With the aid of Table 3-1, identify the stages of mitosis, and sketch each one in the following results and observations section. Label all significant parts of your sketches.

3. Find a clear (good) cell in metaphase, and count the number of chromosomes present. Do the same for a cell in anaphase.

4. Answer the questions pertaining to this portion of the exercise in the following results and observations section.

## RESULTS AND OBSERVATIONS

### *Mitosis*

1. Sketch and label the stages of mitosis.

2. In which stage of mitosis is an equatorial plate evident? _____

3. How many chromosomes were present during metaphase? _____

4. Were the number of chromosomes present during anaphase the same as the number during metaphase? Explain your answer. _____

_____

_____

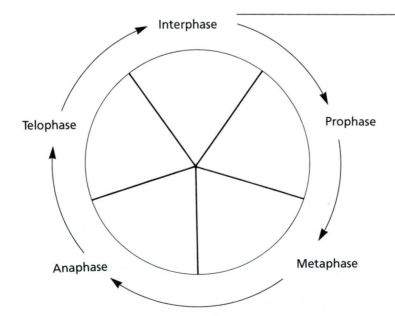

# LABORATORY REVIEW 3

## CELLULAR REPRODUCTION (MITOSIS AND CYTOKINESIS)

### COMPLETION, FILL-IN, AND DEFINITION

1. List and explain the four phases of the cell cycle.

    a. _____

    b. _____

    c. _____

    d. _____

2. Identify all parts of the animal cell undergoing metaphase in mitosis, as shown in Figure Q3-1.

    a. _____

    b. _____

    c. _____

    d. _____

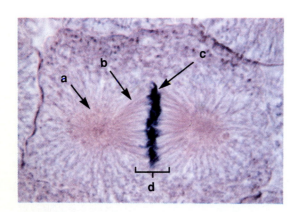

**Figure Q3-1**

3. List and briefly describe the stages of mitosis in their correct sequence.

    a. _____

    b. _____

    c. _____

    d. _____

    e. _____

4. Identify the stages of mitosis shown in Figure Q3-2.

a. _____

b. _____

c. _____

d. _____

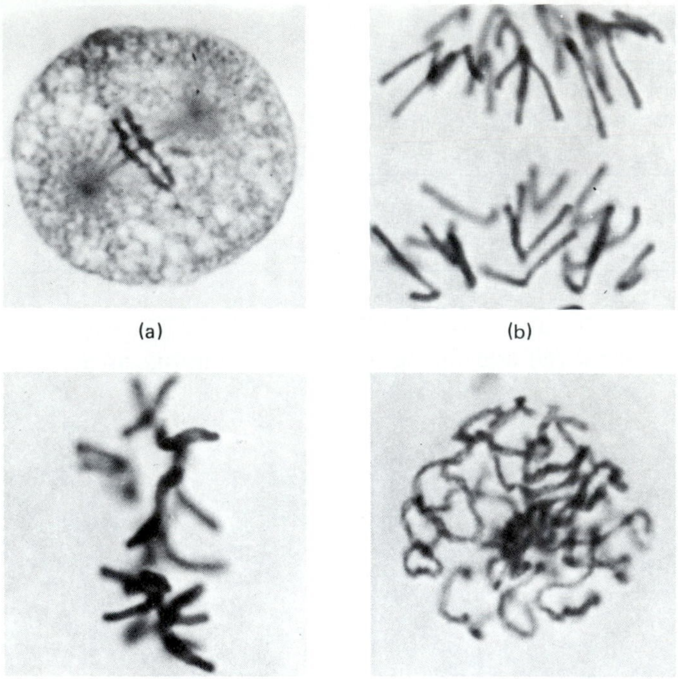

(a)          (b)

(c)          (d)

**Figure Q3-2** Stages of mitosis.

5. Define or explain the following:

a. Nucleolus _____

b. DNA _____

_____

c. Meiosis _____

d. Centromere _____

_____

e. Chromatid _____

## MULTIPLE CHOICE

Insert the correct answers in the spaces provided.

1. Which of the following organelles divides and organizes spindle fibers during mitosis?
   a. nucleus
   b. chromosomes
   c. microtubules
   d. nucleolus
   e. centrioles

2. Which of the following move chromosomes during cell division?
   a. vesicles
   b. nucleus
   c. microtubules
   d. Golgi complex
   e. ribosomes

3. The structures that are condensed forms of DNA, found in the nuclear area and associated with proteins and RNA, are called which of the following?
   a. microtubules
   b. cilia
   c. centrioles
   d. ribosomes
   e. chromosomes

4. During which of the following phases of mitosis does the nuclear membrane of a cell disappear?
   a. interphase
   b. prophase
   c. metaphase
   d. anaphase
   e. telophase

5. During which phase of mitosis do new nuclear membranes form?
   a. interphase
   b. prophase
   c. metaphase
   d. anaphase
   e. telophase

6. The replication of DNA takes place during which of the following phases of mitosis?
   a. interphase
   b. prophase
   c. metaphase
   d. anaphase
   e. telophase

7. Sister chromosomes separate and move to opposite poles of a cell during which phase?
   a. interphase
   b. prophase
   c. metaphase
   d. anaphase
   e. telophase

8. Spindle fibers form in which of the following phase(s)?
   a. interphase
   b. prophase
   c. metaphase
   d. anaphase
   e. telophase

ANSWERS

1. _____
2. _____
3. _____
4. _____
5. _____
6. _____
7. _____
8. _____
9. _____
10. _____

9. Chromosomes become arranged on an equatorial plate in the phase known as what?
   a. interphase
   b. prophase
   c. metaphase
   d. anaphase
   e. telophase

10. During which phase of mitosis do asters form from centrioles?
    a. interphase
    b. prophase
    c. metaphase
    d. anaphase
    e. telophase

# Fluid Transport and Molecular Activity: Brownian Movement, Filtration, Diffusion, and Osmosis

## After completing this exercise, you should be able to

1. List three physical processes by which materials move into and out of cells
2. Perform simple experiments to demonstrate filtration, diffusion, and osmosis
3. Distinguish between the general process of diffusion and the more specific process of osmosis
4. Define or explain *Brownian movement, solution, solute,* and *solvent*

All the chemical reactions in a cellular environment take place in solution. Before a substance can enter a cell, it must dissolve. The term *solution* can refer to the situation in which a substance is dissolved more or less evenly in a liquid. If, for example, the carbohydrate glucose is mixed thoroughly with water, the solid crystals of the substance dissolve and are dispersed throughout the liquid. The resulting mixture—a solution—is composed of the liquid *solvent* and the dissolved *solute.* Substances that dissolve are *soluble,* while those that do not dissolve are *insoluble.* The digestive process in humans and other animals generally involves turning a variety of large compounds into small compounds that are soluble in water and can be carried into and out of cells.

Substances move into and out of cells by several physical processes, including (1) *diffusion,* resulting from a concentration gradient (movement of molecules from an area of higher concentration to a lower one); (2) *osmosis,* the diffusion of water molecules across a semipermeable membrane; and (3) *filtration,* produced by hydrostatic pressure differences.

Within cells, numerous minute particles appear to move in an erratic, zigzag motion known as *Brownian movement.* The movement occurs as small particles suspended within the cytoplasm are bumped about by the molecules of the cell's internal environment.

A cell membrane consists mainly of protein and phospholipid molecules arranged into a delicate, thin layer (see Figure 4-1). Most of its functions are associated with regulating the passage of materials into and out of cells. A membrane that allows molecules to pass through it is called *permeable.* Because cell membranes allow certain ions and certain molecules to pass through them, but do not allow the passage of others, they are said to be *selectively permeable, differentially permeable,* or *semipermeable.*

Movement of dissolved inorganic and organic substances within solutions or through the cell membrane involves a natural physical process termed *diffusion.* Diffusion can be demonstrated easily by placing a small dye crystal (the solute) in a volume of water (the solvent). The solute moves from an area of higher solute concentration to an area of lower concentration.

In *osmosis,* a type of diffusion, a solvent (the water in which substances are dissolved) moves from a solution of lower solute concentration to a solution of higher solute concentration when the solutions are separated by a semipermeable membrane.

*Filtration* is described as the movement of solvents and dissolved substances across a semipermeable membrane, under the influence of mechanical pressure. Such movement is always from an area of higher pressure to an area of lower pressure and continues as long as a difference in pressure exists.

This exercise will demonstrate these various phenomena and processes. (*Dialysis,* a process involving the separation of small molecules from larger ones by diffusion of the smaller molecules through a semipermeable membrane, is discussed in Section XIV.)

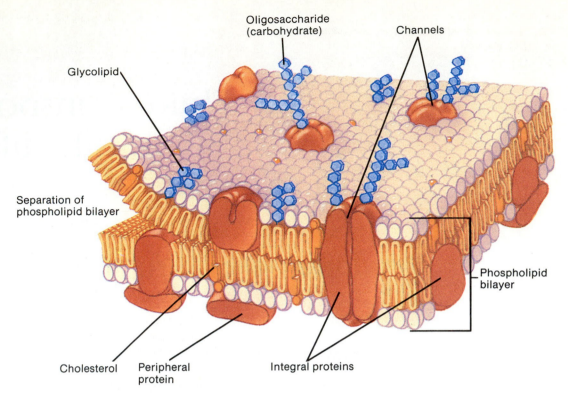

Oligosaccharide
(carbohydrate)

Channels

Glycolipid

Separation of
phospholipid bilayer

Phospholipid
bilayer

Cholesterol  Peripheral
protein

Integral proteins

**Figure 4-1** A diagrammatic enlargement of a cell (plasma) membrane showing the relationship of the phospholipid bilayer and its protein molecules. The bilayer is shown in a separated form, to illustrate the relationships among the various types of molecules found with a membrane. (From G. J. Tortora, *Principles of Human Anatomy*, 6th ed., New York: HarperCollins, 1992.)

## Brownian Movement

### Materials

The following materials should be provided for class use:

1. Dilute solution of India ink
2. Eyedropper and rubber bulb
3. Glass slides and cover slips
4. Microscopes

### Procedure

This procedure is to be performed by students individually.

1. Place a small drop of the dilute India-ink solution on a glass slide, and add a cover slip.
2. Observe the preparation under the high-power objective of your microscope.
3. Reduce the light intensity, and note the direction of movement of the India-ink particles. Indicate the type of movement seen in the results and observations section.

## Filtration

Chemical compounds in solution may behave in one of two ways. The molecules may remain intact—that is, undissociated—or they may dissociate—that is, separate into components termed *ions*. This process is *ionization*. Compounds that behave in this manner are known as *electrolytes;* compounds made up of molecules that do not dissociate are called *nonelectrolytes*.

Ions carry an electrical charge. For example, when the electrolyte sodium chloride is dissolved in water, it provides positively charged sodium ions ($Na^+$) and negatively charged chloride ions ($Cl^-$).

Both nonelectrolytes and electrolytes are used in this exercise, to show how dissolved substances can be forced through physical barriers.

### Materials

1. The following materials should be provided per four students:
   a. One funnel (65 mm diameter)
   b. One ring stand and clamp

c. One 50-ml beaker
d. One sheet of filter paper of appropriate diameter
e. Two test tubes
2. The following materials should be provided for class use.
   a. 5 percent starch solution
   b. 10 percent sodium chloride (NaCl) solution
   c. Iodine reagent
   d. Silver nitrate ($AgNO_3$) solution (0.1 molar)
   e. Powdered wood charcoal
   f. 10-ml graduated cylinders
   g. Wooden tongue depressors

## Procedure

This procedure is to be performed by students in groups of four.
1. Use a graduated cylinder to measure 2 ml of the 10 percent sodium chloride (NaCl) solution. Pour the solution into a test tube. Add 2 drops of the silver nitrate ($AgNO_3$). Observe and record your findings in the results and observations section. This is a test for the presence of chloride ions ($Cl^-$).
2. Put 3 ml of the 5 percent starch solution into the other test tube. Add 3 drops of the iodine reagent. Observe and record your findings in the results and

observations section. This is a test for the presence of starch.
3. Rinse out both test tubes well, and save for a later portion of this exercise.
4. Fold one piece of filter paper in quarters, and open the folded paper to form a cone. Place the cone in a funnel.
5. Set up a filtration system similar to the one shown in Figure 4-2.
6. Prepare a mixture for filtering by mixing the following items in a small beaker:
   a. 4.5 ml starch solution
   b. 2 ml sodium chloride solution
   c. A small amount of the powdered charcoal
   d. 10 ml water
7. Place a test tube under the funnel, and slowly pour 10 ml of water through the system. Collect the water, and discard it. The filter paper should now be firmly in place.
8. Place another test tube temporarily under the funnel, and pour the mixture through the system.
9. Collect the *filtrate* (the solution that passes through) in a beaker.
10. Divide the filtrate between two tubes. Perform the test for starch with one tube and the test for chloride ions with the other.
11. Record your findings, and answer the questions in the results and observations section.

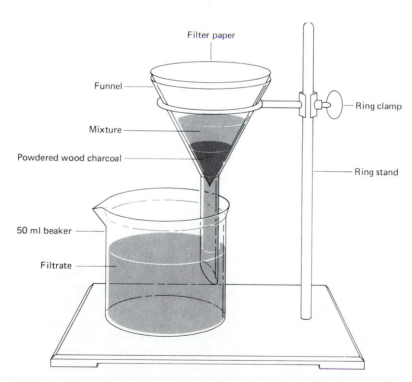

**Figure 4-2** A filtration system. Be certain that all parts of the system are firmly in place.

## Diffusion

### Materials

1. The following materials should be provided per four students:
   a. One plastic Petri plate containing a layer of agar
   b. One pair of forceps
   c. A millimeter ruler
2. The following materials should be provided for class use:
   a. Filter paper discs, approximately 5 mm in diameter
   b. 1 percent crystal violet solution
   c. 1 percent methylene blue solution
   d. 1 percent safranin O solution
   e. Wax pencil or other marking device
   f. Paper towels

### Procedure

This procedure should be performed by students in groups of four.

1. Measure the diameter of the filter paper discs. Enter your measurement in the results and observations section.
2. Obtain a Petri plate containing a thin layer of gelled agar. Mark the bottom of the plate with a wax pencil, to divide the agar preparation into three sectors.
3. Holding a filter paper disc with the forceps, touch it to the surface of the crystal violet solution until the dye penetrates throughout the disc.
4. Remove the Petri plate top, and place the disc in the center of one of the marked sectors. Press the disc gently with the forceps, so that a good contact is made with the agar surface. Close the plate.
5. Wipe the forceps free of the dye solution. Repeat Steps 3 and 4 with the methylene blue and safranin O solutions, placing the discs in the centers of their respective sectors of the Petri plate.
6. Examine the discs on the plate at the beginning of the experiment (0 minutes, time) and at 15, 30, 45, and 60 minutes. Measure the diameters of the developing diffusion zones. (See Color Plate 12.)
7. Enter your findings, and answer the questions in the results and observations section.

## Osmosis

The membrane used in this portion of the exercise is a piece of synthetic material often used as a substitute for a true biological membrane. Such synthetic material contains pores or holes that are approximately the same size as those in biological membranes but are more uniform.

### Materials

1. The following materials should be provided per four students:
   a. One thistle tube (bell-shaped funnel tube)
   b. One dialysis membrane square (about 10 mm square)
   c. One rubber band
   d. One 250-ml beaker and one 500-ml beaker
   e. One ring stand and clamp or similar support system
   f. A millimeter ruler
   g. One marking pen
2. The following materials should be provided for class use:
   a. Four sucrose (cane sugar) solutions having concentrations of 1, 10, 20, and 40 percent sucrose, by weight, respectively
   b. 1 percent methylene blue solution in a dropper bottle
   c. Marking pen or similar device
   d. Transparent tape

### Procedure

This procedure is to be performed by students in groups of four.

An osmometer system will be constructed, with which to determine the rates at which water diffuses through a synthetic membrane separating solutions with differing sucrose concentrations.

1. The instructor will assign a specific sucrose concentration to each student group, such that all four concentrations will be used. Be certain to examine all other systems and the results obtained by other members of the laboratory, so that the significance and influence of different sucrose concentrations can be noted.
2. Using the assigned sucrose concentration, prepare an osmometer system, as follows (see Figure 4-3):
   a. Moisten one dialysis membrane square thoroughly.
   b. Put about 50 ml of the assigned sucrose solution and enough methylene blue to color the preparation into the 250-ml beaker.
   c. While one laboratory partner places a finger firmly over the bottom stem opening of the thistle tube, pour the sucrose solution slowly into the tube until the head (funnel portion) is full.
   d. Allow a small amount of the solution to escape from the stem opening by releasing the finger at the stem opening.
   e. Place the wet membrane square over the funnel end and wind the rubber band around it several

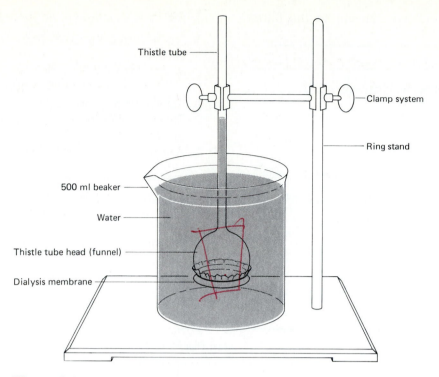

**Figure 4-3** An osmometer system.

times to be sure that the system is secure—that is, that it will not leak.

3. Invert the tube so that the funnel portion is facing down, then attach the thistle tube to a ring stand and clamp or similar support.
4. Fill the 500-ml beaker with water, and place it under the osmometer.
5. Lower the osmometer into the water to the depth as shown in Figure 4-3.
6. Allow 5 minutes for the system to equilibrate. Then locate the top level (meniscus) of the sucrose solution in the tube, and mark it.
7. Using the ruler, measure the level of the meniscus in terms of millimeters moved from the starting point. Measure it again at 15, 30, 45, 60, and 90 minutes.
8. Record your findings, and answer the questions in the results and observations section.
9. Determine the rate of diffusion for each of the sucrose solutions used in the class. (This will require getting the results of others.) To calculate the rate, select a specific time interval (for example, 0–15 minutes), and note the distance covered by the fluid. The rate of diffusion, expressed in millimeters/minute, is equal to the distance divided by the time.

## RESULTS AND OBSERVATIONS

### *Brownian Movement*

1. Use arrows to indicate the type of movement of India-ink particles.

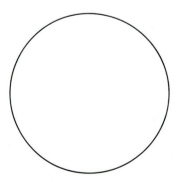

### *Filtration*

1. Indicate the appearance of the solutions after the reagents were added.

   Iodine added to starch: _____

   Silver nitrate added to sodium chloride: _____

2. Which substances were present in the filtrate?

_____

_____

3. What happened to the charcoal? _____

_____

## Diffusion

1. Enter your measurements in Table 4-1.
2. Were the diffusion zones clear and sharp? _____

_____

3. Did the dyes diffuse down through the agar? ____

_____

## Osmosis

1. Enter your measurements in Table 4-2.
2. Enter the diffusion rates of the different solutions in Table 4-3.
3. Which of the solutions had a faster rate of diffusion?

_____

Explain your answer. _____

_____

**Table 4-1**

| Time (in minutes) | Zone Diameters (in mm) | | |
|---|---|---|---|
| | Crystal Violet | Methylene Blue | Safranin O |
| 0 | | | |
| 15 | | | |
| 30 | | | |
| 45 | | | |
| 60 | | | |

**Table 4-2**

| Time | Meniscus Level |
|---|---|
| 0 | |
| 15 | |
| 30 | |
| 45 | |
| 60 | |
| 90 | |

**Table 4-3**

| Sucrose Concentration | Rate of Diffusion (in mm/minute) |
|---|---|
| 1 | |
| 10 | |
| 20 | |
| 40 | |

# LABORATORY REVIEW 4

NAME _____

LAB SECTION _____ DATE _____

## FLUID TRANSPORT AND MOLECULAR ACTIVITY

### COMPLETION AND DEFINITION

1. Distinguish between the items in the following pairs:

   a. Osmosis and diffusion _____

   _____

   _____

   b. Filtration and diffusion _____

   _____

2. What is active transport? _____

   _____

3. Which cellular structures are associated with the following molecular activities?

   a. Osmosis _____

   b. Diffusion _____

   c. Active transport _____

   d. Pinocytosis _____

4. Explain Brownian movement. _____

   _____

5. What is dialysis? _____

   _____

6. Distinguish between electrolytes and nonelectrolytes.

   _____

   _____

## ANSWERS

1. _____
2. _____
3. _____
4. _____
5. _____
6. _____
7. _____

## MULTIPLE CHOICE

Insert the correct answers in the spaces provided.

1. The transport of a solute across a membrane from an area of low concentration to a region of high concentration, using a carrier protein and the expenditure of chemical energy is known as what?
   a. active transport
   b. proton pump
   c. facilitated diffusion
   d. chemiosmosis
   e. Brownian movement

2. The transport of molecules across a cell membrane from a region of high concentration to a region of low concentration, using a carrier protein is known as what?
   a. active transport
   b. diffusion
   c. facilitated diffusion
   d. chemiosmosis
   e. filtration.

3. A term for cell drinking is
   a. active transport
   b. proton pump
   c. facilitated diffusion
   d. chemiosmosis
   e. pinoyctosis.

4. The bulk transport of materials from a cell by extruding from vesicles at the cell surface is called which of the following?
   a. active transport
   b. exocytosis
   c. facilitated diffusion
   d. chemiosmosis
   e. endocytosis

5. The uptake of material into cells by inclusion within an invagination of the plasma membrane forming a vesicle is known as what?
   a. active transport
   b. osmosis
   c. phagocytosis
   d. chemiosmosis
   e. pinocytosis

6. The ingestion of particles, fragments of cells, or even entire cells is called
   a. active transport
   b. proton pump
   c. diffusion
   d. phagocytosis
   e. exocytosis.

7. The channels in the membrane that the cell uses to concentrate metabolites and ions is which of the following?
   a. Na–K pump
   b. proton pump
   c. ATP
   d. coupled channels
   e. chemiosmosis.

# Selective Permeability and the Structural Integrity of Cell Membranes

## After completing this exercise, you should be able to

1. Describe the effects of osmotic imbalances on cell membranes
2. Explain the differences among isotonic, hypotonic, and hypertonic solutions
3. Define *crenation* and *hemolysis*

The cell membrane is an important and dynamic component of cells. In addition to serving as the container for the various cellular structures, compounds, and related materials, it regulates the flow of materials into and out of the cell. All substances, functional or nonfunctional, must pass across the cell's membrane. Several important physical processes associated with such movement were described in Exercise 4.

In Exercise 4, it was shown that when a concentrated sugar solution and water are separated by a membrane permeable to water, the water molecules pass through the membrane to the sugar solution, thereby raising its level. A permeable membrane offers no resistance to the passage of water molecules. Therefore, water molecule movement occurs in both directions. However, water molecules will travel in greater numbers per unit of time from the region where their concentration is highest to where it is lowest.

The force exerted by the water molecules as they flow across a semipermeable membrane is referred to as *osmotic pressure*. Every solution has a potential osmotic pressure. The osmotic pressure of a solution is proportional to the number of solute particles (substance dissolved) per unit volume of a solution. Thus, the more particles in a solution the greater is the osmotic pressure or *osmolality* of the solution. Moreover,

the greater the osmotic pressure of a solution, the greater is its pulling power for water. Water moves toward the area of greater osmolality. It is important to note that the number of particles in solution constitutes the critical factor in determining osmotic pressure. A solution containing an ionizing substance, such as sodium chloride, will exert a greater osmolality than one containing a comparable concentration of a nonionizing substance (refer to the filtration experiment in Exercise 4). The diffusing particles may be ions, molecules, or aggregates of molecules.

From a physiological standpoint, the osmotic properties of solutions can be determined by comparing their effects on red blood cells to the effect of blood plasma (the liquid portion of blood) on the cells. If blood cells are placed in any fluid other than normal plasma, they may remain unaltered, may shrink, or may swell. If they remain unchanged, the solution surrounding them is said to have the same osmotic properties (osmolality) as blood plasma and is called *isotonic* or *isosmotic*. The osmotic pressure of the cell's external environment is equal to that inside the cell (Figure 5-1a, b).

If the cells swell, the solution has lower osmolality than blood plasma and is called *hypotonic* or *hypoosmotic*. The osmotic pressure of the cell's external environment is less than that inside the cell (Figure 5-1c, d). If the pressure does not reach a balance, the cell will continue to swell until it bursts. Such a rupture of red blood cells is known as *hemolysis*.

If the cells shrink, the surrounding solution has higher osmotic properties than plasma and is called *hypertonic* or *hyperosmotic*. The osmotic pressure exerted by the external fluid is greater than that inside the cell. The shrinking and wrinkling of red blood cells in a hypertonic solution is known as *crenation* (Figure 5-1e, f).

In this exercise, yeast and red blood cells are used to demonstrate selected properties of cell membranes.

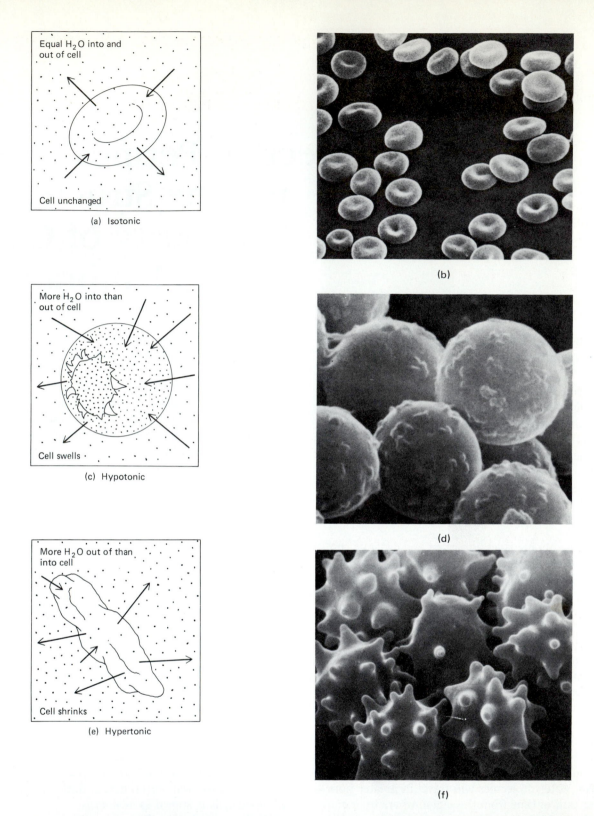

**Figure 5-1** A demonstration of osmotic pressure imbalances. The swelling or shrinkage of a red blood cell (or any cell) depends on the osmotic pressure of the solution in which it is placed. (a) A cell in an isotonic (isosmotic) solution—The osmotic pressure of the cell's external environment is equal to that of its interior. (b) A scanning micrograph of red blood cells in an isotonic solution. (c) A cell in a hypotonic (hypoosmotic) environment—The osmotic pressure is lower outside than inside the cell. Water molecules move into the cell faster than they can leave, diluting the cell's contents and causing it to swell. (d) A scanning micrograph showing swollen cells. (e) A cell in a hypertonic (hyperosmotic) solution—The osmotic pressure is greater outside than inside the cell; water molecules move out of the cell faster than they can enter, causing the cell to shrink. (f) A scanning micrograph showing crenation, one of the effects of a hypertonic environment. (All scanning micrographs from M. P. Sheetz, R. G. Painter, and S. J. Singer. *Journal of Cell Biology*, 70:193–203, 1976.)

## Selective Membrane Permeability

### Materials

The following materials should be provided for class use:

1. Yeast suspension, with Pasteur pipette and rubber bulb
2. 3 percent Congo red solution in dropper bottle
3. Test tubes
4. Water bath set at boiling temperature and equipped with test-tube racks
5. 1-ml pipettes and rubber bulbs
6. Pasteur pipettes and rubber bulbs
7. Glass slides and cover slips
8. Wax marking pencils or similar devices

### Procedure

This procedure is to be performed by students in pairs.

1. Introduce about 1 ml of the yeast suspension into each of three test tubes. Use marking pencils to number these tubes 1, 2, and 3.
2. Add 3 drops of the Congo red solution to Tubes 1 and 2. Shake the tubes gently to mix the contents.
3. Place Tubes 1 and 3 in boiling water bath for 5 minutes. This treatment should kill the yeast cells. Tube 2 should contain living cells.
4. Add 3 drops of the Congo red solution to Tube 3, and shake gently to mix.
5. Allow the heated tubes to reach room temperature (25°C).
6. With separate Pasteur pipettes remove a sample from each tube, and prepare a temporary wet mount of each. (See Exercise 1.)
7. Examine each preparation first under low power and then under high power. Note which preparations contain stained cells.
8. Enter your findings and answer the questions in the results and observations section.

## Structural Integrity of Cell Membranes

### Materials

The following materials should be provided for class use:

1. Individual squeeze bottle containers of the following solutions—
   a. 70 percent alcohol
   b. Distilled water ($H_2O$)
   c. Physiological saline (0.85 percent sodium chloride)
   d. 2 percent sodium chloride (NaCl)
2. Sterile, disposable blood lancets
3. Alcohol prep pads (commercially packaged)
4. Sterile cotton squares
5. Petroleum jelly
6. Wooden applicator sticks
7. Glass depression slides and cover slips
8. Disposable Pasteur pipettes and rubber bulbs
9. Disinfectant solution for disposal of blood-containing materials

### Procedure

This procedure should be performed by students in pairs. It is suggested that one laboratory partner serve as the blood donor. (Refer to Appendix C for alternative sources of blood.)

> ### *CAUTION*
> Working with blood or other body fluids will not pose any danger if the following precautions are observed:
>
> 1. Place any and all blood, or materials containing blood, into the disinfectant containers provided.
> 2. Dispose of any blood-containing materials as indicated by the instructor.
> 3. In cases of blood spillage, wipe the area with a disinfectant-soaked paper towel, and dispose of all materials as indicated by the instructor.
> 4. Always wash your hands after handling blood and associated materials.

1. Obtain three clean, dry glass vials, and label one for each solution, to be used as follows: distilled water, $H_2O$; physiological saline, *sal;* 2 percent sodium chloride, *NaCl.* See Figure 5-2 for steps to be followed in this procedure.
2. Introduce these solutions into their respective vials. The fluid levels should be about 8 mm from the bottom of the vial.
3. Select the finger that will be used, and wipe it with a cotton square that is moistened, but not saturated, with 70 percent alcohol. Do not use a finger that will be involved later in holding a pencil or other such activity.
4. Unwrap the blood lancet. Note that the lancet is to be used only once and then discarded as indicated by the instructor.
5. Hold the finger in a downward position and gently squeeze it.
6. Puncture the ball of the finger with the lancet quickly.
7. Discard the first 2 drops of blood with the aid of a sterile cotton square.

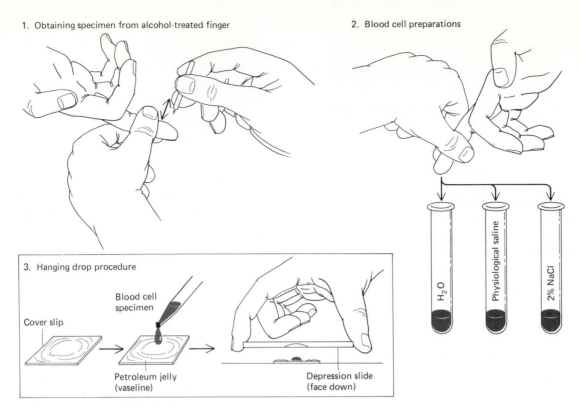

1. Obtaining specimen from alcohol-treated finger

2. Blood cell preparations

$H_2O$

Physiological saline

2% NaCl

3. Hanging drop procedure

Cover slip

Blood cell specimen

Petroleum jelly (vaseline)

Depression slide (face down)

**Figure 5-2** Steps in the laboratory procedure for testing the structural integrity of cell membranes.

8. Keeping the finger in a downward position, place the bleeding site over the open end of the vial containing saline. Hold the vial firmly against the fingertip, and invert it so that the formed blood drop is washed off the finger into the solution.

9. Wipe the finger dry, and repeat Step 8 with each of the other solutions.

10. Examine the vials, and note whether each blood-containing solution is cloudy or clear.

11. With a wooden applicator stick put a *small* amount of petroleum jelly on each corner of three cover slips.

12. Transfer 1 drop of the suspension from the saline vial to the center of one of the prepared cover slips.

13. Place a depression slide face down over the cover slip so that the depression portion is directly over the drop of sample material. Press down gently on the slide so that it will stick firmly to the vaseline on the corners of the cover slip.

14. Turn the slide over in one quick motion so that the drop is suspended in the depression.

15. Examine this preparation first under low power and then under high power.

16. Sketch a number of cells, showing their general appearance, in the results and observations section. Note whether the cells are intact, are round, or show any unusual feature.

17. Repeat Steps 12–16 with the other solutions. (Avoid contamination with the pipette.)

18. Dispose of all blood-containing materials as indicated by your instructor.

19. Wash your hands.

20. Answer the questions in the results and observations section.

## RESULTS AND OBSERVATIONS

### *Selective Membrane Permeability*

1. Which of the preparations contained stained yeast cells? _____

2. Did living or dead yeast cells absorb the dye? _____

3. On the basis of this exercise, how are cell death and cell-membrane control functions related? _____

## Structural Integrity of Cell Membranes

1. Sketch representative fields.

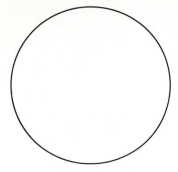

Cells in distilled
water

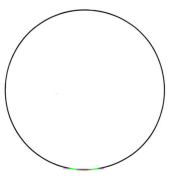

Cells in 0.85
percent saline

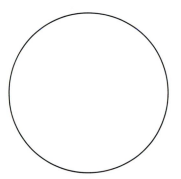

Cells in 2.0
percent sodium
chloride

2. Which of the preparations used was hypotonic to the red blood cells? Explain. _____
   _____
   _____

3. Which of the preparations used was hypertonic to the red blood cells? Explain. _____
   _____
   _____

4. Which of the preparations used was isotonic to the red blood cells? Explain. _____
   _____
   _____

5. How would the different solutions used affect cells with cell walls. (Diagrams can be used.) _____
   _____

6. Which of the preparations exhibited crenation?
   _____
   _____

# LABORATORY REVIEW 5

## SELECTIVE PERMEABILITY AND INTEGRITY OF CELL MEMBRANES

### COMPLETION, FILL-IN, AND DEFINITION

1. a. Do all types of cells have cell membranes? _____

   b. What types of cells lack a cell membrane? _____

   c. What is the chemical composition of a cell membrane? _____

   _____

2. Briefly explain how exposure to the following factors would affect the functioning of a cell membrane:

   a. Heat _____

   b. Lipid solvents _____

3. Figure Q5-1 shows red blood cells suspended in a 2 percent salt solution.

   a. Describe the effect produced. _____

   b. Sketch the effect produced when blood cells are placed in a hypotonic solution.

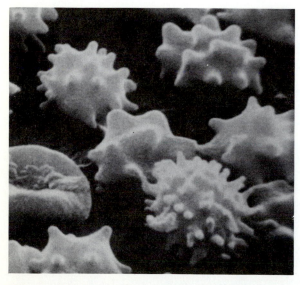

**Figure Q5-1**

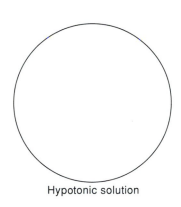

Hypotonic solution

4. Figure Q5-2 shows red blood cells suspended in an isotonic solution.

   a. Describe the effect produced. _____

   b. Sketch the effect produced when blood cells are placed in a hypertonic, not an isotonic solution.

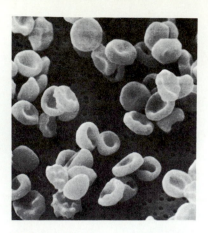

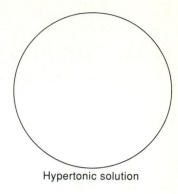

Hypertonic solution

**Figure Q5-2**

5. Define or explain the following:

   a. Osmotic pressure _____

   _____

   b. Crenation _____

   c. Hemolysis _____

   d. Isotonic _____

6. In the system shown in Figure Q5-3, will water molecules pass from A to B or from B to A? Explain your answer. _____

_____

_____

_____

_____

Semipermeable membrane

A       B

Region of low solute concentration

Region of high solute concentration

**Figure Q5-3**

# MULTIPLE CHOICE

## ANSWERS

1. _____
2. _____
3. _____
4. _____
5. _____
6. _____
7. _____
8. _____
9. _____

Insert the correct answers in the spaces provided.

1. Which of the following terms is used to indicate the movement of molecules from a region of high concentration to regions of lower concentration, as a result of spontaneous molecular motions?
   a. diffusion
   b. osmosis
   c. equilibrium
   d. solute
   e. solvent

2. The most common substance in a solution, usually a liquid, is called which of the following?
   a. diffusion
   b. osmosis
   c. equilibrium
   d. solute
   e. solvent

3. Molecules other than those indicated in Question 2, which are dissolved in the liquid in lesser amounts than the liquid, are known as which of the following?
   a. diffusion
   b. osmosis
   c. equilibrium
   d. solute
   e. solvent

4. The diffusion of water across a selectively permeable membrane from an area of greater water concentration to an area of lesser water concentration is called what?
   a. diffusion
   b. osmosis
   c. equilibrium
   d. solute
   e. solvent

5. What is the term that refers to a solution that contains a lower concentration of solute particles (water moves across a membrane out of this solution)?
   a. isotonic
   b. hypertonic
   c. hypotonic
   d. osmotic pressure
   e. equilibrium

6. Which of the following terms refers to a solution with a higher solute concentration (water moves across a membrane into this solution)?
   a. isotonic
   b. hypertonic
   c. hypotonic
   d. osmotic pressure
   e. equilibrium

7. Which term refers to two solutions that have equal concentrations of solute particles?
   a. isotonic
   b. hypertonic
   c. hypotonic
   d. osmotic pressure
   e. equilibrium

8. Which of the following is the term for the potential pressure developed by a solution separated from pure water by a selectively permeable membrane?
   a. isotonic
   b. hypertonic
   c. hypotonic
   d. osmotic pressure
   e. equilibrium

9. Which of the following terms is used to indicate that cytoplasm is pushing out against the cell membrane?
   a. isotonic
   b. hypertonic
   c. hypotonic
   d. osmotic pressure
   e. equilibrium

# Tissues

**After you have completed this exercise, you should be able to**

1. Describe the properties of the four primary tissues of the human body
2. Identify by microscopic examination representatives of primary tissues of the body
3. List the functions of specific tissues
4. Give at least one body location of each type of tissue
5. Identify the component parts of selected tissues

A *tissue* is a group of similar cells that perform a similar function. The study of tissues is called *histology*.

Tissue slides are prepared by fixing part of an organ in a buffered solution of formaldehyde or other fixative, then dehydrating it with an organic solvent and embedding it in paraffin (or epoxyresins for electron microscopy). This embedded preparation is then sectioned with a *microtome* (a device that resembles a very precise meat slicer). All of the tissues shown in this exercise were cut to a thickness of 1 to 8 µm and then stained with appropriate dyes.

## CLASSIFICATION OF TISSUES

The tissues in the human body usually are classified into four main types, based on their structure and function: *epithelial, connective, muscular,* and *nervous* tissues. These four tissue types may be further subdivided into more specific categories (indicated in parentheses).

*Epithelial tissue* lines the inside or the outside of a body, an organ, or a part of an organ (simple, stratified, pseudostratified).

*Connective tissue* connects and supports other tissues in the body (fibrous, adipose, cartilage, bone, blood). Its cells produce a nonliving matrix which becomes located inside (intracellular) or outside (intercellular) the tissue cells.

*Muscular tissue* has the ability to contract (striated skeletal, striated cardiac, and smooth).

*Nervous tissue* (neurons) have the ability to conduct nerve impulses.

Most of the tissue types are discussed in the following parts of this exercise. Each example is accompanied by a photograph of that tissue, and below each photo is a space for your own drawing of any or all of the examples you have the opportunity to observe in the laboratory.[1] If drawings are made, they should be as accurate a representation as possible of the tissue slide you study. If you have no stained slide, you may draw the photograph. Use a hard-lead pencil. Stipple cytoplasmic detail, and label all parts. If you draw more than one cell, label only one. Make careful observations, examine the entire slide to make certain you have found the correct tissue for the drawing, and limit yourself to a small but representative area.

---

[1]Your instructor may assign some drawings and omit others, or ask for no drawings at all. Follow your instructor's directions.

# EPITHELIAL TISSUE

## Materials

The following prepared slides should be provided for class use:

### Epithelial Tissue

1. Cheek cells (squamous)
2. Liver, kidney tubules, or thyroid gland (cuboidal
3. Human skin (stratified squamous)
4. Small intestine cross-section (columnar epithelium)
5. Lining of urinary bladder (stratified transitional epithelium)

### Epithelial Tissue—Exocrine Glands

1. Large intestine (simple tubular glands)
2. Skin (sweat glands)
3. Pyloric region of the stomach (mucous glands)
4. Penile urethra (mucus-secreting glands)
5. Skin containing hair follicles (sebaceous glands)
6. Brunner's glands of duodenum
7. Pancreas (compound acinar glands)
8. Submandibular salivary gland (compound tubulo-acinar gland)

## Simple Epithelium

### Simple Squamous Epithelium

*Squamous* means "flat." These cells often join together at the surface of an organ and give the appearance of pavement tiles (Figure 6-1); hence it is often termed *pavement epithelium*. The cheek cells you studied in Exercise 2 were examples of isolated squamous cells but were actually part of a layered epithelium, to be discussed later in this exercise (stratified epithelium).

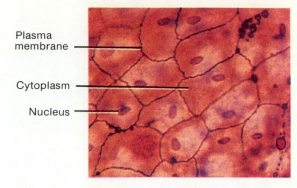

Plasma membrane

Cytoplasm

Nucleus

**Figure 6-1** Squamous epithelium. (© Biophoto, Photo Researchers.)

### Student Drawing

Location of simple squamous epithelium in the body:

1. Peritoneum (lining of the body cavity and its visceral organs)
2. Blood vessel lining (termed *endothelium*)
3. Pleural membranes that surround the lungs
4. Alveoli of lungs
5. Glomerulus of kidney
6. Lens of eye
7. Eardrum (tympanic membrane) layers

## Simple Cuboidal Epithelium

These cells are cube shaped (Figure 6-2) or, if pinched together at one end—as they are in a duct with a small lumen, they are pyramidal (triangular).

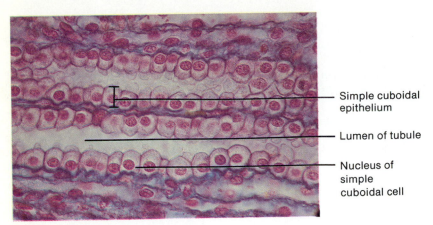

— Simple cuboidal
epithelium

— Lumen of tubule

— Nucleus of
simple
cuboidal cell

**Figure 6-2** Cuboidal epithelium. (Courtesy of Ed Reschke.)

## Student Drawing

Location of cuboidal epithelium in the body:

1. Lining of ducts of kidney tubules and most other tubules.
2. Germinal epithelium (outer layer) of ovary
3. Inner surfaces of the cornea and lens of the eye

## Simple Columnar Epithelium

These cells are elongated (Figure 6-3). The shape of the nucleus usually conforms to the cell shape by becoming somewhat flattened, and it is located at the basal end of the cell (away from the free end). *Goblet cells* are columnar cells that produce mucus. The cells in Figure 6-3 have a great deal of mucus, which is released from the cells' free ends into the digestive tract.

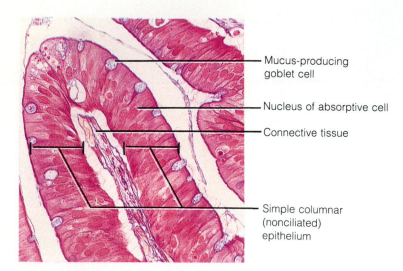

Mucus-producing goblet cell

Nucleus of absorptive cell

Connective tissue

Simple columnar (nonciliated) epithelium

**Figure 6-3** Columnar epithelium. (Courtesy of Lester Bergman and Associates.)

## Student Drawing

Location of simple columnar epithelium in the body:

1. Lining of the digestive tract from stomach to anus
2. Gallbladder lining
3. Ducts of some glands

## Stratified Epithelium

This type of epithelium consists of more than one layer of cells, often combining the three cell types we have just examined: squamous epithelium, cuboidal epithelium, and columnar epithelium. Here, however, all these cells are considered as *one* type of epithelium. The cheek cells previously studied are a part of this tissue.

## *Stratified Squamous Epithelium*

The uppermost layer of this many-layered tissue consists of squamous cells. In Figure 6-4, the cells shown are of the cuboidal type, but they may range into the columnar variety in some tissues.

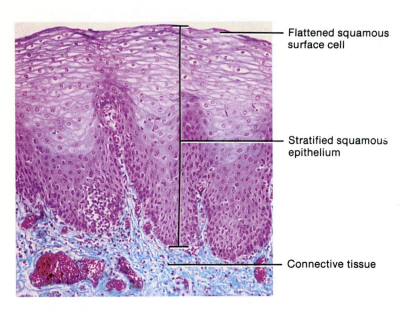

— Flattened squamous surface cell

— Stratified squamous epithelium

— Connective tissue

**Figure 6-4** Stratified squamous epithelium (human cheek cells—430 ×). (Courtesy of Biophoto, SPL, Photo Researchers.)

## *Student Drawing*

Location of stratified squamous epithelium in the body:

1. Epidermis of the skin
2. Lining of the mouth and esophagus
3. Lining of the vagina

## *Stratified Transitional Epithelium*

This type of tissue is similar to stratified squamous except that the layer of cells at the free end is more cube-shaped than squamous (Figure 6-5).

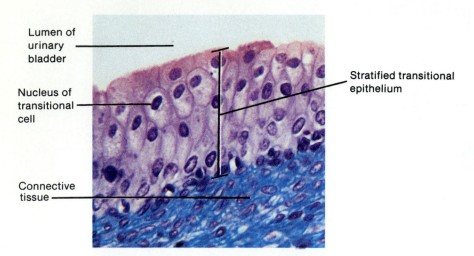

Lumen of urinary bladder

Nucleus of transitional cell

Connective tissue

Stratified transitional epithelium

**Figure 6-5** Stratified transitional epithelium. (Courtesy of Andrew J. Kuntzman.)

## *Student Drawing*

Location of stratified transitional epithelium in the body: lining of urinary bladder

## Epithelium of Exocrine Glands

Glands usually are composed of epithelial cells that have grown inward from the free surface. They are classified structurally as follows and as shown in Figure 6-6.

*Unicellular*—a one-celled gland that produces mucus. An example has already been seen in the *goblet cell* of columnar epithelium.

*Multicellular*—glands composed of many cells.

*Simple*—has a single, nonbranching duct
1. Tubular
2. Branched tubular
3. Coiled tubular
4. Acinar
5. Branched acinar

*Compound*—has a branched duct
1. Tubular
2. Acinar

## CONNECTIVE TISSUE

### Materials

The following prepared slides should be provided for class use:

1. Trachea cross-section (hyaline cartilage)
2. External ear (elastic cartilage)
3. Long-bone cross-section (compact bone)
4. Subcutaneous tissue (adipose or fat cells)
5. Human blood smear (red and white blood cells)
6. Subcutaneous tissue (areolar tissue)
7. Bone marrow (hemopoietic tissue)
8. Tendon (dense connective tissue)
9. Liver (reticular connective tissue)
10. Human blood smear (formal blood elements)

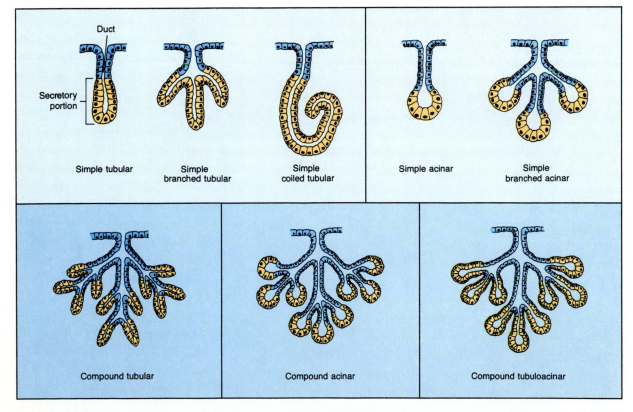

**Figure 6-6** Structural types of multicellular exocrine glands. The secretory portions of the glands are indicated in purple; the blue areas represent the ducts of the glands. (From G. J. Tortora, *Principles of Human Anatomy*, 6th ed., New York: HarperCollins, 1992.)

This large group of diverse tissues has two determining characteristics:

1. The cells produce a nonliving intercellular matrix.
2. The cells are derived from embryonic mesenchyme cells.

There are some exceptions to the production of an intercellular matrix, but in a short classification such as this, exceptions are made for two of the connective tissue types. The special peculiarities of these tissues are noted as they are encountered in this exercise. The connective tissue types you study in this exercise are areolar, adipose, dense, elastic, reticular, cartilage, bone, and vascular (blood).

## Connective Tissue Proper

The many tissues in this category are found in all parts of the body. They bind cells of an organ together, surround and protect major organs, and serve as important links between muscle and bone, as well as between bone and bone.

## *Areolar or Loose Connective Tissue*

The living cells of this tissue (fibroblasts) produce both collagenous and elastic fibers, arranged loosely (Figure 6-7).

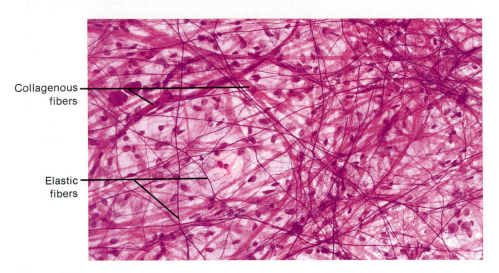

Collagenous fibers

Elastic fibers

**Figure 6-7** Areolar connective tissue. (Courtesy of Biophoto, Photo Researchers.)

## *Student Drawing*

Location of areolar connective tissue in the body:

1. Binding nerves, blood vessels, and body organs to adjacent tissue.
2. Subcutaneous layer of the skin

## Adipose Tissue

These cells are specialized for *intracellular* (within the cells) fat storage (Figure 6-8). This is an exception to the usual intercellular matrix.

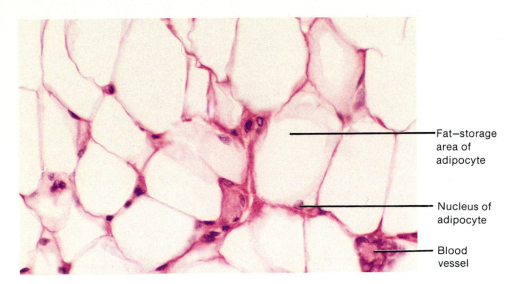

Fat—storage area of adipocyte

Nucleus of adipocyte

Blood vessel

**Figure 6-8** Adipose tissue. (© Bruce Iverson.)

## Student Drawing

Location of adipose tissue in the body:

1. Subcutaneous layer of the skin
2. Surrounding heart, kidneys, and major arteries
3. Yellow bone marrow
4. Various other fatty depots

## Dense (Collagenous) Connective Tissue

The fibroblasts produce strong collagenous (sticky) fibers (Figure 6-9), which are often arranged in parallel fashion, forming tough capsules around organs and forming tendons, ligaments, and aponeuroses (a-po-noo-RO-sees).

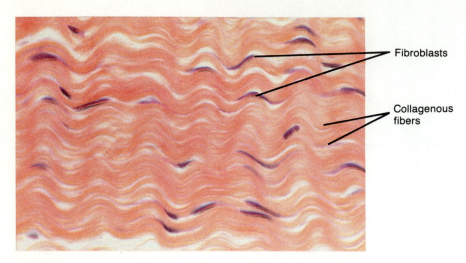

**Figure 6-9** Dense (collagenous) connective tissue. (Courtesy of Andrew J. Kuntzman.)

**Student Drawing**

Location of dense connective tissue in the body:

1. Tendons
2. Ligaments
3. Aponeuroses
4. Capsule surrounding testes
5. Periosteum
6. Perichondrium
7. Fascia

## *Elastic Connective Tissue*

The fibroblasts of this tissue produce both collagenous and elastic fibers (Figure 6-10). The latter lend elasticity to organs.

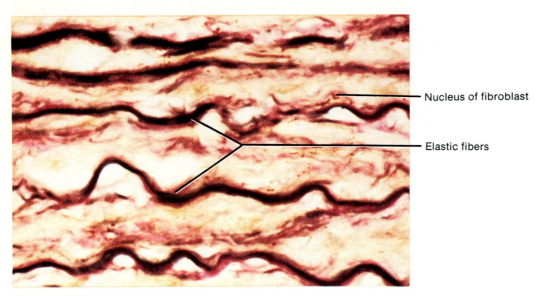

— Nucleus of fibroblast

— Elastic fibers

**Figure 6-10** Elastic connective tissue. (Courtesy of Biophoto, Photo Researchers.)

## *Student Drawing*

Location of elastic connective tissue in the body:

1. Lungs
2. Larynx, trachea, and bronchi
3. Vocal cords

## Reticular Connective Tissue

These very delicate fibers, produced by the fibroblasts, tend to interlace, forming a weblike structure (Figure 6-11).

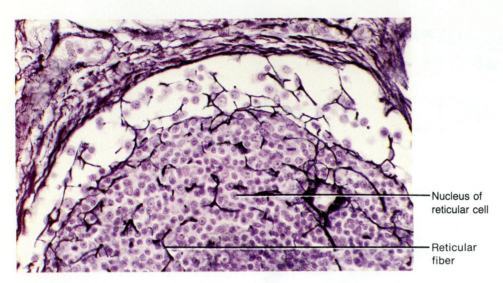

Nucleus of reticular cell

Reticular fiber

**Figure 6-11** Reticular connective tissue. (Courtesy of Biophoto, Photo Researchers.)

## Student Drawing

Location of reticular connective tissue in the body: the infrastructure of many organs, such as the liver and spleen.

## Cartilage

This type of connective tissue is a skeletal tissue. Chondrocytes (cells in cartilage) produce a nonliving matrix, which surrounds and traps the cells within it. There are three types of cartilage: hyaline cartilage, fibrocartilage, and elastic cartilage. Only one, hyaline cartilage, is depicted here.

## Hyaline Cartilage

The matrix surrounding the cells is clear because the fibers within it are so small and fine that they cannot be seen (Figure 6-12).

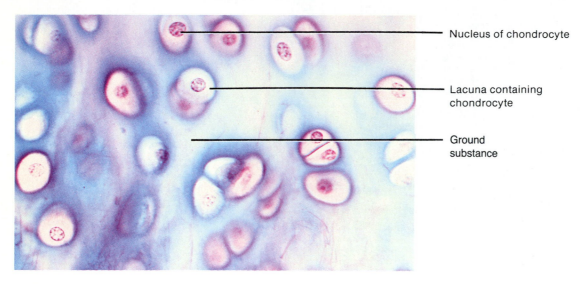

Nucleus of chondrocyte

Lacuna containing chondrocyte

Ground substance

**Figure 6-12** Hyaline cartilage. (Courtesy of Biophoto, Photo Researchers.)

## Student Drawing

Location of hyaline cartilage in the body:

1. Articular cartilages of long bones
2. Earlobe cartilages
3. Larynx, trachea, and bronchi
4. Costal (rib) cartilages

## Bone

The *osteoblasts* (cells that produce the collagenous fibers in this skeletal tissue) become embedded in a hard matrix but retain connections with each other and the blood vessels by means of cytoplasmic extension (canaliculi) (Figure 6-13).

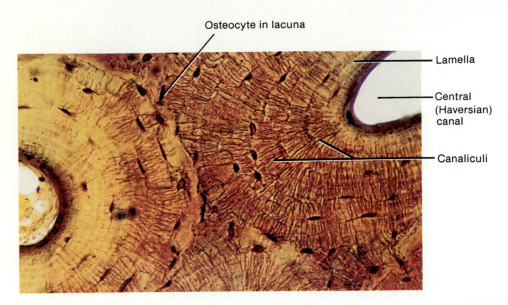

Osteocyte in lacuna

Lamella

Central (Haversian) canal

Canaliculi

**Figure 6-13** Bone. (Courtesy of Biophoto, Photo Researchers.)

*Student Drawing*

Location of bone in the body: the entire framework of bones that makes up the body

## Vascular Tissue (Blood)

This tissue consists of formed elements (erythrocytes, leukocytes, and platelets) and a liquid plasma (Figure 6-14). A stained blood slide will show the pink, disc-shaped *enucleated* (no nucleus) red blood cells and the blue-stained white cells. Examine a blood slide, and pick out as many different white blood cells as you can. (Blood is studied in detail in a later exercise.) If you wish to identify all leukocytes, see Figure 39-2.

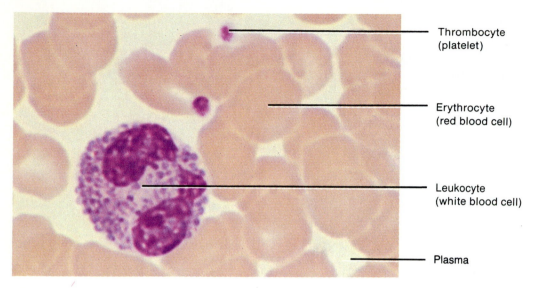

Thrombocyte (platelet)

Erythrocyte (red blood cell)

Leukocyte (white blood cell)

Plasma

**Figure 6-14** Human blood cells: Photo showing erythrocytes and some leukocytes. (Courtesy of Lester Bergman and Associates.)

## Student Drawing

Location of blood in the body: found circulating within arteries, veins, and capillaries of the body's closed circulatory system

## MUSCULAR TISSUE

This highly specialized tissue contains cells that are able to shorten and thicken (contract) upon stimulation.

### Classification of Muscle

Muscle is classified in three ways:

1. *Based on Location in the Body*
   a. *Skeletal*—attached to parts of the skeleton
   b. *Cardiac*—found only in the heart
   c. *Visceral*—found in many visceral organs
2. *Based on Function*
   a. *Voluntary*—skeletal muscle that is under the control of the will
   b. *Involuntary*—muscle that is not under the control of the will (For example, cardiac and visceral muscle are involuntary.)
3. *Based on Microscopic Appearance*
   a. *Striated Skeletal*—cross-striations visible under the light microscope (See text for explanations of this phenomenon.)
   b. *Striated Cardiac*—found only in the heart; cross-striations also visible here
   c. *Nonstriated or Smooth*—found in visceral organs; usually layered; no cross-striations present

Examine each of the aforementioned types of muscle for the given characteristics.

### Striated Skeletal Muscle

In a good preparation, an observer can see that these muscle fibers are *multinucleate* (contain more than one nucleus). This is an unusual characteristic because most cells in the body are *uninucleate* (one nucleus). Seen within the cells are longitudinal muscle fibrils or *myofibrils* and dark-staining masses that appear to be lined up in a row transversely, giving the impression of a line. The myofibrils, which are often difficult to see, run longitudinally, and the dark-staining masses at right angles to them are the cross-striations. A highly specialized cell membrane, the *sarcolemma,* surrounds the cytoplasm, which is referred to as *sarcoplasm.* Find all parts labeled in Figure 6-15 and in the slide given you for laboratory study.

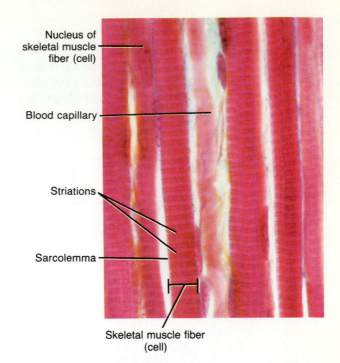

**Figure 6-15** Striated skeletal muscle. (Courtesy of Andrew J. Kuntzman.)

*Student Drawing*

## Striated Cardiac Muscle

The fibers in this type of muscle are all united, the cells are much smaller, and it is difficult to see the myofibrils (Figure 6-16). Notice the dark-staining bars (not seen in striated skeletal muscle) that appear at regular intervals. These bars are *intercalated discs*. Although both cardiac and skeletal muscle are striated, they differ in the following ways:

**STRIATED SKELETAL**

1. Multinucleate
2. Nuclei peripherally located
3. Muscle fibers do not branch
4. There are no intercalated discs

**STRIATED CARDIAC**

1. Uninucleate
2. Nucleus centrally located
3. Muscle fibers are branched, leaving prominent spaces between branches
4. Intercalated discs present

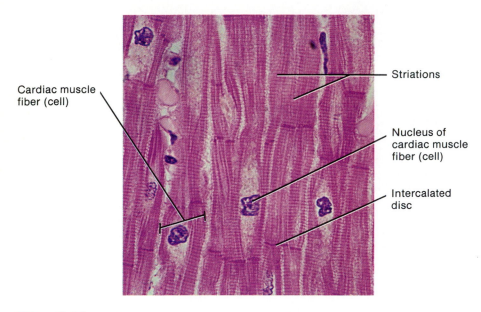

Cardiac muscle fiber (cell)

Striations

Nucleus of cardiac muscle fiber (cell)

Intercalated disc

**Figure 6-16** Cardiac muscle. (Courtesy of Ed Reschke.)

## Student Drawing

## Visceral or Smooth Muscle

*Student Drawing*

These muscle cells are long and tapered and more difficult to see (Figure 6-17). They may be observed *in situ* (in place in the organ where they are located) or teased apart. A good source of smooth muscle is the digestive tract. Look for the myofibrils, the single nucleus, and the cell membrane (sarcolemma).

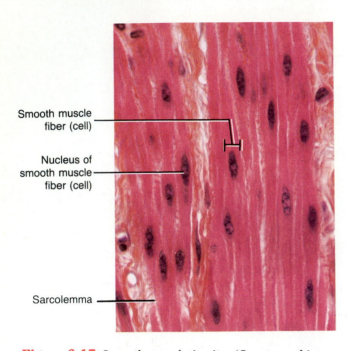

Smooth muscle fiber (cell)

Nucleus of smooth muscle fiber (cell)

Sarcolemma

**Figure 6-17** Smooth muscle *in situ*. (Courtesy of Andrew J. Kuntzman.)

## NERVOUS TISSUE

### Materials

The following prepared slide should be provided for class use: Macerated ox nerve-cord tissue

### Nervous Tissue

The unit of structure and function in nervous tissue is the *neuron*. A neuron is a nerve cell body and its processes. These processes are *dendrites* and *axons*. In a typical nerve cell smear from the spinal cord, you should be able to identify small triangular motor neuron cell bodies and the larger, many-sided association (*internuncial*) neurons. Figure 6-18 shows a single motor neuron, some of its processes, the nucleus, and a nucleolus. Make a drawing of several neurons, showing these parts clearly.

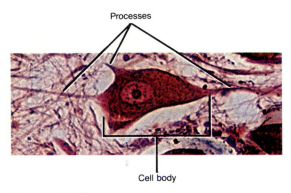

**Figure 6-18** A neuron. (Courtesy of Biophoto Associates.)

# LABORATORY REVIEW *6*

## *TISSUES*

### IDENTIFICATION, COMPLETION, AND DEFINITION

1. Identify the types of tissue shown in Figure Q6-1, and label a nucleus and cytoplasmic membrane of a representative cell in each micrograph.

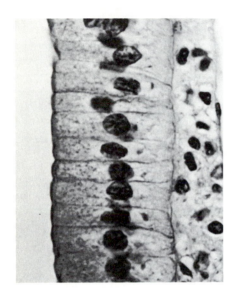

a. _____

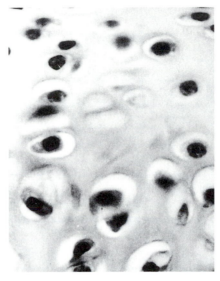

b. _____

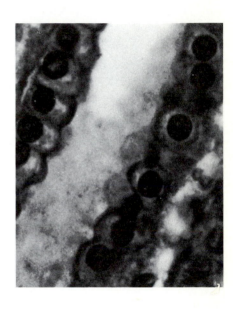

c. _____

d. _____

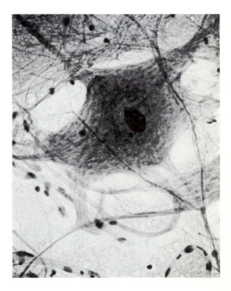

e. _____

f. _____

**Figure Q6-1**

2. List the types and functions of the primary tissues in Table Q6-1.

**Table Q6-1**

| Primary Tissue | Type and Function |
|---|---|
| Connective | a. |
|  | b. |
|  | c. |
|  | d. |
|  | e. |
|  | f. |
| Epithelium | g. |
|  | h. |
|  | i. |
|  | j. |
|  | k. |
| Muscle | l. |
|  | m. |
|  | n. |
| Nervous | o. |
|  | p. |

3. With the aid of your laboratory manual and your text, determine the locations in the body of the following types of tissue.

 a. Smooth muscle _____

 b. Elastic cartilage _____

 c. Compact bone _____

 d. Squamous epithelium _____

 e. Hyaline cartilage _____

 f. Adipose _____

 g. Cardiac muscle _____

 h. Columnar epithelium _____

 i. Hemopoietic _____

4. Place a check (√) in the appropriate space of Table Q6-2 to indicate which muscle type(s) exhibits the property listed.

**Table Q6-2**

| | Cardiac | Skeletal | Visceral (Smooth) |
|---|---|---|---|
| a. Attached to parts of the skeleton | | | |
| b. Found only in the heart | | | |
| c. Found in many visceral organs | | | |
| d. No cross-striations present | | | |
| e. Shows cross-striations | | | |
| f. Has intercalated discs | | | |
| g. Multinucleated | | | |
| h. Uninucleated | | | |
| i. Tapered cells | | | |
| j. Voluntarily controlled | | | |
| k. Involuntarily controlled | | | |

5. Define the word *tissue.* _____

_____

6. What is the name for the study of tissues? _____

_____

7. List below the four main types of tissue, and name one place in the body where each is found:

| TISSUE | LOCATION IN BODY |
| --- | --- |
| a. _____ | _____ |
| b. _____ | _____ |
| c. _____ | _____ |
| d. _____ | _____ |

8. What is the shape of the cell found in the following tissues?

| TISSUE | SHAPE OF CELL |
| --- | --- |
| a. Squamous epithelium | _____ |
| b. Cuboidal epithelium | _____ |
| c. Columnar epithelium | _____ |

9. What does the word *stratified* mean? _____

_____

10. Why is the tissue that makes up the epidermis of the skin called "stratified epithelium"?

_____

11. Connective tissues are made up of cells and a nonliving (usually intercellular) matrix. Describe the appearance of the following:

a. A cartilage cell _____

b. A bone cell _____

c. A fat cell _____

d. What is the matrix composed of in adipose tissue? _____

_____

e. What is the name of the tiny canals that connect bone cells? _____

f. What is the technical name for a bone cell? _____

g. What is the technical name of a cartilage cell? _____

12. What is the function of the neuron? _____

_____

13. Name one organ where each of the following muscle-cell types is located:

a. Cardiac muscle _____

b. Visceral muscle _____

c. Skeletal muscle _____

14. What is the name of the instrument that slices tissue preparations? _____

15. What is the technical name for a red blood cell? _____

16. What is the technical name for a white blood cell? _____

17. Name five different white blood cells: _____

_____

# An Introduction to the Organ and Systems Levels of Organization

**A** division of labor occurs early in the development of higher animals. As Exercise 6 demonstrated, such cells become specialized to perform specific functions and are arranged to form tissues, a process known as *histogenesis*. As the animal body continues to enlarge and develop further, the tissues become organized to produce organs. Later, the organs become clustered to form organ systems, as their functions become more complex.

The exercises of this section emphasize the terms used to specify body position and direction.

### GENERAL REFERENCES

McMinn, R. M. H., and R. T. Hutchings. *Color Atlas of Human Anatomy,* 2nd ed. Chicago: New York Medical Publishers, 1988.

Melloni, J. L., et al. *Melloni's Illustrated Review of Human Anatomy.* Philadelphia: Lippincott, 1988.

Tortora, G. J., and S. R. Grabowski. *Principles of Anatomy and Physiology,* 7th ed. New York: HarperCollins, 1993.

### SPECIFIC REFERENCES

Singer, C. *A Short History of Anatomy and Physiology from the Greeks to Harvey.* New York: Dover Publications, 1957.

Sutton, C. "A Magnetic Window into Bodily Functions." *New Scientist,* 111: 32, 1986.

# Anatomical Orientation and Terminology

*After completing this exercise, you should be able to*

1. Define the anatomical terms used for location and position
2. List the fundamental body planes
3. Identify the major body cavities and their subdivisions
4. Apply proper medical terminology to regions of the body surface for the purpose of indicating location of internal organs
5. Use proper terminology to identify body parts

The purpose of this exercise is to acquaint the student with some of the more fundamental terms used in the study of anatomy. *Anatomy* is a science that describes how an animal is built. One must, therefore, learn the names of body parts, both *macroscopic* (those seen with the naked eye) and *microscopic* (those requiring the use of a microscope). In addition to body parts, there are terms of *location* and *position,* and terms describing *fundamental body planes, surface anatomy, general body regions,* and *body cavities.* A knowledge of these terms is necessary, in order to follow instructions in this lab guide and for comprehension of lecture material and textbook assignments.

## Materials

Models of the human torso should be provided for class use for all the following exercises.

## TERMS OF LOCATION AND POSITION

Terminology for the location and position of body parts of the two-legged human animal varies somewhat from that used for four-legged animals because in four-legged animals the head end is forward in locomotion and the belly surface is parallel to the ground. The differences in terminology are minor, but they can present some difficulty for the beginning student. There is no alternative but to learn all the variations because they are used (often interchangeably) by different authors. These terms are very important, and you should make them a part of your vocabulary as soon as possible (see Figure 7-1).

1. Terms referring to the head end and the rear end fall into three groups.
   *Superior and Inferior*
   a. *Superior*—the head, or upper end; used for humans but not for four-legged animals
   b. *Inferior*—the rear, or lower end, away from the head; used for humans, but not for four-legged animals
   *Cranial and Caudal*
   a. *Cranial*—the head end; used for all animals (including humans) that have a brain case or cranium
   b. *Caudal*—the rear or tail end; used for all animals that have a tail
   *Anterior and Posterior*
   a. *Anterior*—the part of the body that goes forward in locomotion (i.e., the front)
      i. The head end, as in four-legged animals
      ii. The front or belly side of humans

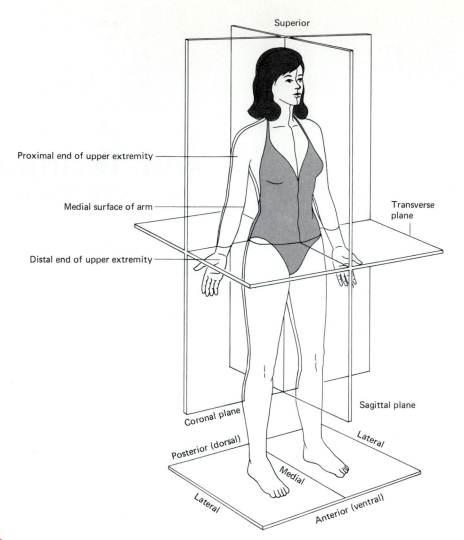

**Figure 7-1** • Anterior view of the human body, including terms of location, position, and body planes.

b. *Posterior*—the part of the body that is not forward in locomotion (i.e., the rear)
   i. The rear end of four-legged animals
   ii. The back side of the human
2. Terms referring to the back side and the belly side usually of a four-legged animal
   *Dorsal and Ventral*
   a. *Dorsal*—the back side of four-legged animals; sometimes also applied to the human back (i.e., the dorsal body cavity)
   b. *Ventral*—the belly side or underside of four-legged animals; sometimes also applied to the human belly side (i.e., the ventral body cavity)
   *Anterior and Posterior*
   a. *Anterior*—the front surface of human
   b. *Posterior*—the back surface of human
3. Terms referring to points on or away from the midline or center
   *Medial and Lateral*
   a. *Medial*—toward or on the midline of the body or an organ
   b. *Lateral*—either away from or to the side of the

midline of the body or an organ
*Central and Peripheral*
a. *Central*—the center of a spheroidal part or organ
b. *Peripheral*—along the outside of a spheroidal part or organ
*Superficial and Deep*
a. *Superficial* (external)—on the outside
b. *Deep* (internal)—on the inside
4. Terms describing appendages or parts of appendages, in reference to the point of attachment to the main axis of the body or to some specific point of reference inside the body
   *Proximal and Distal*
   a. *Proximal*—nearer to the point of attachment to the body axis or trunk in the case of extremities; nearer to the point of reference on an internal organ
   b. *Distal*—farther from the point of attachment to the body axis or trunk in the case of extremities; farther from the point of reference on an internal organ

## TERMS DESCRIBING BODY PLANES

Sections cut through a body or through an organ are very helpful in understanding anatomy. A section or plane is a surface passing through the body that divides it into two parts (see Figure 7-1).

1. *Transverse or cross-section*—a section at right angles to the longitudinal axis of the body or organ, dividing it into superior and inferior parts (in the human)
2. *Sagittal section*—parallel to the longitudinal axis of the body or organ, dividing it into right and left sides
   a. *Midsagittal*—a sagittal section through the midline, resulting in "equal" right and left sides
   b. *Parasagittal*—a sagittal section to the right or left of the midline, resulting in unequal right and left sides
3. *Coronal or frontal section*—a section parallel to the longitudinal axis of the body or organ, dividing it into dorsal and ventral (anterior and posterior in the case of humans) parts

## BODY CAVITIES

The spaces within the body that contain the internal organs are divided into two principal cavities (Figure 7-2), both of which contain further subdivisions (Figure 7-3).

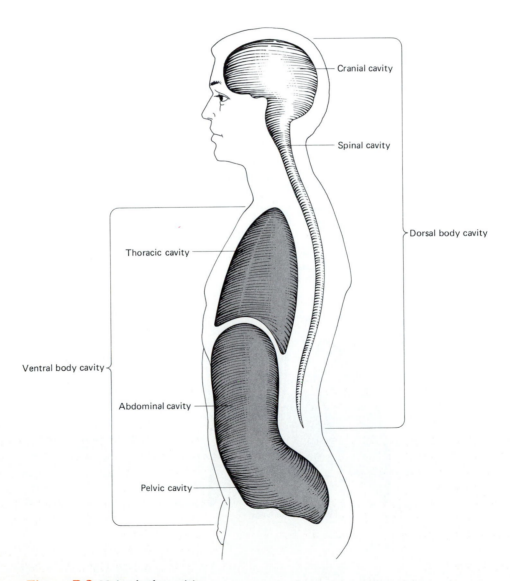

**Figure 7-2** Major body cavities.

1. *Dorsal body cavity*—cavity that has its embryonic origin along the back side of the animal (Notice that the word *dorsal* is used and not *posterior,* for both humans and other animals.)
   a. *Cranial cavity*—cavity formed by the skull, which contains the brain
   b. *Spinal cavity*—cavity formed by the vertebrae, which contains the spinal cord
2. *Ventral body cavity*—cavity that is located in the belly region in both two-legged and four-legged animals
   a. *Thoracic cavity*—located above the diaphragm; the chest cavity (see Figure 7-3)
      i. *Pleural cavities*—cavities between the pleural membranes that surround the lung
      ii. *Pericardial cavity*—cavity between the pericardial membranes that surround the heart
      iii. *Mediastinal cavity*—the cavity between the lungs, containing the heart, trachea, esophagus, and thymus gland
   b. *Abdominal (abdominopelvic) cavity*—cavity located below the diaphragm, containing the viscera
      i. *Abdominal cavity proper*—contains stomach, spleen, liver, gallbladder, pancreas, small intestines, most of large intestine
      ii. *Pelvic cavity*—contains urinary bladder, sigmoid colon, rectum, internal reproductive organs

*Note:* The abdominal cavity, the pleural cavities, and the pericardial cavity are located between two layers of peritoneum: the outer layer is the *parietal peritoneum,* and the inner layer that surrounds the organs is the *visceral peritoneum*. A large cavity, called the abdominal cavity, occupies the space between the parietal and visceral peritoneum below the diaphragm. In the case of both pericardial and pleural cavities, however, the space is very small, and the two layers of peritoneum are closely applied to the respective organs, with only a small fluid-filled space between them.

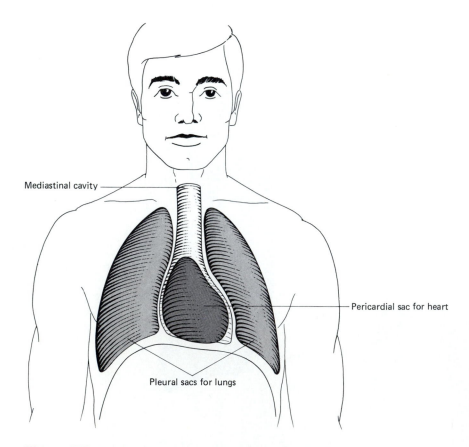

Mediastinal cavity

Pericardial sac for heart

Pleural sacs for lungs

**Figure 7-3** Subdivision of thoracic cavity, anterior view.

## SUBDIVISIONS OF THE ABDOMINAL REGION

The outer surface of the body is marked off into regions
for the purpose of locating and specifying internal parts.
The regions are shown in Figure 7-4.

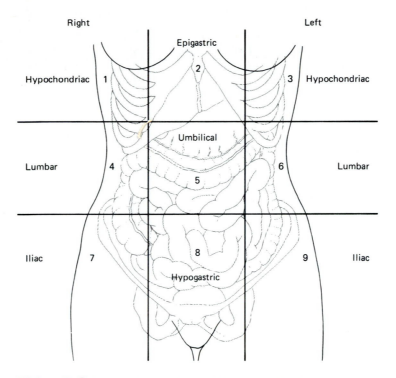

**Figure 7-4** Anatomical regions of the ventral body surface.

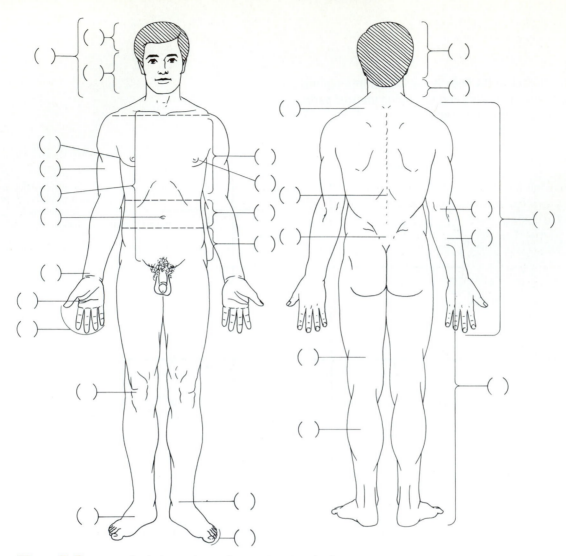

**Figure 7-5** Human body (anterior and posterior views) showing major body regions

## TERMS NAMING BODY PARTS

Identify the body parts on Figure 7-5, using the numerals referring to the common terms and some anatomical terms in the following list. These terms are used in future laboratory exercises, in which more detailed descriptions are given. Insert the numbers from the list on the diagrams of the human body. You may use your textbook.

Head (1)
  Skull, cranium (2)
  Skull, face (3)
Neck (4)
Trunk (5)
  Thorax (chest) (6)
  Breast (7)
  Axilla (armpit) (8)
  Shoulder (9)

Abdomen (10)
Pelvis (11)
Back (12)
Lumbar region (13)
Navel (14)
Upper extremity or arm (15)
  Upper arm (brachium) (16)
  Elbow (17)
  Forearm (antebrachium) (18)
  Wrist (19)
  Hand (20)
  Fingers (21)
Lower extremity or leg (22)
  Thigh (23)
  Knee (24)
  Lower leg or calf (25)
  Ankle (26)
  Foot (27)
  Toes (28)

# LABORATORY REVIEW 7

## ANATOMICAL ORIENTATION AND TERMINOLOGY

### COMPLETION, MATCHING, AND FILL-IN

1. Match the terms on the right with the definitions on the left.
   ( ) A longitudinal cut, dividing body into right and left sides
   ( ) The belly or underside
   ( ) Cavity containing the brain
   ( ) Transverse cut dividing the body into anterior and posterior
   ( ) Space between the membranes that surround the heart
   ( ) Term meaning on the outside of an organ
   ( ) Longitudinal cut dividing body into dorsal and ventral parts
   ( ) Either away from or to the side of the midregion
   ( ) Cavity between the lungs
   ( ) Rear end of an animal

2. In which specific body cavities would the following organs be found? Use your textbook or a chart to locate the organs.

   a. Brain _____

   b. Stomach _____

   c. Heart _____

   d. Thymus gland _____

   e. Spinal cord _____

   f. Pancreas _____

   g. Lungs _____

   h. Esophagus _____

   i. Uterus _____

   j. Liver _____

3. Fill in the correct term for location and position.

   a. The human head is _____ to the neck.

   b. The cat head is _____ to its neck.

   c. The human neck is _____ to the head.

   d. The cat neck is _____ to its head.

   e. The elbow is _____ to the wrist.

   f. The eye is _____ to the nose.

   g. The ankle is _____ to the knee.

   h. The mouth is _____ to the ears.

   i. The tail of the cat is _____ to the trunk.

   j. The vertebral column of the cat is _____ and _____ in position.

### ANSWERS

1. Cranial cavity
2. Coronal section
3. Posterior
4. Lateral
5. Pericardial cavity
6. Sagittal section
7. Mediastinum
8. Ventral
9. Peripheral
10. Cross section

4. On Figure Q7-1, draw the following lines.
   a. A red line to indicate a transverse section
   b. A blue line to indicate a midsagittal section
   c. A green line to indicate a parasagittal section

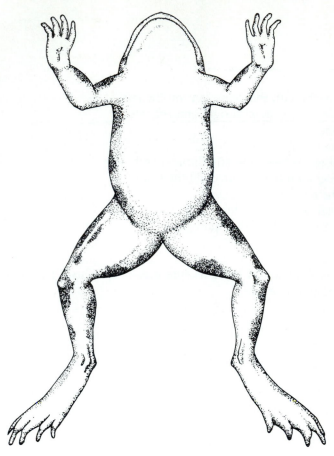

**Figure Q7-1**

5. To which of the five major parts of the body (head, neck, trunk, arm, leg) do the following belong?

   a. Navel _____

   b. Knee _____

   c. Eye _____

   d. Wrist _____

   e. Lumbar area _____

   f. Breast _____

   g. Ankle _____

   h. Chin _____

   i. Pelvis _____

   j. Thigh _____

# The Integumentary System

*T*he skin and several specific derivatives of the skin, such as hair, nails, and various glands, make up the integument of the human body. The skin is only a few millimeters thick, but it is the body's largest organ. Within it, a variety of highly specialized cells are organized into intricate structures and subsystems, to perform many complex functions critical for an individual's survival. For example, the skin (1) provides protection against infection and injury; (2) protects against too little or too much heat loss; (3) serves as the chief means of obtaining vitamin D; (4) plays a vital role in fluid and electrolyte imbalance; and (5) functions as an indicator of changes in the body's external environment. In addition, this organ has been recently recognized to be an integral and active component of the body's immune system.

The organization, structure, and functions of the skin—together with those of its associated derivatives—are presented in this section. The experiments emphasize various properties of the skin, nails, and hair.

### GENERAL REFERENCES

Edelson, T. L., and J. M. Fink. "The Immunologic Function of the Skin." *Scientific American,* 252: 46–53, 1985.

Goldsmith, L. A. (Ed.). *Biochemistry and Physiology of the Skin.* New York: Oxford University Press, 1983.

Mckie, R. *Clinical Dermatology: An Illustrated Textbook,* 3rd ed. New York: Oxford University Press, 1991.

Schaumann, B., and M. Alter. *Dermatoglyphics in Medical Disorders.* New York: Springer-Verlag, 1976.

Tortora, G. J., and S. R. Grabowski. *Principles of Anatomy and Physiology,* 7th ed. New York: HarperCollins, 1993.

### SPECIFIC REFERENCES

Beutner, E. H., T. P. Chorzelski, and S. F. Bean (Eds.). *Immunopathology of the Skin.* New York: Wiley, 1979.

Koss, L. G. *Diagnostic Cytology and Its Histopathologic Bases,* 3rd ed. Philadelphia: Lippincott, 1979.

Pine, D. "Hair! From Personal Statement to Personal Problem." *FDA Consumer,* 25: 20–23, 1991.

Streilein, J. W. "Skin-Associated Lymphoid Tissue (SALT): Origins and Functions." *Journal of Investigative Dermatology,* 80: 12–16, 1983.

# The Skin and Epidermal Derivatives

1. Describe the general structure and organization of skin
2. Identify the major parts of the epidermal derivatives, nails and hair
3. Perform a basic fingerprinting procedure
4. Identify basic digital patterns
5. Measure regional variations in skin temperature

The outer surface of the human body is covered with a protective layer of skin that includes such modified derivatives as nails and hair. The skin is composed of two distinct layers—the superficial layer, known as the *epidermis,* and the deeper layer, termed the *dermis* or *corium.* A subcutaneous layer of tissues is located below the dermis.

The dermis is composed of connective tissue interlaced with blood vessels, fat, glands, hair follicles, and nerves. Fibers from the dermis serve to attach the skin to the subcutaneous layer, which is firmly attached to underlying structures. The subcutaneous layer nourishes, supports, and cushions the skin. The general structure and organization of the skin are shown in Figure 8-1.

The thickness of the skin and of the individual layers varies in different parts of the body. For example, it is thickest on the back, palms, and soles. The thinner skin areas of the body occur on the tympanic (eardrum) membrane and over the eyelids.

In the areas where the epidermis is thickest, it is made up of five *strata* or regions. The outermost region, the *stratum corneum,* consists of numerous layers of dead, scalelike cells composed of proteins termed *ker-*

*atins* (Figure 8-2). Cells of this region are constantly being sloughed or lost by a process termed *exfoliation.* The deepest epidermal region, known as the *stratum basale* or *stratum germinativum,* consists of a single layer of columnar cells undergoing division. Newly formed cells are pushed upward into each successive layer and are eventually sloughed. The epidermis does not contain blood vessels and so depends on the dermis for nutrients.

Skin color depends on the presence of the yellow pigment *carotene* in the surface layer, on the varying concentrations of the brown pigment *melanin* throughout epidermal layers, and on the color of the circulating blood in the dermis. The occasional reddish flush appearance is influenced by the degree of oxygenation of the blood and the extent to which tiny blood vessels open up or close (peripheral blood flow). When the skin is exposed to heat, numerous small blood vessels open up, and more blood flows through the heated tissue. This blood is slightly warmer than usual because it picks up more heat from the warmed tissue. In contrast, the external cooling of the skin causes several arterioles and capillaries to constrict, which reduces the net amount of blood flowing through the tissue.

Of all the human body's organs, none is more exposed to disease and injury than is the skin. Diseases of several internal tissues and organs may be revealed by changes in the skin. These include both color changes or abnormal eruptions (such as rashes) and tumors (Color Plates 13 and 14). The skin is also subject to some disorders and diseases, such as warts and pimples (Color Plate 15).

The entire surface of the skin of humans, in contrast to that of most other mammals, is pitted and grooved by many intersecting lines that form geometric patterns characteristic of each specific area. No area is absolutely free of such markings. These lines, craters, and various microscopic features are stamped indelibly on the skin surface. The palm (palmar) and sole (plan-

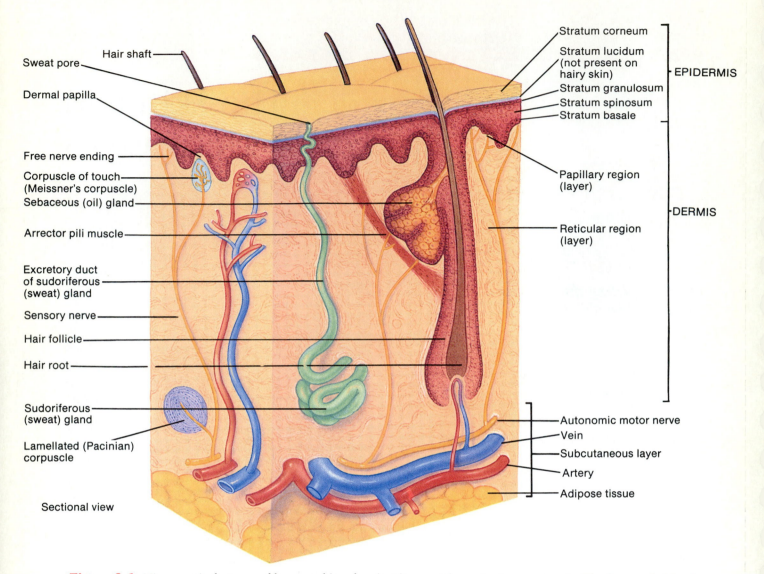

Sweat pore — Hair shaft

Dermal papilla

Free nerve ending

Corpuscle of touch (Meissner's corpuscle)

Sebaceous (oil) gland

Arrector pili muscle

Excretory duct of sudoriferous (sweat) gland

Sensory nerve

Hair follicle

Hair root

Sudoriferous (sweat) gland

Lamellated (Pacinian) corpuscle

Sectional view

Stratum corneum

Stratum lucidum (not present on hairy skin)

Stratum granulosum

Stratum spinosum

Stratum basale

EPIDERMIS

Papillary region (layer)

Reticular region (layer)

DERMIS

Autonomic motor nerve

Vein

Subcutaneous layer

Artery

Adipose tissue

**Figure 8-1** Microscopic features of human skin, showing its most important components. The layers of skin also show a longitudinal section through a hair shaft. (From G. J. Tortora, *Principles of Human Anatomy,* 6th ed., New York: HarperCollins, 1992.)

**Figure 8-2** A scanning micrograph of the skin surface showing the uniform appearance and size of stratum corneum cells. (From F. C. Tring and R. I. Jolly, *Acta Dermatovener (Stockholm),* 53: 343–346, 1973.)

tar) surfaces of an individual characteristically consist of alternating ridges and grooves, which are deeper, closer together, and more clearly seen than those elsewhere on the body. These markings, referred to as the true *dermatoglyphics* (from *derma,* "skin," and *glyphics,* "carvings"), are nearly indestructible and are different in every individual. The ridge patterns on fingertips are divided into three basic digital groups: *arches, loops,* and *whorls* (Figure 8-3). The simplest pattern found on the fingertips is the arch, and the most common is the loop. These markings increase the surface area of the body, which is an adaptation for a more efficient dissipation of heat. In addition, dermatoglyphics are obviously adapted to give the hand and foot a better grip.

Several rules, principles, and definitions have been proposed that allow reliable analyses of dermatoglyphics. In criminology investigation, the intricate details of ridge patterns serve as the bases for fingerprint identifications. In addition, relationships have been reported between unusual dermatoglyphic features and certain genetic abnormalities. Unusual ridge configurations have been shown to exist in individuals with certain chromosomal defects, in persons with single-gene disorders, and in some in whom the genetic basis of disorder is unclear.

Normally, every person has hair on most external areas of the skin surface except the palms and soles. Each individual hair is the product of the epidermis and develops from hair *follicles* (see Color Plates 16a and 16b). The free portion of a hair is referred to as the *shaft,* and the part within the follicle is the *root* (see Figure 8-4). A hair shaft consists of (1) an inner layer, the *medulla,* which is composed of loosely arranged,

soft, horny cells and several air spaces; (2) a layer surrounding the medulla, the *cortex,* which is composed of hard, closely placed, horny cells; and (3) an outer layer, the *cuticle,* which is composed of hard, tightly joined, horny cells (Figure 8-4). Individual hairs have one or several oil (sebaceous) glands, the ducts of which empty into the follicle. Smooth muscles, called *arrector pili,* are attached to the follicle below the sebaceous glands and can move the shaft in response to various emotions and temperatures. "Goose pimples" result from hair-shaft movements.

Adult nails consist of a plate of horny keratinized cells situated on a *nail bed* over the terminal *phalanges* (bones) of the fingers and toes (Figure 8-5). The visible portion of the nail *(nail body)* is bordered laterally and proximally by a fold of skin called the *nail fold.* A *nail groove* lies between this fold and the nail. The portion of the nail hidden under the nail fold is known as the *root.*

Other noticeable features include the *lunula,* a white semicircle at the base of each nail, and the cuticle or *eponychium,* which is the region where the base of the nail connects to the finger or toe. The distal end of the finger or toe under the free edge of the nail is the *hyponychium.* The characteristic pink color of the nail is due to the blood vessels that show through the semi-transparent nail structure.

Nails also are subject to injury, disorders, and diseases. Discoloration of a nail not only represents a major part of the disease or disorder, but may also serve as a clue to a specific diagnosis. Causes of nail discoloration include injury (Color Plate 17), infections, malignancies, and the effects of some specific medications (Color

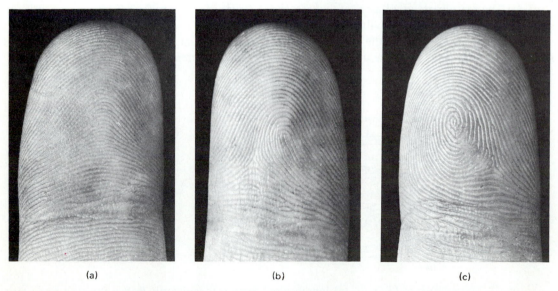

(a)        (b)        (c)

**Figure 8-3** Dermatoglyphics. Photographs demonstrating the three basic digital patterns: (a) *arch,* (b) *loop,* and (c) *whorl.* (From J. R. Miller and J. Giroux. *Journal of Pediatrics,* 69: 302–312, 1966.)

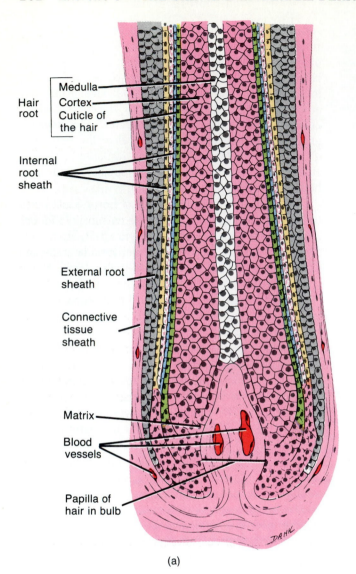

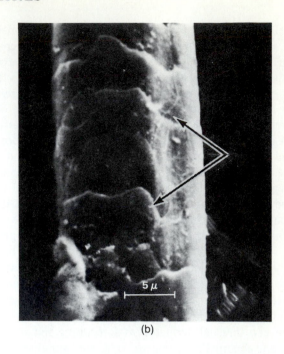

(b)

Figure 8-4 (a) The main parts of a hair root and associated structures. (From G. J. Tortora, *Principles of Human Anatomy,* 6th ed., New York: HarperCollins, 1992. © Leonard Dank.) (b) The surface appearance of hair. A scanning micrograph of the hair shaft showing the cuticular scales (*arrows*) on its surface. (From D. M. Singh, W. B. Green, R. A. Osborne, J. Upshur, and G. R. Hennigar, *Journal of Invest. Dermatology,* 63: 334–336, 1974.)

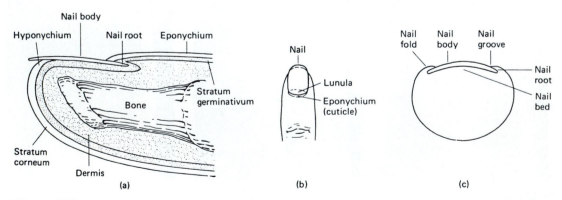

Figure 8-5 Structure of the nail: (a) longitudinal section through the finger showing important features of the nail; (b) top view of the finger; (c) cross-section of nail and nail bed.

Plate 18). Hemorrhaging, nail loosening or detachment, transverse pigmented lines, and white spots or lines are among the common signs of nail changes (Color Plate 19).

The procedures in this exercise emphasize the external structure of nails, the dermatoglyphics of fingertips, and the microscopic properties of skin and hair.

## Examination of Fingers and Fingerprints

### Materials

The following materials should be provided for class use:

1. Printer's ink
2. One ink roller
3. Two glass or metal slabs or ordinary stamp pads
4. Soap (cleanser and paper towels)

### Procedure 1: External Structure of Nails

This procedure is to be performed by students in pairs.

1. Examine the fingers of your partner's right hand.
2. Locate all parts of the nail shown in Figure 8-5.

### Procedure 2: Fingerprints

This procedure is to be performed by students in pairs.

1. Put a small amount of ink on an inking slab. Spread it with a roller into a thin, even film.
2. Place one edge of the thumb of your right hand against the inked slab. Allow your thumb to roll naturally across the slab to its opposite edge, so that the entire area to be printed is covered with ink.
3. Repeat the movement with the thumb on the space provided in the results and observations section. The sheet of paper on which the inked finger is pressed must rest on a firm surface.
4. One at a time, place the remaining fingers of your hand directly down on the inked slab, and press them directly onto the spaces provided in the results and observations section.
5. Repeat Steps 2–4 with your partner's right hand.
6. Compare the fingerprints obtained, and answer the questions in the results and observations section.

## Microscopic Identification of the Skin and Hair

### Materials

The following materials should be provided for class use:

1. Prepared slides of
   a. Human skin showing both epidermis and dermis
   b. Human scalp showing hair follicles
2. Glass slides and cover slips
3. Eyedroppers

### Procedure

This procedure is to be performed by students individually.

1. Examine the prepared slides of human skin.
2. With the aid of Figure 8-1 and your textbook, identify the specific layers of the epidermis and the various glands typically found in the dermis.
3. Examine the preparation of human scalp, and identify the components of a hair follicle (see Color Plate 16).
4. Remove a hair from your scalp and one from your partner's scalp. Prepare a wet mount. (Refer to Exercise 1 for directions.)
5. Examine the hair preparations under both low power and high power.
6. Sketch the appearance of both hairs in the space provided in the results and observations section.
7. Look for any unusual surface features, such as scales and patterns.
8. Enter your findings in the results and observations section, and answer items in the laboratory review pertaining to this exercise.

## RESULTS AND OBSERVATIONS

### *Fingerprints*

1. Student's name: _____

| Thumbprint | Index finger |
|:---:|:---:|

Remaining fingerprints

2. Partner's name: _____

Thumbprint            Index finger

Remaining fingerprints

3. What type of fingerprint pattern did your fingers exhibit? (Refer to Figure 8-3.)

   _____

4. What type of fingerprint pattern did your partner's fingers exhibit?

   _____

5. Were the patterns the same for both individuals?

   _____

6. Were there any unusual features to the patterns obtained? If so, circle them on the specific fingerprint, and describe them here.

   _____

   _____

   _____

*Microscopic Examination of the Hair*

1. Hair sketch

Your hair

Partner's hair

2. Describe any unusual surface features.

   _____

   _____

   _____

   _____

# LABORATORY REVIEW *8*

## *THE SKIN AND EPIDERMAL DERIVATIVES*

### COMPLETION, FILL-IN, AND DEFINITION

1. Identify the parts of the skin in Figure Q8-1.

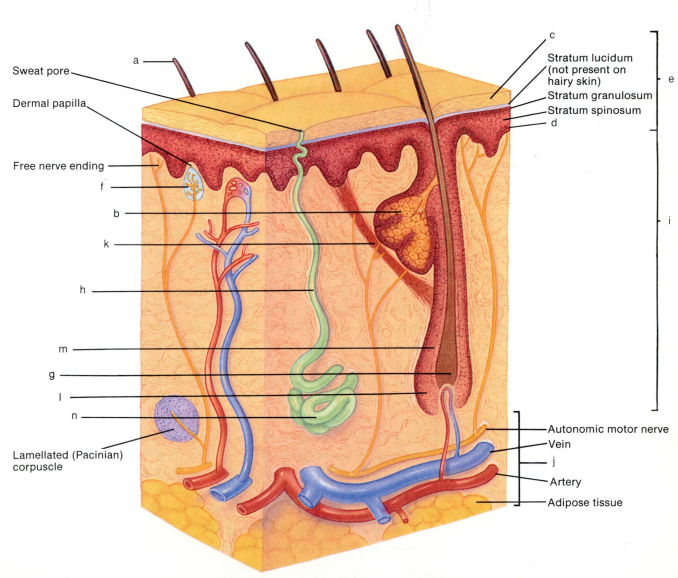

**Figure Q8-1** A view of the skin and associated structure. (From G. J. Tortora and S. R. Grabowski, *Principles of Anatomy and Physiology,* 7th ed., New York: HarperCollins, 1993.)

a. _____ f. _____ k. _____

b. _____ g. _____ l. _____

c. _____ h. _____ m. _____

d. _____ i. _____ n. _____

e. _____ j. _____

2. Identify the parts of the nail in Figure Q8-2.

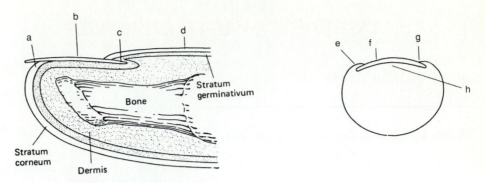

**Figure Q8-2** Parts of the nail.

a. _____    e. _____

b. _____    f. _____

c. _____    g. _____

d. _____    h. _____

3. The fingerprint in Figure Q8-3a was found on a stolen object. The fingerprints of three suspects (b, c, and d) were taken and compared.

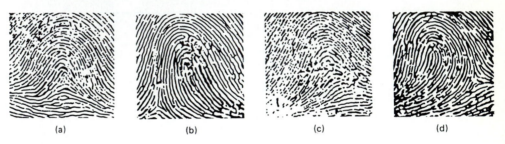

(a)            (b)            (c)            (d)

**Figure Q8-3** Fingerprints.

a. On the basis of this evidence, which of the three individuals would you suspect?

_____

b. What are the digital patterns shown for each fingerprint?

a. _____

b. _____

c. _____

d. _____

4. Define or explain the following:

a. Palmar _____

b. Dermatoglyphics _____

c. Dermis _____

d. Sebaceous gland _____

e. Stratum basale _____

f. Stratum corneum _____

g. Eponychium _____

h. Lunula _____

## MULTIPLE CHOICE

Insert the correct answers in the spaces provided.

1. The outermost layer of epidermis is which of the following?
   a. stratum spinosum
   b. stratum germinativum
   c. hypodermis
   d. stratum corneum
   e. stratum lucidum
2. The hair follicle is which of the following?
   a. an enlarged covering of skin
   b. a shaft of hair
   c. the hair root
   d. an enlarged sebaceous gland
   e. an eccrine gland
3. Sweat is produced by?
   a. sebaceous glands
   b. eccrine glands
   c. cerumenous glands
   d. endocrine glands
   e. papillary skin layer
4. Sebum is secreted by which of the following?
   a. sweat glands
   b. eccrine glands
   c. apocrine glands
   d. papillary skin layer
   e. sebaceous glands
5. Sebaceous glands produce which of the following?
   a. corneum lucidum
   b. sweat
   c. sebum
   d. water
   e. keratin
6. The attached visible part of a nail is known as the
   a. nail root
   b. nail fold
   c. nail bed
   d. nail proper
   e. nail body
7. The cuticle of a nail is known as the
   a. eponychium
   b. hyponychium
   c. matrix
   d. fold
   e. root
8. What are the fingerlike projections of the dermis containing capillaries and receptors called?
   a. pili
   b. polyps
   c. dermal papillae
   d. skin folds
   e. skin roots
9. The waterproofing property of human skin is associated with the presence of
   a. carotene
   b. keratin

ANSWERS

1. _____
2. _____
3. _____
4. _____
5. _____
6. _____
7. _____
8. _____
9. _____
10. _____

c. blood vessels
d. roots
e. melanin

10. Which of the following choices show the proper sequence of the layering of the epidermis, moving from the free outer surface inward to the underlying tissues?
a. corneum, granulosum, lucidium, spinosum
b. corneum, lucidum, granulosum, spinosum, basale
c. spinosum, basale, granulosum, corneum
d. basale, spinosum, granulosum, corneum
e. none of the sequences listed are correct

# The Support System

*I*n vertebrates, the skeleton is associated with three important body functions: support, protection, and locomotion. This internal framework contains both cartilage and bone. As indicated in Exercise 6, cartilage is one of several types of connective tissue. It consists of scattered cells embedded in a flexible, noncellular matrix which the cells secrete.

Vertebrate skeletons generally are divided into two components: the axial skeleton and the appendicular skeleton. The present section contains several exercises that consider the structure, arrangement, and functions of these skeletal components.

### GENERAL REFERENCES

Caplan, A. I. "Cartilage." *Scientific American,* 251: 84–94, 1984.

Platzer, W. "Locomotor System." In W. Kahle, H. Leonhardt, and W. Platzer (Eds.), *Color Atlas and Textbook of Human Anatomy,* Vol. 1. Chicago: Year Book Medical Publishers, 1978.

Roose, C., and D. K. Clawson. *The Musculoskeletal System in Health and Disease.* New York: Harper & Row, 1980.

Tortora, G. J., and S. R. Grabowski. *Principles of Anatomy and Physiology,* 7th ed. New York: HarperCollins, 1993.

Vaughan, J. M. *The Physiology of Bone,* 3rd ed. New York: Oxford University Press, 1981.

### SPECIFIC REFERENCES

Chestnut, C. H., III. "Osteoporosis and Its Treatment." *The New England Journal of Medicine,* 326: 406–407, 1992.

Nadel, E. R. "Physiological Adaptations to Aerobic Training." *American Scientist,* 73: 334–343, 1985.

Ott, S. M. "Bone Density in Adolescents." *The New England Journal of Medicine.* 325: 1646–1647, 1991.

Sharpe, W. D. "Age Changes in Human Bone: An Overview." *Bulletin of The New York Academy of Medicine,* 55: 757, 1979.

Sonstegard. D. A., L. S. Matthews, and H. Kaufer. "The Surgical Replacement of the Human Knee Joint." *Scientific American,* January 1978.

Strauss, R. H. (Ed.). *Sports Medicine and Physiology.* New York: Saunders, 1979.

# Introduction to the Study of the Skeleton

**After completing this exercise, you should be able to**

1. Describe the histology of bone
2. List the parts of bone
3. Be familiar with the major subdivisions of the skeleton

## Materials

The following materials should be provided for class use:

1. Mounted human skeleton
2. Long bone, split lengthwise
3. Prepared slides of bone, for demonstration
4. Microscopes
5. Charts and models of typical bones

## MICROSCOPIC STRUCTURE OF BONE

Examine a demonstration slide showing a section through human compact bone that has been ground by hand. You will see that bone is made up of numerous *Haversian systems* (see Figure 9-1, and review the work done in Exercise 6). These systems are very active. They develop quickly, pushing other older Haversian systems out of the way. As a result you see many Haversian system fragments on the slide. These are *interstitial lamellae.* In the center of each system is the *Haversian canal*—containing blood vessels, lymph vessels, and nerves—and surrounding the canal are concentric rings of bone called *lamellae.* Trapped between the lamellae are the bone cells or *osteocytes,* with their long cytoplasmic extensions in the *canaliculi.*

## MACROSCOPIC APPEARANCE OF BONE

If a long bone that has been split from its proximal to distal ends is available, study the parts. If no specimen is available, use Figure 9-2 to locate the following parts.

1. *Head or epiphysis* (ē-PIF-i-sis)—the end of the bone beyond the epiphyseal (ep-i-FIZ-ēal) line sometimes visible on the outside of the bone. A layer of compact bone covers the inner spongy bone. The *cancellous* (spongy) bone contains the *red bone marrow,* which is the site of blood-cell formation. A layer of *articular cartilage* covers the head, where it is part of a joint.
2. *Shaft* or *diaphysis* (dī-AF-i-sis)—the part of the long bone between the two heads. The bone shaft has an outer layer of compact bone and a hollow center, called the marrow or *medullary cavity,* which contains *yellow bone marrow* in the living specimen.
3. *Periosteum* (pair-ē-OS-tē-um)—a membrane surrounding the outside of a bone (silvery in a freshly killed or living specimen, but absent in the dried bone). This membrane is active in producing more bone on the outer surface.
4. *Endosteum* (end-OS-tē-um)—a membrane lining the medullary cavity. It is similar to the periosteum and produces bone on the inner surface. (It, too, cannot be seen in the dried specimen.)

## MAJOR DIVISIONS OF THE SKELETON

The bones of the human skeleton are grouped into two major divisions, as shown in Figure 9-3.

1. *Axial skeleton* ( *blue* )—the bones located along the midline axis of the body

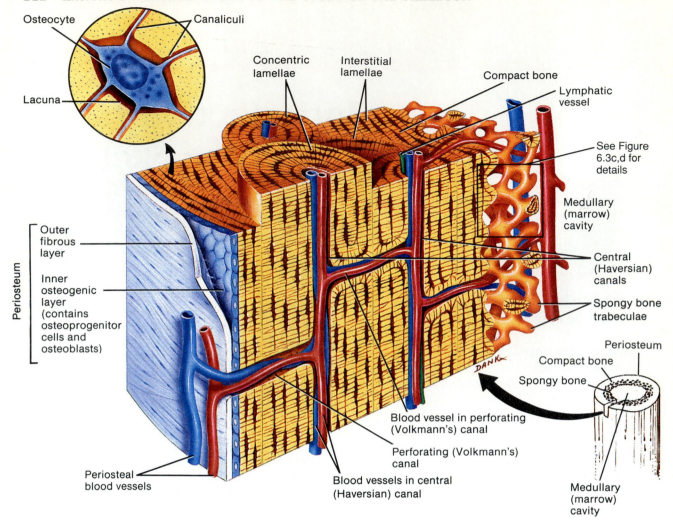

**Figure 9-1** Enlarged aspect of osteons (Haversian systems). (From G. J. Tortora and S. R. Grabowski, *Principles of Anatomy and Physiology,* 7th ed., New York: HarperCollins, 1993.)

2. *Appendicular skeleton* (yellow)—the bones that make up the appendages and the girdles

Each major division is further subdivided into regions. With the help of Figure 9-3, look over the mounted skeleton, and identify as many bones as you can. Be sure you have a clear idea of the bones in each subdivision before beginning detailed study in Exercises 10 through 13.

The subdivisions of the axial skeleton are

1. *Skull—cranial bones,* surrounding the brain, and *facial bones*
2. *Visceral skeleton*—derivatives of the embryonic visceral or branchial (gill) arches; the only bone in this skeletal division which can be seen on the laboratory skeleton is the Hyoid bone
3. *Vertebral column*—the vertebrae, series of bones in the middorsal region inferior to the skull

4. *Sternum*—commonly known as the "breast bone," located in the midventral or anterior thoracic region
5. *Ribs*—12 pairs of widely separated bones in the thoracic region

The subdivisions of the appendicular skeleton are

1. *Pectoral (shoulder) girdle*—two bones located in the shoulder that form the support for the articulation of the arm
2. *Upper extremity*—series of bones, beginning at the pectoral girdle, that make up the arm
3. *Pelvic girdle (os coxae)*—fused bones that form the support for articulation with the leg
4. *Lower extremity*—series of bones, beginning at the pelvic girdle, that make up the leg

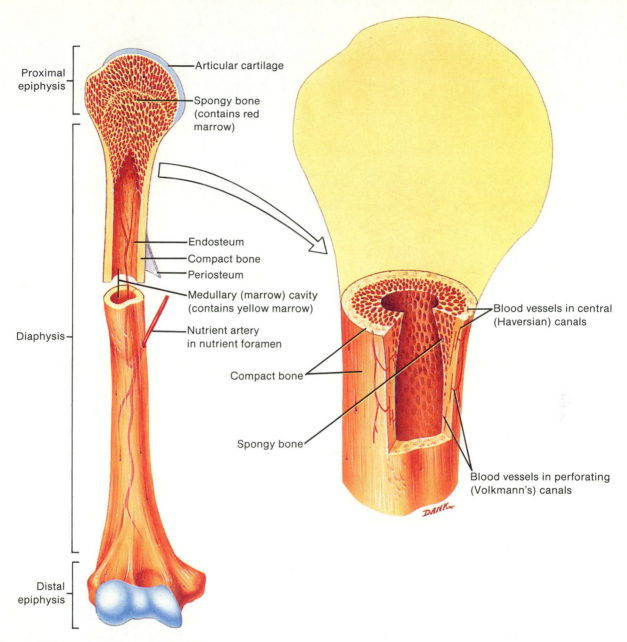

Proximal epiphysis

Articular cartilage

Spongy bone (contains red marrow)

Endosteum

Compact bone

Periosteum

Medullary (marrow) cavity (contains yellow marrow)

Nutrient artery in nutrient foramen

Diaphysis

Compact bone

Spongy bone

Distal epiphysis

Blood vessels in central (Haversian) canals

Compact bone

Blood vessels in perforating (Volkmann's) canals

**Figure 9-2** Longitudinal section through a long bone. (From G. J. Tortora, *Principles of Human Anatomy,* 6th ed., New York: HarperCollins, 1992. © Leonard Dank.)

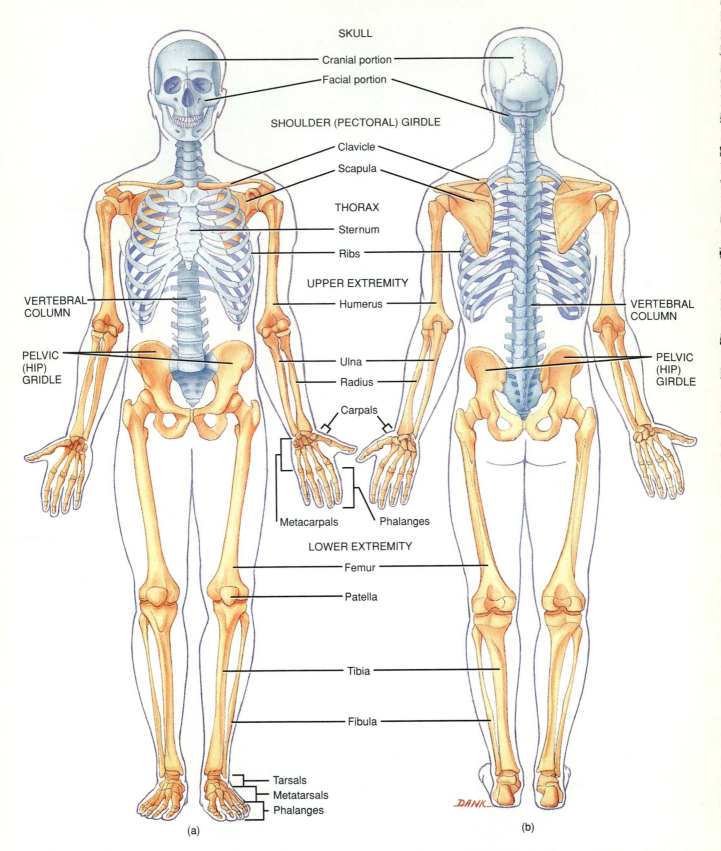

**Figure 9-3** Division of the skeletal system (the axial skeleton is indicated in blue): (a) anterior view; (b) posterior view. (From G. J. Tortora and S. R. Grabowski, *Principles of Anatomy and Physiology*, 7th ed., New York: HarperCollins. 1993. © Leonard Dank. )

# LABORATORY REVIEW 9

## INTRODUCTION TO THE STUDY OF THE SKELETON

### COMPLETION

1. Name the divisions of the axial skeleton.

   a. _____

   b. _____

   c. _____

   d. _____

   e. _____

2. Name the divisions of the appendicular skeleton.

   a. _____

   b. _____

   c. _____

   d. _____

3. What is the key characteristic of any bone in the axial skeleton?

   _____

4. What is the key characteristic of any bone in the appendicular skeleton?

   _____

5. Pronounce the names of the bones described in this exercise at least two times or until you do not have any difficulty with them. Use this method throughout the laboratory exercises whenever large numbers of unfamiliar words are encountered.

# The Skull (Part I)

*After completing this exercise,*
*you should be able to*

1. Identify all major bones of the human skull
2. Locate the major skull sutures and fontanels
3. Correlate the material in this exercise with the chapters in your textbook, with respect to origin and composition of bone
4. Pronounce all names correctly and fluently

## Materials

The following materials should be provided for class use:

1. Mounted human skeleton
2. Human skull
3. Mounted disarticulated skull (Beauchene skull)
4. Disarticulated skull bones

The bones of the skull make up the head of an animal. They are subdivided into *cranial bones* and *facial bones*. Figures 10-1 through 10-6 provide various views and labels. Use them, along with the laboratory specimens, so that you will be able to identify the actual bones and not just pictures of bones.

## THE CRANIAL BONES

Eight bones surround and protect the brain, forming the brain case or *cranium*. Be able to identify these bones on the skulls in the laboratory. Always learn the correct pronunciation and spelling.

1. *Frontal* (FRUN-tal)—large, single bone that forms the forehead
2. *Parietal* (pa-RĪ-i-tal)—paired bones that make up most of the top of the head, posterior to the frontal bone
3. *Occipital* (ok-SIP-i-tal)—large, single bone at the back of the head, posterior to the parietal bones
4. *Temporal* (TEM-por-al)—paired bones at the sides of the skull, inferior to the parietal bones
5. *Sphenoid* (SFĒ-noyd)—large, single bone that forms most of the floor of the cranium
6. *Ethmoid* (ETH-moyd)—single bone on the inner surface of the cranium, anterior to the sphenoid

## THE FACIAL BONES

The remainder of the skull is made up of 14 bones, which together are the *face*.

1. *Nasal* (NĀ-zal)—small, paired bones that constitute the bridge of the nose (Other skeletal structures within the nose are either cartilage and not preserved on this skeleton or parts of other skull bones)
2. *Maxillary* (MAK-si-lair-ē)—large, paired bones of the upper jaw that bear teeth
3. *Lacrimal* (LAK-ri-mal)—small, paired bones located within the orbit of the eye and posterior to the uppermost portion of the maxillary bones
4. *Zygomatic* (zī-gō-MAT-ik)—paired bones that articulate with the maxillaries, commonly known as the *cheek bones*
5. *Palatine* (PAL-a-tīn)—paired bones located in the mouth cavity, posterior to the maxillaries and forming part of the hard palate

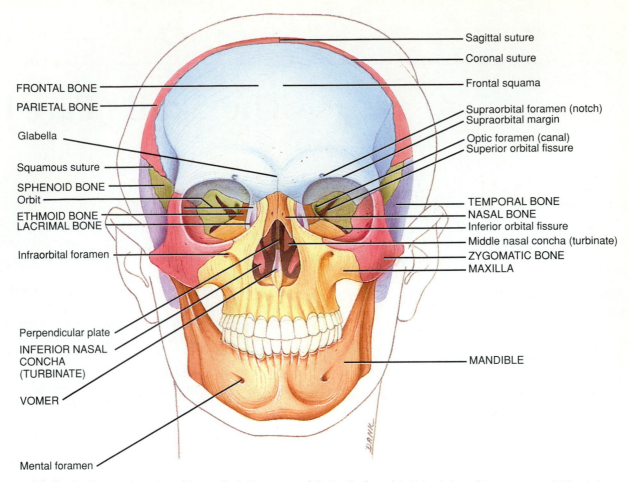

Sagittal suture
Coronal suture
Frontal squama
Supraorbital foramen (notch)
Supraorbital margin
Optic foramen (canal)
Superior orbital fissure
TEMPORAL BONE
NASAL BONE
Inferior orbital fissure
Middle nasal concha (turbinate)
ZYGOMATIC BONE
MAXILLA
MANDIBLE

FRONTAL BONE
PARIETAL BONE
Glabella
Squamous suture
SPHENOID BONE
Orbit
ETHMOID BONE
LACRIMAL BONE
Infraorbital foramen
Perpendicular plate
INFERIOR NASAL CONCHA (TURBINATE)
VOMER
Mental foramen

**Figure 10-1** Skull, anterior view. (From G. J. Tortora and S. R. Grabowski, *Principles of Anatomy and Physiology*, 7th ed., New York: HarperCollins, 1993. © Leonard Dank.)

6. *Vomer* (VŌ-mer)—single, dagger-shaped bone located in the center of the nasal cavity superior to the maxillary bones.
7. *Inferior nasal concha* (KONG-ka)—paired, flat bones that form the lateral walls of the nasal cavity
8. *Mandible* (MAN-di-bul)—single bone that forms the lower jaw, and bears teeth.

## THE SKULL SUTURES

Continue your examination of the skull. Bones come together at lines called "sutures" (SOO-churs). A *suture* is an immovable joint in which bones with serrated edges fit together, separated by a thin layer of fibrous connective tissue. Joints, or articulations, will be taken up in a later exercise, but it is appropriate to name the major skull sutures at this time.

1. *Coronal* (kō-RŌ-nal)—suture between the frontal and parietal bones
2. *Sagittal* (SAJ-i-tal)—suture extending from the coronal suture in the midline and separating the two parietal bones
3. *Squamosal* (skwa-MŌ-sal)—suture extending around the temporal bone and separating it from the parietal and sphenoid bones
4. *Lambdoidal* (lam-DOY-dal)—suture separating the occipital bone from the parietals

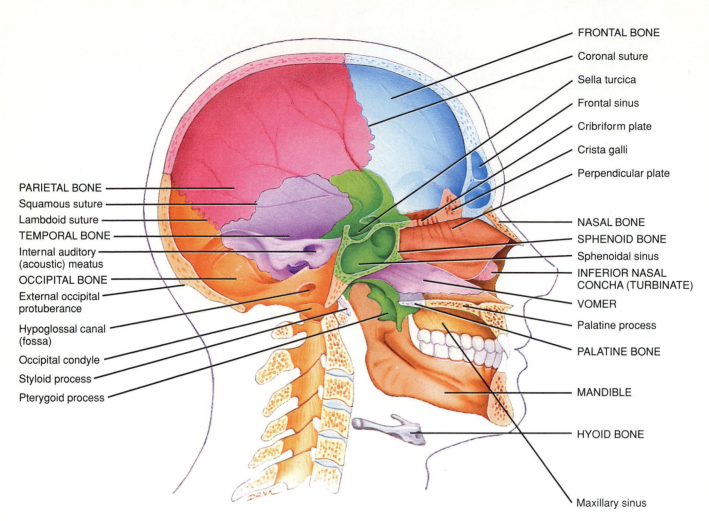

FRONTAL BONE
Coronal suture
Sella turcica
Frontal sinus
Cribriform plate
Crista galli
Perpendicular plate

NASAL BONE
SPHENOID BONE
Sphenoidal sinus
INFERIOR NASAL CONCHA (TURBINATE)
VOMER
Palatine process
PALATINE BONE

MANDIBLE

HYOID BONE

Maxillary sinus

PARIETAL BONE
Squamous suture
Lambdoid suture
TEMPORAL BONE
Internal auditory (acoustic) meatus
OCCIPITAL BONE
External occipital protuberance
Hypoglossal canal (fossa)
Occipital condyle
Styloid process
Pterygoid process

**Figure 10-2** Skull, median view. (From G. J. Tortora and S. R. Grabowski, *Principles of Anatomy and Physiology,* 7th ed., New York: HarperCollins, 1993. © Leonard Dank.)

## THE INFANT SKULL

Examine Figure 10-7 and other demonstration materials. Note the absence of teeth and the general fragility of the bones. It is obvious that the infant skull is still in the process of developing and maturing. Look for the following features:

1. Locate the suture separating the paired frontal bones. In the fully developed skull, the bones will have fused to form a single frontal bone, and the suture will have disappeared.

2. The *fontanels* or "soft spots" are places where the bone has not yet ossified over the membrane that gives rise to it. Locate the *anterior fontanel,* the *posterior fontanel,* and the smaller, paired *posterolateral* and *anterolateral fontanels.* Usually, all of the fontanels will have closed by age 1½ to 2 years.

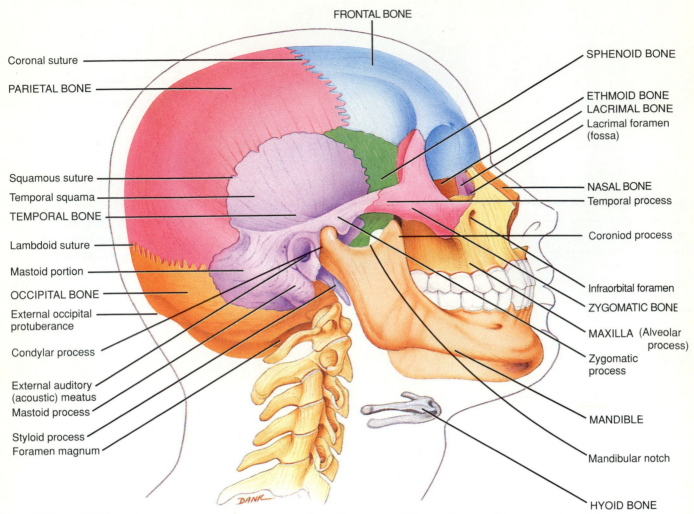

FRONTAL BONE

Coronal suture

PARIETAL BONE

SPHENOID BONE

ETHMOID BONE
LACRIMAL BONE
Lacrimal foramen
(fossa)

Squamous suture

Temporal squama

TEMPORAL BONE

NASAL BONE
Temporal process

Lambdoid suture

Mastoid portion

OCCIPITAL BONE

Coroniod process

External occipital
protuberance

Infraorbital foramen
ZYGOMATIC BONE

Condylar process

MAXILLA (Alveolar
process)
Zygomatic
process

External auditory
(acoustic) meatus

Mastoid process

Styloid process
Foramen magnum

MANDIBLE

Mandibular notch

HYOID BONE

DANK

**Figure 10-3** Skull, right lateral view. (From G. J. Tortora and S. R. Grabowski, *Principles of Anatomy and Physiology,* 7th ed., New York: HarperCollins, 1993. © Leonard Dank.)

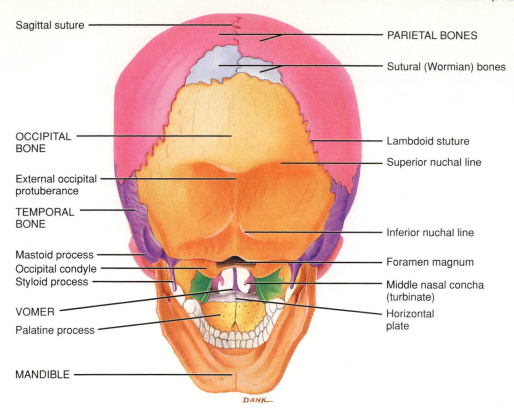

Sagittal suture

PARIETAL BONES

Sutural (Wormian) bones

OCCIPITAL BONE

Lambdoid stuture

Superior nuchal line

External occipital protuberance

TEMPORAL BONE

Inferior nuchal line

Mastoid process

Occipital condyle

Styloid process

Foramen magnum

Middle nasal concha (turbinate)

VOMER

Horizontal plate

Palatine process

MANDIBLE

DANK

**Figure 10-4** Skull, posterior view. (From G. J. Tortora and S. R. Grabowski, *Principles of Anatomy and Physiology,* 7th ed., New York: HarperCollins, 1993. © Leonard Dank.)

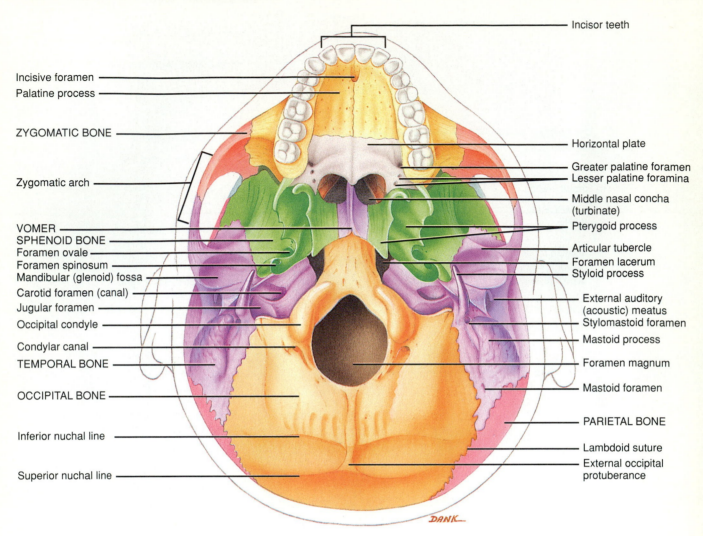

Incisor teeth

Incisive foramen

Palatine process

ZYGOMATIC BONE

Zygomatic arch

VOMER

SPHENOID BONE

Foramen ovale

Foramen spinosum

Mandibular (glenoid) fossa

Carotid foramen (canal)

Jugular foramen

Occipital condyle

Condylar canal

TEMPORAL BONE

OCCIPITAL BONE

Inferior nuchal line

Superior nuchal line

Horizontal plate

Greater palatine foramen

Lesser palatine foramina

Middle nasal concha (turbinate)

Pterygoid process

Articular tubercle

Foramen lacerum

Styloid process

External auditory (acoustic) meatus

Stylomastoid foramen

Mastoid process

Foramen magnum

Mastoid foramen

PARIETAL BONE

Lambdoid suture

External occipital protuberance

**Figure 10-5** Skull in inferior view. (From G. J. Tortora and S. R. Grabowski, *Principles of Anatomy and Physiology*, 7th ed., New York: HarperCollins, 1993. © Leonard Dank.)

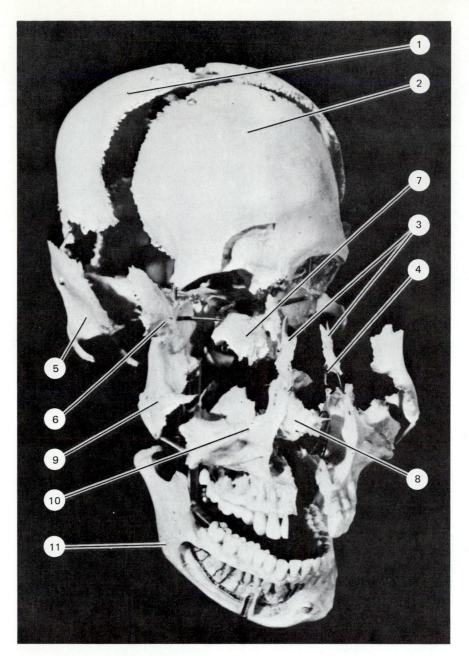

**Figure 10-6** Beauchene skull.
    **1.** Parietal
    **2.** Frontal
    **3.** Nasal
    **4.** Lacrimal
    **5.** Temporal
    **6.** Sphenoid
    **7.** Ethmoid
    **8.** Vomer
    **9.** Zygomatic
    **10.** Maxillary
    **11.** Mandible

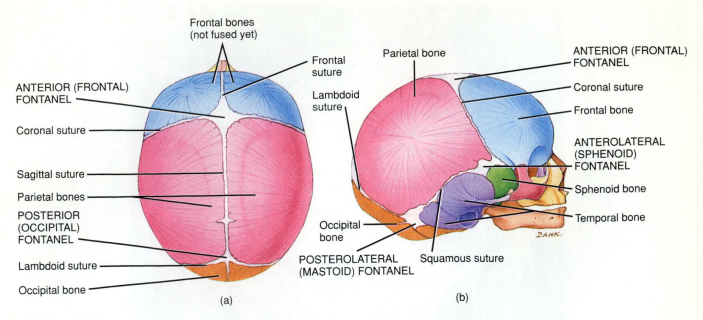

**Figure 10-7** Infant skull: (a) superior view; (b) lateral view. (From G. J. Tortora and S. R. Grabowski, *Principles of Anatomy and Physiology,* 7th ed., New York: HarperCollins, 1993. © Leonard Dank.)

# LABORATORY REVIEW *10*

## *THE SKULL (PART I)*

1. Match the phrase on the left with the numbered choices on the right.

   a. _____ Total number of cranial bones

   b. _____ Dagger-shaped bone in the nasal cavity

   c. _____ Bone that forms the forehead

   d. _____ The cheek bone

   e. _____ Total number of facial bones

   f. _____ Bone that forms the bridge of the nose

   g. _____ Bone that forms the back of the skull

   h. _____ One bone that makes up the hard palate

   i. _____ Another bone that makes up the hard palate

   j. _____ Bone that makes up the lower jaw

## ANSWERS

1. Lacrimal
2. 14
3. Mandible
4. Occipital
5. 20
6. Maxillary
7. Sphenoid
8. Parietal
9. 8
10. Palatine
11. Zygomatic
12. Inferior nasal concha
13. Vomer
14. Frontal
15. Nasal

2. Identify the bones in Figure Q10-1. You may color them if you wish. (If you do use color, be sure to use the same color for each bone in all the drawings.)

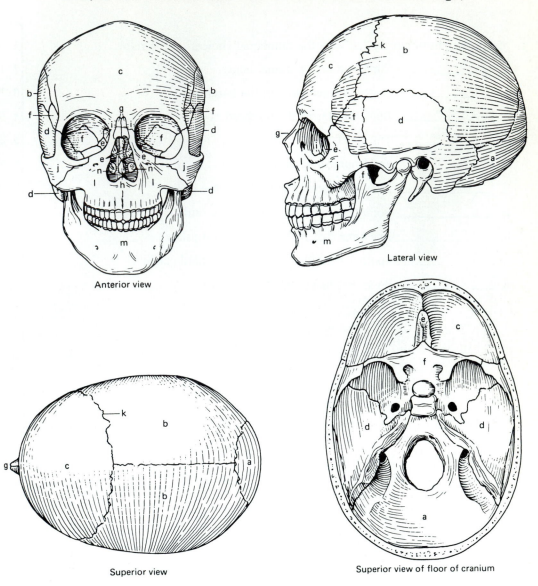

Anterior view

Lateral view

Superior view

Superior view of floor of cranium

**Figure Q10-1**

a. _____    h. _____

b. _____    i. _____

c. _____    j. _____

d. _____    k. _____

e. _____    l. _____

f. _____    m. _____

g. _____    n. _____

# The Skull (Part II) and the Visceral Skeleton

## After completing this exercise, you should be able to

1. Describe the special features of some skull bones
2. Correlate bone markings with their functions
3. Identify the bones of the visceral skeleton
4. Give the correct pronunciation and spelling of each bone and bone marking

### Materials

The following materials should be provided for class use:

1. Mounted human skeleton
2. Human skull
3. Beauchene skull
4. Disarticulated skull bones
5. Demonstration of ear ossicles

## SPECIAL FEATURES OF SKULL BONES

Special surface features of bones are referred to as *markings*. They include prominent or projecting parts, termed *processes* (for example, facets, crests, and spines); depressions, or *fossae;* openings, or *foramina;* grooves, or *sulci;* and enclosed cavities, or *sinuses*. Some of the more important markings of the skull bones are listed here. No attempt is made to name all markings on a bone. If greater detail is sought, the student should consult *Gray's Anatomy* or other similar texts.

## Sphenoid Bone

Situated at the base of the skull and forming most of the base of the cranium, this very complex bone articulates with the temporal, frontal, occipital, and ethmoid bones. Locate the following on a separate sphenoid bone or on the Beauchene skull. The Beauchene skull is a mounted demonstration skull that has been "exploded" so that each bone can be seen readily and its suture lines accurately determined. Consult Figure 11-1.

1. *Sphenoidal sinus*—two large air cavities that drain into the nasal cavity
2. *Sella turcica* (SEL-a TUR-si-ka)—a centrally located depression on the superior surface, which contains the pituitary gland in the living organism
3. *Greater wing*—large, lateral, paired projections that can be seen on the outer surface of the skull
4. *Lesser wing*—small, paired projections superior and medial to the greater wings
5. *Pterygoid process* (TER-i-goyd)—large, paired projections on the inferior surface of the bone
6. *Optic foramen* (OP-tik fō-RĀ-men) or canal—paired holes through which the optic nerves pass
7. *Superior orbital fissure*—paired openings through which pass the cranial nerves III, IV, and parts of V and VI
8. *Foramen ovale* (fō-RĀ-men ō-VA-lē)—paired openings through which part of cranial nerve V passes
9. *Foramen spinosum* (spī-NŌ-sum)—paired openings through which medial meningeal vessels pass
10. *Foramen lacerum* (la-SER-um)—paired slits between the sphenoid and temporal bones, through which the internal carotid arteries pass
11. *Foramen rotundum* (rō-TUN-dum)—paired openings through which part of cranial nerve V passes

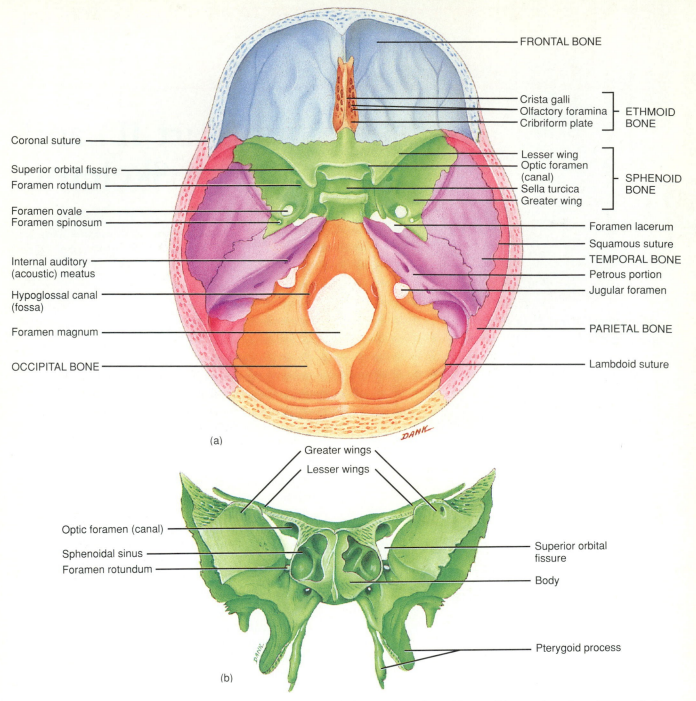

The following labels appear in figure (a):

FRONTAL BONE

Crista galli
Olfactory foramina — ETHMOID BONE
Cribriform plate

Coronal suture

Superior orbital fissure

Foramen rotundum

Lesser wing
Optic foramen (canal)
Sella turcica — SPHENOID BONE
Greater wing

Foramen ovale
Foramen spinosum

Foramen lacerum
Squamous suture
TEMPORAL BONE
Petrous portion
Jugular foramen

Internal auditory (acoustic) meatus

Hypoglossal canal (fossa)

Foramen magnum

PARIETAL BONE

OCCIPITAL BONE

Lambdoid suture

(a)

The following labels appear in figure (b):

Greater wings
Lesser wings

Optic foramen (canal)

Sphenoidal sinus
Foramen rotundum

Superior orbital fissure

Body

Pterygoid process

(b)

**Figure 11-1** Sphenoid bone: (a) viewed in the floor of the cranium from above; (b) anterior view. (From G. J. Tortora and S. R. Grabowski, *Principles of Anatomy and Physiology,* 7th ed., New York: HarperCollins, 1993.)

## Ethmoid Bone

The ethmoid bone is a large but very light spongy bone with a honeycombed structure (see Figures 10-1, 10-2, 10-6, and 11-2). It is the main support of the nasal cavity and forms part of the cranial floor and orbits. The orbit is the eyeball socket. It is made up of several bones.

1. *Perpendicular plate*—forms nasal septum, with the vomer bone
2. *Cribriform plate* (KRIB-ri-form)—porous plate that forms part of the floor of the cranium and contains openings for the olfactory nerve
3. *Crista galli* (KRIS-ta GAL-lē)—triangular process projecting upward from the cribriform plate, which forms part of the floor of the cranium
4. *Superior nasal conchae* (KONG-kē)—paired lateral folds that create air flow for efficient circulation of inhaled air
5. *Middle nasal conchae*—paired lateral folds below the superior nasal conchae, with the same function as the superior nasal conchae

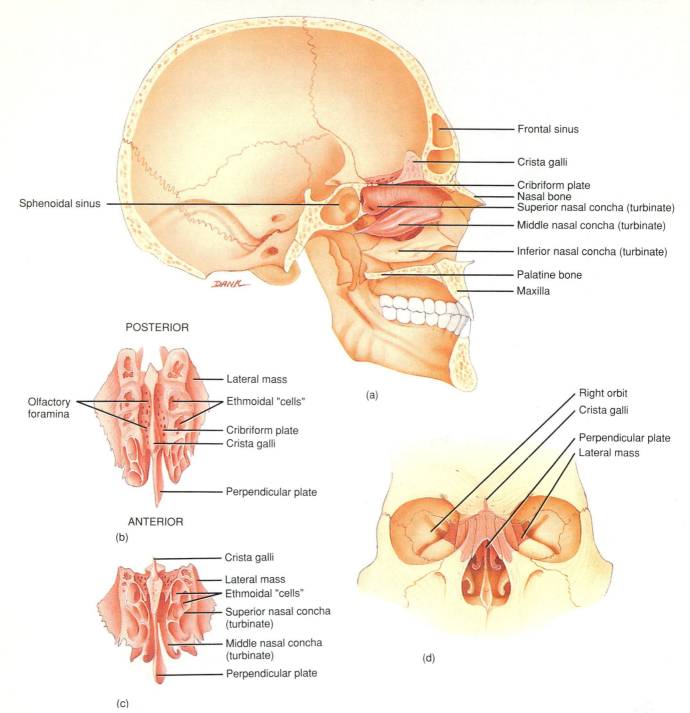

**Figure 11-2** Ethmoid bone. (From G. J. Tortora and S. R. Grabowski, *Principles of Anatomy and Physiology,* 7th ed., New York: HarperCollins, 1993.)

## Maxillary Bone

The paired maxillary bones form the upper jaw and most of the hard palate. They also form part of the orbits and the nasal cavity. (See Figures 10-1, 10-2, 10-3, 10-5, 10-6, and 11-2.)

1. *Maxillary sinus*—air cavity emptying into the nasal cavity (Figure 10-2)
2. *Alveolar process* (al-VĒ-ō-lar)—part of the bone that contains sockets for the teeth (Figure 10-3)

3. *Infraorbital foramen* (in'fra-OR-bi-tal)—paired holes through which the infraorbital nerves and arteries pass (Figure 10-3)

## Temporal Bone

These paired bones form the inferior lateral portion of the cranium and are often referred to as the "temple bones." (See Figures 11-1, 11-2, and 11-3.)

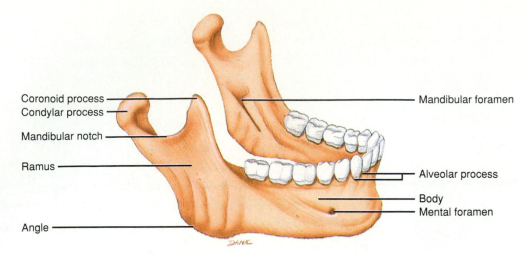

Coronoid process
Condylar process
Mandibular notch
Ramus
Angle
Mandibular foramen
Alveolar process
Body
Mental foramen

**Figure 11-3** Mandible: right lateral view. (From G. J. Tortora and S. R. Grabowski, *Principles of Anatomy and Physiology,* 7th ed., New York: HarperCollins, 1993.)

1. *Zygomatic process*—projection that articulates with the temporal process of the zygomatic bone and forms most of the zygomatic arch
2. *Mandibular fossa* (man-DIB-yoo-ler)—depression into which the lower jaw inserts (Figure 10-5)
3. *Mastoid process* (MAS-toyd)—portion of the bone located behind the earlobe that contains mastoid air cells and serves as the point of attachment for several neck muscles (Figures 10-2, 10-3, 10-4, 10-5)
4. *External auditory meatus* (AW-di-tor-ē mē-Ā-tus)—canal that leads toward the middle ear deep within the temporal bone (Figure 10-3)
5. *Styloid process* (STĪ-loyd)—long, downward-projecting process to which muscles of the tongue and neck are attached (Figures 10-2, 10-3, 10-4, 10-5)

## Frontal Bone

The frontal bone is a bone of both the cranium and the face. It forms the forehead and parts of the orbit and the cranial floor. (See Figure 11-2; see also Figures 10-1, 10-2, 10-3, 10-6, 10-7, and 11-1.)

1. *Glabella* (gla-BELL-a)—flattened area above the nose (Figure 10-1)
2. *Supraorbital foramen*—paired openings through which supraorbital nerve and supraorbital artery pass (Figure 10-1)
3. *Frontal sinus*—air cavity that communicates with the nasal cavity (Figure 10-2, 11-2)

## Mandible

The mandible is the large bone that forms the entire lower jaw and contains all the teeth of the lower jaw (Figures 10-1, 10-2, 10-3, 10-4, 10-6, 10-7, 11-3)

1. *Condylar process* (KON-di-lar)—projection that articulates with the mandibular fossa of the temporal bone (Figures 10-3, 11-3)
2. *Coronoid process* (KOR-ō-noyd)—projection to which temporal muscle attaches (Figures 10-3, 11-3)
3. *Mandibular notch*—depression between the coronoid and condylar processes (Figures 10-3, 11-3)
4. *Mental foramen*—opening through which mental nerve and mental blood vessels pass (Figures 10-1, 11-3)
5. *Alveolar process* (al-VĒ-ō-lar)—part of the bone that contains sockets for the teeth (Figure 11-3)

## THE VISCERAL SKELETON

These and other important skeletal elements (some bone and some cartilage) arise from the embryonic gill arch (visceral arch) region:

## Hyoid Bone

The hyoid (HĪ-oyd) is a U-shaped bone suspended from the styloid process of the temporal bone by ligaments. It serves as point of origin (attachment) for some of the muscles of the tongue. Observe it on the mounted skeleton just inside the mandibular area (Figures 10-2, 10-3).

## Ear Ossicles

Three pairs of tiny bones span the air cavity of the middle ear and serve to transmit sound waves to the inner ear. They are not preserved on the laboratory specimens, but you may have a special preparation to study. Diagrams can be found in your textbook in the chapter on the special senses or in Exercise 3 of this manual. The names of the bones are *malleus* (MAL-ē-us), meaning "hammer"; *incus* (ING-kus), meaning "anvil"; and *stapes* (STĀ-pēz), meaning "stirrup."

## Cartilages of the Larynx

These highly specialized skeletal elements of a visceral organ cannot be seen on the preserved skeleton. They can be observed when the dissection of the cat is made. There are nine cartilages, and they aid in the production of sound (see Figure 11-4). Three are single, and three are paired:

Thyroid cartilage (1)

Epiglottis (1)

Cricoid (1)

Arytenoid (2)

Corniculate (2)

Cuneiform (2)

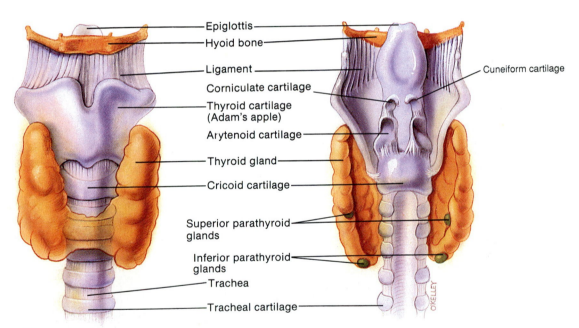

**Figure 11-4** Larynx: (a) diagram of anterior view; (b) diagram of posterior view. (From G. J. Tortora, *Principles of Human Anatomy,* 6th ed., New York: HarperCollins, 1992.)

# LABORATORY REVIEW *11*

## THE SKULL (PART II) AND THE VISCERAL SKELETON

### MATCHING AND COMPLETION

1. Match the terms on the left with the bones listed on the right. This question reviews the markings of the bones of the skull.

   a. _____ Sella turcica

   b. _____ Mastoid process

   c. _____ Zygomatic process

   d. _____ Mandibular notch

   e. _____ Cribriform plate

   f. _____ Greater wing

   g. _____ Mental foramen

   h. _____ Middle nasal concha

   i. _____ Maxillary sinus

   j. _____ Pterygoid processes

   k. _____ Foramen spinosum

   l. _____ Supraorbital foramen

   m. _____ Infraorbital foramen

   n. _____ Perpendicular plate

   o. _____ Condylar process

   p. _____ Foramen lacerum

   q. _____ Glabella

   r. _____ Coronoid process

   s. _____ Styloid process

   t. _____ Superior nasal concha

   u. _____ Foramen ovale

   v. _____ Crista galli

   w. _____ Sphenoidal sinus

   x. _____ Optic foramen

   y. _____ External auditory meatus

2. Answer the following questions briefly.

   a. Why is the "visceral skeleton" so named? _____

   _____

   _____

## ANSWERS

1. Frontal
2. Parietal
3. Occipital
4. Temporal
5. Sphenoid
6. Ethmoid
7. Nasal
8. Maxillary
9. Lacrimal
10. Zygomatic
11. Palatine
12. Vomer
13. Inferior nasal concha
14. Mandible

    b. What is the function of the laryngeal cartilages? _____
_____
_____

    c. What is the function of the middle-ear ossicles? _____
_____
_____

    d. Which skeletal muscle originates on the hyoid bone, and what is the function
of this muscle? _____
_____
_____

    e. Distinguish between the terms *visceral skeleton* and *viscera.* _____
_____

# The Vertebral Column, Sternum, and Ribs

## After completing this exercise, you should be able to

1. List the major subdivisions of the vertebral column
2. Describe the special features of an individual vertebra
3. Correlate markings on a single vertebra with its functions
4. Identify the bones of the sternum
5. Divide ribs into groups, according to their type of articulation

## Materials

The following materials should be provided for class use:

1. Mounted human skeleton
2. Disarticulated skeleton
3. Strung vertebral columns
4. Individual vertebrae

## THE VERTEBRAL COLUMN

The bones of the vertebral column surround and protect the spinal cord in the living organism. They are named for their location along the body axis. Study them on the laboratory specimens, and compare them with Figure 12-1.

Continue your observations by locating the parts of a vertebra shown in Figure 12-2. This happens to be a *thoracic vertebra,* but it is possible to find the same parts on any typical vertebra.

1. *Spinous process*—posterior projection that serves as point of attachment for muscles (It can be felt on your own body in the midback region.)
2. *Vertebral foramen*—central canal through which the spinal cord passes
3. *Body* or *centrum*—large circular area anterior to the vertebral foramen, which forms the joints at the intervertebral discs
4. *Transverse processes*—lateral projections that serve as points of attachment for muscles
5. *Articular processes*—smooth surfaces that articulate with other vertebrae
   a. *Superior articular processes*—pair of smooth articular surfaces facing toward the spinous process
   b. *Inferior articular processes*—pair of smooth articular surfaces facing away from the spinous process
6. *Vertebral arch*—arch over the vertebral foramen, between the centrum and the spinous process
   a. *Pedicles*—lateral portions of arch
   b. *Laminae* or *roof plates*—posterior portion of arch

## Cervical Vertebrae

The seven vertebrae in the neck region are the *cervical vertebrae* (see Figure 12-3). They differ from other vertebrae, in that each transverse process contains a lateral opening termed the *transverse foramen*. The first two vertebrae are highly specialized to form the type of joint that enables you to turn your head (to pivot the head on its axis). The first cervical vertebra is the *atlas,* and it supports the head. The second is the *axis*. The centrum, or body of the axis, is the peglike *odontoid process,* or *dens*. It projects into the ring of the atlas, which lacks a centrum, and forms a pivot, which enables us to turn our heads. Blood vessels pass through the *transverse foramina* of the cervical group.

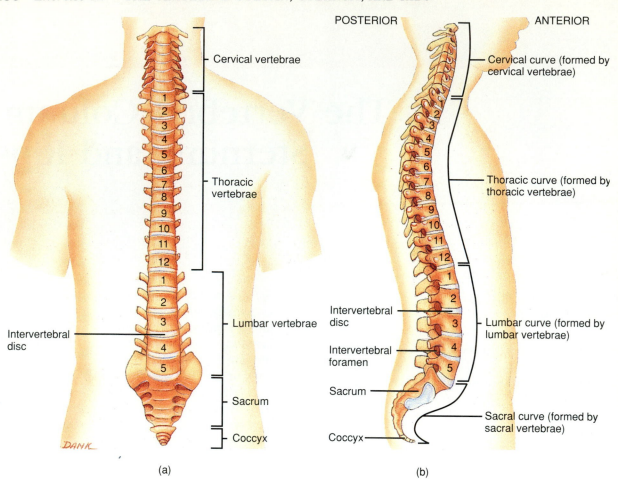

POSTERIOR        ANTERIOR

Cervical vertebrae

Cervical curve (formed by cervical vertebrae)

Thoracic vertebrae

Thoracic curve (formed by thoracic vertebrae)

Intervertebral disc

Lumbar vertebrae

Intervertebral disc

Intervertebral foramen

Lumbar curve (formed by lumbar vertebrae)

Sacrum

Sacrum

Sacral curve (formed by sacral vertebrae)

Coccyx

Coccyx

(a)

(b)

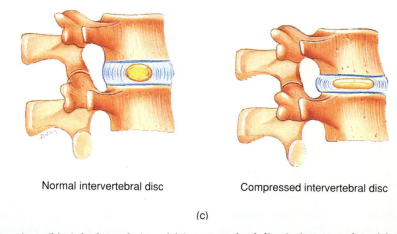

Normal intervertebral disc       Compressed intervertebral disc

(c)

**Figure 12-1** Vertebral column: (a) anterior view; (b) right lateral view; (c) intervertebral disc in its normal position (left) and under compression (right). The relative size of the intervertebral disc has been enlarged for emphasis. An apparent window has been cut in the annulus fibrosus so that the nucleus pulposus can be seen. (From G. J. Tortora and S. R. Grabowski, *Principles of Anatomy and Physiology,* 7th ed., New York: HarperCollins, 1993.)

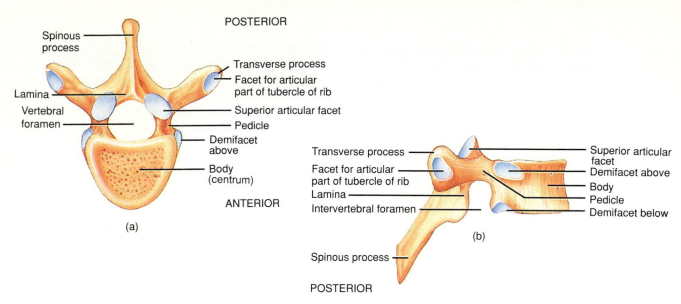

**Figure 12-2** Vertebra, as illustrated by a thoracic vertebra: (a) superior view; (b) right lateral view. (From G. J. Tortora and S. R. Grabowski, *Principles of Anatomy and Physiology,* 7th ed., New York: HarperCollins, 1993.)

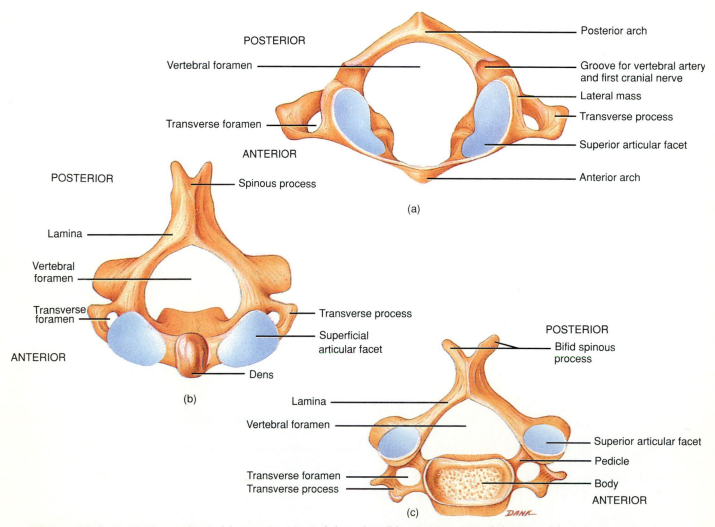

**Figure 12-3** Cervical vertebrae: (a) superior view of the atlas; (b) superior view of the axis; (c) superior view of a typical cervical vertebra. (From G. J. Tortora and S. R. Grabowski, *Principles of Anatomy and Physiology,* 7th ed., New York: HarperCollins, 1993.)

## Thoracic Vertebrae

The 12 thoracic vertebrae are characterized by long spinous processes. Each vertebra articulates with a pair of ribs. On Figure 12-2, locate the four smooth facets (two pairs) on each centrum that are the points of articulation with the head of a rib. These smooth surfaces, termed *costal demifacets,* are found on all thoracic vertebrae. In addition, there are paired *facets* on the transverse processes of all but the last two thoracic vertebrae for articulation with the *tubercle* of the rib. (See Figure 12-4.)

## Ribs

The ribs are divided into two groups (Figure 12-4). The first seven pairs of ribs articulate with the sternum by means of their own *costal cartilages*. They are *true ribs*.

The last five pairs of ribs do not articulate directly with the sternum by means of their own costal cartilages, but indirectly, if at all, through the cartilages of the seventh true rib. These are *false ribs*. The cartilages of the eighth, ninth, and tenth false ribs join each other and then articulate with the cartilage of the seventh true rib.

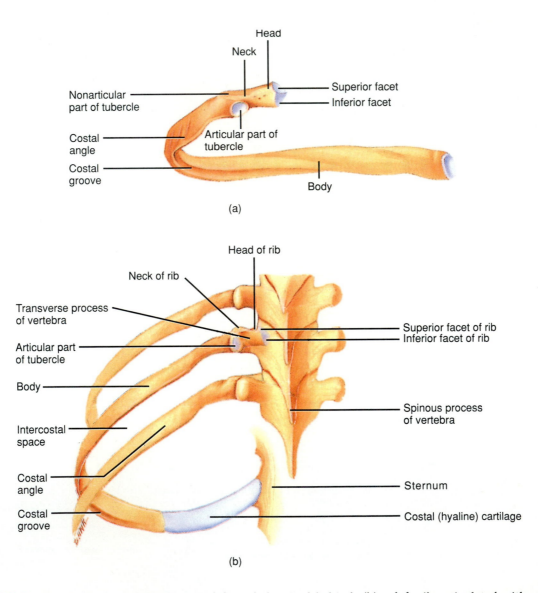

**Figure 12-4** Typical rib: (a) a left rib viewed from below and behind; (b) a left rib articulated with a thoracic vertebra and the sternum. (From G. J. Tortora and S. R. Grabowski, *Principles of Anatomy and Physiology,* 7th ed., New York: HarperCollins, 1993.)

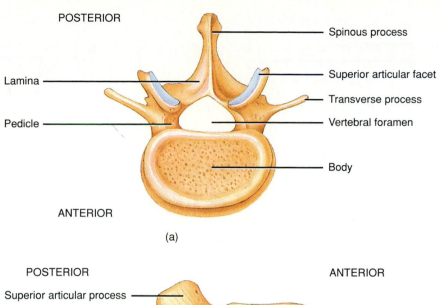

POSTERIOR

Spinous process

Lamina

Superior articular facet

Transverse process

Pedicle

Vertebral foramen

Body

ANTERIOR

(a)

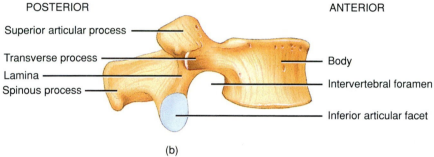

POSTERIOR

ANTERIOR

Superior articular process

Transverse process

Lamina

Spinous process

Body

Intervertebral foramen

Inferior articular facet

(b)

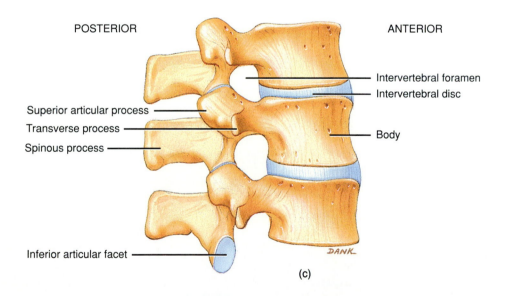

POSTERIOR

ANTERIOR

Intervertebral foramen

Intervertebral disc

Superior articular process

Transverse process

Spinous process

Body

Inferior articular facet

DANK

(c)

**Figure 12-5** Lumbar vertebrae: (a) superior view; (b) right lateral view; (c) diagram of lumbar vertebrae articulated in right lateral view. (From G. J. Tortora and S. R. Grabowski, *Principles of Anatomy and Physiology,* 7th ed., New York: Harper Collins, 1993.)

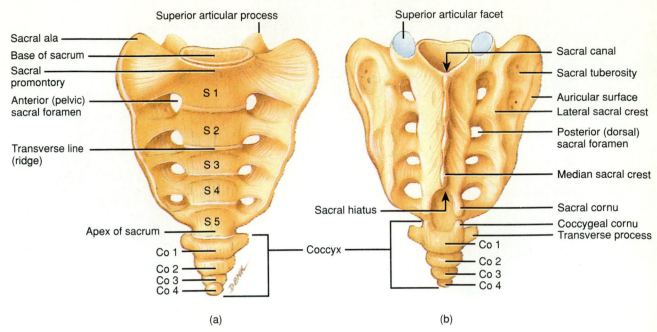

Figure 12-6 labels (left, anterior view):
- Sacral ala
- Base of sacrum
- Sacral promontory
- Anterior (pelvic) sacral foramen
- Transverse line (ridge)
- Apex of sacrum
- Superior articular process
- S 1
- S 2
- S 3
- S 4
- S 5
- Co 1
- Co 2
- Co 3
- Co 4
- Coccyx
- (a)

Figure 12-6 labels (right, posterior view):
- Superior articular facet
- Sacral canal
- Sacral tuberosity
- Auricular surface
- Lateral sacral crest
- Posterior (dorsal) sacral foramen
- Median sacral crest
- Sacral cornu
- Coccygeal cornu
- Transverse process
- Sacral hiatus
- Co 1
- Co 2
- Co 3
- Co 4
- (b)

**Figure 12-6** Sacrum and coccyx: (a) anterior view; (b) posterior view. (From G. J. Tortora and S. R. Grabowski, *Principles of Anatomy and Physiology,* 7th ed., New York: HarperCollins, 1993.)

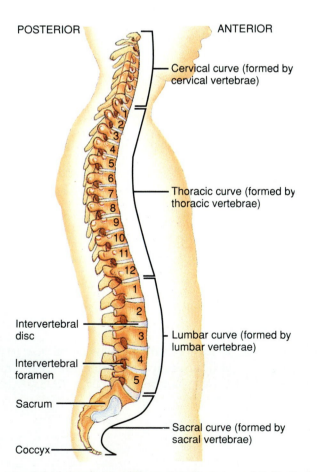

Figure 12-7 labels:
- POSTERIOR
- ANTERIOR
- Cervical curve (formed by cervical vertebrae)
- Thoracic curve (formed by thoracic vertebrae)
- Lumbar curve (formed by lumbar vertebrae)
- Sacral curve (formed by sacral vertebrae)
- Intervertebral disc
- Intervertebral foramen
- Sacrum
- Coccyx

**Figure 12-7** Vertebral column in right lateral view. (From G. J. Tortora and S. R. Grabowski, *Principles of Anatomy and Physiology,* 7th ed., New York: HarperCollins, 1993.)

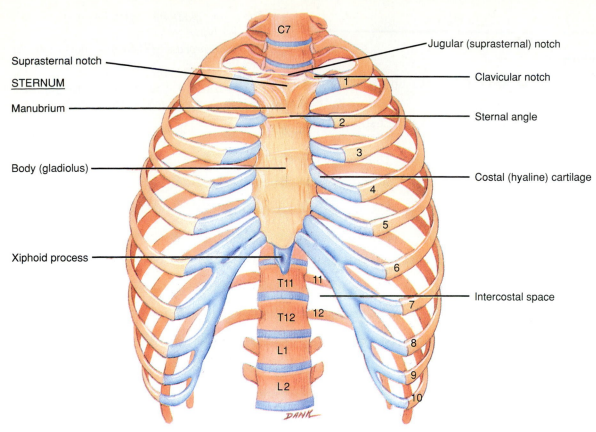

**Figure 12-8** Sternum and ribs: anterior view. (From G. J. Tortora and S. R. Grabowski, *Principles of Anatomy and Physiology*, 7th ed., New York: HarperCollins, 1993.)

The last two pairs of the false rib group do not have costal cartilages and do not attach to the sternum at all. They are *floating ribs*.

## Lumbar Vertebrae

The five lumbar vertebrae (Figure 12-5) are generally larger and heavier than the other vertebrae and are located in an area of the back that bears a considerable amount of the body's weight (see Figure 12-2). They have no costal demifacets on their centra, and their spines, which project backward almost horizontally, are not as long and pointed as those of the thoracic group (see Figure 12-2).

## Sacral Vertebrae (Sacrum) and Coccyx

In this region of the vertebral column, five vertebrae have fused (for added strength) to form one large triangular bone termed the *sacrum* (Figures 12-1 and 12-6). It supports the pelvic girdle at the *sacroiliac* joint. The coccyx is a vestigial (nonfunctional) part of the vertebral column that represents the tail (Figure 12-6). There may be three, four, or five of these small bones, and they may or may not be fused.

## Spinal Curvatures

The spinal column exhibits four normal curves. When viewed from the side, two are concave, and two are convex (see Figure 12-7):

Cervical curve (convex)

Thoracic curve (concave)

Lumbar curve (convex)

Sacral curve (concave)

A convex curve, when viewed from the side, has its convexity anteriorly. A concave curve when viewed from the side, has its concavity posteriorly.

## THE STERNUM AND RIBS

### Sternum

Identify the three bones that make up the sternum (Figure 12-8):

1. *Manubrium* (ma-NOO-brē-um)—most superior in position

2. *Body* or *gladiolus* (glad-ē-Ō-lus)—middle and longest bone

3. *Xiphoid process* (ZĪ-foyd)—most inferior and smallest

With the aid of the diagram, locate the *jugular notch* and the *clavicular notch* on the manubrium.

# LABORATORY REVIEW *12*

## *THE VERTEBRAL COLUMN, STERNUM, AND RIBS*

### COMPLETION

a. _____ Name the part of a vertebra that forms the point of articulation for a rib.

b. _____ How many cervical vertebrae does the human have?

c. _____ What is the special name given to the first cervical vertebra?

d. _____ What is the special name given to the second cervical vertebra?

e. _____ Which bone is formed from five fused sacral vertebrae?

f. _____ Which part of the human vertebral column is considered vestigial?

g. _____ What is the name for the cavity in a vertebra through which the spinal cord passes?

h. _____ Which vertebrae are the largest and heaviest?

i. _____ On which part of the vertebral column does walking place the most stress?

j. _____ What is the function of the transverse processes?

k. _____ Do spinal nerves exit through vertebrae or between them?

l. _____ Name the superior portion of the sternum.

m. _____ Name the middle portion of the sternum.

n. _____ Name the inferior portion of the sternum.

o. _____ Which part of a vertebra is most posterior in position?

2. What criterion makes a rib a *true rib?* _____

_____

3. Name and state the location of the spinal curvatures.

a. _____

b. _____

c. _____

d. _____

**ANSWERS**

1. Vertebral column
2. Ribs
3. Sternum

4. Match the terms on the right with the part of the skeleton (listed on the left) in which they belong.

a. _____ Transverse process

b. _____ Tubercle

c. _____ Atlas

d. _____ Gladiolus

e. _____ Centrum

f. _____ Manubrium

g. _____ Inferior articular process

h. _____ Spinous process

i. _____ Xiphoid process

j. _____ Jugular notch

k. _____ Coccyx

# The Appendicular Skeleton

1. List the major bones and markings of the appendicular skeleton
2. Give the correct pronunciation and spelling of all major bones and markings
3. Identify all parts on laboratory specimens

## Materials

The following materials should be provided for class use:

1. Mounted human skeleton
2. Disarticulated skeleton
3. Individual pelvis and scapula bones
4. Articulated hand-wrist and foot-ankle bones

## SUBDIVISIONS OF THE APPENDICULAR SKELETON

The bones of the appendicular skeleton form the *appendages* (arms and legs) and the *girdles* (pectoral and pelvic). The *girdles* are bony structures to which the arms and legs attach and join the *torso*. The bones are grouped into four subdivisions.

1. *Upper extremity* or *arm* (Figure 13-1)
   a. *Humerus* (HYOO-mer-us)—bone of the upper arm
   b. *Radius* (RĀ-dē-us)—lateral bone of the lower arm or forearm
   c. *Ulna* (UL-na)—medial bone of the lower arm or forearm
   d. *Carpals* (KAR-puls)—bones of the wrist
   e. *Metacarpals*—bones of the hand
   f. *Phalanges* (fa-LAN-jēz; singular is *phalanx*, FĀ-lanks)—bones of the fingers
2. *Pectoral girdle* (Figure 13-1)
   a. *Scapula* (SKAP-yoo-la)—posterior component of girdle, commonly known as the shoulder blade
   b. *Clavicle* (KLAV-i-kul)—anterior component, commonly known as the collarbone
3. *Lower extremity* or *leg* (Figure 13-2)
   a. *Femur* (FĒ-mer)—bone of the thigh
   b. *Patella* (pa-TEL-a)—sesamoid bone (which develops in a tendon) that forms the kneecap
   c. *Tibia* (TIB-ē-a)—medial bone of the lower leg
   d. *Fibula* (FIB-yoo-la)—lateral bone of the lower leg
   e. *Tarsals* (TAR-sals)—bones of the ankle
   f. *Metatarsals*—bones of the foot
   g. *Phalanges*—bones of the toes
4. *Pelvic girdle* (Figures 13-2 and 13-3)
   a. *Coxal bone*—commonly known as the hipbone, made of three fused bones
      i. *Ilium* (ILL-ē-um)
      ii. *Ischium* (IS-kē-um)
      iii. *Pubis* (PYOO-bis)

## BONES OF THE UPPER EXTREMITY

Having learned the names of the bones of the appendicular skeleton, the student should now examine each bone for some of the more important markings. We consider the bones of the upper extremity first.

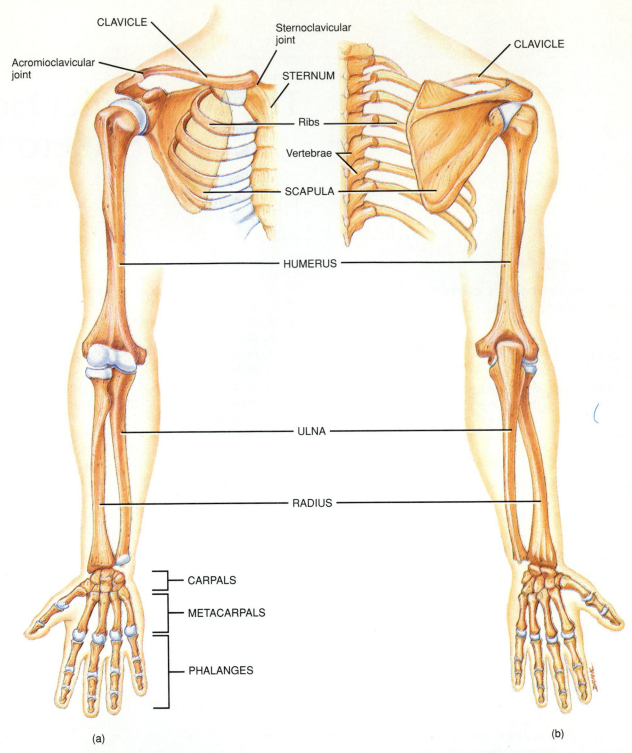

**Figure 13-1** Right pectoral (shoulder) girdle and upper extremity: (a) anterior view; (b) posterior view. (From G. J. Tortora and S. R. Grabowski, *Principles of Anatomy and Physiology,* 7th ed., New York: HarperCollins, 1993.)

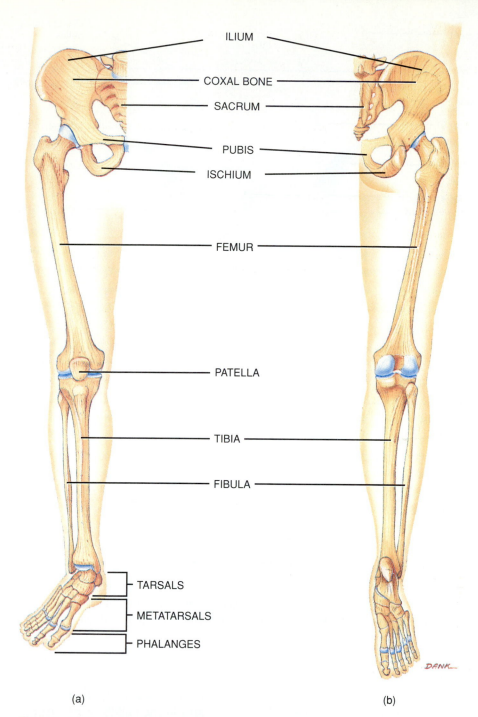

ILIUM

COXAL BONE

SACRUM

PUBIS

ISCHIUM

FEMUR

PATELLA

TIBIA

FIBULA

TARSALS

METATARSALS

PHALANGES

(a)

(b)

**Figure 13-2** Right pelvic girdle and lower extremity: (a) anterior view; (b) posterior view. (From G. J. Tortora and S. R. Grabowski, *Principles of Anatomy and Physiology,* 7th ed., New York: HarperCollins, 1993.)

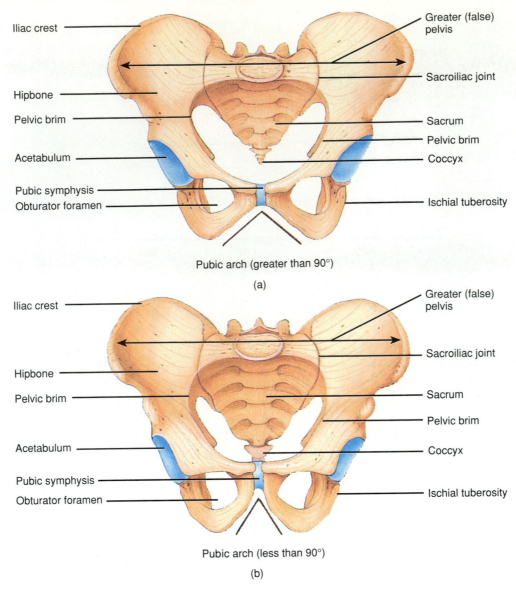

**Figure 13-3** Pelvis, anterior views: (a) female pelvis; (b) male pelvis. (From G. J. Tortora and S. R. Grabowski, *Principles of Anatomy and Physiology,* 7th ed., New York: HarperCollins, 1993.)

## Humerus

The longest bone of the arm, the *humerus,* articulates with the pectoral girdle (at the scapula) proximally and with the radius and ulna distally at the elbow. Locate the following parts on a laboratory specimen (Figure 13-4):

1. *Greater tubercle* (TOO-ber-kul)—lateral projection distal to the neck of the humerus
2. *Lesser tubercle*—anterior projection distal to the neck of the humerus
3. *Intertubercular sulcus* or *bicipital groove* (in'-ter-too-BER-kyoo-lar SUL-kus) (bī-SIP-i-tal)—furrow between the two tubercles

4. *Deltoid tuberosity* (too-ber-OS-i-tē)—roughened area in the middle of the shaft
5. *Capitulum* (ka-PIT-yoo-lum)—knoblike process articulating with the radius
6. *Radial fossa*—depression for the head of the radius
7. *Trochlea* (TRŌK-lē-a)—articular surface for the ulna
8. *Coronoid fossa*—anterior depression for the ulna when forearm is flexed
9. *Olecranon fossa* (ō-LEK-ra-non)—posterior depression for the olecranon process of the ulna when forearm is extended
10. *Medial epicondyle* (ep'-i-KON-dīl)—roughened projection on the medial distal surface
11. *Lateral epicondyle*—roughened projection on the lateral distal surface

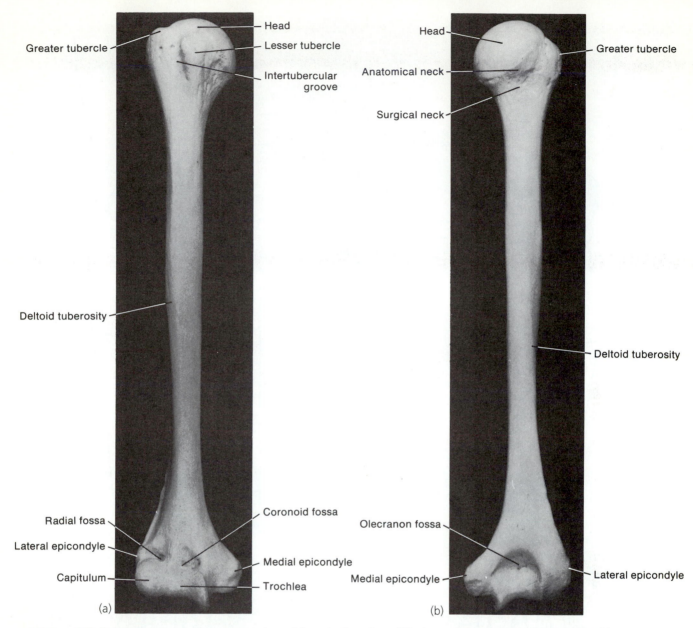

**Figure 13-4** The humerus: (a) anterior view; (b) posterior view. (Photographs courtesy of Lenny Patti.)

## Radius and Ulna

The *radius* is the lateral bone of the forearm. It forms a radial or pivotal joint with the ulna at the proximal ends of each bone (Figure 13-5):

## Carpals, Metacarpals, and Phalanges

The *carpus* or wrist consists of eight small bones closely held together by ligaments. They are arranged in two rows, each with four bones. Using Figure 13-6 and a human skeleton or separate hand–wrist preparation, locate all the carpal bones.

The four bones of the proximal row, from medial to lateral, are

1. *Pisiform* (PĪ-si-form)
2. *Triquetral* (trī-KWĒ-tral) or *triangular*
3. *Lunate* (LOO-nāt)
4. *Scaphoid* (SKAF-oyd) or *navicular* (na-VIK-yoo-lar)

The four bones of the distal row, from medial to lateral, are

1. *Hamate* (HAM-āt)
2. *Capitate* (KAP-i-tāt)
3. *Trapezoid* (TRAP-i-zoyd) or *lesser multangular* (mul-TANG-gyoo-lar)
4. *Trapezium* (tra-PĒ-zē-um) or *greater multangular*

The *metacarpus,* the region between the wrist and the fingers, consists of five bones that form the palm of the

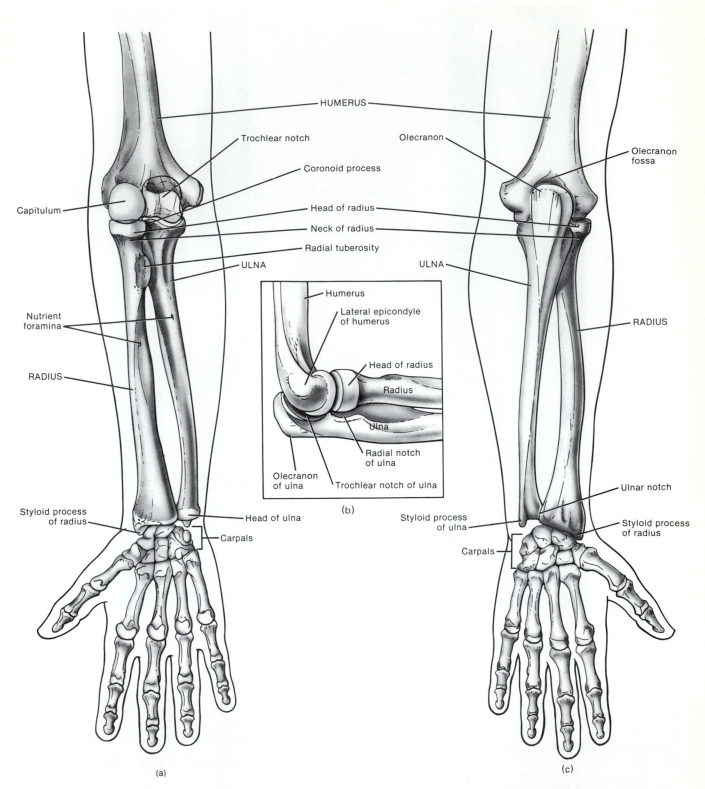

**Figure 13-5** Right radius and ulna: (a) anterior view; (b) posterior view; (c) right elbow, lateral view. (From G. J. Tortora, *Principles of Human Anatomy,* 3rd, 4th, 5th eds., New York: HarperCollins, 1983, 1986, 1989.)

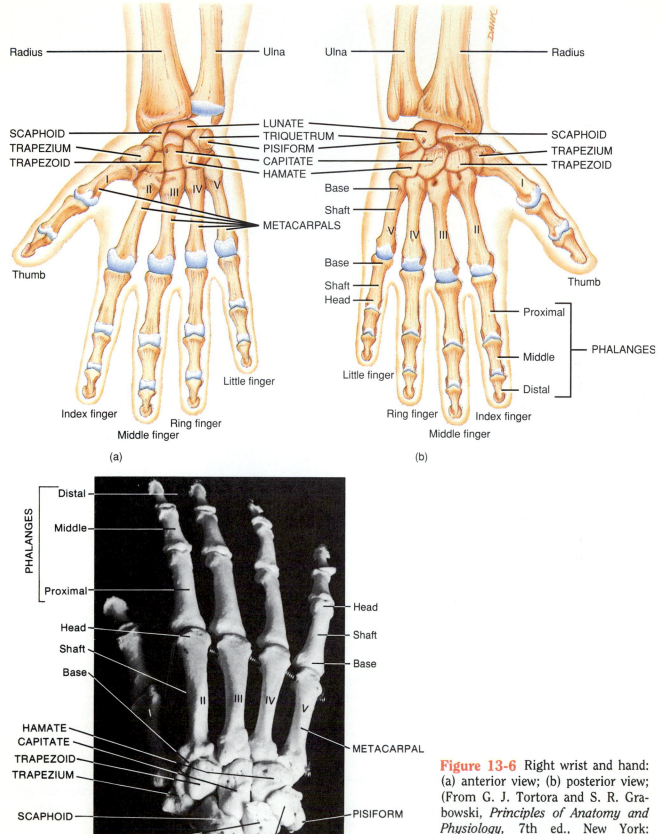

Radius — Ulna
Ulna — Radius

SCAPHOID
TRAPEZIUM
TRAPEZOID

LUNATE
TRIQUETRUM
PISIFORM
CAPITATE
HAMATE

SCAPHOID
TRAPEZIUM
TRAPEZOID

I II III IV V

METACARPALS

Thumb

Base
Shaft

Base
Shaft
Head

I

Thumb

V IV III II

Index finger

Little finger

Proximal

PHALANGES

Middle
Distal

Middle finger
Ring finger

Little finger

Ring finger Index finger
Middle finger

(a)

(b)

PHALANGES
Distal
Middle
Proximal

Head
Shaft
Base

HAMATE
CAPITATE
TRAPEZOID
TRAPEZIUM

SCAPHOID

LUNATE

TRIQUETRAL

(c)

II III IV V

Head
Shaft
Base

METACARPAL

PISIFORM

**Figure 13-6** Right wrist and hand: (a) anterior view; (b) posterior view; (From G. J. Tortora and S. R. Grabowski, *Principles of Anatomy and Physiology,* 7th ed., New York: HarperCollins, 1993.) (c) posterior view (courtesy of Lenny Patti); (d) anteroposterior roentgenogram (courtesy of Daniel Sorrentino.)

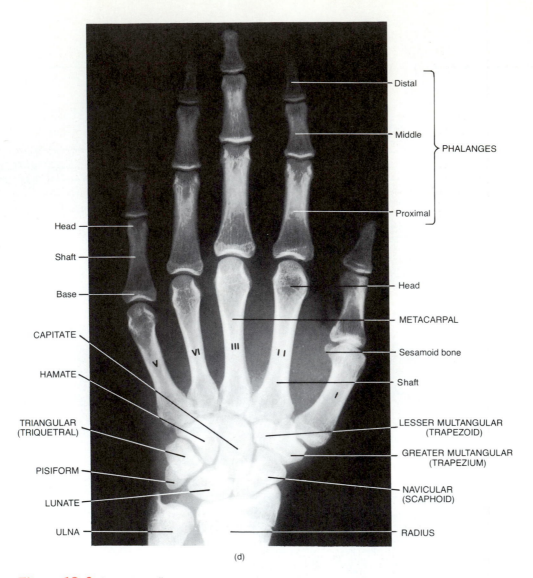

**Figure 13-6** *(Continued)*

hand. They are numbered from I to V, beginning with the metacarpal that articulates with the thumb. Assign the proper numbers to the metacarpals on your lab specimens.

The *phalanges* are the bones of the digits, or fingers. There are 14 in each hand, and 3 in each finger except the thumb, which has only 2.

## BONES OF THE PECTORAL GIRDLE

The pectoral girdle is the point of attachment of the arm to the torso (Figures 13-7 and 13-8). It is made up of only two bones: the clavicle and the scapula. The anterior bone is the *clavicle,* a long, slender, horizontally placed bone. It lies just above the first rib and articulates with the sternum and the scapula. The pos-

terior bone is the *scapula,* an essentially flat triangular bone with many markings for muscle attachment. Identify the following parts on an individual bone.

1. *Spine*—large, sharp ridge that runs transversely across the posterior surface
2. *Acromion* (a-KRŌ-mē-on)—large projection at the end of the spine that articulates with the clavicle
3. *Coracoid process*—(KOR-a-koyd)—projection on the anterior surface
4. *Glenoid fossa* or *cavity*—depression below the acromion and coracoid process that articulates with the proximal head of the humerus to form the shoulder joint
5. *Supraspinous fossa* (soo'-pra-spī-NOUS)—depression above the spine on the posterior surface, point of attachment for supraspinatus muscles
6. *Infraspinous fossa* (in'-fra-spī-NOUS)—large con-

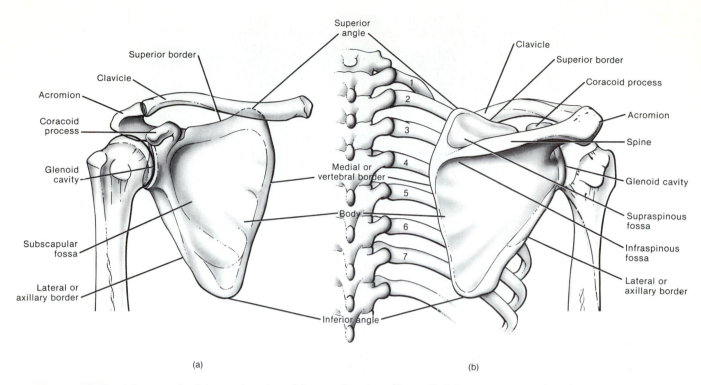

**Figure 13-7** Right scapula: (a) anterior view; (b) posterior view. (From G. J. Tortora, *Principles of Human Anatomy,* 3rd, 4th, 5th eds., New York: HarperCollins, 1983, 1986, 1989.)

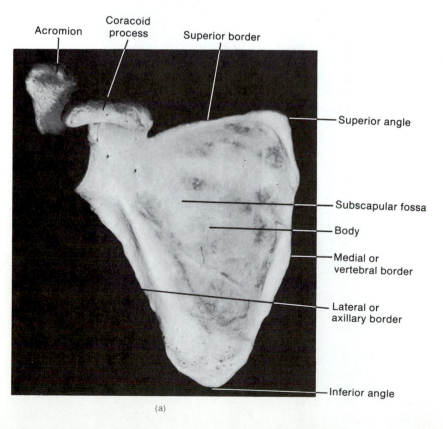

**Figure 13-8** Right scapula: (a) anterior view; (b) posterior view; (c) lateral view. (Courtesy of Lenny Patti.)

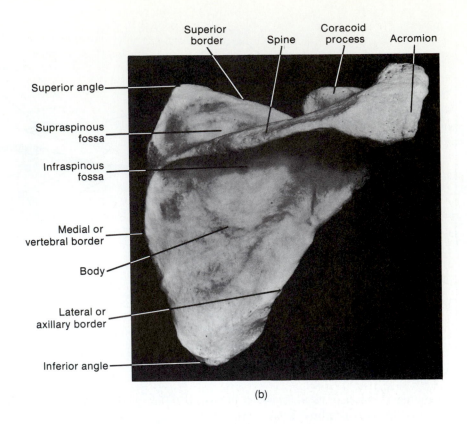

(b)

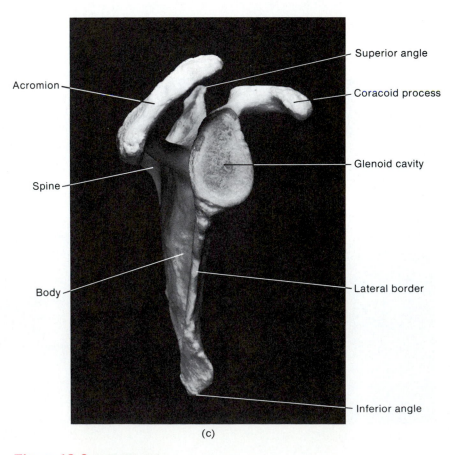

(c)

**Figure 13-8** (Continued)

cavity below the spine on the posterior surface, point of attachment for infraspinatus muscles

7. *Subscapular fossa* (sub-SKAP-yoo-lar)—slight concavity occupying most of the anterior surface, point of attachment for shoulder muscles

8. *Vertebral* or *medial border*—thin edge of the body near the vertebral column

9. *Axillary* or *lateral border*—thickened edge of the body near the arm

10. *Superior border*—superior edge of the body

11. *Inferior angle*—angle formed where medial and lateral borders meet, most inferior part of the bone

12. *Superior angle*—angle formed where superior and medial borders meet

## BONES OF THE LOWER EXTREMITY

Review the names of the bones of the leg, and then study their important markings.

### Femur

The *femur,* or thigh bone, is the longest and heaviest bone in the body. Using Figure 13-9 and a laboratory specimen, locate the following parts.

1. *Head*—rounded proximal end that articulates with the hip socket in the pelvic girdle

2. *Neck*—constricted area distal to the head

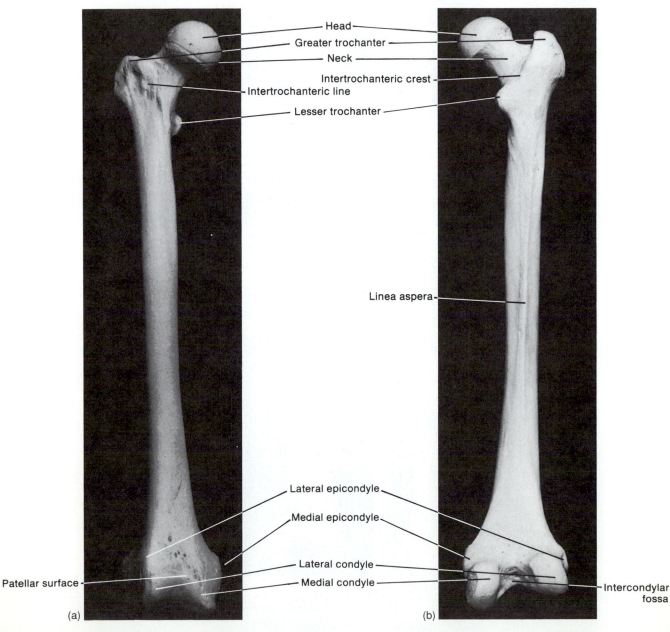

**Figure 13-9** The femur: (a) anterior view; (b) posterior view. (Photographs courtesy of Lenny Patti.)

3. *Greater trochanter* (trō-KAN-ter)—large, roughened lateral projection at the proximal end
4. *Lesser trochanter*—smaller, roughened medial projection
5. *Linea aspera* (LIN-ē-a AS-per-a)—roughened vertical ridge on the shaft
6. *Medial condyle* (KON-dīl)—projection at the distal end for articulation with the tibia
7. *Lateral condyle*—projection at the distal end for articulation with the tibia
8. *Intercondylar fossa*—depression between the medial and lateral condyles

9. *Medial epicondyle*—projection superior to the medial condyle
10. *Lateral epicondyle*—projection superior to the lateral condyle

## Tibia and Fibula

The *tibia* is the larger of the two bones of the lower leg. It is located medially. Identify the following parts, using Figure 13-10 and the laboratory specimen.

1. *Lateral condyle*—lateral projection at the proximal end that articulates with the femur

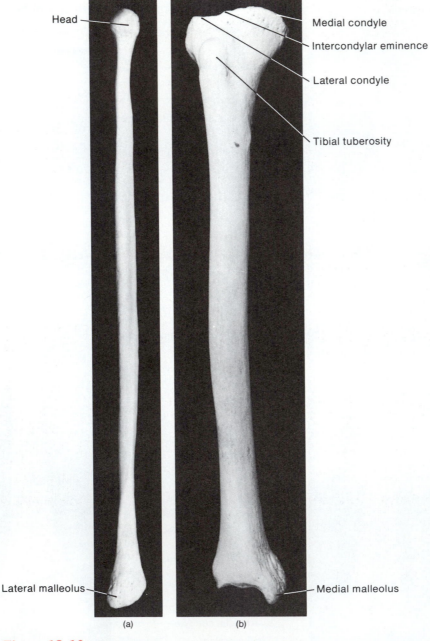

Head

Medial condyle

Intercondylar eminence

Lateral condyle

Tibial tuberosity

Lateral malleolus

Medial malleolus

(a)                (b)

**Figure 13-10** Fibula and tibia: (a) Right fibula, anterior view; (b) right tibia, anterior view. (Photographs courtesy of Lenny Patti.)

2. *Medial condyle*—medial projection at the proximal end that articulates with the femur
3. *Intercondylar eminence* (in'-ter-KON-di-lar)—small projection between the condyles
4. *Tibial tuberosity*—roughened projection on the anterior surface just below the head
5. *Medial malleolus* (mal-LĒ-ō-lus)—medial prominence at the distal end
6. *Fibular notch*—lateral surface on the lateral condyle that articulates with the fibula

The *fibula* is the lateral bone of the lower leg. It is much smaller than the tibia. Locate the following parts.

1. *Head*—proximal end that articulates with the tibia
2. *Lateral malleolus*—long projection at the distal end, which articulates with the talus of the ankle

### Tarsals, Metatarsals, and Phalanges

The *tarsus,* or ankle, consists of seven bones. Using Figure 13-11 and the laboratory skeleton, locate the following tarsal bones.

1. *Calcaneus* (kal-KĀ-nē-us)—largest of the tarsals, forming the heel

2. *Cuboid* (KYOO-boyd)
3. *Navicular*
4. *First (medial) cuneiform* (kyoo-NĒ-i-form)
5. *Second (intermediate) cuneiform*
6. *Third (lateral) cuneiform*
7. *Talus*

The *metatarsals,* the bones of the foot, are numbered from I to V (Figure 13-11), starting with I, which articulates with the big toe.

The *phalanges* of the foot have the same number and arrangement as those of the hand. There are 14 toe bones, 2 in the big toe, and 3 each in the other digits (Figure 13-11).

## BONES OF THE PELVIC GIRDLE

The pelvic girdle serves as the point of attachment of the leg to the main axis of the body. It is made up of a total of six fused bones, three on each side (see Figure 13-12). The three bones on the right and left sides form the *coxal bones* or *os coxae* (formerly termed the *innominate bones*). It is not possible to see the lines of

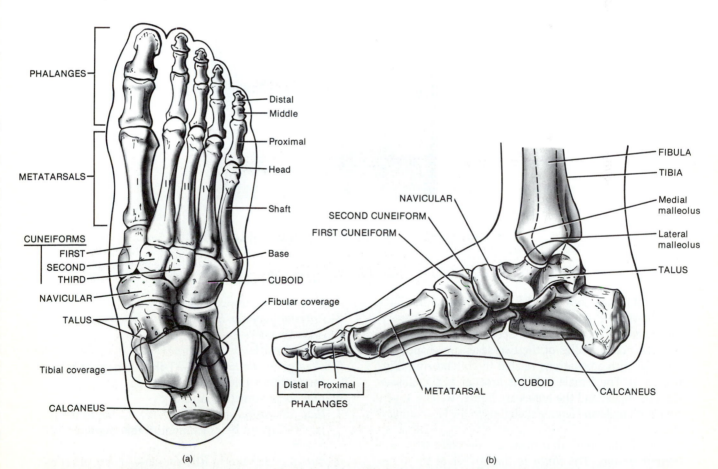

(a)                                                                 (b)

**Figure 13-11**  Right foot and ankle: (a) superior view; (b) medial view; (From G. J. Tortora and N. P. Anagnostakos, *Principles of Anatomy and Physiology*, 3rd, 4th, 5th, 6th eds., New York: HarperCollins, 1981, 1984, 1987, 1990.)

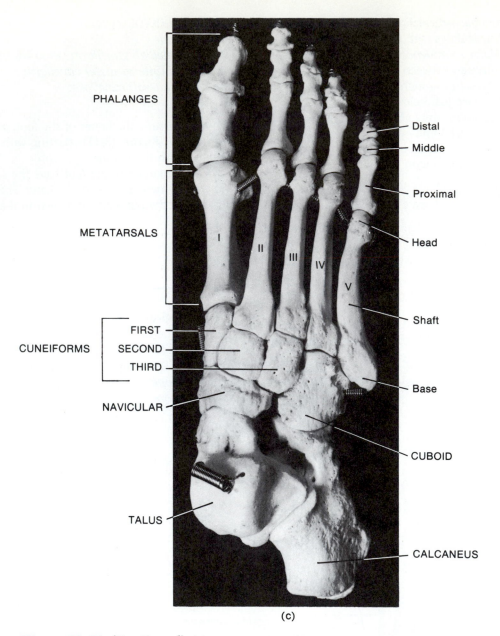

PHALANGES

METATARSALS

CUNEIFORMS
FIRST
SECOND
THIRD

NAVICULAR

TALUS

Distal
Middle
Proximal
Head
Shaft
Base
CUBOID
CALCANEUS

I  II  III  IV  V

(c)

**Figure 13-11** *(Continued)* (c) superior view. (Courtesy of Lenny Patti.)

fusion of these bones, but the approximate suture lines are shown in Figure 13-12. Try to locate the approximate lines in your laboratory specimen.

The pelvic girdles of the two sexes are shaped differently, with adaptations related to childbearing (Figure 13-3). The female pelvis is broader and shallower than the male, and the bones are lighter weight. If you have both male and female skeletons in the lab, compare them.

Study the bones that make up one coxal bone and their markings. The *ilium* is the largest of the three bones and the most superior in position. Locate the following parts (see Figures 13-3 and 13-12).

1. *Iliac crest*—convex superior border
2. *Anterior superior iliac spine*
3. *Anterior inferior iliac spine*
4. *Posterior inferior iliac spine*
5. *Greater sciatic notch* (sī-AT-ik)—concavity below the posterior inferior iliac spine
6. *Iliac fossa*—medial concavity
7. *Iliac tuberosity*—roughened area medial to the iliac fossa, point of attachment of the sacroiliac ligament
8. *Auricular surface* (o-RIK-yoo-lar)—point of articulation with the sacrum

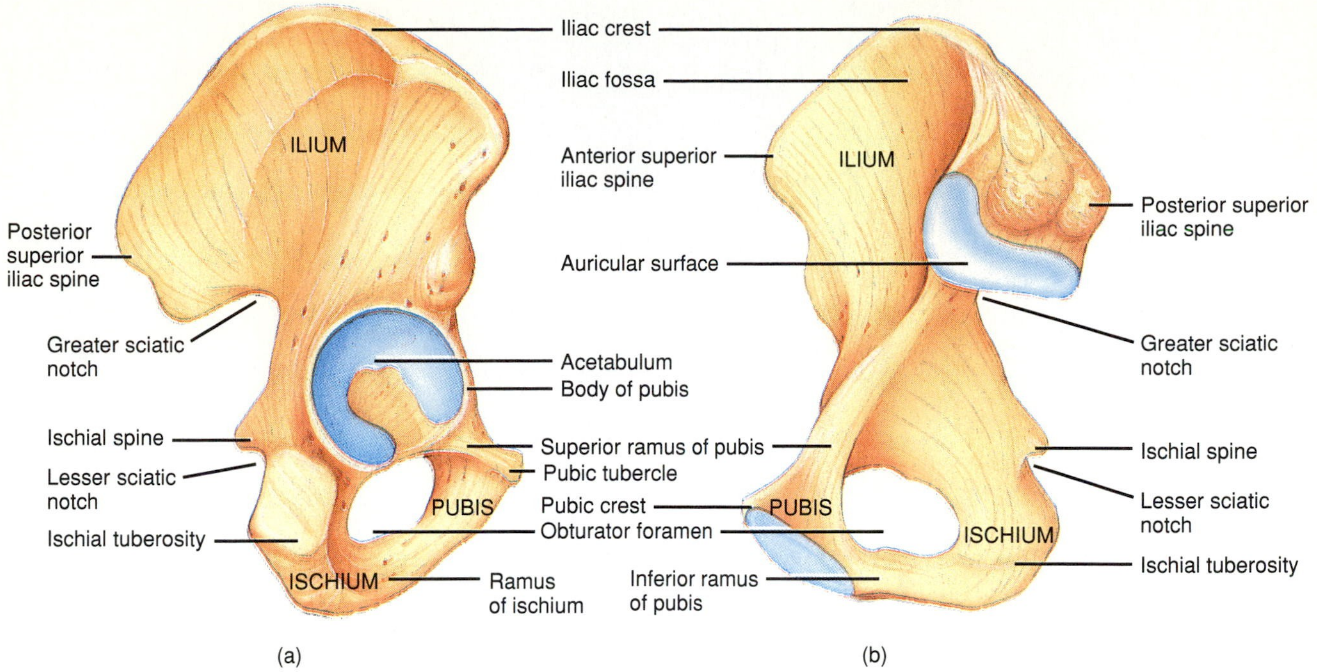

**Figure 13-12** Right coxal bone: (a) lateral view (fusion lines are shown but are probably not seen on your specimen); (b) medial view. (From G. J. Tortora and S. R. Grabowski, *Principles of Anatomy and Physiology,* 7th ed., New York: HarperCollins, 1993.)

The *ischium* is the inferior and posterior part of the coxal bone. It is the part of the pelvic girdle you sit on. Locate the following parts.

1. *Ischial tuberosity*—roughened area at the most posterior and inferior point
2. *Ischial spine*—sharp projection above the ischial tuberosity
3. *Lesser sciatic notch*—concavity below the ischial spine

The *pubis* is the most anterior part of the coxal bone. At the midline, where right and left pubic bones meet,

a joint is formed, which is called the *pubic symphysis* (symphysis pubis). It is discussed in the exercise on articulations.

Other markings on the pelvic girdle are

1. *Acetabulum* (as'-e-TAB-yoo-lum)—fossa formed by the ilium, ischium, and pubis, which articulates with the proximal head of the femur; commonly called the hip socket
2. *Obturator foramen* (OB-tyoo-rā'-ter)—large oval hole surrounded by parts of the ischium and the pubis

# LABORATORY REVIEW 13

NAME _____

LAB SECTION _____ DATE _____

## THE APPENDICULAR SKELETON

1. Match the bones on the left with the major areas to which they belong.

   a. _____ Metatarsal

   b. _____ Radius

   c. _____ Clavicle

   d. _____ Phalanx

   e. _____ Tibia

   f. _____ Humerus

   g. _____ Carpal

   h. _____ Fibula

   i. _____ Femur

   j. _____ Ilium

   k. _____ Scapula

   l. _____ Tarsal

   m. _____ Ischium

   n. _____ Ulna

   o. _____ Metacarpal

   p. _____ Pubis

2. Match the markings on the left with the bones on which they appear.

   a. _____ Ulnar notch

   b. _____ Coracoid process

   c. _____ Semilunar notch

   d. _____ Glenoid fossa

   e. _____ Deltoid tuberosity

   f. _____ Coronoid process

   g. _____ Acromion process

   h. _____ Radial fossa

   i. _____ Subscapular fossa

   j. _____ Olecranon fossa

   k. _____ Olecranon process

   l. _____ Inferior angle

   m. _____ Radial tuberosity

   n. _____ Spine

   o. _____ Coronoid fossa

   p. _____ Supraspinous fossa

## ANSWERS

1. Upper arm
2. Lower arm
3. Wrist
4. Hand
5. Fingers
6. Pectoral girdle
7. Upper leg
8. Lower leg
9. Ankle
10. Foot
11. Toes
12. Pelvic girdle

1. Humerus
2. Radius
3. Ulna
4. Scapula

**ANSWERS**

q. _____ Radial notch

r. _____ Intertubercular sulcus

s. _____ Infraspinous fossa

t. _____ Lesser tubercle

3. Write the names of the eight carpal bones. Try to do it from memory.

a. _____

b. _____

c. _____

d. _____

e. _____

f. _____

g. _____

h. _____

4. Match the markings on the right with the bones on which they appear.

1. Femur

2. Tibia

3. Fibula

4. Pelvic girdle

a. _____ Ischial tuberosity

b. _____ Linea aspera

c. _____ Lateral malleolus

d. _____ Iliac crest

e. _____ Lateral epicondyle

f. _____ Acetabulum

g. _____ Intercondylar eminence

h. _____ Greater trochanter

i. _____ Greater sciatic notch

j. _____ Medial epicondyle

k. _____ Obturator foramen

l. _____ Medial malleolus

m. _____ Fibular notch

n. _____ Lesser trochanter

5. Name the seven bones of the ankle. Try to do it from memory.

a. _____

b. _____

c. _____

d. _____

e. _____

f. _____

g. _____

# Articulations

1. List the terminology associated with joints
2. Classify joints
3. Demonstrate joint movements on your own body

## Materials

The following materials should be provided for class use:

1. Mounted human skeleton
2. Disarticulated vertebral column

An *articulation,* or *joint,* is a place where two or more skeletal elements come together. *Arthrology* is the study of joints.

In the laboratory, joint movements can be demonstrated by referring to the mounted human skeletons or by making specific movements on your own body. Often, the latter is preferable, because your bones are not wired together, and you do not have a metal rod running the length of your vertebral column. The knowledge you have gained in the study of the skeleton will now be applied to good advantage. In order to study the joints intelligently, a familiarity with the names and parts of bones is essential.

It is a common misconception that a joint is always movable. As a matter of fact, most joints are only slightly movable (often, the movement is impossible to detect) or totally immovable.

There are many ways in which joints are classified, but we restrict the classification systems to two: a *functional classification,* based on the degree of movement at a joint, and a *structural classification,* based on the anatomical features of the joint area.

## FUNCTIONAL CLASSIFICATION

This classification divides joints into three categories: synarthroses, amphiarthroses, and diarthroses.

A *synarthrosis* (sin-ar-THRŌ-sis) is an immovable joint. There are four general types of synarthroses.

1. *Suture*—bones joined by a thin layer of dense connective tissue, found only in the skull
   **Example:** Coronal suture between the frontal bone and the parietal bones
2. *Synchondrosis* (sin-kon-DRŌ-sis)—two bones separated by hyaline cartilage
   **Example:** Epiphyseal plate region of a child's bone, with epiphysis and diaphysis separated by cartilage
3. *Synostosis* (sin-os-TŌ-sis)—two bones separated by another plate of bone (ossification of connective tissue or cartilage between two bones)
   **Example:** a. End of a long bone after growth has stopped, when the cartilaginous epiphyseal plate turns to bone
   b. Joint between right and left frontal bones, after they have fused into one bone
4. *Gomphosis*—a cone-shaped peg that fits into a socket
   **Example:** Roots of teeth in the alveolar processes of skull

An *amphiarthrosis* (am-fē-ar-THRO-sis) is a slightly movable joint (see Figure 14-1). The degree of movement is almost imperceptible. The two types of amphiarthroses are

1. *Symphysis* (SIM-fi-sis)—two bones separated by white fibrocartilage (Figure 14-1)
   **Example:** a. *Pubic symphysis* between the two pubic bones of the pelvic girdle

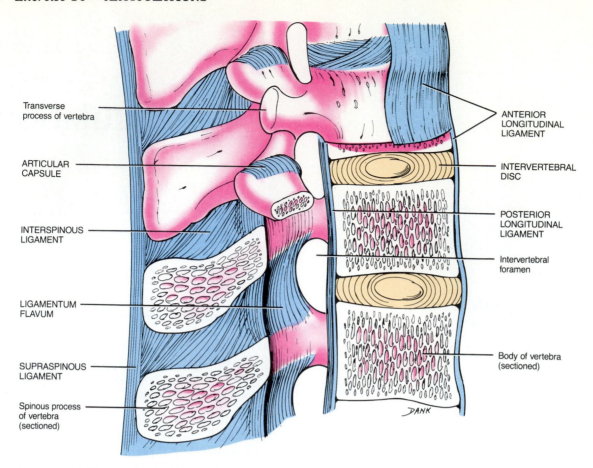

**Figure 14-1** A slightly movable joint (amphiarthrosis). The intervertebral discs of the spinal column separate the centra of adjacent vertebrae. (From G. J. Tortora, *Principles of Human Anatomy,* 6th ed., New York: HarperCollins, 1992. © Leonard Dank.)

  b. *Intervertebral discs* between the centra of adjacent vertebrae
2. *Syndesmosis* (sin-dez-MŌ-sis)—two bones separated by relatively thick, dense fibrous connective tissue
   **Example:** Distal ends of tibia and fibula

A *diarthrosis* (dī-ar-THRŌ-sis) is a freely movable joint (Figure 14-2). These joints are characterized by the presence of a space between the bones involved. The space is termed the *synovial* or *joint cavity.* This separation allows great freedom of movement. Six types of diarthrotic joints, as shown in Figure 14-3, are as follows:

1. *Hinge joint*—convex surface of one bone fits into concave surface of another in such a way that movement in only one plane (monaxial) is allowed (see Figure 14-3b)
   **Example:** a. *Elbow* between the distal end of the humerus and the proximal end of the ulna
   b. *Knee* between the distal end of the femur and the proximal end of the tibia

  c. *Interphalangeal joints* between adjacent phalanges of fingers or toes
2. *Condyloid* (KON-di-loyd) *joint* or *ellipsoidal* (ē-lip-SOYD-al) *joint*—spherical head of one bone fits into an elliptical cavity, allowing movement in two planes (biaxial) (see Figure 14-3d)
   **Example:** a. Between the distal end of the metatarsals 2 to 5 and the proximal end of the first phalanx
   b. Between the atlas of the vertebral column and the occipital condyles of the skull
   c. Between radius and ulna and adjacent carpals (wrist joint)
3. *Ball and socket joint*—rounded head of a bone fits into a capsule of another, allowing movement in three planes (triaxial) (see Figure 14-3f)
   **Example:** a. *Hip* between the proximal end of the femur and the acetabulum
   b. *Shoulder* between the proximal end of the humerus and the glenoid fossa (the most freely movable joint in the body)
4. *Pivotal* or *rotary joint*—one bone turns on an-

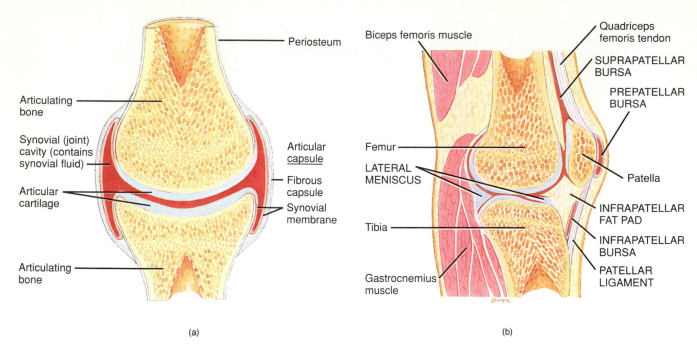

**Figure 14-2** Synovial joint: (a) generalized synovial joint in frontal section; (b) sagittal section of the knee joint, showing articular discs (menisci) and bursae. (From G. J. Tortora and S. R. Grabowski, *Principles of Anatomy and Physiology,* 7th ed., New York: HarperCollins, 1993.)

other, allowing monaxial movement (see Figure 14-3c)

**Example:** a. Between the atlas and the axis
b. Between the proximal and distal ends of the radius and the ulna

5. *Plane* or *gliding joint*—flat articulating surfaces move side to side or forward and back on one another, allowing highly restricted biaxial movement (see Figure 14-3a)

**Example:** a. Between adjacent carpal bones of the wrist
b. Between adjacent tarsal bones of the ankle
c. Between superior and inferior articular surfaces of the vertebrae
d. Between costal demifacets of thoracic vertebrae and ribs

6. *Saddle joint*—concave surface of one bone fits over a convex surface of another, allowing biaxial movement (much like the ellipsoidal joint) (see Figure 14-3e)

**Example:** Between the first metacarpal (thumb) and the greater multangular of the wrist

## STRUCTURAL CLASSIFICATION

On the basis of structure, joints may be classified according to the presence or absence of a joint cavity and the kind of connective tissue between the bones. There are three structural classifications.

1. *Fibrous joint*—bones held together by fibrous connective tissue; no joint cavity; may be synarthrotic or amphiarthrotic
2. *Cartilaginous joint*—bones are held together by cartilage; no joint cavity; may be synarthrotic or amphiarthrotic
3. *Synovial joint*—bones not bound together; separated by a synovial cavity; diarthrotic

### Examples of a Fibrous Joint

Suture—immovable (review skull sutures); syndesmosis—slightly movable (review distal ends of tibia and fibula); and gomphosis—immovable (review teeth in bony sockets).

### Examples of a Cartilaginous Joint

Synchondrosis—immovable (Figure 14-4); symphysis—slightly movable.

### Examples of Synovial Joints

Hinge, condyloid (ellipsoidal), ball and socket pivotal (rotary), plane (gliding), and saddle joints have been shown in Figures 14-2 and 14-3, and Figure 14-5 shows a photograph of a partially dissected diarthrotic synovial joint.

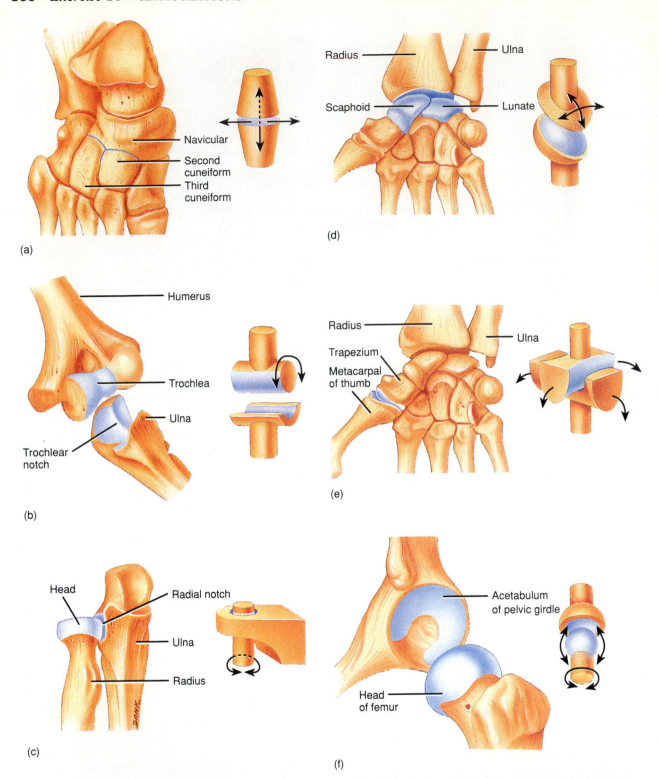

(a)

(b)

(c)

(d)

(e)

(f)

**Figure 14-3** Simplified representations of types of synovial joints. (From G. J. Tortora and S. R. Grabowski, *Principles of Anatomy and Physiology,* 7th ed., New York: HarperCollins, 1993.)

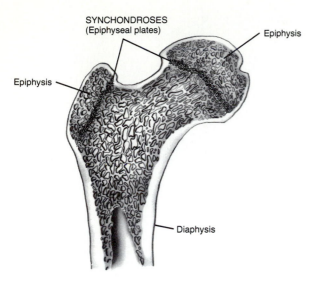

**Figure 14-4** An immovable joint *(synchondrosis).* The epiphyseal plates (cartilage) separate the heads of this young femur from the shaft. (From G. J. Tortora and N. P. Anagnostakos, *Principles of Anatomy and Physiology,* 3rd, 4th, 5th, 6th eds., New York: HarperCollins, 1981, 1984, 1987, 1990.)

## JOINT MOVEMENT

The types of movement allowed at freely movable joints are many and varied. There are always factors that prevent dislocation or restrict movement, such as (a) large muscle masses that limit bending, (b) tight tendons and ligaments, and (c) muscle tension. However, most movements can be clearly defined, and the student should try to demonstrate them on his or her own body. Try to make the following movements, using Figure 14-6 as a guide.

### Types of Movements

1. *Plane or gliding*—Make a simple sliding movement from side to side between adjacent carpals of the wrist.
2. *Flexion*—Decrease the angle between two bones: Fold the elbow.
3. *Extension*—Increase the angle between two bones: Straighten the elbow (arm).
4. *Dorsiflexion*—Bend the foot at the ankle region: With your foot on the ground, bring your toes upward.
5. *Hyperextension*—Straighten one part on another beyond its normal position: Bend the head backward.
6. *Plantar flexion*—Extend the foot on the ankle as far as possible: Stand on your toes.
7. *Abduction*—Move an appendage away from the

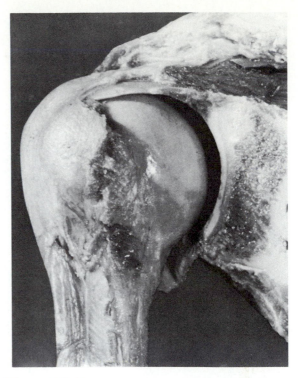

**Figure 14-5** A freely movable joint (diarthrosis). A photograph of the shoulder joint partially dissected to expose the glenoid fossa between the humerus and scapula.

midline: Move your arm or leg out and away from the body; spread your fingers.
8. *Adduction*—Move an appendage toward the midline: Bring your arm or leg back to the midline from an abducted position.
9. *Circumduction*—While the proximal end of the appendage remains stationary, make a circle or arc with the distal end: Extend the arm, and make the arc with your hand.
10. *Rotation*—Turn one bone on another: Turn the atlas on the axis (turn the head); rotate the radius on the ulna to turn the palm of the hand upward (*supination*) or downward (*pronation*); turn the leg inward or outward by rotating at the hip.

### Axes of Movement

Further, joints may be classified as

1. *Monaxial*—movement in one plane only
2. *Biaxial*—movement in two planes
3. *Triaxial*—a wide range of movements

Examples of the aforementioned joints are

### Monaxial

*Hinge joint*—movement is flexion and extension.
*Pivotal joint*—movement is primarily rotation.

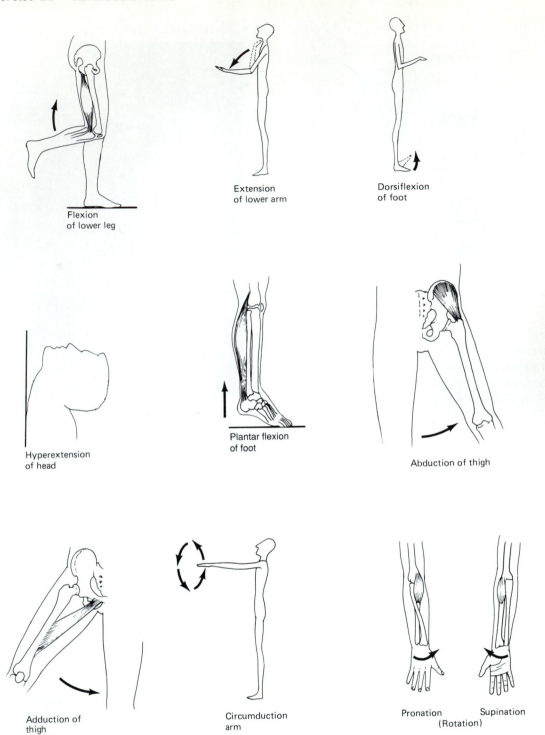

**Figure 14-6** Joint movements.

## Biaxial

*Condyloid or ellipsoidal*—movement is side-to-side and back and forth.

*Saddle*—a modified condyloid joint with movement side-to-side and back and forth.

## Triaxial

*Ball and socket*—movement is flexion–extension, abduction–adduction, and rotation

# LABORATORY REVIEW *14*

## *ARTICULATIONS*

### MATCHING AND COMPLETION

1. Match the functional classifications of joints on the left with the structural classifications to which they belong.

    a. _____ Suture

    b. _____ Synchondroses

    c. _____ Synostoses

    d. _____ Symphyses

    e. _____ Syndesmoses

    f. _____ Diarthroses

2. Give an example of a specific joint for each functional classification listed in Question 1.

    a. _____

    b. _____

    c. _____

    d. _____

    e. _____

    f. _____

3. List the types of diarthrotic joints, and give an example of each.

    a. _____

    b. _____

    c. _____

    d. _____

    e. _____

    f. _____

4. Match the joints on the left with the term describing its action.

    a. _____ Shoulder

    b. _____ Between atlas and axis

    c. _____ Pubic symphysis

    d. _____ Between adjacent skull bones

    e. _____ Between radius and ulna

    f. _____ Between articular surfaces of vertebrae

    g. _____ Knee

    h. _____ Hip

    i. _____ Child's epiphyseal plate region in the long bone

    j. _____ Elbow

## ANSWERS

1. Fibrous

2. Cartilaginous

3. Synovial

<br>

1. Monaxial

2. Biaxial

3. Triaxial

4. Partially movable

5. Immovable

# The Muscle System

*T*he human body has more than 600 skeletal muscles that account for approximately 40–45 percent of an individual's body weight. These muscles are used to move and manipulate body parts.

This section contains exercises on the dissection of cat muscles and their comparison with human muscles. The muscles are divided into groups so that a logical progression can be made to dissect the superficial muscles and the more important deep muscles. No attempt is made to cover all the cat muscles. The cat muscles should be compared with human muscles, as well as references made to the text drawings and to charts and models.

With each dissection sequence, there are the following study aides: (1) a black and white photo of the specific muscle group, (2) a line drawing corresponding to the black and white photo, (3) a table summarizing the origins, insertions, and actions of cat muscles, (4) a table summarizing the origins, insertions, and actions of corresponding human muscles, and (5) color plates of cat muscle dissections.

Because of their importance in body movements, skeletal muscles have long been the subject of studies concerned with their chemical composition, ultra-structure, and mechanism of contraction. Exercises in this section also present several aspects of muscle physiology, including the microscopic and submicroscopic properties of skeletal muscle fibers, the biochemistry of muscular contraction, the demonstration and measurement of the degree of muscular contraction (using a laboratory animal), isotonic and isometric muscle contractions, muscle fatigue, and electromyography.

### GENERAL REFERENCES

Berger, R. A. *Applied Exercise Physiology.* Philadelphia: Lea & Febiger, 1982.

Eisenberg, E., and L. E. Greene. "The Relation of Muscle Biochemistry to Muscle Physiology." *Annual Review of Physiology,* 42:293, 1980.

Kapit, W., R. I. Macey, and E. Meisami. *The Physiology Coloring Book.* New York: Harper & Row, 1987.

Tortora, G. J., and S. R. Grabowski. *Principles of Anatomy and Physiology,* 7th ed. New York: HarperCollins, 1993.

Wilson, F. C. *Musculoskeletal System: Basic Processes and Disorders,* 2nd ed. Philadelphia: Lippincott, 1983.

## SPECIFIC REFERENCES

Basmajian, J. V. "Electromyography—Dynamic Gross Anatomy: A Review." *American Journal of Anatomy,* 159: 245–260, 1980.

Edwards, R. H. T. "Human Muscle Function and Fatigue." In *Human Muscle Fatigue: Physiological Mechanisms.* Ciba Foundation Symposium, Vol. 82, pp. 1–18. London: Pittman Medical, 1981.

Hamilton, W. G. "Surgical Anatomy of the Foot and Ankle." *Clinical Symposia,* 37(3), 1985.

Lampe, E. W. "Surgical Anatomy of the Hand." *Clinical Symposia,* 40(3), 1988.

LeVeau, B. *William's & Lissner's Biomechanics of Human Motion,* 3rd ed. Philadelphia: Saunders, 1991.

Nadel, E. R. "Physiological Adaptations to Aerobic Training." *American Scientist,* 73: 334–343, 1985.

Shepard, R. J. *Physiology and Biochemistry of Exercise.* New York: Praeger Publishers, 1982.

Veikko, A. K., III. "The Physiology of Marathon Running." *Science Progress,* 70: 109–127, 1986.

# Removal of the Skin and Muscle Review

## After completing this exercise, you should be able to

1. Remove the skin of the cat in one piece
2. Prepare the specimen for future dissection by removing all fat, fascia, and fur from the muscle surface so that muscle outlines and fiber direction can be seen

## Materials

The following materials should be provided for class use:

1. Embalmed cats in plastic bags
2. Dissecting tools
   a. Heavy-duty scissors
   b. Scalpel
   c. Probe
   d. Two dissecting needles
   e. Forceps
3. Bone forceps
4. String
5. Rubber bands
6. Name tags

## INTRODUCTION

This portion of the manual contains exercises on the dissection of cat muscles and their comparison to human muscles. The cat muscles can be compared to human muscles using the drawings that follow each group of cat muscles.

With each dissection sequence there are five study aides: (1) a black and white photo of the specific muscle group in the cat, (2) a line drawing corresponding to the black and white photo, (3) color plates of the dissections done on the cat, (4) tables summarizing the origins, insertions, and actions of cat muscles and of corresponding human muscles, and (5) color photos of human muscle dissections. The labeled color plates of cat-muscle dissections are found after Exercise 21. Compare them with the black and white photos while making each dissection.

The ideal way to make a dissection of the cat muscles is for each student to work on a specimen alone. However, the expense involved and the difficulty in procuring that large number of cats usually results in two or more students sharing an animal. If this is the case, it is to be understood that each student does an equal amount of work and that each uses his or her own dissecting tools.

If laboratory grades are given, the dissection grade must be shared by those responsible for the specimen.

Complete instructions, photographs, and line drawings are provided for each step in the dissection. Laboratory review material should be completed as the dissection progresses.

For the safety of the student, it is recommended that a heavy-duty, nonrazor scalpel be used, especially in the skinning procedure. If the student wishes, a razor scalpel for use on softer tissues is acceptable. Always point scissors away from you when cutting tissue, and be careful to keep fingers away from the scalpel blade.

## Removal of the Skin

Read the *entire section* carefully before you begin. Removing the skin in one piece is not difficult, but care must be taken to ensure that no muscles are damaged or destroyed in the process. In many places along the surface of the body, the skin is closely applied to the underlying muscle and tightly held to that muscle by tough fascia. In other places, the skin is loosely attached to the body and is separated from the muscle by sub-cutaneous fat. Cats do not ordinarily store large masses of fat under the skin, as humans do, but occasionally a fat cat will be encountered. A lactating or pregnant female is a case in point.

Another thing to keep in mind before starting the skinning procedure is that the cat has large, thin muscle masses that attach directly to the skin, which, when they contract, move the skin. In the head and neck regions, this muscle is called *platysma*. In the dorsal and ventral body regions posterior to the neck, it is called the *cutaneous maximus* (see Figure 15-1). These muscle strands may be removed with the skin, or they may be removed later, after all the skin has been taken off. Humans have a small *platysma* but no *cutaneous maximus*.

To begin skin removal, place the cat face down on a large dissecting tray. Using Figure 15-2 as a guide, insert the scissors in the midline just above the tail, and cut forward, separating skin from muscle and fascia as you go along, using your fingers wherever possible. Extend your first incision line to a point just above the neck. Now separate with your hands as much skin as possible to the right and to the left. Cut off the tail at this point (see Figure 15-2a), using either a saw or bone forceps. Discard the tail.

When you have loosened all the skin on the dorsal surface, begin your next incision lines along the backside of each arm to the wrist, along the backside of each leg to the ankle, and around the neck. When these incision lines have been made, once again loosen as much skin as possible with your fingers, using scissors or scalpel as cutting tools only when necessary.

The next step is to cut the skin around each wrist and around each ankle, and all around the neck onto the ventral side (see Figure 15-2b).

Turn the cat onto its back, and begin pulling the skin from the neck downward (see Figure 15-2c) by holding the cat with one hand and pulling the skin with the other. If the skin does not come off easily, you will have to free it by cutting.

The last stage in the skin removal should be done very carefully if you have a male cat. In Figure 15-2d, note the two large ducts leading up from the testes between the legs (a probe lies under the left duct). These ducts are an important part of the reproductive system and must *not* be cut off.

Continue pulling the skin down until you reach the external genitalia. Cut a circle around the genitalia, and remove the remaining skin on the legs. The skin around the external genitalia is not removed until you study the reproductive system.

If you have a female cat, the precautions suggested above need not be taken. Cut a circle around the anus, and simply pull off the remaining skin. Occasionally, a lactating female is encountered. The mammary glands may be left attached to the skin as it is being removed, or they may be removed later. Mammary glands are well developed in cats that have just given birth (see Figure 15-3).

Now you have removed the skin in one piece. It may be placed under the cat, or you may keep it in the plastic bag in which the cat was originally stored during

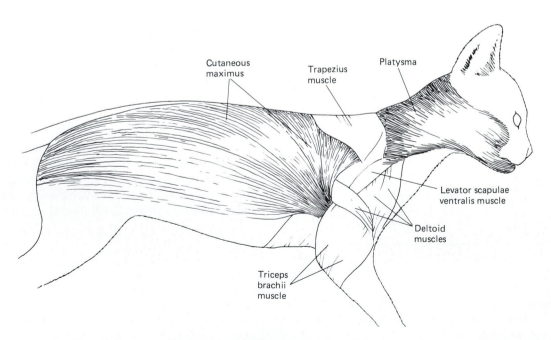

**Figure 15-1** Cutaneous maximus and platysma muscles of the cat.

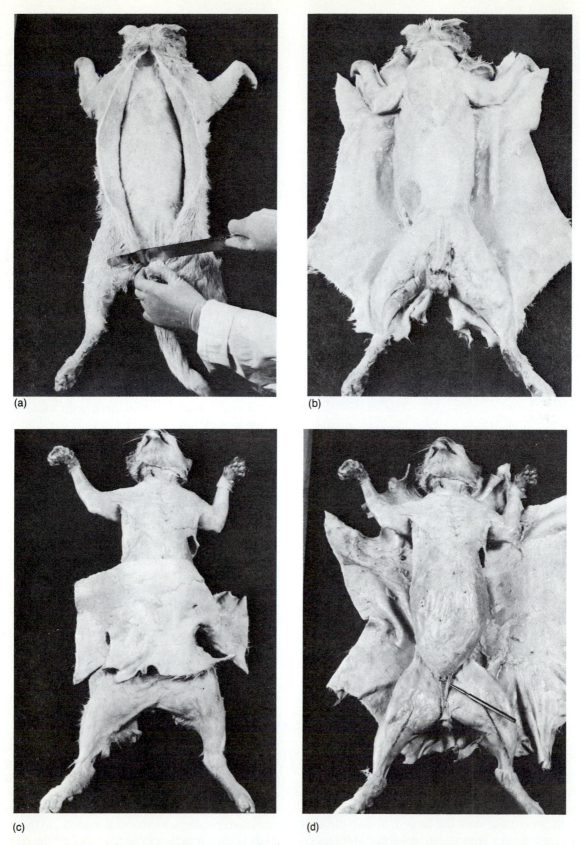

**Figure 15-2** (a) First incision, exposing back musculature and removal of the tail; (b) skin laid back in one piece, exposing musculature of back, arms, and legs; (c) partial freeing of the skin, shown from the ventral surface; (d) skin completely removed in one piece, and showing a probe under the spermatic cord.

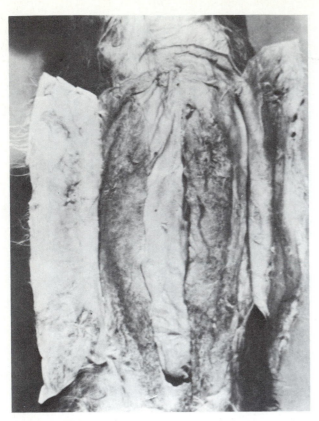

**Figure 15-3** Fully developed mammary glands in the female cat.

the laboratory period. In any event, the skin is to be retained and used to wrap the cat at the end of the lab period before storing it in its plastic bag. Wrapping the body in its skin helps to keep the cat moist between laboratory periods. Tie the skin to the body with string.

One must be careful not to let the cat dry out because it is impossible to do a good muscle dissection on a dry animal. If, during the course of the semester, you find that the cat is losing too much moisture, add a solution of 3 percent phenol from time to time. This solution may be placed directly on the musculature, used to remoisten the skin, or placed in small quantities in the plastic bag before it is closed with string or a rubber band at the end of the lab period.

The last step before beginning the dissection of the muscles is the removal of all extraneous material from the surface, such as bits of fat, hair, loose fascia, and remnants of cutaneous maximus muscle and platysma. If you have any doubts about the removal of any type of tissue, ask the instructor for advice so that you do not cut off something essential to your future dissection. You can always remove an unwanted substance from the surface at a later time, but you can't reattach something you have cut off and thrown away. Always

proceed slowly, after giving each move careful thought. When the surface is properly cleaned off, you should be able to make out some muscle outlines and see the directions of the muscle fibers. Make certain that all cutaneous maximus and platysma have been removed. Have your instructor check your cat before you proceed further.

## OVERVIEW OF MUSCLES AND MUSCLE TISSUE

Although muscular tissue can be divided into three categories on the basis of its location, the cells in muscle tissue all have one thing in common. The fibrils within them shorten and thicken and thus effect movement. Review the histology of the following muscles in Exercise 6.

1. *Visceral muscle*—found in visceral organs, largely involuntary action
2. *Cardiac muscle*—found only in the heart, involuntary
3. *Skeletal muscle*—associated with skeletal parts, voluntary (*Note:* Skeletal muscle is the subject for the study of muscles in this unit.)

### The Structure of a Muscle

Each muscle has three main parts:

1. *Origin*—end attached to the part of the skeleton that does not move when the muscle contracts
2. *Insertion*—end attached to the part of the skeleton that moves when the muscle contracts
3. *Body* or *belly*—center of the muscle between the origin and the insertion

Most muscles are subdivided into bundles or *fasciculae*. This is an important concept to keep in mind because there is often a tendency for the student to try to subdivide a muscle into several parts simply because it seems possible to do so. The entire extent, size, and shape of a muscle must be determined *before* beginning to dissect it, and its origin and insertion should be located on the skeleton. If you have a cat skeleton available, identify the bones. You will find that they are similar to those of the human. Locate each origin and insertion as you go along. This manual provides photographs, line drawings, and tables indicating the origin, insertion, and action of each muscle, so that careful study of each muscle can be made *before* the dissection is begun. A drawing of the cat skeleton (see Figure 15-4) will help you locate muscle origins and insertions.

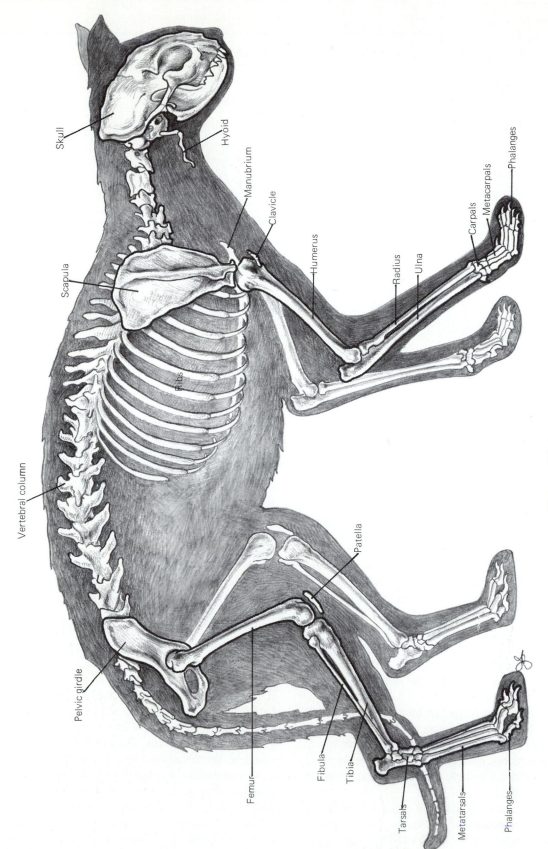

Skull

Hyoid

Manubrium

Clavicle

Humerus

Radius

Ulna

Carpals

Metacarpals

Phalanges

Scapula

Ribs

Vertebral column

Pelvic girdle

Patella

Femur

Fibula

Tibia

Tarsals

Metatarsals

Phalanges

**Figure 15-4** Cat skeleton.

**Table 15-1**

| Muscle Action | Description |
|---|---|
| Flexion | *Flexor muscle* bends a joint or reduces the angle between two bones. |
| Extension | *Extensor muscle* straightens a joint or increases the angle between two bones. |
| Abduction | *Abductor muscle* carries an appendage or a part away from the midline. |
| Adduction | *Adductor muscle* brings an appendage or a part back to the midline. |
| Elevation | *Levator muscle* raises a part. |
| Depression | *Depressor muscle* lowers a part. |
| Rotation | *Rotator muscle* turns one part on another. |

## Muscle Actions

Use your text for demonstrations of muscle actions. Some of the most frequently encountered are listed in Table 15-1.

## Dissection

To *dissect* a muscle does not mean to cut it, but merely to separate it cleanly from the adjacent muscles between its origin and insertion. *Never cut an origin or an insertion.* If deeper muscles are to be viewed and dissected, it will be necessary to make *transections.* To transect a muscle, find the midpoint between its origin and insertion, and cut it in half (or as close to half as possible in the case of odd-shaped muscles), perpendicular to the direction of its fibers. Fold each half back to origin or insertion. You will then be able to observe the underlying muscles. Never transect a muscle unless you are specifically instructed to do so in this manual or by your instructor.

Each muscle is surrounded by connective tissue called *fascia.* Extensions of fascia around and into the muscle are differentiated into *epimysium, perimysium,* and *endomysium,* but because we do not separate muscles into their bundles in these exercises, the term *fascia* is used exclusively.

Most muscles do not attach directly to the bone, but rather by extensions of fascia. Connective tissue that binds a muscle to a bone is a *tendon.* Connective tissue that connects one bone with another bone is a *ligament.* The name *aponeurosis* (ap'-ō-nyoo-RŌ-sis) is often applied to a broad, flat tendon.

The connective tissue that binds muscle together may be very tough and require cutting with scissors or a scalpel, or it may be so soft and fragile that it can be broken by running a finger between the muscles. Do whatever you can with your fingers, resorting to scissors or scalpel only when necessary. The older the animal is, the tougher the fascia. If you have an old cat to dissect, you will find that you have to work harder to separate muscles than your neighbor who has a younger specimen.

In order to save time, instructions for dissection are given for one side of the cat. Corresponding photographs and line drawings in each exercise show the left side.

Because this is probably the first time you have made a dissection of this kind, it would be wise to take your time, think carefully before dissecting or transecting, and remember that once a wrong cut is made, it cannot be corrected (except with needle and thread). One's laboratory skill often enters into the determination of the course grade. This is the only cat you will get, so take precautions, and do not make mistakes that could be avoided simply by reading the instructions.

Immediately following are diagrams and photographs of human musculature. Refer to these (Figures 15-5, 15-6, 15-7, and 15-8) and other diagrams throughout all your work on human muscles, and compare the cat muscles you have dissected with their human counterparts.

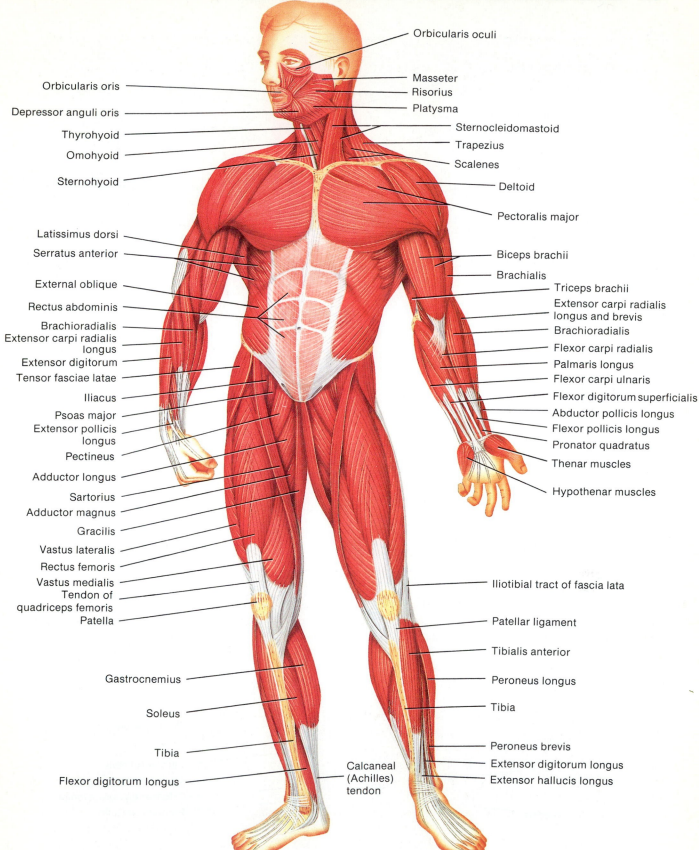

Orbicularis oculi

Masseter
Risorius
Platysma

Orbicularis oris
Depressor anguli oris
Thyrohyoid
Omohyoid
Sternohyoid

Sternocleidomastoid
Trapezius
Scalenes

Deltoid

Pectoralis major

Latissimus dorsi
Serratus anterior

Biceps brachii
Brachialis

External oblique
Rectus abdominis
Brachioradialis
Extensor carpi radialis longus
Extensor digitorum
Tensor fasciae latae
Iliacus
Psoas major
Extensor pollicis longus
Pectineus
Adductor longus
Sartorius
Adductor magnus
Gracilis
Vastus lateralis
Rectus femoris
Vastus medialis
Tendon of quadriceps femoris
Patella

Triceps brachii
Extensor carpi radialis longus and brevis
Brachioradialis
Flexor carpi radialis
Palmaris longus
Flexor carpi ulnaris
Flexor digitorum superficialis
Abductor pollicis longus
Flexor pollicis longus
Pronator quadratus
Thenar muscles

Hypothenar muscles

Iliotibial tract of fascia lata

Patellar ligament

Tibialis anterior

Peroneus longus

Tibia

Gastrocnemius

Soleus

Tibia

Flexor digitorum longus

Calcaneal (Achilles) tendon

Peroneus brevis
Extensor digitorum longus
Extensor hallucis longus

**Figure 15-5** Principal superficial skeletal muscles: anterior view. (From G. J. Tortora and S. R. Grabowski, *Principles of Anatomy and Physiology,* 7th ed., New York: HarperCollins, 1993. © Leonard Dank.)

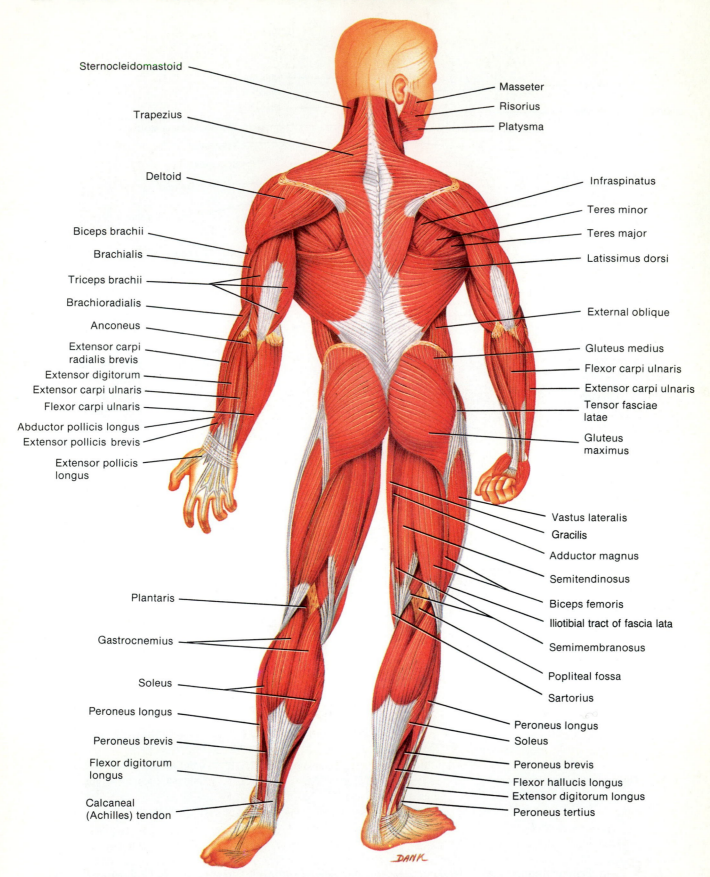

Sternocleidomastoid

Trapezius

Deltoid

Biceps brachii

Brachialis

Triceps brachii

Brachioradialis

Anconeus

Extensor carpi radialis brevis

Extensor digitorum

Extensor carpi ulnaris

Flexor carpi ulnaris

Abductor pollicis longus

Extensor pollicis brevis

Extensor pollicis longus

Plantaris

Gastrocnemius

Soleus

Peroneus longus

Peroneus brevis

Flexor digitorum longus

Calcaneal (Achilles) tendon

Masseter

Risorius

Platysma

Infraspinatus

Teres minor

Teres major

Latissimus dorsi

External oblique

Gluteus medius

Flexor carpi ulnaris

Extensor carpi ulnaris

Tensor fasciae latae

Gluteus maximus

Vastus lateralis

Gracilis

Adductor magnus

Semitendinosus

Biceps femoris

Iliotibial tract of fascia lata

Semimembranosus

Popliteal fossa

Sartorius

Peroneus longus

Soleus

Peroneus brevis

Flexor hallucis longus

Extensor digitorum longus

Peroneus tertius

DANK

**Figure 15-6** Principal superficial skeletal muscles: posterior view. (From G. J. Tortora and S. R. Grabowski, *Principles of Anatomy and Physiology*, 7th ed., New York: HarperCollins, 1993. © Leonard Dank.)

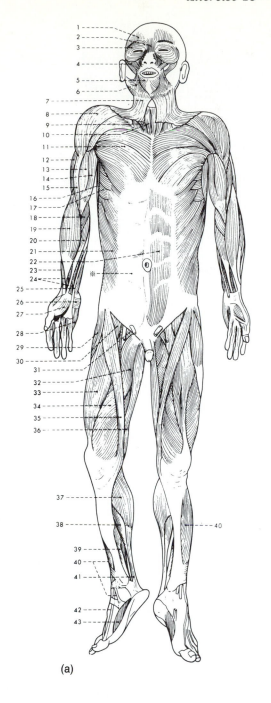

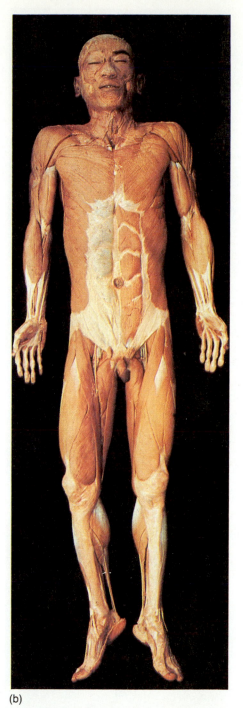

(a)

(b)

**Figure 15-7** Superficial muscles—anterior aspect: 1. frontal belly of occipito-frontalis; 2. procerus; 3. orbicularis oculi; 4. zygomaticus major, minor, levator anguli oris, levator labii superioris alaeque nasi; 5. depressor labii inferioris; 6. risorius, depressor anguli oris; 7. platysma; 8. deltoid; 9. sternocleidomastoid; 10. sternohyoid; 11. pectoralis major; 12. lateral head of triceps; 13. long head of biceps; 14. short head of biceps; 15. brachialis; 16. latissimus dorsi; 17. serratus anterior; 18. pronator teres; 19. brachioradialis; 20. flexor carpi radialis; 21. external abdominal oblique; 22. rectus abdominis; 23. abductor pollicis longus; 24. flexor digitorum superficialis; 25. palmaris longus; 26. palmaris brevis; 27. thenar eminence; 28. abductor digiti minimi; 29. tensor fasciae latae; 30. iliopsoas; 31. pectineus; 32. adductor longus; 33. rectus femoris; 34. vastus medialis; 35. sartorius; 36. gracilis; 37. gastrocnemius; 38. soleus; 39. flexor digitorum longus; 40. tibialis anterior; 41. tendo calcaneus; 42. extensor hallucis longus; 43. abductor hallucis aponeurosis of external oblique (rectus sheath). (From Yokochi/Rohen *Photographic Anatomy of the Human Body,* 2nd ed., Baltimore: University Park Press, and Tokyo: Igaku-Shoin Ltd., 1978.)

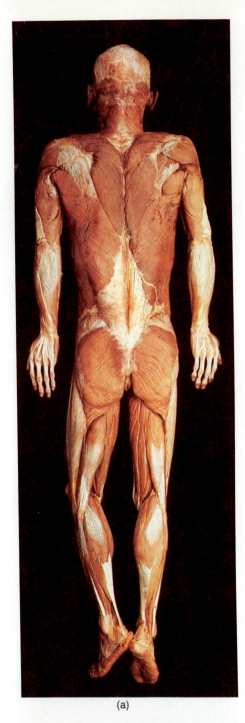

(a)

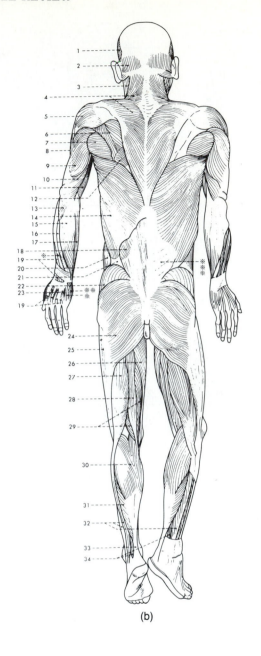

(b)

**Figure 15-8** Superficial muscles, posterior aspect: 1. temporalis; 2. occipital belly of occipito-frontalis; 3. sternocleidomastoideus; 4. trapezius; 5. deltoid; 6. infraspinatus; 7. teres major; 8. lateral head of triceps; 9. long head of triceps; 10. medial head of triceps; 11. brachialis; 12. rhomboideus major; 13. anconeus; 14. latissimus dorsi; 15. flexor carpi ulnaris; 16. extensor carpi ulnaris; 17. flexor digitorum superficialis; 18. serratus posterior inferior; 19. extensor digitorum (communis) 20. extensor abdominal oblique; 21. lumbar triangle; 22. gluteus medius; 23. dorsa interosseous; 24. gluteus maximus; 25; iliotibial tract of fascia lata; 26. gracilis; 27. biceps femoris. (From Yokochi/Rohen *Photographic Anatomy of the Human Body,* 2nd ed., Baltimore: University Park Press, and Tokyo: Igaku-Shoin Ltd., 1978.)

# LABORATORY REVIEW *15*

NAME _____

LAB SECTION _____ DATE _____

## REMOVAL OF THE SKIN AND MUSCLE REVIEW

### COMPLETION AND DEFINITIONS

1. For each of the following terms, state what it is, where it is located, and what function, if any, it has.

   a. Cutaneous maximus _____

   _____

   _____

   b. Platysma _____

   _____

   _____

   c. Fascia _____

   _____

   _____

2. How is the origin of a muscle different from the insertion? _____

   _____

   _____

3. What is the difference between dissecting and transecting?

   _____

   _____

   _____

4. Under what circumstances would it be necessary to transect a muscle?

   _____

5. Define these terms:

   a. Tendon _____

   _____

   b. Ligament _____

   _____

   c. Aponeurosis _____

   _____

# Muscles of the Back and Shoulder

*After completing this exercise, you should be able to*

1. Dissect the muscles of the back of the cat
2. Identify comparable human muscles

## Materials

The following materials should be provided for class use:

1. Embalmed cats
2. Dissecting tools
3. Dissecting trays
4. String
5. 3 percent phenol solution

The muscles dissected in Exercise 16 are listed here, according to their specific exercises.

### SUPERFICIAL BACK AND SHOULDER MUSCLES

Spinotrapezius

Acromiotrapezius

Clavotrapezius

Levator scapulae ventralis

Latissimus dorsi

Spinodeltoid

Acromiodeltoid

Clavodeltoid (clavobrachialis)

### MUSCLES ASSOCIATED WITH THE SCAPULA

Rhomboideus major

Rhomboideus minor

Rhomboideus capitis

Supraspinatus

Infraspinatus

Teres major

Teres minor

### DEEP BACK MUSCLES

Splenius

Spinalis dorsi

Longissimus dorsi

    Longissimus thoracis

    Longissimus cervicis

    Longissimus capitis

Iliocostalis

Serratus dorsalis

Multifidus spinae

Serratus ventralis

Subscapularis

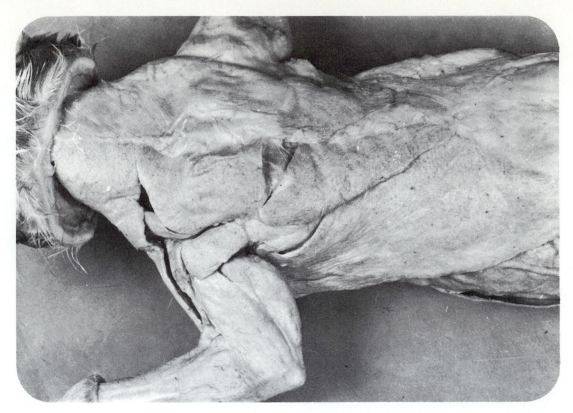

**Figure 16-1** (a) Superficial back muscles of cat (photograph).

## SUPERFICIAL BACK AND SHOULDER MUSCLES

### Instructions for Dissection

Begin your dissection in the upper back region between the two arms. Make sure that the skin has been removed from the neck as far forward as the ears (see Figure 16-1). Determine the midline *raphe* (RĀ-fē), and locate the *trapezius group* of muscles, which have their origins on this raphe. Humans have only one trapezius muscle, but the cat has three, as shown in Figure 16-1b.

Start with the most posterior of the group, the *spinotrapezius*. This is a triangular muscle that lies over the latissimus dorsi at its posterior end and joins the acromiotrapezius anteriorly. Dissect it completely, being careful not to loosen either its origin or insertion.

Next dissect the *acromiotrapezius* and finally the *clavotrapezius*, both four-sided muscles. The clavotrapezius inserts on the very small clavicle bone embedded in the muscle between clavotrapezius and clavodeltoid. Try to feel the clavicle with your fingertips and locate it on the cat skeleton.

Underneath the acromiotrapezius and running craniad is the *levator scapulae ventralis*. Free it as much as possible. It will be seen in greater length in a later dissection.

Posterior to the spinotrapezius is the very large *latissimus dorsi*. Free it between origin and insertion. Because it originates by a broad, tough aponeurosis along the midline of the back, it will be difficult to loosen it all the way to its origin, but it can be done at a later time.

The *deltoid* group of muscles is represented in the

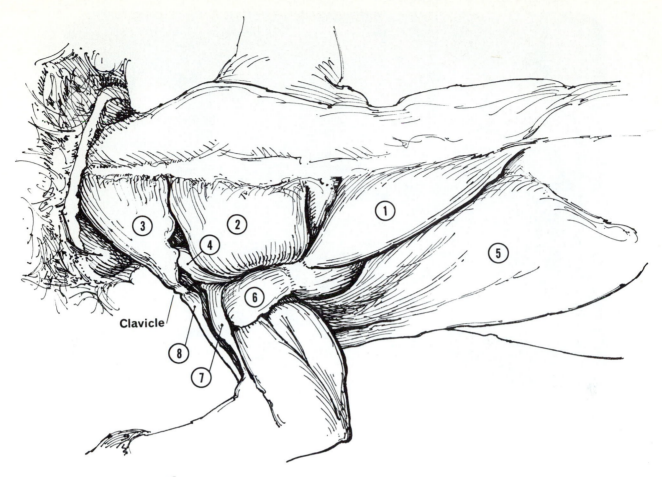

1. Spinotrapezius (spī-nō-tra-PĒ-zē-us)
2. Acromiotrapezius (a-KRŌ-mē-ō)
3. Clavotrapezius (KLAV-ō)
4. Levator scapulae ventralis (le-VĀ-tor SKAP-yoo-lē ven-TRAL-is)
5. Latissimus dorsi (la-TISS-i-mus DOR-sī)
6. Spinodeltoid (spī-nō-DEL-toyd)
7. Acromiodeltoid
8. Clavodeltoid or clavobrachialis (klav-ō-brak-ēAL-is)

**Figure 16-1** (b) Superficial back muscles of cat (line drawing).

cat by three muscles, while in the human, there is only one. The *clavodeltoid* extends down the arm from the clavicle bone and is sometimes termed the *clavobrachialis*. Lying posterior to the clavodeltoid is the triangular *acromiodeltoid*. Located just over the shoulder tip, it is difficult to loosen and so should just be outlined. The most posterior of the group is the *spinodeltoid*. This is a rectangular muscle and, like the acromiodeltoid, should be outlined rather than dis-

sected away from the bone. You may emphasize the muscle outline by pressure from a dissecting needle or a fine-pointed dissecting scissors.

Take note of the fact that each of the muscles in these two different groups have identical prefixes, such as *acromio*trapezius and *acromio*deltoid. Making yourself aware of things like this will facilitate memorization of names.

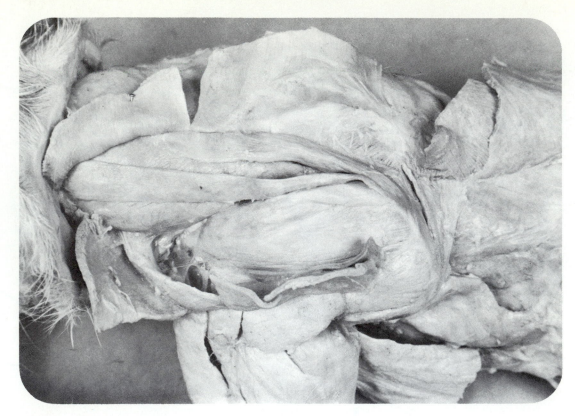

**Figure 16-2** (a) Muscles associated with scapula of cat (photograph).

## MUSCLES ASSOCIATED WITH THE SCAPULA

### Instructions for Dissection

At this time, it will be necessary to transect the trapezius group of muscles and the latissimus dorsi. Locate the *spinotrapezius.* It is triangular in shape, so that the transection line will have to be angled. The *acromiotrapezius* and *clavotrapezius* should be transected midway between their origins and insertions. Pull back the cut ends. The *latissimus dorsi* is irregular in shape, and it will not be possible to get two equal halves. Angle your transection line. In order to pull the muscle to its origin on the midline, you will have to use scissors to complete the cut in the fascia near the posterior end of the muscle. Once this has been done, the muscle can be pulled to the midline quite easily, and this reveals beneath it a thick silvery aponeurosis, about which more is written in a later exercise.

Pull the scapula away from the body, remove any excess fascia and fat, and observe the *rhomboideus major, rhomboideus minor,* and *rhomboideus capitis* muscles (Figure 16-2a). The rhomboideus minor is larger than the rhomboideus major and lies anterior to it (Figure 16-2b). The *rhomboideus capitis* is the most anterior of the group of rhomboid muscles, and it extends to the head. It is ribbonlike. On the lateral surface of the scapula, locate the *supraspinatus, infraspinatus,* and *teres major.* The supraspinatus and infraspinatus muscles need not be dissected away from the bony scapula, but be certain that the spine of the scapula is visible, showing the separation of the two muscles. Separate the *teres major* from the infraspinatus, and you will see the *teres minor,* which runs parallel to the teres major. Observe it, but do not dissect it further.

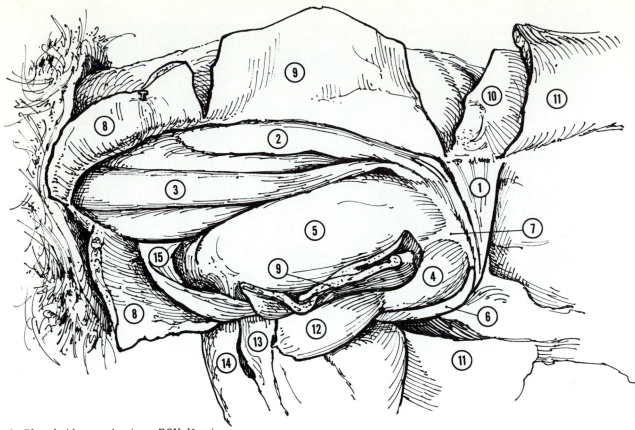

1. Rhomboideus major (rom-BOY-dē-us)
2. Rhomboideus minor
3. Rhomboideus capitis or occipitoscapularis (KA-pi-tis; ok-sip′-i-tō-skap′-yoo-LAR-is)
4. Infraspinatus (in′-fra′spī-NA-tus)
5. Supraspinatus (soo′-pra)
6. Teres major (TER-ēz)
7. Scapular spine (bone)
8. Clavotrapezius (reflected)
9. Acromiotrapezius (reflected)
10. Spinotrapezius (reflected)
11. Latissimus dorsi
12. Spinodeltoid
13. Acromiodeltoid
14. Clavodeltoid or clavobrachialis
15. Levator scapulae ventralis

**Figure 16-2** (b) Muscles associated with scapula of cat (line drawing).

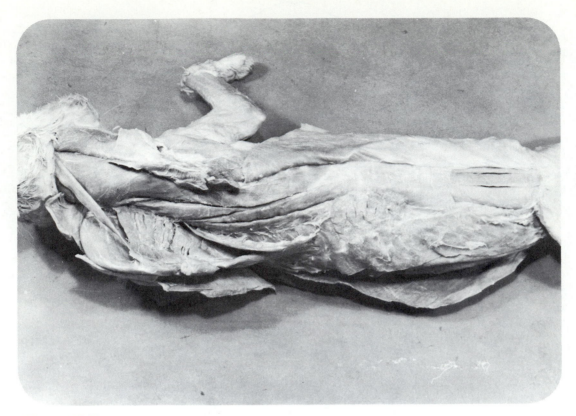

**Figure 16-3** (a) Deep back muslces of cat (photograph)

## DEEP BACK MUSCLES

### Instructions for Dissection

In order to observe the deeper muscles of the back, it is necessary to transect the *rhomboideus major* and *rhomboideus minor* muscles at this time. The rhomboideus muscles hold the dorsal part of the scapula to the body, and when they are transected, the scapula can be pulled away from the body to the left side of the cat (Figure 16-3a). Do not transect the rhomboideus capitis.

Anterior to the rhomboids, near the head, is a broad, flat muscle, the *splenius*. Do not dissect it, but observe its extent. Posterior to the splenius is the *spinalis dorsi* (Figure 16-3b). Lateral to the *spinalis dorsi* is a very long muscle, the *longissimus dorsi*, with three distinct parts, and lateral to it is the *iliocostalis*. Lying over the *iliocostalis* is a thin fascia, which is the origin of the *serratus dorsalis*, a serrated muscle that must be pulled down in order to reveal the *iliocostalis*. Carefully cut the origin of the serratus dorsalis for about 5 to 6 inches and pull the muscle down toward its insertion, thus

revealing the entire *iliocostalis* muscle. It is unusual to cut an origin, but it is necessary in this case, to make the deep muscle dissection complete.

Locate the three anterior parts of the *longissimus dorsi:* (1) *longissimus capitis* (the only part that extends to the head), (2) *longissimus cervicis,* and (3) *longissimus thoracis.* The separations are not easy to see, but a little probing with a blunt dissecting tool will soon locate the lines of separation. Follow the *longissimus* posteriad, and you will see that it disappears beneath a tough fascia, from which it cannot be separated. In order to see the posterior end of the muscle, which divides into three bundles, cut off the tough fascia in a rectangular block (see Figures 16-3a and b). This will expose four muscle bundles. The one closest to the midline is the *multifidus spinae,* and the other three are the *longissimus dorsi.*

Between the scapula and the ventral body region is another serrated muscle, the *serratus ventralis*. The muscle occupying the subscapular fossa is the *subscapularis.* Refer to Table 16-1 for the origins, insertions, and actions of the cat's back muscles; Table 16-2 shows these three aspects for human back muscles.

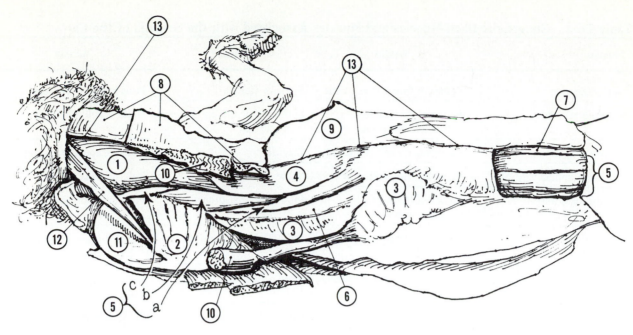

1. Splenius
2. Serratus ventralis
3. Serratus dorsalis (partially reflected)
4. Spinalis dorsi
5. Longissimus dorsi (posterior part)
   a. Longissimus thoracis
   b. Longissimus cervicis
   c. Longissimus capitis
6. Iliocostalis
7. Multifidus spinae
8. Trapezius group (reflected)
9. Latissimus dorsi (reflected)
10. Rhomboideus (cut end)
11. Subscapularis
12. Rhomboideus capitis
13. Midline

**Figure 16-3**   (b) Deep back muscles of cat (line drawing).

**Table 16-1  Superficial Back Muscles and Muscles Associated with the Scapula of the Cat**

| Muscle | Origin | Insertion | Action |
| --- | --- | --- | --- |
| Spinotrapezius | Spines of thoracic vertebrae | Spine of scapula | Pulls scapula dorsocaudad |
| Acromiotrapezius | Spines of cervical and thoracic vertebrae | Spine and acromion process of scapula | Holds scapula in place |
| Clavotrapezius | Lambdoidal crest and axis | Clavicle | Draws scapula craniodorsad |
| Levator scapulae ventralis | Occipital bone and atlas | Spine of scapula | Pulls scapula craniad |
| Latissimus dorsi | Spines of thoracic and lumbar vertebrae | Intertubercular groove of humerus | Pulls arm dorsad and caudad |
| Spinodeltoid | Spine of scapula | Deltoid ridge of humerus | Flexes humerus and rotates it outward |
| Acromiodeltoid | Acromion process of scapula | Deltoid ridge of humerus | Flexes humerus and rotates it outward |
| Clavodeltoid | Clavotrapezius muscle and clavicle | Humerus | Flexes forearm |
| Rhomboideus major | Spines of thoracic vertebrae | Vertebral border of scapula | Pulls scapula dorsally and medially |
| Rhomboideus minor | Spines of thoracic vertebrae | Vertebral border of scapula | Pulls scapula dorsally and medially |
| Rhomboideus capitis | Lambdoidal ridge | Vertebral border of scapula | Pulls scapula craniad |
| Supraspinatus | Supraspinous fossa of scapula | Greater tubercle of humerus | Extends humerus |
| Infraspinatus | Infraspinous fossa of scapula | Greater tubercle of humerus | Rotates humerus outward |
| Teres major | Axillary border of scapula | Humerus | Flexes humerus and rotates it inward |

**Table 16-2  Superficial Back Muscles of the Human**

| Muscle | Origin | Insertion | Action |
|---|---|---|---|
| Trapezius | Occipital bone and cervical and thoracic vertebrae | Spine and acromion process of scapula | Raises clavicle; adducts, elevates, and depresses scapula |
| Deltoid | Clavicle and spine and acromion process of scapula | Humerus | Abducts arm |
| Levator scapulae | Cervical vertebrae | Vertebral border of scapula | Raises scapula |
| Latissimus dorsi | Spines of thoracic and lumbar vertebrae, crests of sacrum and ilium, last four ribs | Intertubercular groove of humerus | Extends, adducts, and rotates arm medially; pulls shoulder down and back |
| Rhomboideus major | Spines of second through fifth thoracic vertebrae | Vertebral border of scapula | Moves scapula back and up and rotates it downward |
| Rhomboideus minor | Spines of seventh cervical and first thoracic verbebrae | Superior angle of scapula | Adducts scapula |
| Supraspinatus | Supraspinous fossa of scapula | Greater tubercle of humerus | Adducts arm |
| Infraspinatus | Infraspinous fossa of scapula | Greater tubercle of humerus | Rotates humerus laterally |
| Teres major | Inferior angle of scapula | Proximal end of humerus | Extends humerus; adducts and rotates arm |
| Spinalis capitis | Semispinalis thoracis | Spinalis thoracis | Extends vertebral column |
| Spinalis cervicis | Ligamentum nuchae and spinous process of seventh cervical vertebra | Spinous process of axis | Extends vertebral column |
| Spinalis thoracis | Spinous processes of upper lumbar and lower thoracic vertebrae | Spinous processes of upper thoracic vertebrae | Extends vertebral column |
| Multifidus | Sacrum, ilium, transverse processes of lumbar, thoracic, and lower four cervical vertebrae | Spinous process of a higher vertebra | Extends vertebral column and rotates it to opposite side |

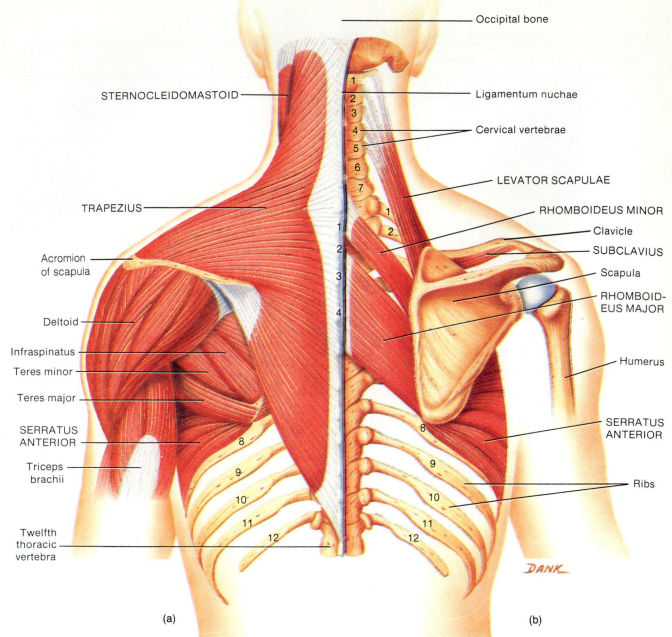

- Occipital bone
- STERNOCLEIDOMASTOID
- Ligamentum nuchae
- Cervical vertebrae
- TRAPEZIUS
- LEVATOR SCAPULAE
- RHOMBOIDEUS MINOR
- Clavicle
- SUBCLAVIUS
- Acromion of scapula
- Scapula
- RHOMBOID-EUS MAJOR
- Deltoid
- Infraspinatus
- Teres minor
- Teres major
- Humerus
- SERRATUS ANTERIOR
- Triceps brachii
- SERRATUS ANTERIOR
- Ribs
- Twelfth thoracic vertebra

(a)                     (b)

**Figure 16-4**  Human back muscles: (a) posterior superficial view; (b) posterior deep view. (From G. J. Tortora and S. R. Grabowski, *Principles of Anatomy and Physiology,* 7th ed., New York: HarperCollins, 1993. © Leonard Dank.)

## COMPARABLE HUMAN MUSCLES

The muscles associated with the scapula in the human, which are comparable to those you dissected in the cat, are the *trapezius,* the *rhomboids* (*major, minor,* and *capitis*), and the *levator scapulae.* In addition, there are the *serratus anterior* and the *pectoralis minor,* which you will dissect in a later exercise. Locate as many of these muscles on your own body as possible, and make movements that demonstrate their actions (see Figures 16-4 and 16-5).

The human *supraspinatus, infraspinatus, subscapularis,* and the *teres minor* form what is termed a *musculotendinous cuff* over the head of the humerus. They reinforce the joint while allowing a great deal of movement. Make as many shoulder joint movements as you can, and you will see why it is called the most freely movable joint in the body. The subscapularis is described in a later exercise on the cat.

The deep muscles of the human back stabilize the bones of the vertebral column. Many of them are short and span the distance between two vertebrae. Others

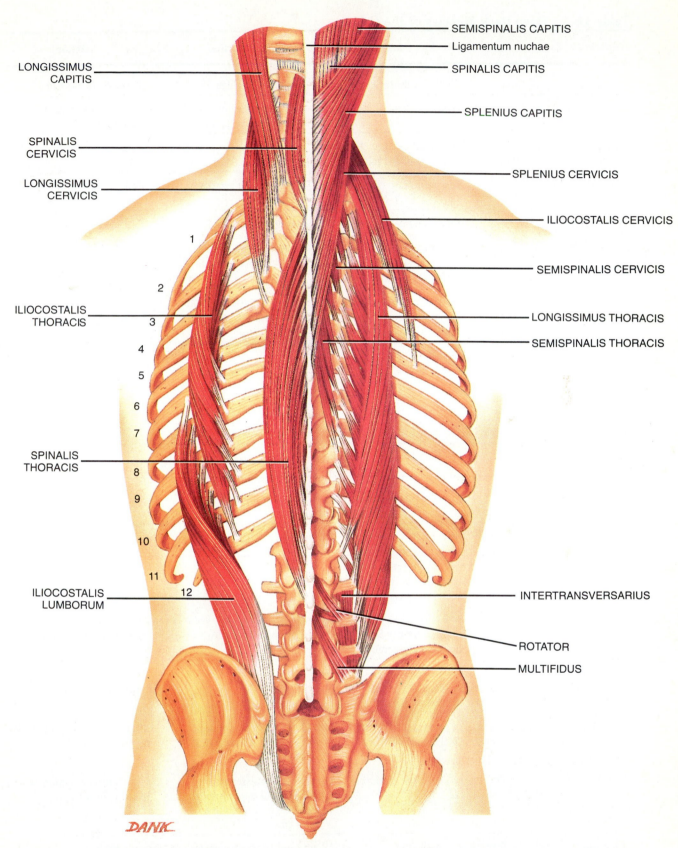

**Figure 16-5** Human muscles that move the vertebral column, as seen in posterior view. (From G. J. Tortora and S. R. Grabowski, *Principles of Anatomy and Physiology,* 7th ed., New York: HarperCollins, 1993. © Leonard Dank.)

**Table 16-3   Deep Back Muscles of the Cat**

| Muscle | Origin | Insertion | Action |
|---|---|---|---|
| Splenius | Spines of thoracic vertebrae | Lambdoidal ridge | Turns or raises head |
| Serratus ventralis | First ten ribs | Medial surface of scapula near the vertebral border | Pulls scapula forward and depresses it |
| Serratus dorsalis | Middorsal cervical, thoracic, and lumbar raphe | On each rib | Pulls ribs craniad |
| Spinalis dorsi | Last four thoracic vertebrae | Thoracic and cervical vertebrae | Extends vertebral column |
| Longissimus (capitis, cervicis, and dorsai) | Sacral and caudal vertebrae | Cervical vertebrae | Extends vertebral column |
| Iliocostalis | Angles of posterior thoracic ribs, by separate bundles | 3 to 4 ribs anterior to origin of each bundle | Draws ribs together |
| Multifidus spinae | Transverse processes of lumbar vertebrae | Spines of vertebrae anterior to origin of each slip | Extends back; turns vertebral column sideways |

**Table 16-4   Deep Back Muscles of the Human**

| Muscle | Origin | Insertion | Action |
|---|---|---|---|
| Splenius capitis | Transverse processes of last four cervical vertebrae | Occipital bone and mastoid process | Flexes head; rotates face |
| Semispinalis capitis | Articular process of cervical vertebrae 4, 5, and 6 and transverse processes of the seventh cervical vertebra and the first six thoracic vertebrae | Occipital bone | Extends head; rotates face |
| Spinalis thoracis | Spines of upper lumbar and lower thoracic vertebrae | Spines of upper thoracic vertebrae | Extends vertebral column |
| Longissimus Capitus | Transverse processes of last four cervical vertebrae | Mastoid process | Extends head, rotates face |
| Cervicis | Transverse processes of fourth and fifth thoracic vertebrae | Transverse processes of second through sixth cervical vertebrae | Extends cervical region of vertebral column |
| Thoracis | Transverse processes of lumbar vertebrae | Transverse processes of thoracic and upper lumbar vertebrae and ninth and tenth ribs | Extends thoracic region of vertebral column |
| Iliocostalis cervicis and iliocostalis lumborum | Iliac crest and all ribs | All ribs and transverse processes of fourth through sixth cervical vertebrae | Extends back |

are relatively long. They are postural muscles and not only maintain the upright posture, but also influence the curvatures of the spinal column. They extend, flex, and rotate parts of the vertebral column.

The location of these postural muscles is between the angle of the ribs and the vertebral spines. Refer to Tables 16-3 and 16-4 for the origins, insertions, and actions of the back muscles of the cat and human, respectively.

Locate some human muscle on your own body that can be compared with those dissected in the cat. Refer to the tables in Exercises 15 and 16, which list origins, insertions, and actions of human back muscles. Learn as much as you can about the muscles listed here by performing the following tests.

1. *Trapezius* (Figure 16-6)
   The human trapezius muscle elevates and depresses the clavicle and the scapula, adducts the scapula, and extends the head. Perform each of the actions of the trapezius muscle on your own body. Shrug your left shoulder with your right hand on first the clavicle, then the scapula:

   a. Describe the movement of the clavicle. _____

   _____

   b. Describe the movement of the scapula. _____

   _____

   c. What is the name of the muscle action in Question (a)? _____

   _____

   What is the name of the muscle action in Question (b)? _____

   _____

   d. Move the left shoulder toward the midline and back several times. What effect does this have on the scapula? _____

   _____

   e. Bow your head, then slowly raise it. What effect is the trapezius muscle having on the head?

   _____

   _____

2. *Rhomboideus major* (Figure 16-7)
   The rhomboideus major adducts the scapula and rotates the scapula upward, slightly outward, and downward.

   Contract the rhomboideus muscle by trying to pull the scapula closer to the midline. Have your lab partner lie face down with arms behind the back, palms up. Ask your partner to raise his or her arm while you place your hand between the medial border of the scapula and the midline of the body (over the rhomboideus major). Observe the origins and insertion of the muscle. What movement can you detect in the scapula? _____

   _____

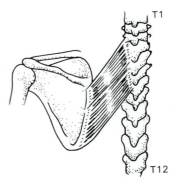

**Figure 16-7** Rhomboideus (posterior view). (T1 and T12 refer to the first and twelfth thoracic vertebrae, respectively.) (From P. J. Donnelly and G. A. Wistreich, *Exploring the Human Body,* New York: Harper & Row, 1989.)

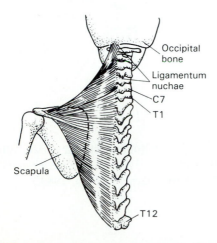

**Figure 16-6** Trapezius (posterior view). (From P. J. Donnelly and G. A. Wistreich, *Exploring the Human Body,* New York: Harper & Row, 1989.)

3. *Levator scapulae* (Figure 16-8)
   This muscle elevates the scapula and bends the neck laterally when the scapula is fixed.

   Shrug the shoulders to elevate the scapulae. Next, try to keep the scapulae fixed while contracting the levator scapulae on the left side of the body only. Do you feel any movement of the neck? Are there any significant differences in the distance the shoulder can be raised? Make observations and comments: _____

   _____

   _____

   _____

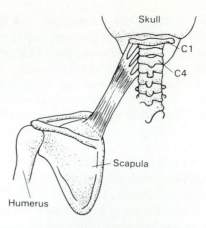

**Figure 16-8** Levator scapula (posterior view). (From P. J. Donnelly and G. A. Wistreich, *Exploring the Human Body,* New York: Harper & Row, 1989.)

4. *Latissimus dorsi* (Figure 16-9)
   This muscle extends, adducts, and rotates the arm medially. It also draws the shoulder downward and backward. It is a large triangular muscle covering the lower half of the back. Along with the trapezius, rhomboideus major, and levator scapulae muscles, the latissimus dorsi connects the arm to the vertebral column.

   With elbow fixed, arm at your side, turn the arm medially in the glenoid fossa and extend the arm slowly. Draw the shoulder downward and backward. Where on the back do you feel muscle contraction? _____

   _____

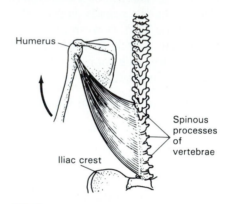

**Figure 16-9** Latissimus dorsi (posterior view). (From P. J. Donnelly and G. A. Wistreich, *Exploring the Human Body,* New York: Harper & Row, 1989.)

5. *Deltoid* (Figure 16-10)
   The deltoid muscle has an anterior and a posterior portion. If the entire muscle contracts, the arm is abducted. The anterior portion alone flexes the arm, and the posterior portion alone extends the arm.

   Perform the following three actions:
   a. Hold your arm at your side, elbow straight, and move your arm away from your body (abduction). Return your arm to its resting position at the side of your body.
   b. Move your arm forward (contracting the anterior fibers).
   c. Move your arm backward (contracting the posterior fibers). This is the action you would make if you were lifting, then rolling a bowling ball. Name other activities that involve the use of the deltoid muscle: _____

   _____

   _____

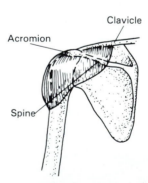

**Figure 16-10** Deltoid (posterior view). (From P. J. Donnelly and G. A. Wistreich, *Exploring the Human Body,* New York: Harper & Row, 1989.)

6. *Supraspinatus* (Figure 16-11)
   This muscle abducts the arm, along with the deltoid muscle, and slightly rotates the arm.

   Move your arm away from your body laterally. With your elbow fixed, rotate your arm laterally. Describe how you can palpate contraction in the supraspinous region of the scapula. _____

   _____

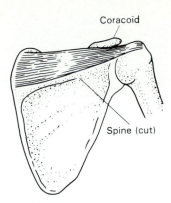

**Figure 16-11** Supraspinatus (posterior view). (From P. J. Donnelly and G. A. Wistreich, *Exploring the Human Body,* New York: Harper & Row, 1989.)

7. *Infraspinatus* (Figure 16-12)
   The infraspinatus rotates the arm laterally and both abducts and adducts the humerus. The upper part of the muscle abducts, and the lower part adducts.

   With your arm at the side of the body, and the elbow fixed, rotate the arm laterally. Describe how you can palpate contraction in the infraspinous region of the scapula. _____

   _____

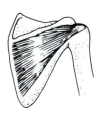

**Figure 16-12** Infraspinatus (posterior view). (From P. J. Donnelly and G. A. Wistreich, *Exploring the Human Body,* New York: Harper & Row, 1989.)

8. *Teres minor* (Figure 16-13)
   Together with the subscapularis, supraspinatus, and infraspinatus, this muscle forms the *rotator cuff* of the shoulder. By itself, it rotates the arm laterally. Which sports actitivities might aggravate this group of muscles? _____

   _____

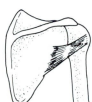

**Figure 16-13** Teres minor (posterior view). (From P. J. Donnelly and G. A. Wistreich, *Exploring the Human Body,* New York: Harper & Row, 1989.)

9. *Subscapularis* (Figure 16-14)
   The subscapularis (a rotator-cuff muscle) rotates the arm medially. It is often injured in sports activities.

   Name some everyday activities that involve the rotator cuff muscles. _____

   _____

   _____

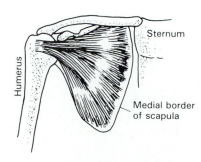

**Figure 16-14** Subscapularis (anterior view). (From P. J. Donnelly and G. A. Wistreich, *Exploring the Human Body,* New York: Harper & Row, 1989.)

# LABORATORY REVIEW *16*

## *MUSCLES OF THE BACK AND SHOULDER*

### IDENTIFICATION

1. Identify the muscles in Figure Q16-1 by matching the letters in the figure with the names in the following list. You may color the diagram if you wish. Use your text as a guide.

_____ Spinalis thoracis

_____ Iliocostalis thoracis

_____ Semispinalis capitis

_____ Iliocostalis cervicis

_____ Splenius capitis

_____ Longissimus thoracis

_____ Longissimus cervicis

_____ Longissimus capitis

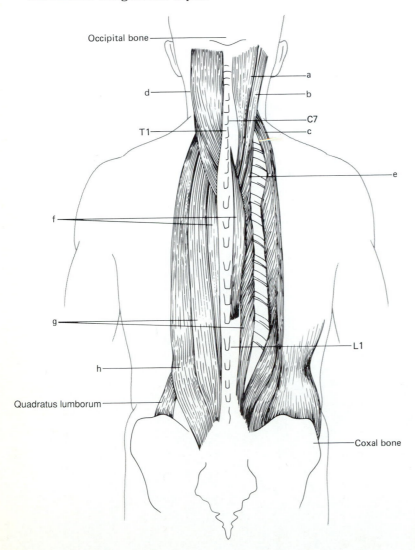

**Figure Q16-1**

2. Name the muscles indicated in Figure Q16-2. Color the diagram if you wish.

a. _____

b. _____

c. _____

d. _____

e. _____

f. _____

g. _____

h. _____

i. _____

j. _____

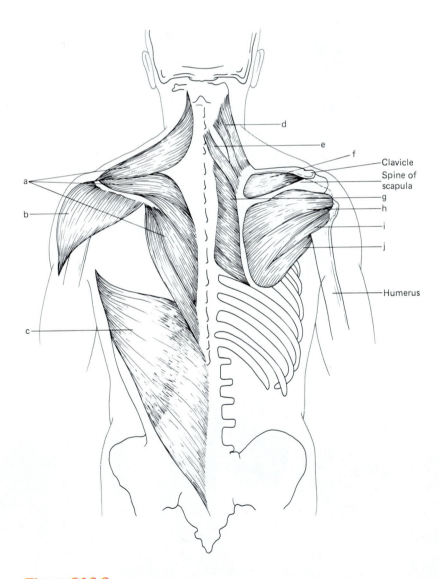

**Figure Q16-2**

3. Figure Q16-3 is a posterior view of the skeleton. Draw in and label the following
   muscles. Review origins and insertions in Table 16-2.
   a. Trapezius
   b. Deltoid
   c. Latissimus dorsi
   d. Levator scapulae
   e. Rhomboideus major
   f. Rhomboideus minor
   g. Supraspinatus
   h. Infraspinatus

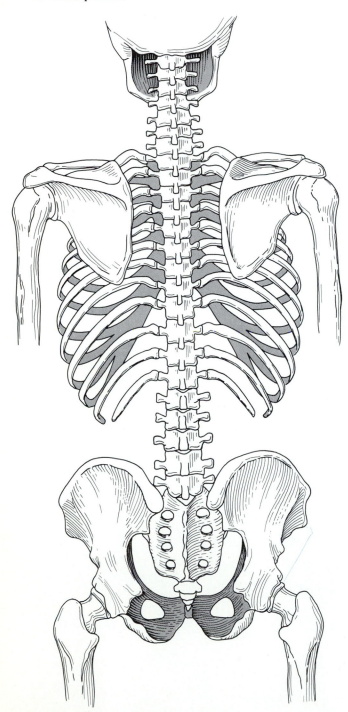

**Figure Q16-3**

4. Figure Q16-4 shows a posterior view of the human skeleton. Draw in and label the following muscles on Side A.
   a. Splenius capitis
   b. Spinalis thoracis
   c. Longissimus thoracis
   d. Iliocostalis thoracis

Draw in and label the following muscles on Side B.
   e. Semispinalis capitis
   f. Longissimus capitis
   g. Iliocostalis cervicis
   h. Longissimus cervicis

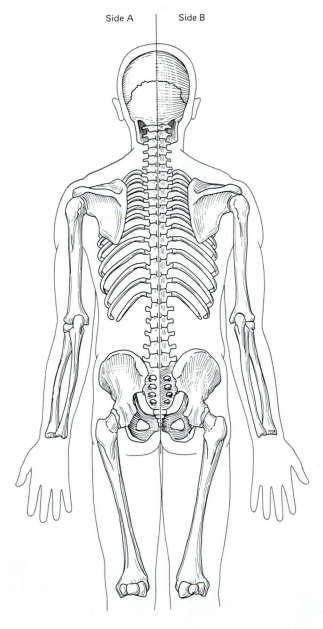

**Figure Q16-4**

# Muscles of the Chest and Abdomen

---

**After completing this exercise,
you should be able to**

1. Dissect the muscles of the chest and abdomen of the cat
2. Identify comparable human muscles

---

## Materials

The following materials should be provided for class use:

1. Embalmed cats
2. Dissecting tools
3. Dissecting trays
4. String
5. 3 percent phenol solution

All cat muscles dissected in Exercise 17 are listed here, according to specific laboratory exercises.

**PECTORAL MUSCLES**

Pectoantebrachialis

Pectoralis major

Pectoralis minor

Xiphihumeralis

**ABDOMINAL MUSCLES**

External oblique

Internal oblique

Rectus abdominis

Transversus abdominis

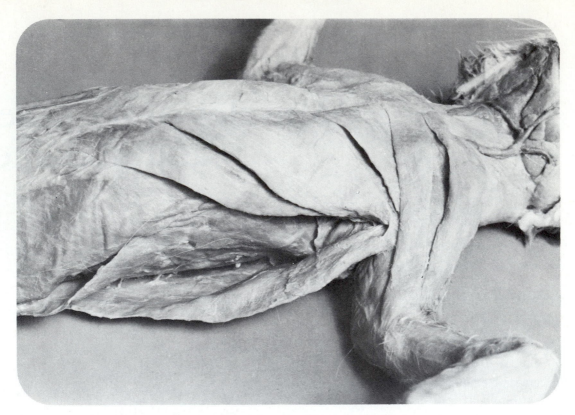

**Figure 17-1** (a) Pectoral muscles of cat (photograph).

## PECTORAL MUSCLES

### Instructions for Dissection

Turn the cat on its back, as shown in Figure 17-1, and find the general outline of the pectoral group of four muscles. Begin with the *pectoantebrachialis,* a long, thin, ribbonlike muscle. In most cats, this is a difficult muscle to dissect because the fascia is tough. Be certain that you see its outlines before you begin, and do not dissect too deeply or you will damage the muscle beneath it. Separate the pectoantebrachialis between its origin and its insertion.

Next, find the posterior border of the *pectoralis major,* and loosen it at that point. This muscle is also thin, but it is much broader than the pectoantebrachialis, which overlies it (Figure 17-1b). With your finger or a probe, loosen the pectoralis major from the underlying muscle, the *pectoralis minor.*

Now, find the posterior border of the pectoralis minor. It is larger than the pectoralis major (despite its name) and runs beneath both the pectoralis major and the pectoantebrachialis. The pectoralis minor is the thickest muscle of this group, as well as the broadest.

The last muscle of the group is the *xiphihumeralis.* Its origin is on the xiphoid process of the sternum on the midventral line. Its insertion on the humerus is a fragile tendon. Be careful not to tear this tendon. There are neither pectoantebrachialis nor xiphihumeralis muscles in the human.

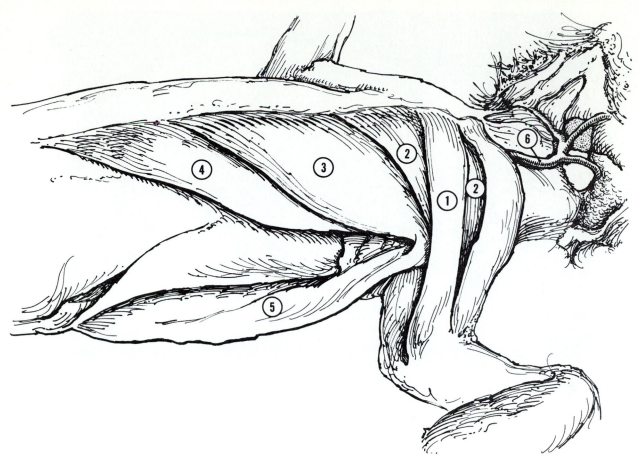

1. Pectoantebrachialis (pek′-tō-an-ti-brāk-ē-AL-is)
2. Pectoralis major (pek-tō-RAL-is)
3. Pectoralis minor
4. Xiphihumeralis (zif′-ē-hyoo′-mer-AL-is)
5. Latissimus dorsi
6. Jugular vein and tributaries

**Figure 17-1** (b) Pectoral muscles of cat (line drawing).

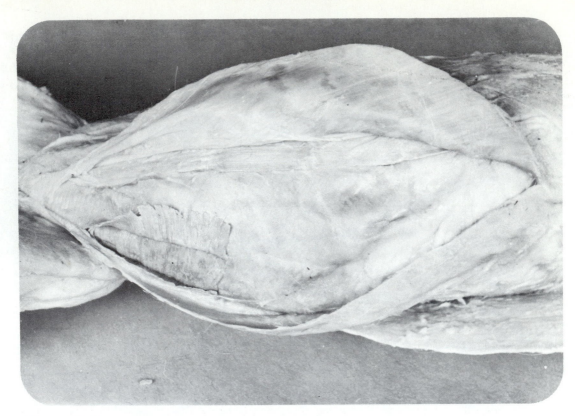

**Figure 17-2** (a) Abdominal muscles of cat (photograph)

## ABDOMINAL MUSCLES

### Instructions for Dissection

There are three thin layers of muscle covering the abdominal cavity (Figure 17-2). You must be careful in this dissection not to cut too deeply. Determine origin and insertion of the *external oblique,* the first layer, and mark off the midpoint between origin and insertion with a pencil or other tool. This muscle must be transected to see the one below it, and the transection line will be slightly curved, due to the shape of the muscle.* Begin the cut with the tip of the scissors and *gradually* extend it until you can get a probe between the external oblique and the *internal oblique,* the second layer (see Figure 17-2). When you are sure that you have not gone

too deep, extend the transection line, and separate with your fingers as you go along. Bring each half back to the origin and the insertion.

Near the insertion on the *linea alba* (midline) you will see a long, ribbonlike muscle, the *rectus abdominis.* Expose as much of it as you can. You cannot see the anterior part of the muscle at this time, but you will see it in a later dissection.

To separate the internal oblique from the *transversus abdominis,* the third or innermost layer, use extreme caution, as both muscles are very thin. Sometimes, the transversus abdominis is so thin that its fibers separate, and you can see the *parietal peritoneum* (lining of the body cavity) between them. Separate as much of the internal oblique as you can, and reflect the cut ends. Do not make your transection line too close to the broad aponeurosis by which the muscle inserts. Note that the direction of the fibers of the external oblique and internal oblique are almost at right angles to one another. Refer to Table 17-1 for the origins, insertions, and actions of the cat's abdominal muscles.

---

*Do not use the aponeurosis, which forms the insertion of the muscle, when measuring its width. Transect midway between borders of the muscular portion.

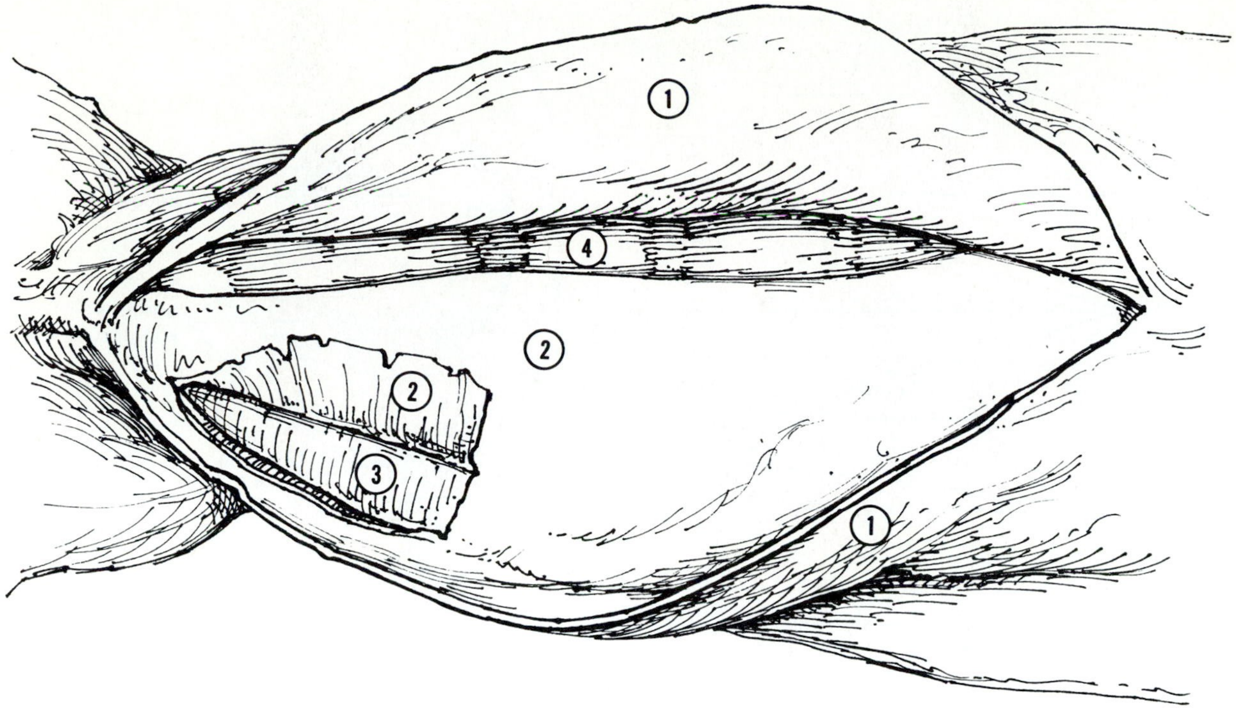

1. External oblique (reflected) (o-BLĒK)
2. Internal oblique (reflected)
3. Transversus abdominis
4. Rectus abdominis (REK-tus ab-DOM-i-nis)

**Figure 17-2** (b) Abdominal muscles of cat (line drawing).

## Table 17-1   Pectoral and Abdominal Muscles of the Cat

| Muscle | Origin | Insertion | Action |
|---|---|---|---|
| Pectoantebrachialis | Manubrium | Proximal end of forearm by fascia | Adducts arm |
| Pectoralis major | Anterior third of sternum | Proximal two-thirds of humerus | Adducts arm |
| Pectoralis minor | Almost entire length of sternum | Proximal half of humerus | Adducts arm |
| Xiphihumeralis | Xiphoid process | Proximal end of humerus | Adducts arm |
| External oblique | Last nine ribs | Linea alba from sternum to pubis | Compresses abdomen |
| Internal oblique | Lumbodorsal fascia | Linea alba | Compresses abdomen |
| Transversus abdominis | Costal cartilage of false ribs, transverse processes of lumbar vertebrae, and ilium | Linea alba | Compresses abdomen |
| Rectus abdominis | Pubic bone | Costal cartilage of first through fourth ribs | Compresses abdomen |

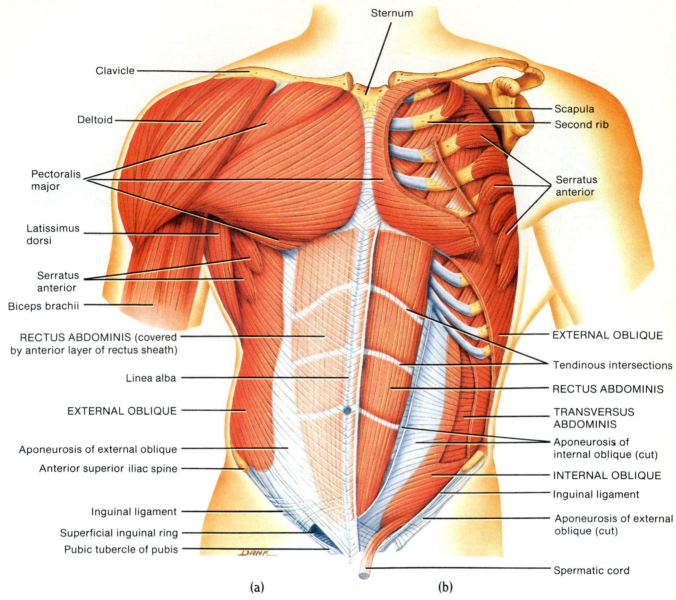

**Figure 17-3** Human muscles of the anterior abdominal wall: (a) superficial view; (b) deep view. (From G. J. Tortora and S. R. Grabowski, *Principles of Anatomy and Physiology,* 7th ed., New York: HarperCollins, 1993.)

## COMPARABLE HUMAN MUSCLES

The human pectoral group consists of a *pectoralis major* and a *pectoralis minor* (see Figures 17-3 and 17-4). There is no pectoantebrachialis or xiphihumeralis. The function of the pectoral muscles in humans is the same as in the cat—that is, to adduct the arm, and to perform the other actions listed in Table 17-2.

The abdominal group of human muscles is also similar to those in the cat. The *transversus abdominis*

lies over the parietal peritoneum. The middle layer is the *internal oblique,* and the outer layer is the *external oblique.* The *rectus abdominis* arises in segments connected by flat tendons and is enclosed in a sheath formed by the aponeuroses of the other three abdominal muscles. The segmentation of the rectus abdominis is seldom seen in the cat. The abdominal muscles compress the contents of the abdominal cavity and also contribute to flexion and rotation of the torso. In the next section try to demonstrate as many of these muscles as possible on your own body.

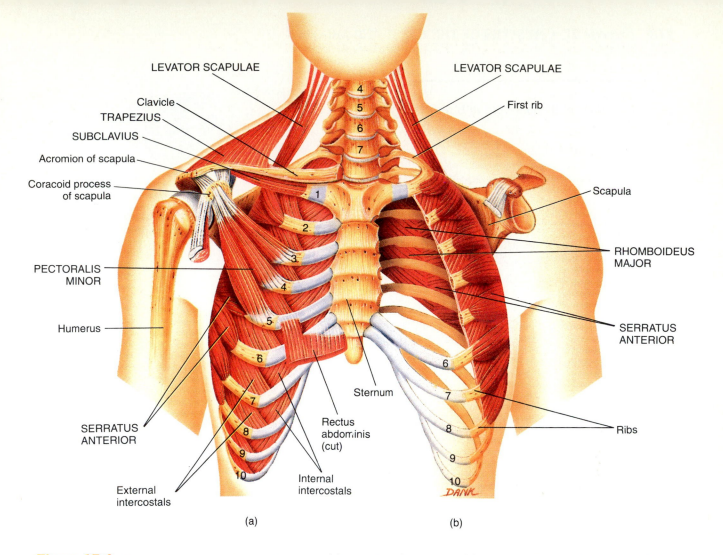

LEVATOR SCAPULAE

LEVATOR SCAPULAE

Clavicle

First rib

TRAPEZIUS

SUBCLAVIUS

Acromion of scapula

Coracoid process
of scapula

Scapula

PECTORALIS
MINOR

RHOMBOIDEUS
MAJOR

Humerus

SERRATUS
ANTERIOR

Sternum

SERRATUS
ANTERIOR

Rectus
abdominis
(cut)

Ribs

External
intercostals

Internal
intercostals

(a)

(b)

**Figure 17-4** The human pectoral muscle group: (a) anterior deep view; (b) anterior deeper view. (From G. J. Tortora and S. R. Grabowski, *Principles of Anatomy and Physiology,* 7th ed., New York: HarperCollins, 1993.)

## Table 17-2    Pectoral and Abdominal Muscles of the Human

| Muscle | Origin | Insertion | Action |
|---|---|---|---|
| Pectoralis major | Clavicle, sternum, and cartilage of second through sixth ribs | Greater tubercle of humerus | Flexes, adducts, and rotates arm medially |
| Pectoralis minor | Third through fifth ribs | Coracoid process of scapula | Depresses scapula; rotates shoulder upward |
| External oblique | Lower eight ribs | Iliac crest and linea alba | Compresses abdomen |
| Internal oblique | Iliac crest, inguinal ligament, and thoracolumbar fascia | Costal cartilage of last three or four ribs | Compresses abdomen |
| Transversus abdominis | Iliac crest, inguinal ligament, lumbar fascia, and cartilage of last six ribs | Xiphoid process, linea alba, and pubis | Compresses abdomen |
| Rectus abdominis | Pubic crest and symphysis pubis | Costal cartilage of fifth to seventh ribs and xiphoid process | Flexes vertebral column |

## HUMAN CHEST MUSCLES

Study the actions of the following with respect to your own body.

1. *Pectoralis major* (Figure 17-5)
   This muscle adducts, flexes, and rotates the arm medially. It occupies most of the chest area above the diaphragm.
   a. With your right hand on the left chest, near the arm, hold your left arm straight out in front of you. Then move the arm as far medially as possible. Do you feel the contraction of the pectoralis major muscle? _____

   b. With your right hand in the same place, flex and rotate the left arm. Describe the contraction of the pectoralis major. Is it greater or less than during adduction? _____

2. *Pectoralis minor* (Figure 17-6)
   This muscle depresses the scapula and rotates the shoulder joint anteriorly. It also elevates the third through fifth ribs during forced inhalation when scapulae are fixed.
   a. To demonstrate the pectoralis minor, keep the shoulders in a fixed position, place your hands over the superior rib cage, and force inhale. Describe the movement you feel in the rib cage.

   _____

   b. Relax the left shoulder, then lower it. Slowly rotate the shoulder anteriorly without shrugging it. Describe how you can feel the muscle contract. _____

The abdominal muscles compress the abdomen and flex the vertebral column.

1. *External oblique* (Figure 17-7), *internal oblique* (Figure 17-8), and *transversus abdominus* (Figure 17-9)
   These thin sheets of muscle, the fibers of which run in different directions from one another, compress the abdomen and bend the vertebral column laterally.
   a. Contract this group of muscles in order to pull in your abdomen. Try to do it without holding your breath. Can you hold this position without holding your breath? _____

   b. Contract the same three muscles on the left side, only to the extent that you notice a slight bending of your torso to the left. Repeat this on the right side. Can you do this without holding your breath or using your shoulder muscles? _____

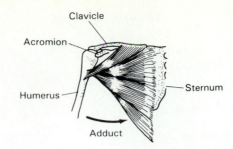

**Figure 17-5** Pectoralis major (anterior view). (From P. J. Donnelly and G. A. Wistreich, *Exploring the Human Body*, New York: Harper & Row, 1989.)

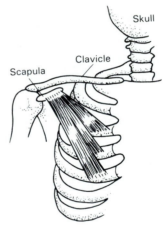

**Figure 17-6** Pectoralis minor (anterior view). (From P. J. Donnelly and G. A. Wistreich, *Exploring the Human Body*, New York: Harper & Row, 1989.)

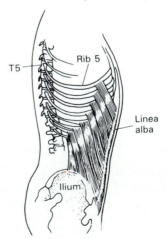

**Figure 17-7** External oblique. (From P. J. Donnelly and G. A. Wistreich, *Exploring the Human Body*, New York: Harper & Row, 1989.)

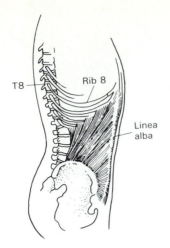

**Figure 17-8** Internal oblique. (From P. J. Donnelly and G. A. Wistreich, *Exploring the Human Body,* New York: Harper & Row, 1989.)

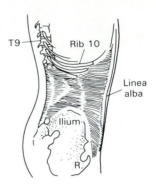

**Figure 17-9** Transversus abdominis. (From P. J. Donnelly and G. A. Wistreich, *Exploring the Human Body,* New York: Harper & Row, 1989.)

2. *Rectus abdominis* (Figure 17-10)
   The rectus abdominis plays a role in compressing the abdomen, along with the muscles above it, but it also flexes the vertebral column.
   a. To demonstrate the action of this muscle, first compress your abdomen, then contract your rectus abdominis. Can you feel a slight bending forward of your torso? _____

   b. Lie flat on your back, with hands behind your head. Raise your head and shoulders about 2 inches. Can you feel your rectus abdominus contracting? _____

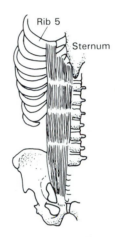

**Figure 17-10** Right rectus abdominis (anterior view). (From P. J. Donnelly and G. A. Wistreich, *Exploring the Human Body,* New York: Harper & Row, 1989.)

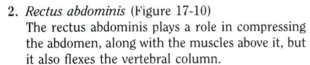

# LABORATORY REVIEW *17*

NAME _____

LAB SECTION _____ DATE _____

## MUSCLES OF THE CHEST AND ABDOMEN

### IDENTIFICATION

1. Identify the muscles in Figure Q17-1, by matching the letters in the figure with the names listed after the figure.

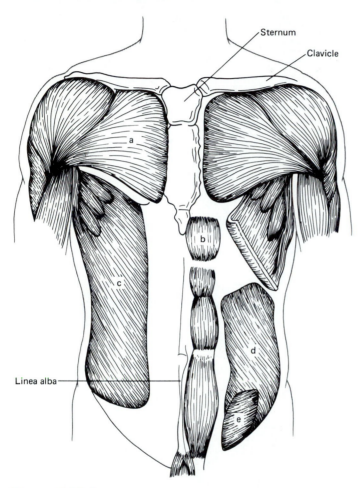

**Figure Q17-1**

_____ Internal oblique

_____ Transversus abdominis

_____ Pectoralis major

_____ External oblique

_____ Rectus abdominis

**215**

2. Figure Q17-2 is an anterior view of the skeleton. Refer to Table 17-2. Draw in and label the following muscles on Side A.
   a. Pectoralis major
   b. External oblique
   Draw in and label the following muscle on Side B.
   c. Rectus abdominis

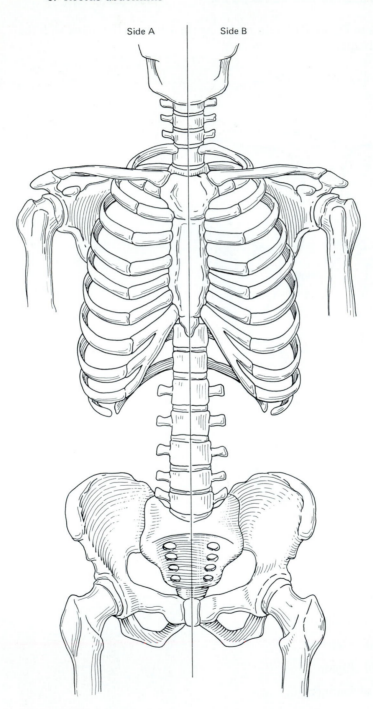

Side A     Side B

**Figure Q17-2**

# Muscles of the Neck, Jaw, Face, and the Salivary Glands

## After completing this exercise, you should be able to

1. Dissect the muscles of the neck and jaw of the cat
2. Identify comparable human muscles

## Materials

The following materials should be provided for class use:

1. Embalmed cats
2. Dissecting tools
3. Dissecting trays
4. String
5. 3 percent phenol solution

All muscles dissected in Exercise 18 are listed here, according to specific laboratory exercises.

### SUPERFICIAL NECK AND JAW MUSCLES

Masseter

Sternomastoid

Sternohyoid

Mylohyoid

Digastric

### DEEP NECK MUSCLES AND SALIVARY GLANDS

Stylohyoid

Cleideomastoid

Sternothryoid

Thyrohyoid

Cricothyroid

Geniohyoid

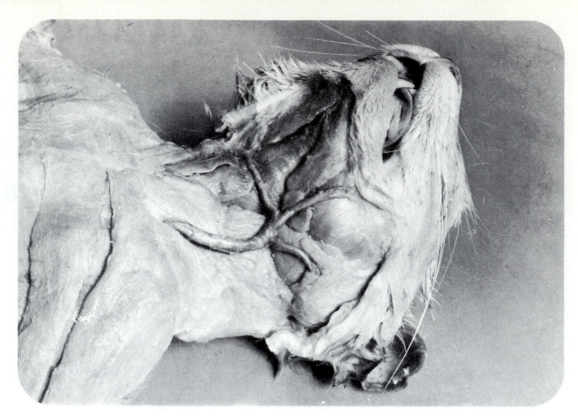

**Figure 18-1** (a) Superficial neck and jaw muscles of cat (photograph).

## SUPERFICIAL NECK AND JAW MUSCLES

### Instructions for Dissection

The skin must be removed from the entire neck region and the left side of the face, as shown in Figure 18-1a. Remove the skin carefully, in order not to destroy the *jugular vein* and its tributaries and the *parotid duct,* which runs across the cheek, as shown in Figure 18-1b. You may also see *cranial nerve VII* passing over the cheek, just below and parallel to the parotid duct. Prominent *lymph nodes* will be seen beneath the veins, and the large *submandibular gland* lies below the *parotid gland.* Locate all of these structures before beginning the identification of the muscles.

The muscle of the cheek is called the *masseter.* It is large and convex and is partially covered by the lymph nodes and salivary glands.

The large V-shaped muscle between the sternum and the head is the *sternomastoid.* (In the human, this muscle is united with another, the cleidomastoid, to form the sternocleidomastoid. The cleidomastoid of the cat is observed in the next dissection.)

In the middle of the neck, between the two sternomastoids, are the paired *sternohyoids,* and above them, with fibers running transversely, are the *mylohyoids.*

Attached to the mandible on both sides are the *digastric* muscles.

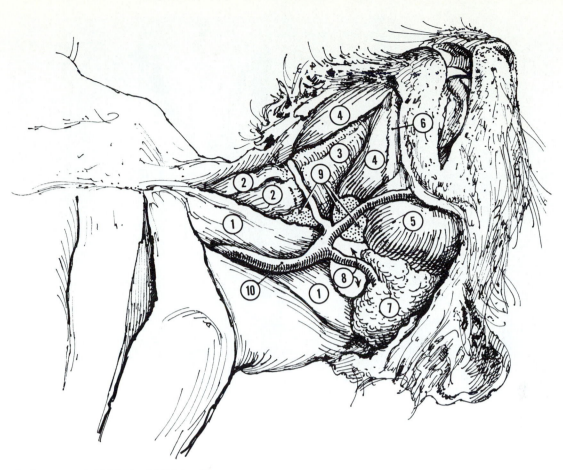

1. Sternomastoid (ster′-nō-MAS-toyd)
2. Sternohyoid (ster′-nō-HĪ-oyd)
3. Mylohyoid (mī′-lō-HĪ-oyd)
4. Digastric (dī-GAS-trik)
5. Masseter (MASS-e-ter)
6. Mandible (bone)
7. Parotid gland (pa-ROT-id)
8. Submandibular gland
9. Lymph nodes
10. External jugular vein and tributaries

**Figure 18-1** (b) Superficial neck and jaw muscles of cat (line drawing).

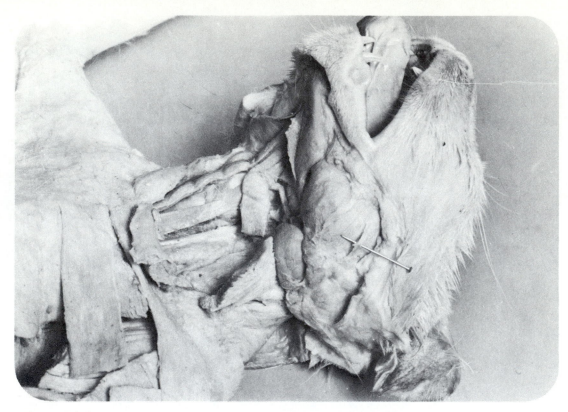

**Figure 18-2** (a) Deep neck muscles and salivary glands of cat (photograph).

## DEEP NECK MUSCLES AND SALIVARY GLANDS

### Instructions for Dissection

Carefully dissect the veins and lymph nodes on the left side of the neck (see Figure 18-2). If you want to preserve them (they are damaged on the right side, as a result of the injection procedure), free the entire mass in one piece, clip the veins at their anterior ends, and pull the whole mass posteriorly, so that it may be put back in place once the dissection is completed. If you do not want to preserve them, cut them off. Identify again the muscles previously named.

Lying beneath the jugular vein at the junction between sternohyoid and mylohyoid is a small muscle, the *stylohyoid,* the fibers of which run transversely (see Figure 18-2b).

Transect the sternomastoid, and reflect the cut ends as far as possible, to see the underlying structures. Also, transect the left sternohyoid, and reflect its ends. Under the sternomastoid, laterally, is the *cleidomastoid,* extending from the clavicle to the mastoid process.

Under the cut ends of the sternohyoid, you can see two parallel muscles. Posterior in position is the *sternothyroid,* which extends upward only as far as the thyroid cartilage of the larynx. Just above the sternothyroid is the *thyrohyoid,* which spans the short distance between the thyroid cartilage of the larynx posteriorly and the hyoid cartilage of the larynx anteriorly.

Lying directly on the larynx is the *cricothyroid* muscle, which is not to be dissected, but merely observed. Cut the *mylohyoid* up the midline as far as possible, and reflect the cut ends laterally. This reveals paired *geniohyoid* muscles. Refer to Table 18-1 for the origins, insertions, and actions of the neck muscles of the cat.

The salivary glands consist of three paired glands. You have already seen the parotid gland and its duct and the submandibular gland. In order to see the sublingual gland, loosen the submandibular gland, and pull it to the cat's left side. This will reveal a much smaller cone-shaped gland—the sublingual gland—and also the submandibular and sublingual ducts.

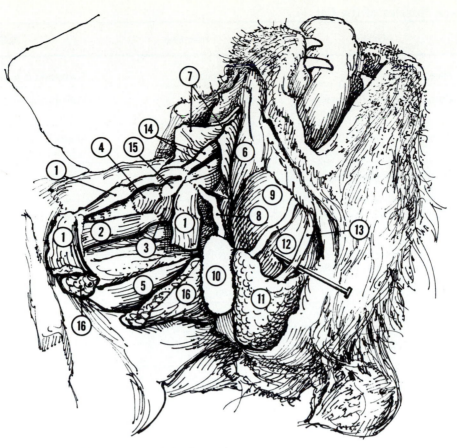

1. Sternohyoid (reflected on left side only)
2. Sternothyroid (ster'-nō-THĪ-royd)
3. Thyrohyoid (thī-rō-HĪ-oyd)
4. Cricothyroid (krī-kō-THĪ-royd)
5. Cleidomastoid (klī-dō-MAS-toyd)
6. Digastric
7. Mylohyoid (reflected)
8. Stylohyoid (stī'-lō-HĪ-oyd)
9. Masseter
10. Submandibular
11. Parotid gland
12. Cranial nerve VII (facial)
13. Parotid duct
14. Geniohyoid
15. Larynx
16. Sternomastoid (reflected)

**Figure 18-2** (b) Deep neck muscles and salivary glands of cat (line drawing).

**Table 18-1   Neck Muscles of the Cat**

| Muscle | Origin | Insertion | Action |
| --- | --- | --- | --- |
| Sternomastoid | Manubrium | Mastoid process | Turns and depresses head |
| Sternohyoid | Costal cartilage of first rib | Hyoid bone | Pulls hyoid caudally; raises ribs and sternum |
| Mylohyoid | Cricoid cartilage of larynx | Hyoid bone | Raises floor of mouth; pulls hyoid forward |
| Digastric | Occipital bone and mastoid process | Ventral border of mandible | Depresses lower jaw |
| Masseter | Zygomatic bone | Lateral margin near the ventral border of the mandible at the coronoid fossa | Elevates lower jaw |
| Sternothyroid | Costal cartilage of first rib | Thyroid cartilage of larynx | Pulls larynx caudad |
| Thyrohyoid | Thyroid cartilage of larynx | Hyoid bone | Raises larynx |
| Cricothyroid | Cricoid cartilage of larynx | Thyroid cartilage of larynx | Tenses vocal cords |
| Cleidomastoid | Clavicle | Mastoid process | Pulls head laterally; pulls clavicle craniad |
| Stylohyoid | Stylohyal bone of hyoid | Body of hyoid bone | Raises hyoid |
| Geniohyoid | Medial surface of mandible | Body of hyoid bone | Pulls hyoid craniad |

## COMPARABLE HUMAN MUSCLES

The largest muscle of the neck in the human is the *sternocleidomastoid.* In the cat, this muscle is divided into two distinct muscles, the sternomastoid and the cleidomastoid. In the human, this muscle's position between the clavicle, sternum, and mastoid process of the skull gives it complex capability of movement. If only the right or the left sternocleidomastoid contracts, it turns the head to the opposite side. If both right and left sternocleidomastoids contract, the head is flexed on the chest.

Many of the deep muscles of the human neck are considered to be in the chest and deep back of the cat.

They are the *scalenus* muscles, which turn the vertebral column to one side when they contract, and the *semispinalis, longissimus,* and *splenius* muscles, which extend the head and neck. They are found in the *posterior triangle* of the neck.

The muscles associated with the hyoid bone are grouped into the *suprahyoid* muscles and the *infrahyoid* muscles, depending on whether they are found above or below the hyoid bone. The suprahyoid muscles lie in the floor of the mouth, and the infrahyoid muscles are found lower in the neck, around the larynx and trachea. The following drawings (Figures 18-3 through 18-6 and tables (Tables 18-2 through 18-5) include the human facial muscles of the lower jaw and of the tongue and larynx, and the muscles that move the head.

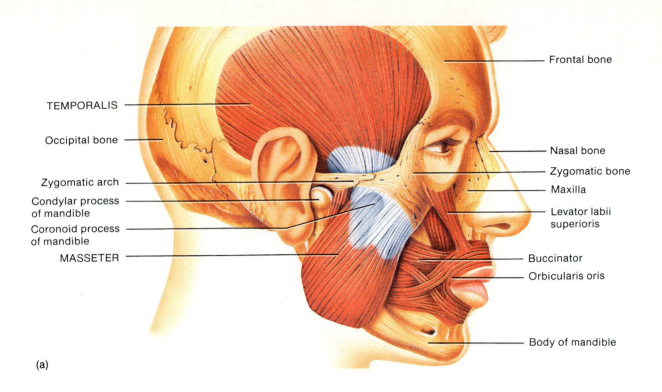

(a)

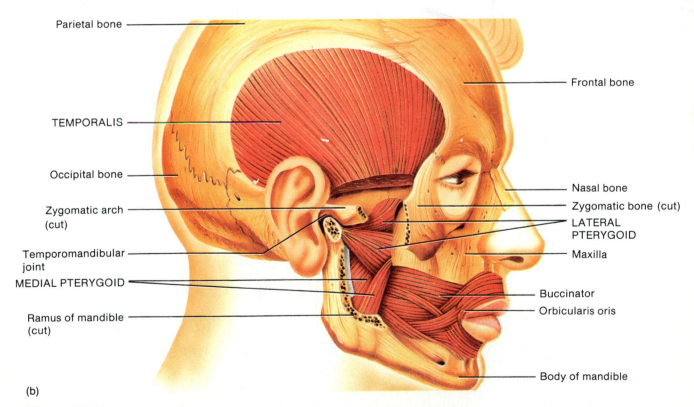

(b)

**Figure 18-3** Human muscles that move the lower jaw: (a) superficial right lateral view; (b) deep right lateral view. (From G. J. Tortora and S. R. Grabowski, *Principles of Anatomy and Physiology,* 7th ed., New York: HarperCollins, 1993. © Leonard Dank.)

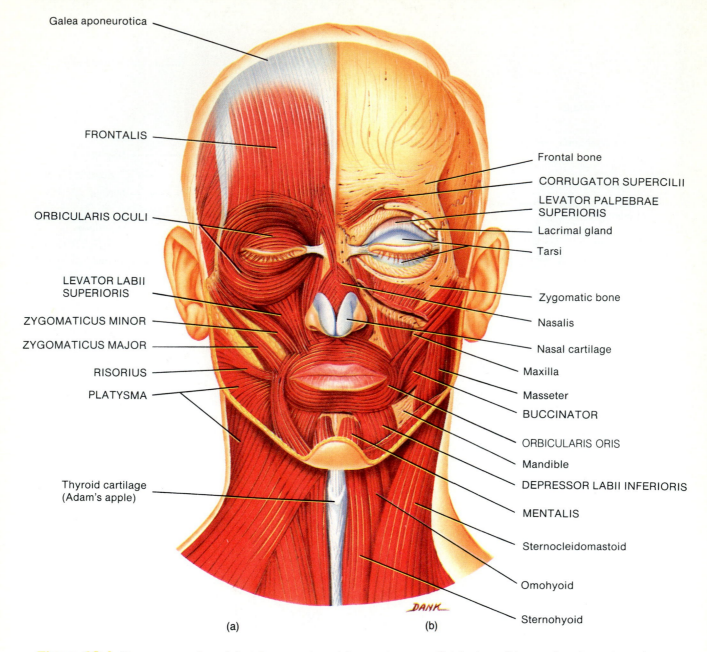

Galea aponeurotica

FRONTALIS

ORBICULARIS OCULI

LEVATOR LABII
SUPERIORIS

ZYGOMATICUS MINOR

ZYGOMATICUS MAJOR

RISORIUS

PLATYSMA

Thyroid cartilage
(Adam's apple)

Frontal bone

CORRUGATOR SUPERCILII

LEVATOR PALPEBRAE
SUPERIORIS

Lacrimal gland

Tarsi

Zygomatic bone

Nasalis

Nasal cartilage

Maxilla

Masseter

BUCCINATOR

ORBICULARIS ORIS

Mandible

DEPRESSOR LABII INFERIORIS

MENTALIS

Sternocleidomastoid

Omohyoid

Sternohyoid

(a)                              (b)

**Figure 18-4** Human muscles of facial expression: (a) anterior superficial view; (b) anterior deep view. (From G. J. Tortora and S. R. Grabowski, *Principles of Anatomy and Physiology,* 7th ed., New York: HarperCollins, 1993. © Leonard Dank.)

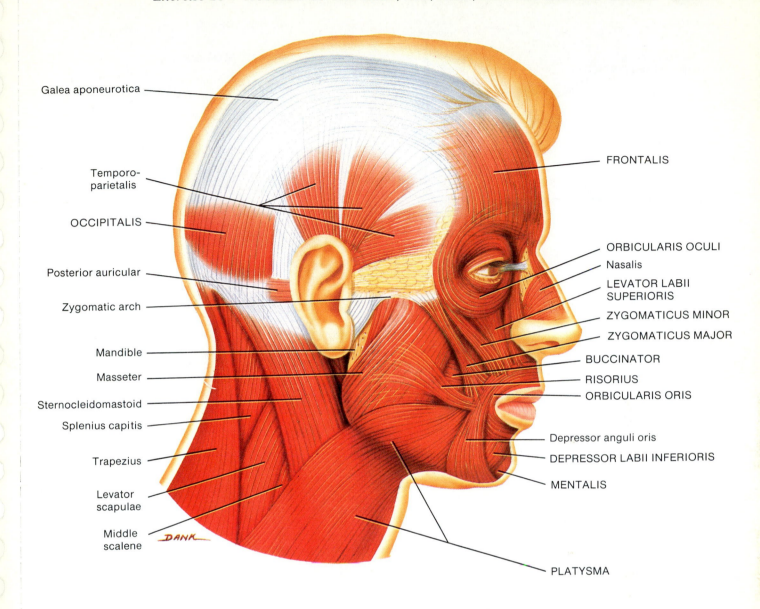

**Figure 18-5** Human muscles of the scalp, face, and neck (lateral view). (From G. J. Tortora and S. R. Grabowski, *Principles of Anatomy and Physiology,* 7th ed., New York: HarperCollins, 1993. © Leonard Dank.)

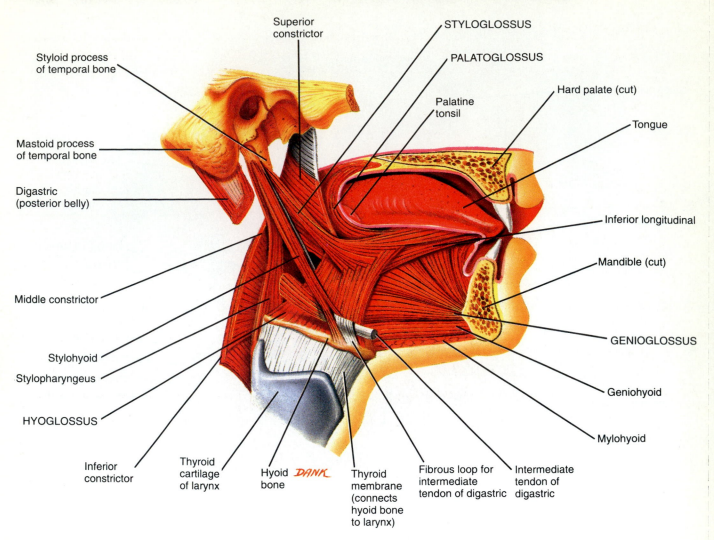

**Figure 18-6** Human muscles that move the tongue (viewed from the right side). (From G. J. Tortora and S. R. Grabowski, *Principles of Anatomy and Physiology,* 7th ed., New York: HarperCollins, 1993. © Leonard Dank.)

**Table 18-2   Muscles That Move the Lower Jaw**

| Muscle | Origin | Insertion | Action | Innervation |
|---|---|---|---|---|
| Masseter (*masseter* = chewer) | Maxilla and zygomatic arch | Angle and ramus of mandible | Elevates mandible, as in closing the mouth, and protracts (protrudes) mandible | Mandibular branch of trigeminal nerve (V) |
| Temporalis (*tempora* = temples) | Temporal bone | Coronoid process of mandible | Elevates and retracts mandible | Temporal nerve from mandibular division of trigeminal nerve (V) |
| Medial pterygoid (*medial* = closer to midline; *pterygoid* = like a wing) | Medial surface of lateral pterygoid plate of sphenoid; maxilla | Angle and ramus of mandible | Elevates and protracts mandible and moves mandible from side to side | Mandibular branch of trigeminal nerve (V) |
| Lateral pterygoid (*lateral* = farther from midline) | Greater wing and lateral surface of lateral pterygoid plate of sphenoid | Condyle of mandible; temporomandibular articulation | Protracts mandible, opens mouth, and moves mandible from side to side | Mandibular branch of trigeminal nerve (V) |

**Table 18-3   Muscles of Facial Expression**

| Muscle | Origin | Insertion | Action | Innervation |
|---|---|---|---|---|
| Epicranius (*epi* = over; *crani* = skull) | This muscle is divisible into two portions: the frontalis over the frontal bone and the occipitalis over the occipital bone. The two muscles are united by a strong aponeurosis, the galea aponeurotica, which covers the superior and lateral surfaces of the skull. | | | |
| Frontalis (*front* = forehead) | Galea aponeurotica | Skin superior to supraorbital line | Draws scalp forward, raises eyebrows, and wrinkles forehead horizontally | Facial nerve (VII) |
| Occipitalis (*occipito* = base of skull) | Occipital bone and mastoid process of temporal bone | Galea aponeurotica | Draws scalp backward | Facial nerve (VII) |
| Orbicularis oris (*orb* = circular; *or* = mouth) | Muscle fibers surrounding opening of mouth | Skin at corner of mouth | Closes lips, compresses lips against teeth, protrudes lips, and shapes lips during speech | Facial nerve (VII) |
| Zygomaticus major (*zygomatic* = cheek bone; *major* = greater) | Zygomatic bone | Skin at angle of mouth and orbicularis oris | Draws angle of mouth upward and outward, as in smiling or laughing | Facial nerve (VII) |

**Table 18-3** *(Continued)*

| Muscle | Origin | Insertion | Action | Innervation |
|---|---|---|---|---|
| Levator labii superioris (*levator* = elevates; *labii* = lip; *superioris* = upper) | Superior to infraorbital foramen of maxilla | Skin at angle of mouth and orbicularis oris | Elevates upper lip | Facial nerve (VII) |
| Depressor labii inferioris (*depressor* = depresses; *inferioris* = lower) | Mandible | Skin of lower lip | Depresses lower lip | Facial nerve (VII) |
| Buccinator (*bucc* = cheek) | Alveolar processes of maxilla and mandible and pterygomandibular raphe (fibrous band extending from the pterygoid hamulus to the mandible) | Orbicularis oris | Major cheek muscle; compresses cheek, as in blowing air out of mouth, and causes the cheeks to cave in, producing the action of sucking | Facial nerve (VII) |
| Mentalis (*mentum* = chin) | Mandible | Skin of chin | Elevates and protrudes lower lip and pulls skin of chin up, as in pouting | Facial nerve (VII) |
| Platysma (*platy* = flat, broad) | Fascia over deltoid and pectoralis major muscles | Mandible, muscles around angle of mouth, and skin of lower face | Draws outer part of lower lip downward and backward, as in pouting; depresses mandible | Facial nerve (VII) |
| Risorius (*risor* = laughter) | Fascia over parotid (salivary) gland | Skin at angle of mouth | Draws angle of mouth laterally as in tenseness | Facial nerve (VII) |
| Orbicularis oculi (*ocul* = eye) | Medial wall of orbit | Circular path around orbit | Closes eye | Facial nerve (VII) |
| Corrugator supercilii (*corrugo* = to wrinkle; *supercilium* = eyebrow) | Medial end of superciliary arch of frontal bone | Skin of the eyebrow | Draws eyebrow downward as in frowning | Facial nerve (VII) |
| Levator palpebrae superioris (*palpebrae* = eyelids) | Roof of orbit (lesser wing of sphenoid bone) | Skin of the upper eyelid | Elevates the upper eyelid | Oculomotor nerve (III) |

**Table 18-4   Muscles of the Larynx**

| Muscle | Origin | Insertion | Action | Innervation |
|---|---|---|---|---|
| EXTRINSIC<br>Omohyoid (*omo* = relationship to the shoulder; *hyoedies* = U-shaped, pertaining to the hyoid bone) | Superior border of scapula and superior transverse ligament | Body of hyoid bone | Depresses hyoid bone | Branches of ansa cervicalis nerve (C1–C3) |
| Sternohyoid (*sterno* = sternum) | Medial end of clavicle and manubrium of sternum | Body of hyoid bone | Depresses hyoid bone | Branches of ansa cervicalis nerve (C1–C3) |
| Sternothyroid (*thyro* = thyroid gland) | Manubrium of sternum | Thyroid cartilage of larynx | Depresses thyroid cartilage | Branches of ansa cervicalis nerve (C1–C3) |
| Thyrohyoid | Thyroid cartilage of larynx | Greater cornu of hyoid bone | Elevates thyroid cartilage and depresses hyoid bone | Cervical nerves C1–C2 and descending hypoglossal nerve (XIII) |
| Stylopharyngeus<br>Palatopharyngeus<br>Inferior constrictor<br>Middle constrictor | | | | |
| INTRINSIC<br>Cricothyroid (*crico* = cricoid cartilage of larynx) | Anterior and lateral portion of cricoid cartilage of larynx | Anterior border of inferior cornu of thyroid cartilage of larynx and posterior part of inferior border of lamina of thyroid cartilage | Produces tension and elongation of vocal folds | External laryngeal branch of vagus nerve (X) |

**Table 18-5   Muscles That Move the Tongue**

| Muscle | Origin | Insertion | Action | Innervation |
|---|---|---|---|---|
| Genioglossus (*geneion* = chin; *glossus* = tongue) | Mandible | Undersurface of tongue and hyoid bone | Depresses and thrusts tongue forward (protraction) | Hypoglossal nerve (XII) |
| Styloglossus (*stylo* = stake or pole) | Styloid process of temporal bone | Side and undersurface of tongue | Elevates tongue and draws it backward (retraction) | Hypoglossal nerve (XII) |
| Stylohyoid (*hyoeides* = U-shaped; pertaining to hyoid bone) | Styloid process of temporal bone | Body of hyoid bone | Elevates and retracts tongue | Facial nerve (VII) |
| Hyoglossus | Body of hyoid bone | Side of tongue | Depresses tongue and draws down its sides | Hypoglossal nerve (XII) |

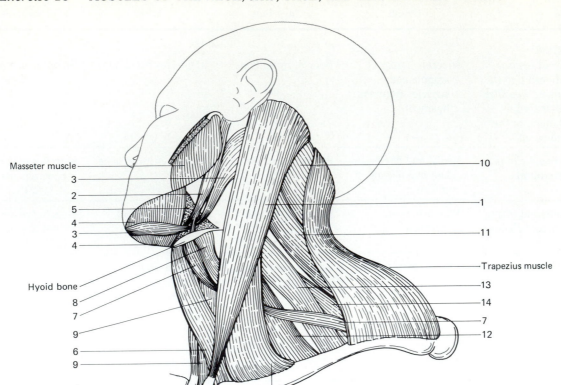

Masseter muscle
3
2
5
4
3
4
Hyoid bone
8
7
9
6
9
1
Sternum
Clavicle

10
1
11
Trapezius muscle
13
14
7
12
1

Anterior triangle muscle
 1. Sternocleidomastoid

Suprahyoid muscles
 2. Stylohyoid
 3. Digastric
 4. Mylohyoid
 5. Hyoglossus

Infrahyoid muscles
 6. Sternohyoid
 7. Omohyoid
 8. Thyrohyoid
 9. Sternothyroid

Posterior triangle muscles
10. Splenius capitis
11. Levator scapulae
12. Scalenus anterior
13. Scalenus medius
14. Scalenus posterior

**Figure 18-7** Muscles of the neck and jaw (left lateral view).

The neck is a complicated area. The muscles in Figure 18-7 are neck and jaw muscles, shown in side view. The large *sternocleidomastoid* forms the *anterior triangle* of the neck. Using the hyoid bone as a point of reference, the muscles of the anterior triangle are divided into two groups: the *suprahyoid* muscles and the *infrahyoid* muscles. The muscles of the *posterior triangle* are those located posterior to the sternocleidomastoid. Muscles included in these groups are listed in the caption for Figure 18-7. Color the muscles, and compare their functions and locations with their counterparts in the cat. (Note that the muscles of the posterior triangle of the cat are discussed in relation to the cat's deep back and chest muscles.) An anterior view of the muscles of the neck is shown in Figures 18-8 and 18-9. Color the muscles. Note their relationship to the larynx and the trachea.

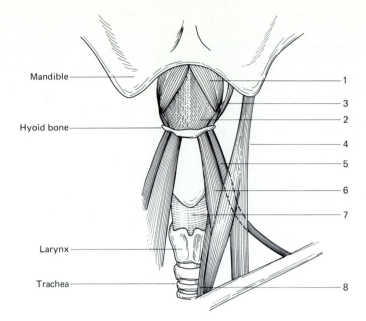

Mandible

Hyoid bone

Larynx

Trachea

1
3
2
4
5
6
7
8

1. Digastric
2. Mylohyoid
3. Stylohyoid
4. Sternocleidomastoid
5. Omohyoid
6. Sternohyoid
7. Cricothyroid
8. Sternothyroid

**Figure 18-8** Muscles of the neck, which move the tongue (anterior view).

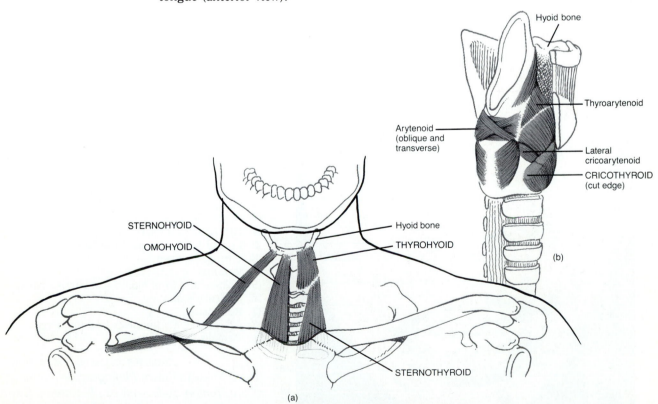

Hyoid bone

Thyroarytenoid

Arytenoid
(oblique and
transverse)

Lateral
cricoarytenoid

CRICOTHYROID
(cut edge)

(b)

STERNOHYOID

OMOHYOID

Hyoid bone

THYROHYOID

STERNOTHYROID

(a)

**Figure 18-9** Muscles of the larynx: (a) anterior view; (b) right lateral view. (From G. J. Tortora, *Principles of Human Anatomy,* 3rd, 4th, 5th eds., New York: HarperCollins, 1983, 1986, 1989.)

## HUMAN NECK, JAW, AND FACE MUSCLES

The muscles discussed here are some of the more important facial muscles of the human. These were not dissected in the cat, but because they are so important as a means of silent communication in the human, they are studied here.

1. *Epicranius or occipitofrontalis* (Figure 18-10.)
   The epicranius is divisible into two parts: (1) the *frontalis,* which covers the frontal bone, and (2) the *occipitalis,* which covers the superior and lateral surfaces of the skull. The two muscles are united by an aponeurosis (galea aponeurotica) that covers the superior and lateral surfaces of the skull. They raise the eyebrows and wrinkle the skin of the forehead horizontally.
   a. Look in a mirror. Wrinkle your forehead. What do you observe? How many horizontal lines appear? Are there differences among your classmates?

Observations: _____

_____

_____

_____

b. Comparisons: _____

_____

_____

c. When you raise your eyebrows or wrinkle your forehead, can you feel movement over the hairy portion of the skull? Does your hair move?

_____

2. *Orbicularis oculi* (Figure 18-10)
   This circular muscle surrounds the opening of the eye. It closes the eye.
   Close your eye, to feel the contraction of this muscle. Have your lab partner do the same, so that you can observe the varying degrees of contracture and closure of the eye. What are your observations?

_____

_____

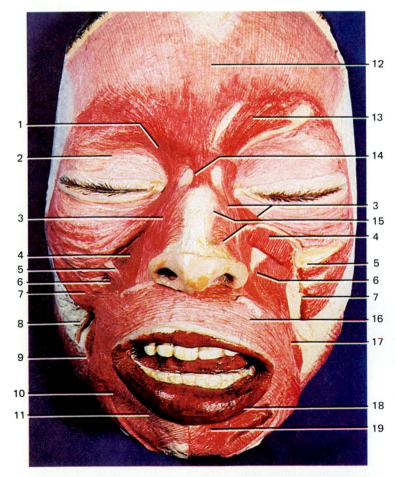

1. Depressor supercilli
2. Orbicularis oculi
3. Levator labii superioris alaeque nasi
4. Levator labii superioris
5. Zygomaticus minor
6. Levator anguli oris
7. Zygomaticus major
8. Risorius
9. Platysma
10. Depressor anguli oris
11. Depressor labii inferioris
12. Frontal belly of occipito-frontalis
13. Corrugator supercilli
14. Procerus
15. Transverse part of nasalis (compressor naris)
16. Orbicularis oris
17. Buccinator
18. Orbicularis oris (incisivus labii inferioris)
19. Mentalis

**Figure 18-10** Muscles of facial expression. (From Yokochi/Rohen *Photographic Anatomy of the Human Body,* 2nd ed., Baltimore: University Park Press and Tokyo: Igaku-Shoin Ltd., 1978.)

3. *Zygomaticus major* (Figure 18-10)
   This muscle draws the angle of the mouth upward and outward, as in smiling or laughing.

   Smile broadly, using muscles on both sides of your face. Make as many facial expressions as you can involving this muscle, and identify these expressions. Use a mirror or your lab partner. Describe these expressions. _____

   _____

   _____

4. *Platysma* (Figure 18-10)
   The platysma draws the outer part of the lower lip downward and backward, as in pouting or crying. It also depresses (lowers) the mandible. This muscle was probably removed in the cat dissection because it sticks so closely to the skin.
   a. Demonstrate the platysma by pouting for your lab partner. With your fingers at the corners of your mouth, feel the contraction.
   b. Make other facial expressions similar to pouting, using this muscle. Describe these expressions.

   _____

   _____

5. *Orbicularis oris* (Figure 18-10)
   The orbicularis oris produces many facial expressions. Perform each of the four actions listed next, and describe in your own words the resulting facial expressions. Perhaps you can associate these expressions with a particular mood.

   a. Close the lips. _____

   _____

   b. Compress the lips against the teeth. _____

   _____

   c. Protrude the lips. _____

   _____

   d. Prepare your mouth to say "oh." _____

   _____

   e. Prepare your mouth to kiss. _____

   _____

6. *Buccinator* (Figure 18-10)
   The buccinator compresses the cheek, as in blowing air out of the mouth. It causes the cheeks to cave in as in sucking, and it is used a great deal of the time to manipulate the lips.

Make both blowing and sucking movements with your mouth, and describe the configuration of your cheeks. _____

_____

_____

Two of the three muscles that move the lower jaw and head (masseter and temporalis) are discussed next. They are used in eating and chewing.

1. *Masseter* (Figure 18-11)
   The masseter raises the mandible as the mouth is closed. It causes the protrusion (protraction) of the mandible. It is used in speech, as the mandible is moved up and down, forward and back, to achieve not only a particular sound, but also particular facial posturing that often accompanies these sounds.
   a. Demonstrate the muscle by creating a phrase or sentence in which the masseter is involved with both elevation and protrusion of the mandible. Try slowly reading this aloud, and describe the movement of the mandible. _____

   _____

   _____

   b. The masseter is also used while eating. Make chewing movements with your mouth closed. Place your fingers at the angles of your jaw to feel the contraction.

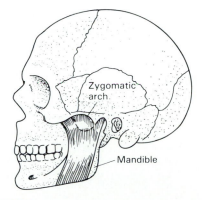

**Figure 18-11** Masseter muscle. (From P. J. Donnelly and G. A. Wistreich, *Exploring the Human Body,* New York: Harper & Row, 1989.)

2. *Temporalis* (Figure 18-12)

The temporalis elevates and retracts the mandible. Pull your lower jaw back to demonstrate it. Keep your mouth closed, and let your teeth be your guide as to how much movement can be achieved. Place your fingers on your temples. Describe what you

feel. _____

_____

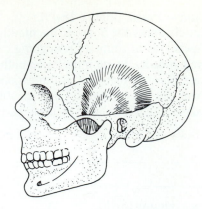

**Figure 18-12** Temporal muscle. (From P. J. Donnelly and G. A. Wistreich, *Exploring the Human Body,* New York: Harper & Row, 1989.)

3. *Sternocleidomastoid* (Figure 18-13)

The large muscle in the human neck is the *sternocleidomastoid.* It is complicated and has two origins, one on the manubrium and the other on the clavicle. By contracting both muscles of the pair together, you can flex the cervical vertebrae, draw the head forward, and elevate the chin. By contracting one muscle of the pair only, you can rotate your face toward the side opposite the contracting muscle.

a. To demonstrate both muscles of the pair, stand straight, eyes looking forward. Bend your head forward, and feel the muscle contraction below your ears. If your muscles are strong, another test for this muscle is to be able to raise your head while lying in a supine position. Do not assist this action by raising your shoulders.

b. To demonstrate only one muscle at a time, contract only the left sternocleidomastoid. Turn your head to the right, and look up. Now contract only the right muscle. Turn your head to the left, and look up. Do this exercise while looking in a mirror, or make observations on your lab

partner. _____

_____

_____

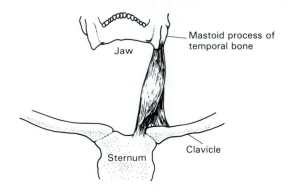

**Figure 18-13** Sternocleidomastoid (anterior view). (From P. J. Donnelly and G. A. Wistreich, *Exploring the Human Body,* New York: Harper & Row, 1989.)

# LABORATORY REVIEW *18*

## MUSCLES OF THE NECK, JAW, FACE, AND THE SALIVARY GLANDS

### COMPLETION

Which human muscles perform the following functions?

a. _____ Compress lips against teeth

b. _____ Depress lower lip

c. _____ Close the eye

d. _____ Raise the eyebrows

e. _____ Elevate upper lip

f. _____ Elevate mandible to close the mouth

g. _____ Move mandible from side to side

h. _____ Draw tongue backward

i. _____ Draw soft palate down on the tongue

j. _____ Depress thyroid cartilage of larynx

k. _____ Elongate the vocal folds

l. _____ Thrust tongue forward

# Muscles of the Deep Chest and Arm

## After completing this exercise, you should be able to

1. Dissect the muscles of the deep chest and arm of the cat
2. Identify comparable human muscles

## Materials

The following materials should be provided for class use:

1. Embalmed cats
2. Dissecting tools
3. Dissecting trays
4. String
5. 3 percent phenol solution

The muscles dissected in Exercise 19 are listed here, according to specific laboratory exercises.

### DEEP CHEST AND RELATED MUSCLES

Serratus ventralis

Subscapularis

Teres major

Coracobrachialis

Transversus costarum

Scalenus

   Scalenus anterior

   Scalenus medius

   Scalenus posterior

External intercostal

Internal intercostal

### SUPERFICIAL ARM MUSCLES, LATERAL SURFACE

Triceps brachii

   Triceps lateral head

   Triceps medial head

   Triceps long head

Brachialis

Anconeus

Brachioradialis

Extensor carpi radialis

Extensor digitorum communis

Extensor digitorum lateralis

Extensor carpi ulnaris

### SUPERFICIAL ARM MUSCLES, MEDIAL SURFACE

Epitrochlearis

Biceps brachii

Extensor carpi radialis brevis

Extensor carpi radialis longus

Pronator teres

Flexor carpi radialis

Palmaris longus

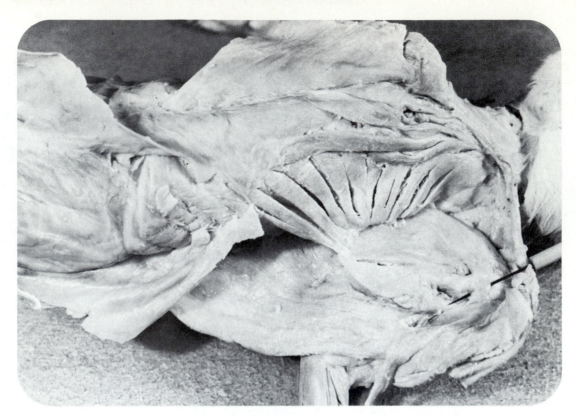

**Figure 19-1** (a) Deep chest and related muscles of the cat (photograph).

# DEEP CHEST AND RELATED MUSCLES

## Instructions for Dissection

With the cat on its back (Figure 19-1), transect the pectoral muscles—the pectoantebrachialis, the pectoralis major (which must be cut all the way to the humerus in order to reflect it), and finally the pectoralis minor. Do not transect the xiphihumeralis.

Observe once again the serratus ventralis, the subscapularis, and the teres major, as shown in Figure 19-1. At the shoulder is a very small muscle, the *coracobrachialis*. It is no more than ⅓ inch long and tightly binds the scapula to the arm. (In the human, it is much larger.)

Near the midline, find a small oblique muscle, the *transversus costarum*, and lateral and somewhat above it is the *scalenus*. The most prominent part of the three-part scalenus is the *scalenus medius*, which extends farthest posteriorly. On either side of the scalenus medius are shorter slips of muscle, the *scalenus anterior* and the *scalenus posterior*, but all three unite anteriorly to form one muscle. (In the human, the scalenus is located higher in the neck region than it is in the cat.) Separate each slip. Note also in this area the anterior end of the rectus abdominis.

Between the ribs, under the external oblique muscle, clear a space of all fat and fascia. Between two adjacent ribs observe the *external intercostal* muscle, which has fibers that run obliquely from the posterior border of one rib to the anterior border of the next rib. Under the external intercostal lies the *internal intercostal*. Separate these two muscles, keeping in mind that they are very thin. The internal intercostal can be recognized because its fibers, also oblique, run at a different angle from those of the external intercostal. Refer to Table 19-1 for the origins, insertions, and actions of the cat's deep chest muscles.

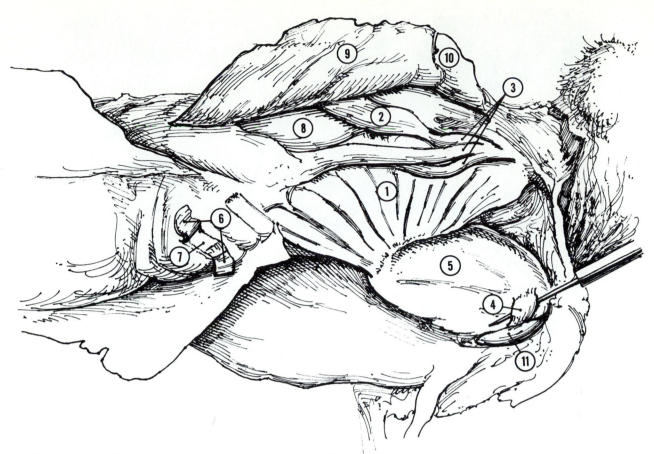

1. Serratus ventralis
2. Transversus costarum (kos-TAR-um)
3. Scalenus (skā-LĒ-nus)
4. Coracobrachialis (kor'-a-kō-brāk-e-AL-is)
5. Subscapularis (sub-skap'-yoo-LAR-is)
6. External intercostal (reflected)
7. Internal intercostal
8. Rectus abdominis
9. Pectoralis minor (reflected)
10. Pectoralis major (reflected)

**Figure 19-1**  (b) Deep chest and related muscles of the cat (line drawing).

## Table 19-1   Deep Chest Muscles of the Cat

| Muscle | Origin | Insertion | Action |
|---|---|---|---|
| Serratus ventralis | First ten ribs | Medial surface of scapula | Pulls scapula forward and depresses it |
| Transversus costarum | Lateral border of sternum | First rib | Pulls sternum craniad |
| Scalenus | Transverse processes of cervical vertebrae | Ribs | Raises ribs; flexes neck |
| Coracobrachialis | Coracoid process of scapula | Proximal end of humerus | Adducts humerus |
| Subscapularis | Subscapular fossa | Lesser tubercle of humerus | Pulls humerus medially |
| External intercostals | Caudal border of one rib | Cranial border of next rib caudad | Raise ribs at inhalation and thereby increase size of thoracic cavity |
| Internal intercostals | Cranial border of one rib | Caudal border of the next rib craniad | Draw adjacent ribs together and lower ribs |

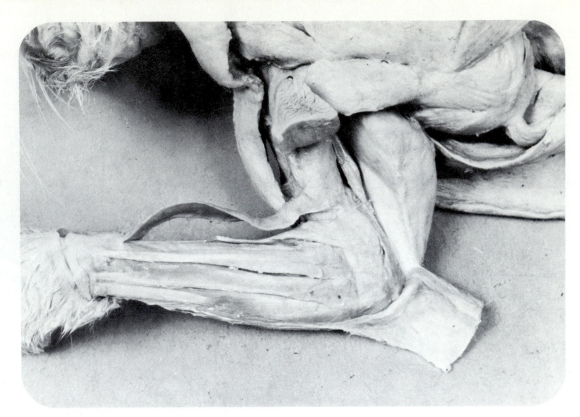

**Figure 19-2** (a) Superficial arm muscles (lateral surface) of the cat (photograph).

# SUPERFICIAL ARM MUSCLES, LATERAL SURFACE

## Instructions for Dissection

Begin with the upper arm (Figure 19-2). You have already dissected the clavodeltoid, which, although not an arm muscle, does cover the anterior surface of the arm. The largest superficial muscle in the upper arm is the *triceps brachii*. It is made up of three distinct muscles, with three heads of origin, but it has only one point of insertion. The large mass over the posterior surface is the *long head* of the triceps. Lying next to it on the lateral surface is the *lateral head* of the triceps. Free each of these muscles between origin and inser-

tion, and then transect the lateral head of the triceps. Pull back the cut ends to reveal the smaller medial head of the triceps and next to it the *brachialis,* as seen in Figure 19-2.

Distal to the triceps medial head is an L-shaped muscle surrounding the elbow, the *anconeus.* You need not dissect the anconeus but merely observe it. It is usually darker in color than the surrounding muscles, which helps in locating it (it is not found in humans).

In the lower arm, dissect only the superficial muscles. An extension from the brachialis of the upper arm is the *brachioradialis.* The muscles next to the brachioradialis—in order—are the *extensor carpi radialis,* the *extensor digitorum communis,* the *extensor digitorum lateralis,* and the *extensor carpi ulnaris.*

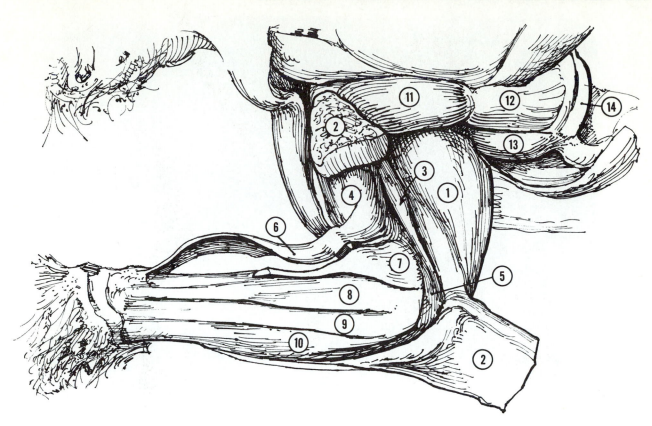

1. Triceps brachii, long head (BRĀK-ē-ē)
2. Triceps brachii, lateral head (reflected)
3. Triceps brachii, medial head
4. Brachialis (brāk-ē-AL-is)
5. Anconeus (an-CŌN-ē-us)
6. Brachioradialis (brāk-ē-ō-rā′-dē-AL-is)
7. Extensor carpi radialis (KAR-pē)
8. Extensor digitorum communis (dij-i-TOR-um kom-MYOO-nis)
9. Extensor digitorum lateralis
10. Extensor carpi ulnaris (ul-NAR-is)
11. Spinodeltoid
12. Infraspinatus
13. Teres major
14. Rhomboideus major

**Figure 19-2** (b) Superficial arm muscles (lateral surface) of the cat (line drawing).

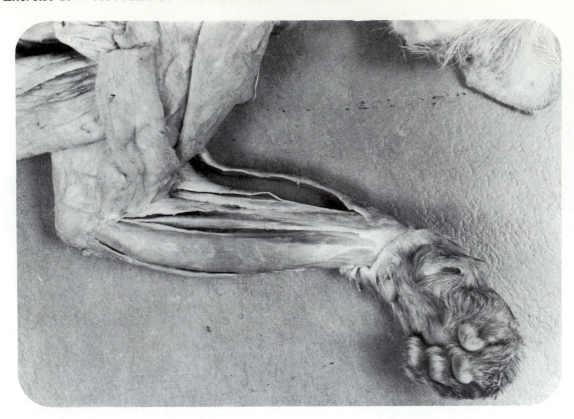

**Figure 19-3** (a) Superficial arm muscles (medial surface) of the cat (photograph).

## SUPERFICIAL ARM MUSCLES, MEDIAL SURFACE

### Instructions for Dissection

The medial surface of the upper arm is almost entirely covered by a broad, flat sheet of muscle, the *epitrochlearis* (absent in humans). Dissect the epitrochlearis, and you will find the somewhat convex *biceps brachii* muscle beneath the pectoral muscle.

In the lower arm, begin at the anterior border, with the previously observed brachioradialis. Lying next to it is the *extensor carpi radialis,* which has two parts: a shorter, somewhat triangular *extensor carpi radialis brevis* and beneath it the *extensor carpi radialis longus.*

The next superficial muscles to separate are the narrow *pronator teres* and next to it the *flexor carpi radialis.* The last muscle on the medial surface is the widest of the group: the *palmaris longus.* Refer to Table 19-2 for the origins, insertions, and actions of the cat's arm muscles.

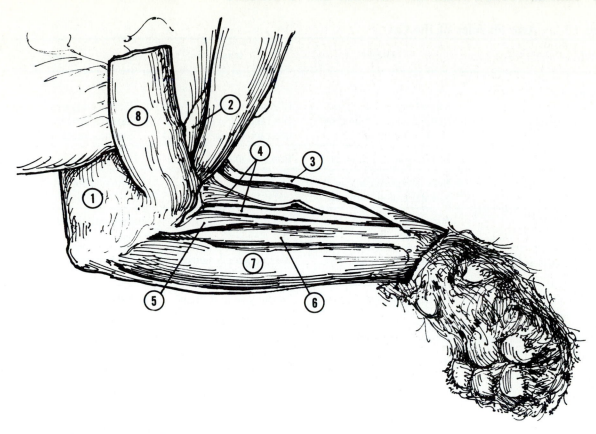

1. Epitrochlearis (ep-i-trōk-lē-AR-is)
2. Biceps brachii
3. Brachioradialis
4. Extensor carpi radialis (longus and brevis)
5. Pronator teres (prō-NĀ-tor TER-ēz)
6. Flexor carpi radialis
7. Palmaris longus (pal-MAR-is)
8. Pectoantebrachialis

**Figure 19-3** (b) Superficial arm muscles (medial surface) of the cat (line drawing).

**Table 19-2  Arm Muscles of the Cat**

| Muscle | Origin | Insertion | Action |
|---|---|---|---|
| Triceps brachii | | | |
|   Long head | Lateral border of scapula | Olecranon process of ulna | Extends forearm |
|   Lateral head | Deltoid ridge of humerus | Olecranon process of ulna | Extends forearm |
|   Medial head | Shaft of humerus | Olecranon process of ulna | Extends forearm |
| Brachialis | Lateral side of humerus | Proximal end of ulna | Flexes forearm |
| Anconeus | Near lateral epicondyle at distal end of humerus | Proximal end of ulna | Tenses elbow; rotates ulna |
| Brachioradialis | Middorsal border of humerus | Distal end of radius | Supinates hand |
| Extensor carpi radialis longus | Supracondylar ridge of humerus | Base of second metacarpal | Extends hand |
| Extensor digitorum communis | Lateral surface of humerus | Tendons on four digits | Extends digits |
| Extensor digitorum lateralis | Lateral surface of humerus | Digits by joining tendons of extensor digitorum communis | Extends digits |
| Extensor carpi ulnaris | Distal end of humerus and lateral border of ulna | Base of fifth metacarpal | Extends ulnar side of wrist and fifth digit |
| Extensor carpi radialis brevis | Lateral epicondyle of humerus | Base of third metacarpal | Extends hand |
| Pronator teres | Distal end of humerus | Medial border of radius | Rotates radius for pronation |
| Flexor carpi radialis | Distal end of humerus | Second and third metacarpals | Flexes wrist |
| Palmaris longus | Distal end of humerus | Tendons on all digits | Flexes digits |
| Flexor carpi ulnaris | Distal end of humerus | Tendons on all digits | Flexes digits |
| Biceps brachii | Tuberosity near glenoid fossa and coracoid process of scapula | Radial tuberosity | Flexes and supinates forearm |
| Epitrochlearis | Ventral border of latissimus dorsi near insertion | Fascia on medial surface of elbow | Extends forearm and rotates ulna to supinate hand |

## COMPARABLE HUMAN MUSCLES

As noted in Exercise 18, the muscles of the deep chest of the cat are not easily related to those of the human body. The *scalenus* is primarily a neck muscle in the human, and the *coracobrachialis* is much larger. There is no *transversus costarum* in the human.

In contrast to the deep chest, there is a great deal of similarity between cats and humans in the upper and lower arm regions. One exception is the *epitrochlearis,* a conspicuous cat muscle not found in the human. The *biceps brachii* (covered by the epitrochlearis in the cat) is more prominent in the human. The following drawings of human muscles (Figures 19-4 through 19-6) will show you both the differences and the similarities. Study as many as time permits, learning their origins, insertions, and actions from Tables 19-3 and 19-4).

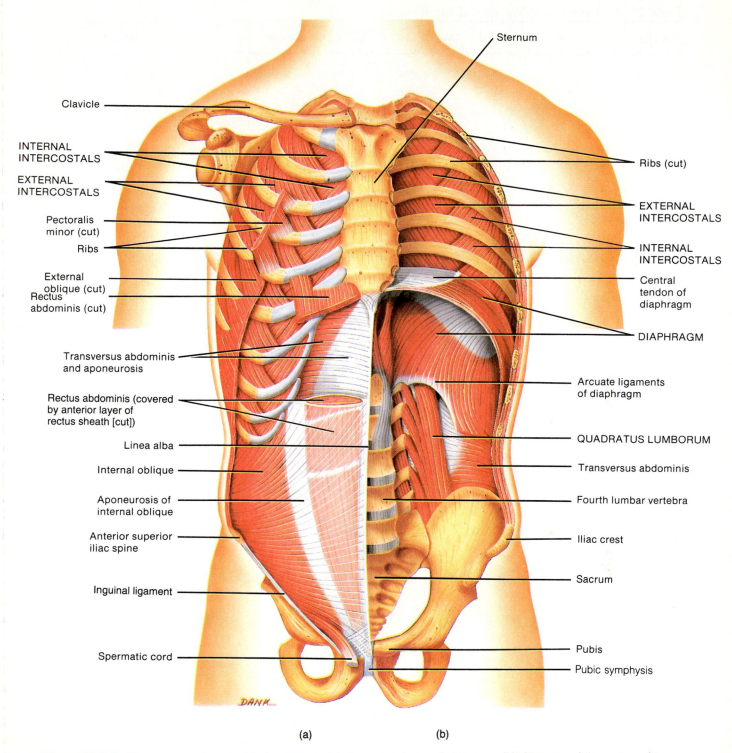

Sternum

Clavicle

INTERNAL
INTERCOSTALS

EXTERNAL
INTERCOSTALS

Pectoralis
minor (cut)

Ribs

External
oblique (cut)
Rectus
abdominis (cut)

Transversus abdominis
and aponeurosis

Rectus abdominis (covered
by anterior layer of
rectus sheath [cut])

Linea alba

Internal oblique

Aponeurosis of
internal oblique

Anterior superior
iliac spine

Inguinal ligament

Spermatic cord

Ribs (cut)

EXTERNAL
INTERCOSTALS

INTERNAL
INTERCOSTALS

Central
tendon of
diaphragm

DIAPHRAGM

Arcuate ligaments
of diaphragm

QUADRATUS LUMBORUM

Transversus abdominis

Fourth lumbar vertebra

Iliac crest

Sacrum

Pubis

Pubic symphysis

DANK

(a)                    (b)

**Figure 19-4** Human muscles used in breathing: (a) diagram of superficial view; (b) diagram of deep view. (From G. J. Tortora and S. R. Grabowski, *Principles of Anatomy and Physiology,* 7th ed., New York: HarperCollins, 1993. © Leonard Dank.)

**Table 19-3   Deep Chest Muscles of the Human**

| Muscle | Origin | Insertion | Action |
|---|---|---|---|
| Serratus anterior | First eight or nine ribs | Vertebral border and inferior angle of scapula | Pulls scapula forward and rotates it upward |
| Coracobrachialis (not shown) | Coracoid process of scapula | Humerus | Adducts, flexes, and rotates arm |
| External intercostals | Inferior border of rib above | Superior border of rib below | Raise rib at inhalation and thereby increase size of thoracic cavity |
| Internal intercostals | Superior border of rib below | Inferior border of rib above | Draw adjacent ribs together and lower ribs at exhalation |

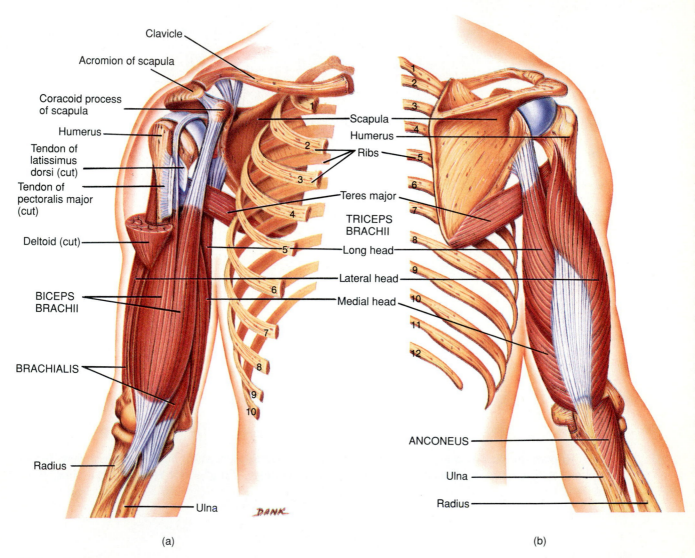

(a)          (b)

**Figure 19-5** Human muscles of the shoulder that move the forearm: (a) anterior view; (b) posterior view. (From G. J. Tortora and S. R. Grabowski, *Principles of Anatomy and Physiology,* 7th ed., New York: HarperCollins. 1993. © Leonard Dank.)

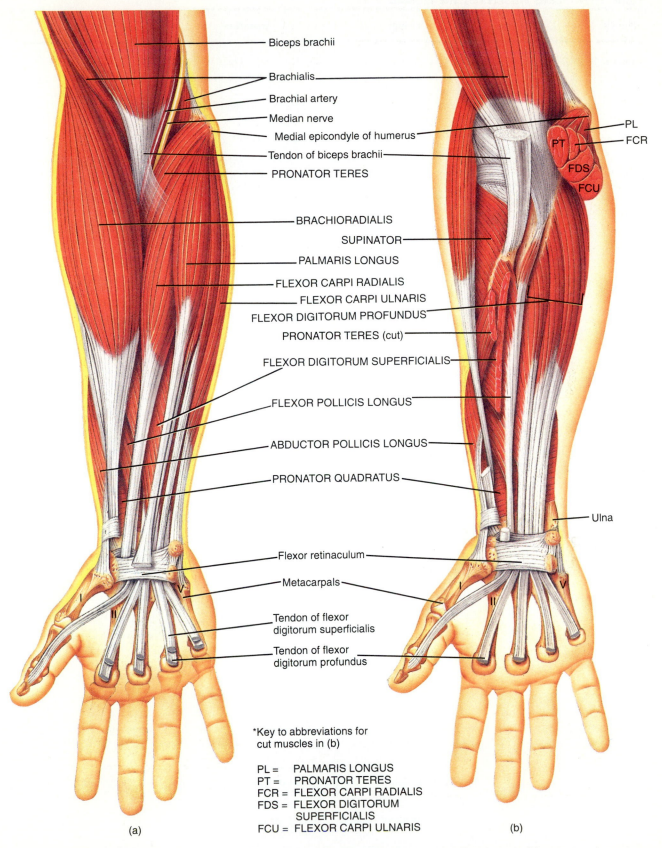

Biceps brachii
Brachialis
Brachial artery
Median nerve
Medial epicondyle of humerus
Tendon of biceps brachii
PRONATOR TERES
BRACHIORADIALIS
SUPINATOR
PALMARIS LONGUS
FLEXOR CARPI RADIALIS
FLEXOR CARPI ULNARIS
FLEXOR DIGITORUM PROFUNDUS
PRONATOR TERES (cut)
FLEXOR DIGITORUM SUPERFICIALIS
FLEXOR POLLICIS LONGUS
ABDUCTOR POLLICIS LONGUS
PRONATOR QUADRATUS
Flexor retinaculum
Metacarpals
Tendon of flexor digitorum superficialis
Tendon of flexor digitorum profundus

PL
PT
FCR
FDS
FCU

Ulna

*Key to abbreviations for cut muscles in (b)

PL = PALMARIS LONGUS
PT = PRONATOR TERES
FCR = FLEXOR CARPI RADIALIS
FDS = FLEXOR DIGITORUM SUPERFICIALIS
FCU = FLEXOR CARPI ULNARIS

(a)                    (b)

**Figure 19-6** Human muscles that move the wrist and fingers; (a) anterior superficial view; (b) anterior deep view. (From G. J. Tortora and S. R. Grabowski, *Principles of Anatomy and Physiology,* 7th ed., New York: HarperCollins, 1993. © Leonard Dank.)

**Table 19-4   Arm Muscles of the Human**

| Muscle | Origin | Insertion | Action |
|---|---|---|---|
| Triceps brachii | | | |
|    Long head | Infraglenoid tuberosity of scapula | Olecranon process of ulna | Extends forearm |
|    Lateral head | Above radial groove of humerus | Olecranon process of ulna | Extends forearm |
|    Medial head | Below radial groove of humerus | Olecranon process of ulna | Extends forearm |
| Brachialis | Anterior surface of humerus | Tuberosity and coronoid process of ulna | Flexes forearm |
| Biceps brachii | | | |
|    Long head | Near glenoid fossa of scapula | Radial tuberosity | Flexes and supinates forearm |
|    Short head | Coracoid process of scapula | Radial tuberosity | Flexes and supinates forearm |
| Brachioradialis | Supracondyloid ridge of humerus | Above styloid process of radius | Flexes forearm |
| Flexor carpi radialis | Medial epicondyle of humerus | Second and third metacarpals | Flexes and abducts wrist |
| Flexor carpi ulnaris | Medial epicondyle of humerus and upper dorsal border of ulna | Pisiform, hamate, and fifth metacarpal | Flexes and adducts wrist |
| Extensor carpi radialis longus | Lateral epicondyle of humerus | Second metacarpal | Extends and abducts wrist |
| Extensor carpi ulnaris | Lateral epicondyle of humerus and dorsal border of ulna | Fifth metacarpal | Extends and adducts wrist |
| Flexor digitorum profundus | Body of ulna | Bases of distal phalanges | Flexes all distal phalanges |
| Flexor digitorum superficialis | Medial epicondyle of humerus, coronoid process of ulna, and oblique line of radius | Middle phalanges | Flexes middle phalanges of each finger |
| Extensor digitorum | Lateral epicondyle of humerus | Middle and distal phalanges | Extends phalanges |
| Extensor indicis | Dorsal surface of ulna | Tendon of extensor digitorum of index finger | Extends index finger |
| Supinator | Lateral epicondyle of humerus and ridge on ulna | Oblique line of radius | Supinates forearm |
| Pronator teres | Medial epicondyle of humerus and coronoid process of ulna | Lateral surface of radius | Pronates forearm |
| Palmaris longus | Medial epicondyle of humerus | Radial tuberosity | Flexes and supinates forearm |
| Coracobrachialis | Coracoid process of scapula | Middle of medial border of humerus | Flexes and adducts arm |

It is not easy to demonstrate the deep chest muscles, but some of the muscles in the human arm are easily available. Some arm muscles easily experimented on are listed:

1. *Triceps brachii* (Figure 19-7)
   The triceps has three heads of origin in the human, just as it does in the cat. It extends the forearm.

   With the elbow flexed, hand touching the shoulder, slowly straighten (extend) the elbow joint. Feel the triceps contract. (a) Name two sports activities during which the triceps brachii is used extensively, and (b) describe the specific phase of the activity during which it is contracted.

   a. _____

   _____

   b. _____

   _____

2. *Biceps brachii* (Figure 19-8)
   The biceps has two distinct heads of origin on the scapula by tendons that pass over the proximal head of the humerus. It flexes the forearm and supinates the forearm.

   Demonstrate the action of this muscle by flexing the elbow joint. Name some specific activities that you perform regularly, which require the use of the biceps. _____

   _____

   _____

3. *Supinator* (Figure 19-9)
   The supinator turns the forearm and hand so that the palm faces upward.

   To demonstrate its action, move your arm (with the hand relaxed) away from your body in front of you. Slowly turn the palm of your hand upward (supination). Place your fingers on the forearm below the elbow, to feel the action. Name two specific activities in which this muscle is used. _____

   _____

   _____

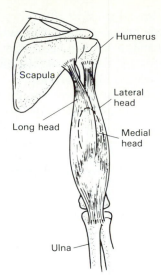

**Figure 19-7** Human triceps brachii (posterior view). (From P. J. Donnelly and G. A. Wistreich, *Exploring the Human Body,* New York: Harper & Row, 1989.)

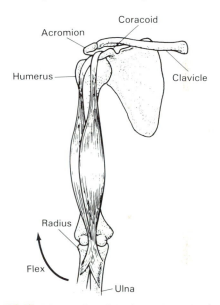

**Figure 19-8** Biceps brachii (anterior view). (From P. J. Donnelly and G. A. Wistreich, *Exploring the Human Body,* New York: Harper & Row, 1989.)

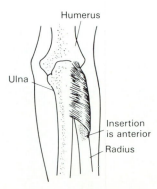

**Figure 19-9** Supinator (posterior view). (From P. J. Donnelly and G. A. Wistreich, *Exploring the Human Body,* New York: Harper & Row, 1989.)

4. *Pronator teres* (Figure 19-10)

This muscle pronates (turns downward) the forearm and hand. It has two heads of origin, one on the humerus and the other on the ulna. Its fibers pass obliquely over the forearm.

With your arm outstretched, in the palm-up position, turn your palm down. Make the same motion with your arm at your side. What specific action does a baseball player make when contracting this muscle? _____

_____

Some of the muscles that move the wrist and fingers, which perform some of the most intricate movements of which the human is capable, are as follows:

1. *Flexor carpi radialis* (Figure 19-11)

This flexor muscle bends and abducts the wrist and hand.

With your arm in front of you, palm in the pronated position, flex your hand as far as possible. Then turn your wrist laterally (abduction). Name an activity in which you would move your hand in this way. _____

_____

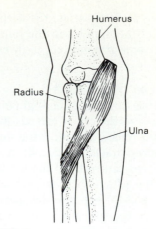

**Figure 19-10** Pronator teres (anterior view). (From P. J. Donnelly and G. A. Wistreich, *Exploring the Human Body,* New York: Harper & Row, 1989.)

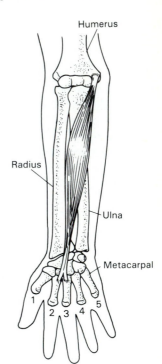

**Figure 19-11** Flexor carpi radialis (posterior view). (From P. J. Donnelly and G. A. Wistreich, *Exploring the Human Body,* New York: Harper & Row, 1989.)

2. *Flexor carpi ulnaris* (Figure 19-12)

This muscle also has two heads of origin, one on the humerus, the other on the ulna.

To demonstrate this muscle, hold your arm in front of you, palm down. Flex your hand, and turn your hand inward (adduction). Name an activity in which the hand and wrist would be moved in this way. _____

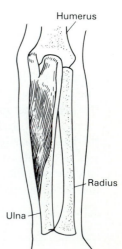

**Figure 19-12** Flexor carpi ulnaris (posterior view). (From P. J. Donnelly and G. A. Wistreich, *Exploring the Human Body,* New York: Harper & Row, 1989.)

3. *Palmaris longus* (Figure 19-13)

This is a slender muscle with a long tendon of insertion. It frequently sends a small tendon to the thumb muscles (not shown in Figure 19-13).

To demonstrate it, flex the wrist with the fingers curled as if you were about to strike piano keys. Feel the palmer aponeurosis tense.

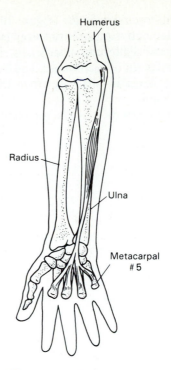

**Figure 19-13** Palmaris longus (anterior view). (From P. J. Donnelly and G. A. Wistreich, *Exploring the Human Body,* New York: Harper & Row, 1989.)

4. *Flexor digitorum superficialis* (Figure 19-14)

This flexor muscle bends the phalanges and the hand.

Demonstrate this muscle's action by flexing fingers two through five while keeping the interphalangeal joints extended. What purposeful activity could be accomplished with this muscle? ____

_____

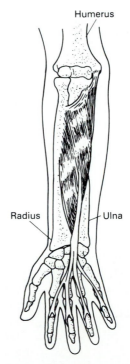

**Figure 19-14** Flexor digitorum superficialis (anterior view). (From P. J. Donnelly and G. A. Wistreich, *Exploring the Human Body,* New York: Harper & Row, 1989.)

5. *Flexor digitorum profundis* (Figure 19-15)
The tendon of this muscle runs deep to the tendon of the flexor digitorum superficialis. It flexes the distal phalanges of each finger, and by continued action, it flexes all phalanges (two through five) and the hand.

To demonstrate it, make the hand into a fist.

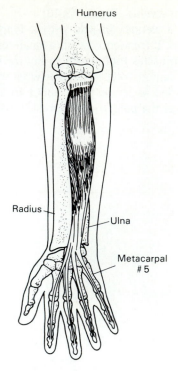

**Figure 19-15** Flexor digitorum profundis (anterior view). (From P. J. Donnelly and G. A. Wistreich, *Exploring the Human Body,* New York: Harper & Row, 1989.)

6. *Extensor carpi radialis longus* (Figure 19-16)
This muscle extends and abducts the wrist and hand.

To demonstrate, hold your hand out, palm down. Bend the hand backward (hyperextension). Relax the hand and turn it away from the body (abduction). In what specific activities does this

muscle play a part? _____

_____

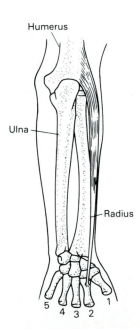

**Figure 19-16** Extensor carpi radialis longus (posterior view). (From P. J. Donnelly and G. A. Wistreich, *Exploring the Human Body,* New York: Harper & Row, 1989.)

7. *Extensor carpi ulnaris* (Figure 19-17)
This extensor straightens and adducts the wrist and hand.

     To demonstrate, hold your hand in front of you. Bend the hand backward (hyperextension). Relax again. Then move the hand and wrist inward (toward the midline). Describe an activity during

which this muscle is used. _____

_____

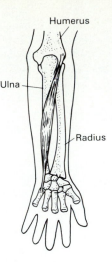

**Figure 19-17** Extensor carpi ulnaris (posterior view). (From P. J. Donnelly and G. A. Wistreich, *Exploring the Human Body,* New York: Harper & Row, 1989.)

# LABORATORY REVIEW *19*

## *MUSCLES OF THE DEEP CHEST AND ARM*

### COMPLETION

1. Complete Table Q19-1. Note how differences in location affect the action of the muscles. If the actions are similar, note that too.

**Table Q19-1**

| Muscle | Location | Action |
|---|---|---|
| Coracobrachialis (cat) | | |
| Coracobrachialis (human) | | |
| Serratus ventralis (cat) | | |
| Serratus anterior (human) | | |
| Intercostal muscles (cat) | | |
| Intercostal muscles (human) | | |

2. Study Figure Q19-1, showing the deep chest and neck muscles of the human. Compare them, as to size and location, with the muscles just dissected in the cat. Color this diagram if you wish.

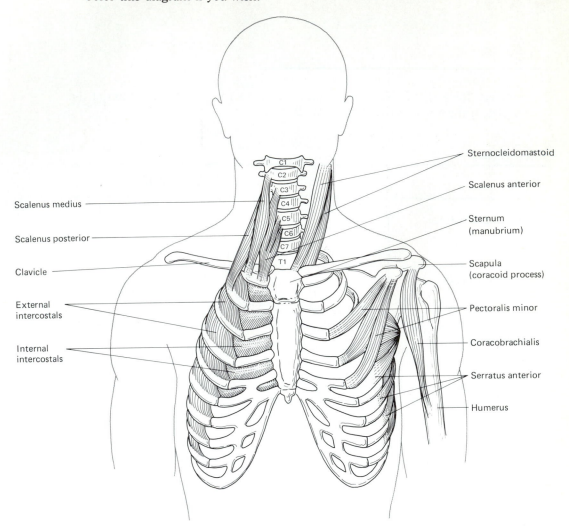

**Figure Q19-1** (From G. J. Tortora, *Principles of Human Anatomy*, 3rd, 4th, 5th eds., New York: HarperCollins, 1983, 1986, 1989.)

3. Fill in the blanks.

a. _____ Name a muscle that flexes the forearm.

b. _____ Name a muscle that extends and adducts the wrist.

c. _____ Name a muscle that flexes and abducts the wrist.

d. _____ Name a muscle that extends the forearm.

e. _____ Name a muscle that extends a phalanx.

f. _____ Name another muscle that extends a phalanx.

g. _____ Name a muscle that pronates the forearm.

h. _____ Name a muscle that supinates the forearm.

i. _____ What is the position of the hand in pronation?

j. _____ What is the position of the hand in supination?

k. _____ Name a muscle that flexes the forearm.

l. _____ Name another muscle that flexes the forearm.

4. Study Figure Q19-2, showing the human arm muscles. Relate each muscle to one dissected in the cat. Color the drawing for clarity.

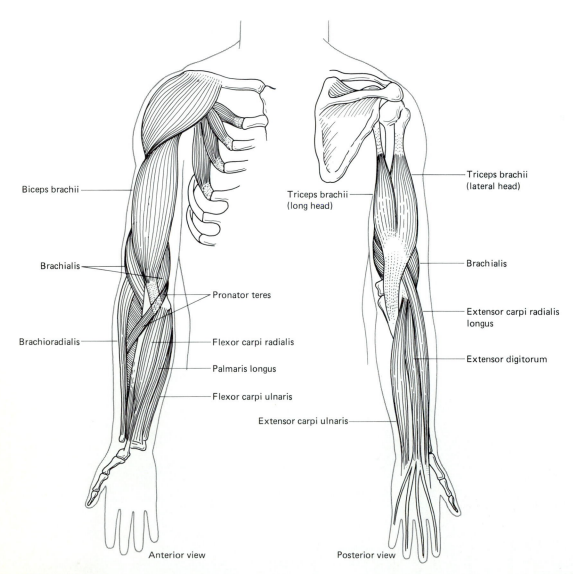

**Figure Q19-2**

# Muscles of the Hip and Thigh

**After completing this exercise, you should be able to**

**After completing this exercise, you should be able to**

1. Dissect the muscles of the hip and thigh of the cat
2. Identify comparable human muscles

## Materials

The following materials should be provided for class use:

1. Embalmed cats
2. Dissecting tools
3. Dissecting trays
4. String
5. 3 percent phenol solution

The muscles dissected in Exercise 20 are listed here, according to specific laboratory exercises.

### SUPERFICIAL THIGH MUSCLES, MEDIAL VIEW

Sartorius

Gracilis

### DEEP THIGH MUSCLES, MEDIAL VIEW

Quadriceps femoris
   Vastus lateralis
   Vastus intermedius
   Vastus medialis
   Rectus femoris

Iliopsoas

Pectineus

Adductor longus

Adductor femoris

Semimembranosus

### SUPERFICIAL THIGH AND HIP MUSCLES, LATERAL VIEW

Gluteus medius

Gluteus maximus

Tensor fasciae latae

Caudofemoralis

Biceps femoris

### DEEP THIGH MUSCLES, LATERAL VIEW

Tenuissimus

Semitendinosus

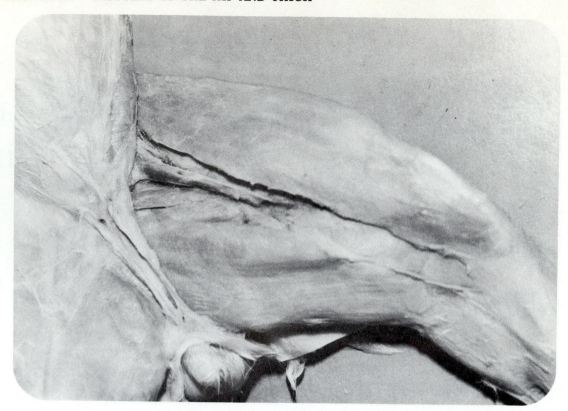

**Figure 20-1** (a) Superficial thigh muscles (medial surface) of the cat (photograph).

## SUPERFICIAL THIGH MUSCLES, MEDIAL VIEW

### Instructions for Dissection

Almost the entire medial surface of the thigh is covered by two ribbonlike muscles separated proximally by a triangular area (the *femoral triangle*), in which you see blood vessels and a nerve, as displayed in Figure 20-1. Remove any fat and fascia that obscures these vessels and the underlying muscles. The blood vessels (Figure 20-1) are the *femoral artery* (injected red) and the *femoral vein* (injected blue). The large nerve running parallel to the blood vessels is the *femoral nerve*.

The thin, wide muscle located laterally is the *sartorius*. Loosen it between its origin and its insertion. The thin, wide muscle located medially is the *gracilis*. The muscles lying in the triangular area are considered in the next dissection.

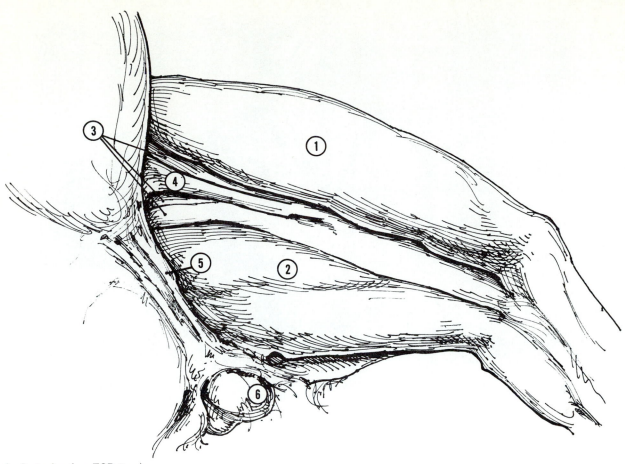

1. Sartorius (sar-TOR-ē-us)
2. Gracilis (gra-SIL-is)
3. Adductor muscles (femoral triangle)
4. Blood vessels (femoral triangle)
5. Spermatic cord (male cats only)
6. Scrotum (male cats only)

**Figure 20-1** (b) Superficial thigh muscles (medial surface) of the cat (line drawing).

**Figure 20-2** (a) Deep thigh muscles (medial view) of the cat (photograph).

## DEEP THIGH MUSCLES, MEDIAL VIEW

### Instructions for Dissection

In order to see this important group of muscles, it is necessary to transect both the sartorius and the gracilis muscles. Lay their ends back as far as possible, and begin your observations at the lateral border.

Covering about one-half of the area of the thigh is a large muscle named the *quadriceps femoris*. It is so large that the four muscles that make it up must be considered separately, as they are in the human. The quadriceps separates readily into three distinct bundles: (1) a large lateral muscle, the *vastus lateralis*, (2) a middle muscle, the *rectus femoris*, and (3) a medial muscle, the *vastus medialis*. The fourth muscle of the group, the *vastus intermedius*, may be seen by separating the rectus femoris from either adjacent muscle, or by transecting the rectus femoris and laying back its cut ends. (In Figure 20-2, the rectus femoris is not transected, so the vastus intermedius is not shown.)

Dissect away the *femoral artery, femoral vein,* and *femoral nerve,* so that you are better able to see the group of muscles beneath them. Most lateral of the group is the *iliopsoas* (composed of two muscles, the *iliacus* and the *psoas major,* but seen as one muscle in this dissection). Medial to the iliopsoas is the *pectineus,* and medial to it is the *adductor longus*.

The remaining two medial muscles are considerably larger, the *adductor femoris* and medial to it the *semimembranosus,* as shown in Figure 20-2.

1. Vastus lateralis (VAS-tus la'-ter-AL-is)
2. Rectus femoris (REK-tus FEM-or-is)
3. Vastus medialis
4. Vastus intermedius (not shown)
5. Iliopsoas (il'-ē-ō-SŌ-as)
6. Pectineus (pek-TIN-e-us)
7. Adductor longus
8. Adductor femoris
9. Semimembranosus (sem'-ē-mem-bre-NŌ-sus)
10. Sartorius (reflected)
11. Gracilis (reflected)

**Figure 20-2**  (b) Deep thigh muscles (medial view) of the cat (line drawing).

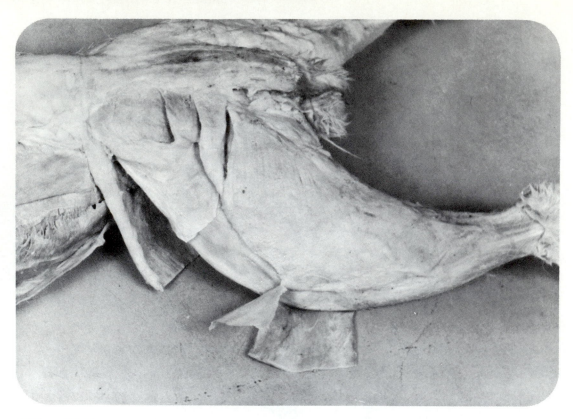

**Figure 20-3** (a) Superficial thigh muscles and hip muscles (lateral view) of the cat (photograph).

## SUPERFICIAL THIGH AND HIP MUSCLES, LATERAL VIEW

### Instructions for Dissection

Turn the cat over so that the lateral surface is facing you, as in Figure 20-3. In the hip region, which is covered with heavy fascia, are the *gluteal muscles*. Remove this heavy fascia by cutting as much of it off as is necessary to expose the *gluteus medius* and posterior to it the *gluteus maximus*. Note that the gluteus maximus is the smaller of the two, contrary to their relative sizes in the human.

Lying within the same mass of fascia that you have just removed, and below the gluteus medius, is the *tensor fasciae latae* (see Figure 20-3). The tensor is an irregularly shaped muscle that not only originates in a heavy fascia, but also inserts by a long broad fascia (fascia lata), almost completely covering the *vastus lateralis*. Loosen the fascia lata, and transect it so that you can see the entire vastus lateralis.

Posterior to the fascia lata the gluteus maximus is the *caudofemoralis* (a muscle not found in the human), which tapers to its insertion by a very fine tendon. Separate it from the large *biceps femoris*, whose mass covers most of the lateral surface of the thigh.

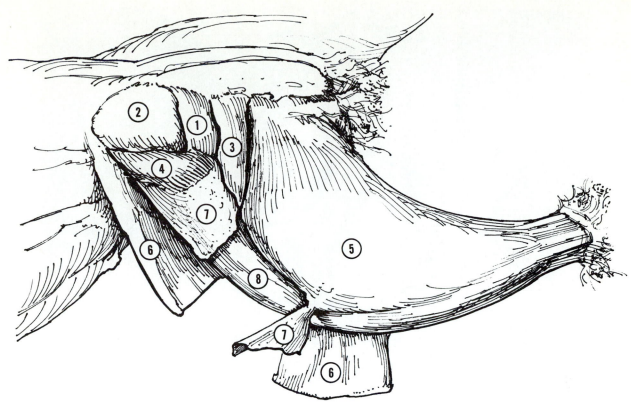

1. Gluteus maximus (GLOO-tē-us MAX-i-mus)
2. Gluteus medius (GLOO-tē-us MĒ-dē-us)
3. Caudofemoralis (kaw'-dō-fem-or-AL-is)
4. Tensor fasciae latae
5. Biceps femoris (BĪ-seps FEM-or-is)
6. Sartorius
7. Fascia lata (transected to expose vastus lateralis)
8. Vastus lateralis

**Figure 20-3** (b) Superficial thigh muscles and hip muscles (lateral view) of the cat (line drawing).

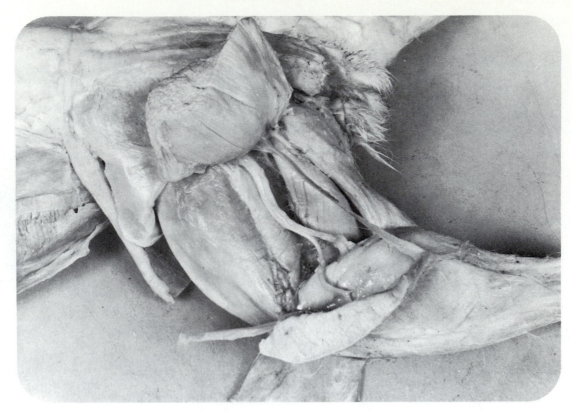

**Figure 20-4** (a) Deep thigh muscles (lateral view) of the cat (photograph).

## DEEP THIGH MUSCLES, LATERAL VIEW

### Instructions for Dissection

Separate the biceps femoris from the tissue below it (see Figure 20-4) with your finger and a probe. Before making a transection, locate and dissect a very small muscle, the *tenuissimus,* which runs the length of the biceps femoris but is only about ¼ inch wide (not found in the human leg). Also, locate the *sciatic nerve,* lying

in the same general area (see Figure 20-4). When you have done this, carefully transect the biceps femoris, and draw each end to its origin and insertion.

In addition to the tenuissimus and the *sciatic nerve,* observe the semimembranosus already seen in the medial view. Medial to it is the *semitendinosus.* The vastus lateralis and the adductor femoris are also prominent in this view. Refer to Table 20-1 for the origins, insertions, and actions of the cat's hip and thigh muscles.

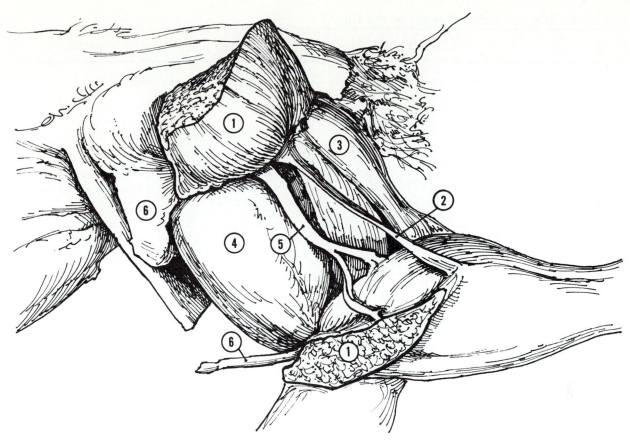

1. Biceps femoris (reflected)
2. Tenuissimus (ten'-yoo-ISS-i-mus)
3. Semitendinosus
4. Vastus lateralis
5. Sciatic nerve
6. Fascia lata (reflected)

**Figure 20-4** (b) Deep thigh muscles (lateral view) of the cat (line drawing).

**Table 20-1   Hip and Thigh Muscles of the Cat**

| Muscle | Origin | Insertion | Action |
|---|---|---|---|
| Gluteus maximus | Transverse processes of last sacral and first caudal vertebrae | Greater trochanter of femur | Abducts thigh |
| Gluteus medius | Fascia from sacral and caudal vertebrae and iliac crest | Greater trochanter of femur | Abducts thigh |
| Caudofemoralis | Transverse processes of second and third caudal vertebrae | Patella by tendon that pierces fascia lata | Abducts thigh; extends shank |
| Tensor fasciae latae | Ventral border of ilium and fascia covering gluteus medius | Patella and tibia by fascia lata | Extends thigh; aids in abduction and rotation |
| Biceps femoris | Ischial tuberosity | Patella and lateral border of tibia | Adducts thigh; flexes shank |
| Sartorius | Ilium | Patella | Adducts and rotates femur; extends tibia |
| Gracilis | Border between ischium and pubis | Proximal end of tibia | Adducts thigh and pulls in caudad |
| Quadriceps femoris | | | |
|     Vastus lateralis | Greater trochanter and shaft of femur | Patella | Extends lower leg |
|     Rectus femoris | Ventral border of ilium near acetabulum | Patella | Extends lower leg |
|     Vastus medialis | Shaft of femur | Patella | Extends lower leg |
|     Vastus intermedius | Almost entire length of shaft of femur | Patella | Extends lower leg |
| Iliopsoas | Lumbar vertebrae and ventral border of ilium | Lesser trochanter of femur | Flexes thigh and rotates it, turning foot outward |
| Pectineus | Cranial border of pubis | Near lesser trochanter of femur | Adducts thigh |
| Adductor longus | Cranial border of pubis | Linea aspera of femur | Adducts thigh |
| Adductor femoris | Pubis and ischium | Ventral surface and shaft of femur | Extends thigh |
| Semimembranosus | Ischial tuberosity | Medial surface and shaft of femur | Extends thigh |
| Semitendinosus | Ischial tuberosity | Proximal end of tibia | Flexes shank |
| Tenuissimus | Transverse process of second caudal vertebra | Inserting fascia of biceps femoris | Adducts thigh |

## COMPARABLE HUMAN MUSCLES

The gluteus muscles in the human are divided into three groups. Externally, there is the large posterior *gluteus maximus,* and above and lateral to it is the *gluteus medius.* The smaller *gluteus minimus* lies beneath the other two (Figure 20-5).

The muscles of the posterior thigh are equally effective at both hip and knee joints as extensors and flexors. The tendons of these muscles can be felt at the back of the knee when the knee is partly flexed. In addition to the *biceps femoris,* there are the *semimembranosus* and the *semitendinosus.* These muscles, called the "hamstrings," enable you to stand in a relaxed condition.

The muscles of the anterior thigh extend the knee joint. They are the quadriceps group: the vastus lateralis, rectus femoris, vastus medialis, and vastus intermedius. Flexors of the hip joint are the *iliopsoas, rectus femoris, tensor fasciae latae, pectineus, gracilis,* and to some extent the *sartorius.*

There is a great ideal of similarity between cat and human in these groups of muscles. Study as many human hip and thigh muscles as you can. Refer to Table 20-2 for the muscles' origin, insertion, and action. Perform as many of the suggested movements on pages 275–278 as time permits.

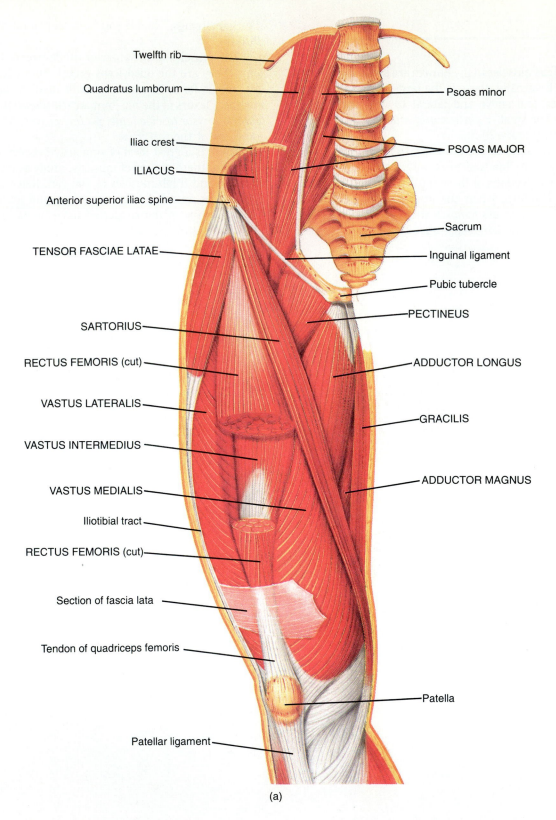

Twelfth rib

Quadratus lumborum

Iliac crest

ILIACUS

Anterior superior iliac spine

TENSOR FASCIAE LATAE

SARTORIUS

RECTUS FEMORIS (cut)

VASTUS LATERALIS

VASTUS INTERMEDIUS

VASTUS MEDIALIS

Iliotibial tract

RECTUS FEMORIS (cut)

Section of fascia lata

Tendon of quadriceps femoris

Patellar ligament

Psoas minor

PSOAS MAJOR

Sacrum

Inguinal ligament

Pubic tubercle

PECTINEUS

ADDUCTOR LONGUS

GRACILIS

ADDUCTOR MAGNUS

Patella

(a)

**Figure 20-5** Muscles that act on the leg: (a) anterior superficial view; (b) anterior deep view;

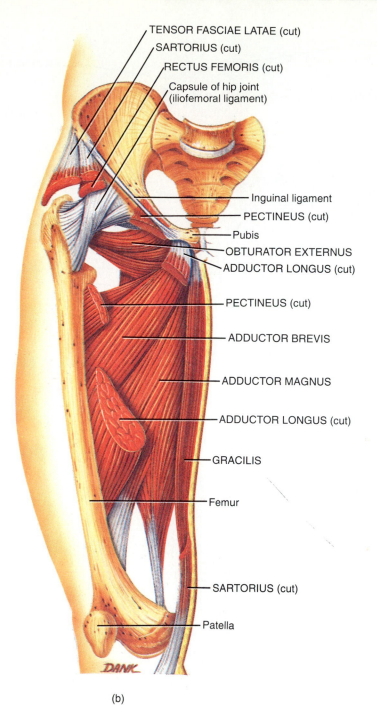

TENSOR FASCIAE LATAE (cut)

SARTORIUS (cut)

RECTUS FEMORIS (cut)

Capsule of hip joint
(iliofemoral ligament)

Inguinal ligament

PECTINEUS (cut)

Pubis

OBTURATOR EXTERNUS

ADDUCTOR LONGUS (cut)

PECTINEUS (cut)

ADDUCTOR BREVIS

ADDUCTOR MAGNUS

ADDUCTOR LONGUS (cut)

GRACILIS

Femur

SARTORIUS (cut)

Patella

DANK

(b)

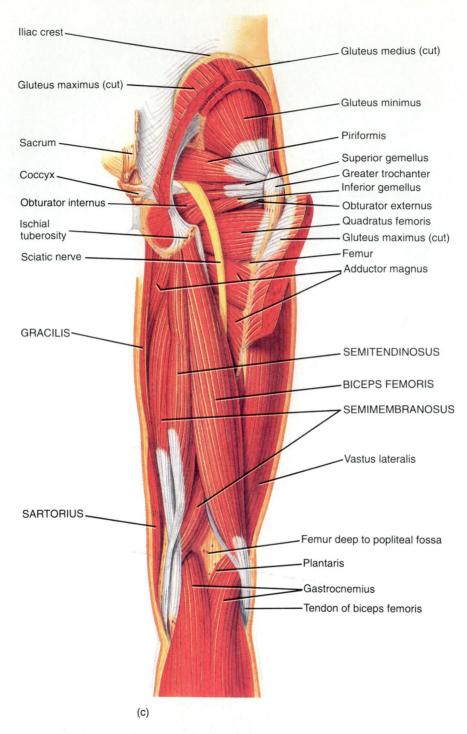

Iliac crest

Gluteus maximus (cut)

Sacrum

Coccyx

Obturator internus

Ischial tuberosity

Sciatic nerve

GRACILIS

SARTORIUS

Gluteus medius (cut)

Gluteus minimus

Piriformis

Superior gemellus

Greater trochanter

Inferior gemellus

Obturator externus

Quadratus femoris

Gluteus maximus (cut)

Femur

Adductor magnus

SEMITENDINOSUS

BICEPS FEMORIS

SEMIMEMBRANOSUS

Vastus lateralis

Femur deep to popliteal fossa

Plantaris

Gastrocnemius

Tendon of biceps femoris

(c)

**Figure 20-5** *(Continued)* (c) posterior superficial view; (d) posterior deeper view. (From G. J. Tortora and S. R. Grabowski, *Principles of Anatomy and Physiology*, 7th ed., New York: HarperCollins, 1993. © Leonard Dank.)

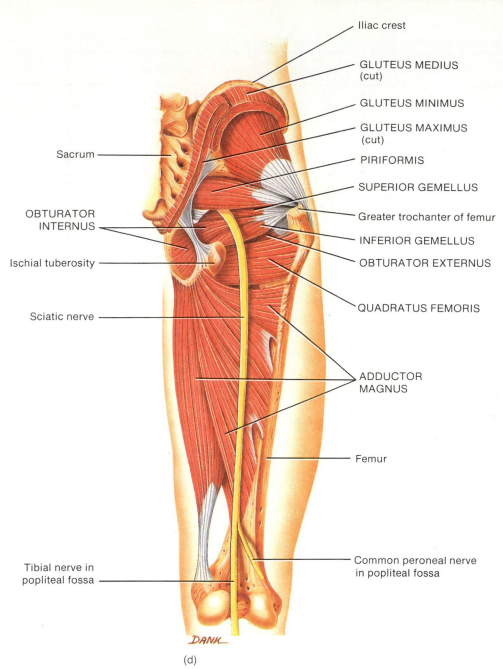

Iliac crest

GLUTEUS MEDIUS
(cut)

GLUTEUS MINIMUS

GLUTEUS MAXIMUS
(cut)

PIRIFORMIS

SUPERIOR GEMELLUS

Greater trochanter of femur

INFERIOR GEMELLUS

OBTURATOR EXTERNUS

QUADRATUS FEMORIS

Sacrum

OBTURATOR
INTERNUS

Ischial tuberosity

Sciatic nerve

ADDUCTOR
MAGNUS

Femur

Common peroneal nerve
in popliteal fossa

Tibial nerve in
popliteal fossa

DANK

(d)

**Table 20-2   Hip and Thigh Muscles of the Human**

| Muscle | Origin | Insertion | Action |
|---|---|---|---|
| Gluteus maximus | Iliac crest, sacrum, and coccyx | Iliotibial tract of fascia lata and gluteal tuberosity of femur | Extends and rotates femur laterally |
| Gluteus medius | Ilium | Greater trochanter of femur | Abducts and rotates femur laterally |
| Gluteus minimus | Ilium | Greater trochanter of femur | Abducts and rotates femur medially |
| Tensor fasciae latae | Iliac crest | Iliotibial tract of fascia lata | Flexes and abducts femur |
| Adductor longus | Pubic crest and symphysis | Linea aspera of femur | Adducts, rotates, and flexes femur |
| Adductor magnus (femoris) | Pubis and ischium | Linea aspera of femur | Adducts, flexes, and extends femur |
| Psoas major | Transverse processes and centra of lumbar vertebrae | Lesser trochanter of femur | Flexes and rotates femur laterally; flexes vertebral column |
| Iliacus | Iliac fossa | Tendon of psoas major | Flexes and rotates femur laterally; flexes vertebral column |
| Pectineus | Fascia of pubis | Pectineal line of femur | Flexes, adducts, and rotates thigh laterally |
| Biceps femoris | Ischial tuberosity and linea aspera of femur | Head of fibula and lateral condyle of tibia | Flexes knee; extends thigh |
| Semitendinosus | Ischial tuberosity | Proximal end of tibia | Flexes knee; extends thigh |
| Semimembranosus | Ischial tuberosity | Medial condyle of tibia | Flexes knee; extends thigh |
| Gracilis | Pubic symphysis and pubic arch | Medial surface of tibia | Flexes knee; adducts thigh |
| Sartorius | Anterior superior spine of ilium | Medial surface of tibia | Flexes knee; flexes thigh and rotates it laterally |
| Quadriceps femoris | | | |
| Rectus femoris | Anterior inferior spine of ilium | Upper border of patella | Extends knee; flexes thigh |
| Vastus lateralis | Greater trochanter and linea aspera of femur | Patella and tibial tuberosity | Extends knee |
| Vastus medialis | Linea aspera of femur | Patella and tibial tuberosity | Extends knee |
| Vastus intermedius | Anterior and lateral surface of femur | Patella and tibial tuberosity | Extends knee |

## HIP AND THIGH MUSCLES

Two large groups of muscles move the leg. They are the quadriceps femoris group and the hamstrings. The quadriceps flexes the thigh and extends the leg. The hamstring group flexes the leg and extends the thigh. In addition there are abductors, adductors and rotators. Perform the movements discussed below.

### *The Quadriceps Femoris*

1. *Rectus femoris* (Figure 20-6) extends the leg and flexes the thigh.
2. *Vastus medialis* (Figure 20-7) extends the leg.
3. *Vastus lateralis* (Figure 20-8) extends the leg.
4. *Vastus intermedius* (Figure 20-9) extends the leg.
   To demonstrate leg extension, sit on a chair. Slowly raise your legs off the floor until the upper and lower legs form a straight line (knee extended). Place your hands on the anterior thigh to feel the contraction of the quads. These are powerful extensors and are indispensable to walking.

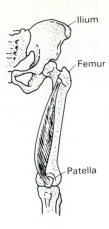

**Figure 20-7** Vastus medialis (anterior view). (From P. J. Donnelly and G. A. Wistreich, *Exploring the Human Body,* New York: Harper & Row, 1989.)

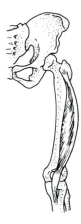

**Figure 20-8** Vastus lateralis (anterior view). (From P. J. Donnelly and G. A. Wistreich, *Exploring the Human Body,* New York: Harper & Row, 1989.)

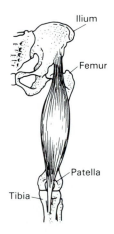

**Figure 20-6** Rectus femoris (anterior view). (From P. J. Donnelly and G. A. Wistreich, *Exploring the Human Body,* New York: Harper & Row, 1989.)

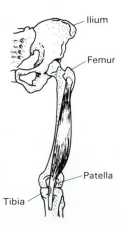

**Figure 20-9** Vastus intermedius (anterior view). (From P. J. Donnelly and G. A. Wistreich, *Exploring the Human Body,* New York: Harper & Row, 1989.)

## *The Hamstrings*

1. *Biceps femoris* (Figure 20-10) flexes the leg at the hip and extends the thigh.
2. *Semitendinosus* (Figure 20-11) flexes the leg at the hip and extends the thigh.
3. *Semimembranosus* (Figure 20-12) flexes the leg at the hip and extends the thigh.

   The hamstring group of muscles, along with the quads, is used in walking, running, and jumping. To demonstrate the hamstrings, flex the leg and place your hand on the posterior surface of the thigh above the knee. Feel the powerful contraction. To test for strength of the hamstrings, lie down in a supine position. Have the instructor or another student hold one leg down while you raise the other.

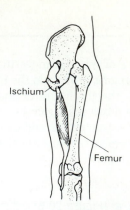

**Figure 20-11** Semitendinosus (posterior view). (From P. J. Donnelly and G. A. Wistreich, *Exploring the Human Body*, New York: Harper & Row, 1989.)

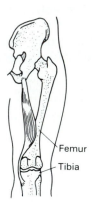

**Figure 20-12** Semimembranosus (posterior view). (From P. J. Donnelly and G. A. Wistreich, *Exploring the Human Body*, New York: Harper & Row, 1989.)

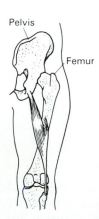

**Figure 20-10** Biceps femoris (posterior view). (From P. J. Donnelly and G. A. Wistreich, *Exploring the Human Body*, New York: Harper & Row, 1989.)

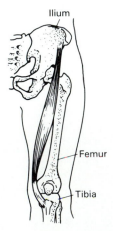

**Figure 20-13** Sartorius (anterior view). (From P. J. Donnelly and G. A. Wistreich, *Exploring the Human Body*, New York: Harper & Row, 1989.)

### *Abductors, Adductors, and Rotators*

1. *Sartorius* (see Figure 20-13)
   This muscle enables you to rotate the leg laterally so that you can sit in a cross-legged (tailor) fashion. It originates on the ilium of the pelvic girdle and inserts on the tibia, running across the leg from lateral to medial. Sit tailor fashion, and feel the muscle contract. It flexes the leg, rotates the leg laterally, and flexes the thigh.

   a. Name one synergist of the sartorius. _____

   b. Name one antagonist of the sartorius. _____

2. *Tensor fasciae latae* (Figure 20-14)
   The tensor flexes the thigh, rotates the thigh medially, and abducts the thigh. It also tenses the fascia latae.
      Name other muscles in the leg, which perform these same functions:

   Flexes thigh _____

   Rotates thigh medially _____

   Abducts thigh _____

3. *Adductor magnus (femoris)* (Figure 20-15)
   This is the largest and most powerful muscle of the adductor group. Its anterior part adducts and flexes the thigh and also rotates the thigh medially. The posterior part extends the thigh and rotates it laterally.

   a. Flex the thigh of one leg a short distance. Adduct and rotate medially simultaneously. This action is controlled by the anterior portion of the muscle. Comments? _____

   _____

   b. Straighten the thigh, and rotate it laterally. This is accomplished by the posterior portion. Comments? _____

   _____

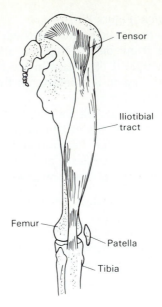

**Figure 20-14** Tensor fasciae latae (lateral view). (From P. J. Donnelly and G. A. Wistreich, *Exploring the Human Body,* New York: Harper & Row, 1989.)

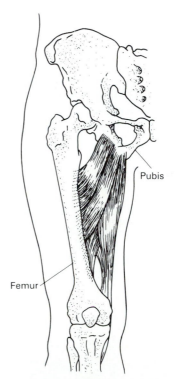

**Figure 20-15** Adductor magnus (anterior view). (From P. J. Donnelly and G. A. Wistreich, *Exploring the Human Body,* New York: Harper & Row, 1989.)

4. *Psoas major* (Figure 20-16)

The psoas flexes and rotates the thigh laterally and flexes the trunk on the thigh. It also bends the trunk laterally. It originates in the body itself in the lumbar region and inserts on the lesser trochanter of the femur.

To demonstrate the muscle, lie on your back. Have a student or the instructor place a hand on the left iliac crest for stabilization. Slightly abduct the right thigh, then flex it and rotate it laterally. Are you able to feel the contraction between the

sartorius and the adductor tendons? _____

5. *Iliacus* (Figure 20-17)

This muscle flexes the thigh and rotates it laterally. It also flexes the vertebral column. Its origin is high on the pelvic girdle, and it inserts on the lesser trochanter of the femur.

To demonstrate it, sit on the edge of a table, with hands grasping the table edge. Flex the hips, one leg at a time. Can you feel its contraction? ____

6. *Gluteus maximus* (Figure 20-18)

This pair is the largest of the gluteus group of muscles. Each muscle is very thick, and the pair forms the human buttocks. These two muscles are instrumental in maintaining the upright posture. They are in a state of contraction if you are standing.

To demonstrate the muscles, rotate one leg at a time laterally while in a standing position. Bend over a table, elbows on the table and feet on the floor. Extend one leg at a time. Feel the contraction and compare it with the intensity of contraction in the standing position. _____

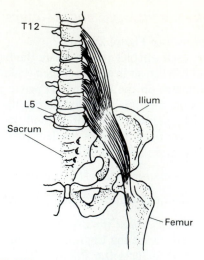

**Figure 20-16** Psoas major (anterior view). (From P. J. Donnelly and G. A. Wistreich, *Exploring the Human Body*, New York: Harper & Row, 1989.)

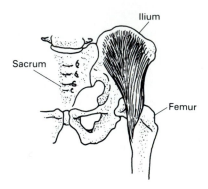

**Figure 20-17** Iliacus (anterior view). (From P. J. Donnelly and G. A. Wistreich, *Exploring the Human Body*, New York: Harper & Row, 1989.)

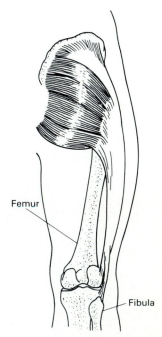

**Figure 20-18** Gluteus maximus (posterior view). (From P. J. Donnelly and G. A. Wistreich, *Exploring the Human Body*, New York: Harper & Row, 1989.)

# LABORATORY REVIEW *20*

## *MUSCLES OF THE HIP AND THIGH*

1. The gluteus muscles and the tensor fasciae latae are shown diagrammatically in Figure Q20-1. Color them, and review their origins and insertions.

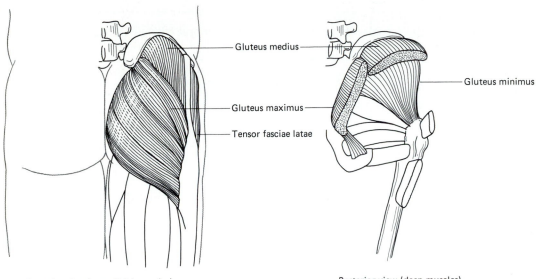

Gluteus medius

Gluteus maximus

Tensor fasciae latae

Gluteus minimus

Posterior view (superficial muscles)

Posterior view (deep muscles)

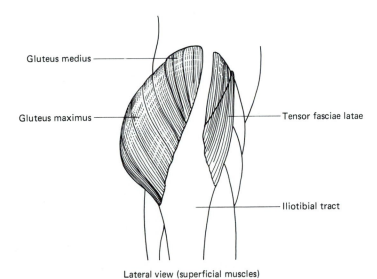

Gluteus medius

Gluteus maximus

Tensor fasciae latae

Iliotibial tract

Lateral view (superficial muscles)

**Figure Q20-1**

2. What are the actions of the human gluteal muscles?

   a. Gluteus maximus _____

   _____

   b. Gluteus medius _____

   _____

   c. Gluteus minimus _____

   _____

3. Name the muscles that can be seen in the femoral triangle of the cat. _____

   _____

4. List the three muscles often designated as the "hamstrings."

   a. _____

   b. _____

   c. _____

# Muscles of the Lower Leg

*After completing this exercise,
you should be able to*

1. Dissect the muscles of the lower leg of the cat
2. Identify comparable human muscles

## Materials

The following materials should be provided for class use:

1. Embalmed cats
2. Dissecting tools
3. Dissecting trays
4. String
5. 3 percent phenol solution

The muscles dissected in Exercise 21 are listed here, according to specific laboratory exercises.

### SUPERFICIAL LOWER LEG MUSCLES, MEDIAL SURFACE

Gastrocnemius and Achilles tendon

Flexor digitorum longus

Tibialis anterior

### SUPERFICIAL LOWER LEG MUSCLES, LATERAL SURFACE

Gastrocnemius and Achilles tendon

Soleus

Peroneus

Extensor digitorum longus

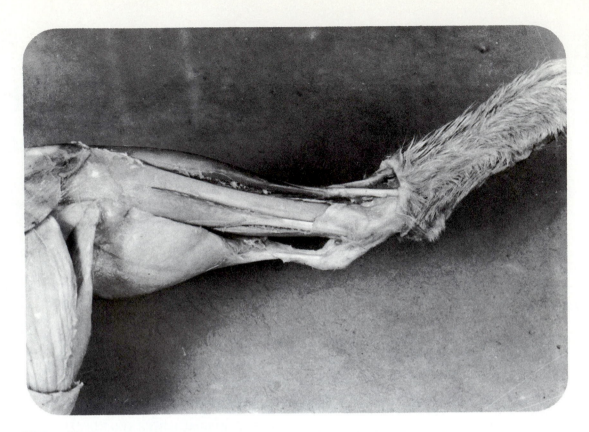

**Figure 21-1** (a) Superficial lower leg muscles (medial surface) of the cat (photograph).

## SUPERFICIAL LOWER LEG MUSCLES, MEDIAL SURFACE

### Instructions for Dissection

In the lower leg, there is a large, prominent muscle, the *gastrocnemius* (see Figure 21-1). It can be seen on both the medial and the lateral surfaces. It has two distinct heads of origin, but the two muscle masses unite to form the large and very tough *Achilles tendon* or *calcaneal tendon,* which inserts on the calcaneus bone. Lying between the gastrocnemius and the tibial bone is the *flexor digitorum longus*. This muscle has two heads of origin. Locate them. On the anterior surface of the tibial bone is the *tibialis anterior* muscle, as shown in Figure 21-1.

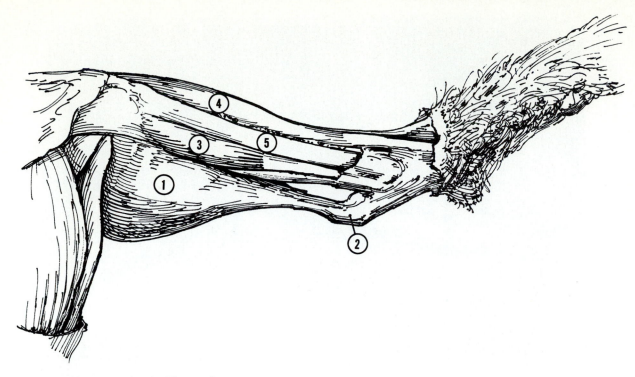

1. Gastrocnemius (gas'-trok-NE-mē-us)
2. Achilles tendon (a-KIL-ēz)
3. Flexor digitorum longus (dij-i-TŌ-rum)
4. Tibialis anterior (tib'-ē-AL-is)
5. Tibia (bone)

**Figure 21-1** (b) Superficial lower leg muscles (medial surface) of the cat (line drawing).

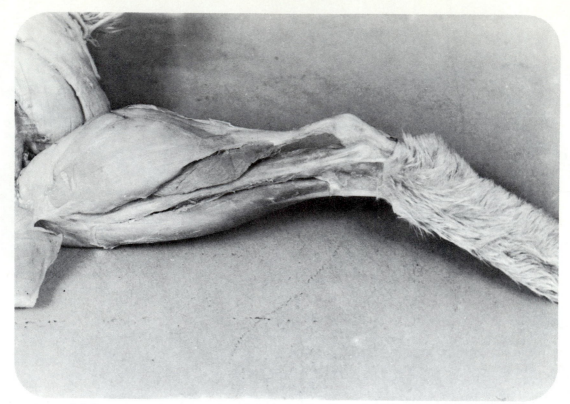

**Figure 21-2** (a) Superficial muscles of the lower leg (lateral surface) of the cat (photograph).

## SUPERFICIAL LOWER LEG MUSCLES, LATERAL SURFACE

### Instructions for Dissection

On the lateral surface of the leg, note once again the gastrocnemius muscle and the Achilles tendon (see Figure 21-2).

Lying next to the gastrocnemius and prominent along its distal end, is the small *soleus* muscle. Next to the soleus is the *peroneus,* a long, thin muscle, as shown in Figure 21-2.

On the anterior border, locate the *extensor digitorum longus,* which inserts by long tendons on each of the five digits. Refer to Table 21-1 for the origins, insertions, and actions of the cat's lower leg muscles.

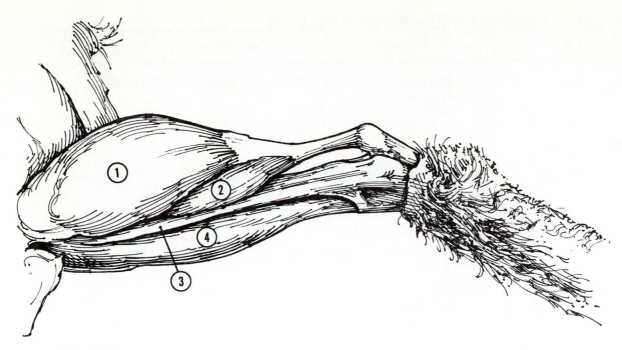

1. Gastrocnemius
2. Soleus (SŌ-lē-us)
3. Peroneus (per'-ō-NĒ-us)
4. Extensor digitorum longus

**Figure 21-2** (b) Superficial muscles of the lower leg (lateral surface) of the cat (line drawing).

**Table 21-1  Superficial Lower Leg Muscles of the Cat**

| Muscle | Origin | Insertion | Action |
|---|---|---|---|
| Gastrocnemius | Knee fascia and distal end of femur | Calcaneus by Achilles tendon | Extends foot |
| Flexor digitorum longus | Distal end of tibia and head and shaft of fibula | Tendons on all digits | Flexes digits |
| Tibialis anterior | Proximal ends of tibia and fibula | First metatarsal | Flexes foot |
| Soleus | Proximal end of fibula | Bases of metatarsals | Extends foot |
| Peroneus | Proximal end of fibula | Bases of metatarsals | Flexes foot |
| Extensor digitorum longus | Distal end of femur | All phalanges | Extends toes |

**Table 21-2  Superficial Lower Leg Muscles of the Human**

| Muscle | Origin | Insertion | Action |
|---|---|---|---|
| Gastrocnemius | Lateral and medial condyles of femur and capsule of knee | Calcaneus by Achilles tendon | Plantar flexes (extends) foot |
| Soleus | Head of fibula and medial border of tibia | Calcaneus by Achilles tendon | Plantar flexes foot |
| Peroneus longus | Head and body of fibula and lateral condyle of tibia | First metatarsal and first cuneiform | Plantar flexes and everts foot |
| Peroneus brevis | Body of fibula | Fifth metatarsal | Plantar flexes and everts foot |
| Tibialis anterior | Lateral condyle and body of tibia | First metatarsal and first cuneiform | Flexes foot dorsally and inverts it |
| Tibialis posterior | Membrane between tibia and fibula | Second, third, and fourth metatarsals; navicular; third cuneiform; cuboid | Plantar flexes and inverts foot |
| Flexor digitorum longus | Tibia | Distal phalanges of four outer toes | Flexes toes; plantar flexes and inverts foot |
| Extensor digitorum longus | Lateral condyle of tibia and anterior surface of fibula | Middle and distal phalanges of four outer toes | Extends toes, flexes foot dorsally and everts it |

## COMPARABLE HUMAN MUSCLES

The muscles of the anterior lower leg (Figure 21-3) are dorsiflexors of the ankle and extensors of the toes. They are the *tibialis anterior, extensor digitorum longus,* and *peroneus tertius.*

On the lateral surface of the leg are the peroneal muscles, which evert the foot. They may also protect against excessive inversion. Specifically, they are the *peroneus longus* and the *peroneus brevis.*

The *gastrocnemius* muscle covers the posterior surface of the lower leg and has two heads. It ends in the calcaneal tendon (Achilles tendon) and inserts on the calcaneus bone. It plantar flexes (extends) the foot. Beneath it is the *soleus,* which inserts on the calcaneus and also plantar flexes the foot. Stand on your toes and feel these muscles contract.

Muscles of this region are similar in cat and human. Learn as many human muscles as you can, and check all origins, insertions, and actions with Table 21-2. Demonstrate the action of as many muscles of the leg as you can. Try to see or feel the contraction as you make a specific movement.

The color plates on the pages following this exercise should be used, along with the figures in Exercises 13 through 18.

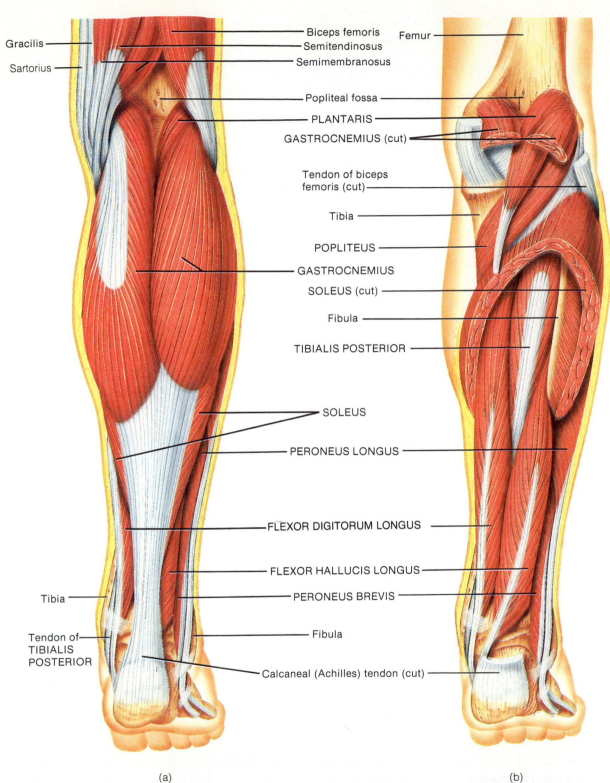

Gracilis

Sartorius

Biceps femoris

Semitendinosus

Semimembranosus

Femur

Popliteal fossa

PLANTARIS

GASTROCNEMIUS (cut)

Tendon of biceps femoris (cut)

Tibia

POPLITEUS

GASTROCNEMIUS

SOLEUS (cut)

Fibula

TIBIALIS POSTERIOR

SOLEUS

PERONEUS LONGUS

FLEXOR DIGITORUM LONGUS

FLEXOR HALLUCIS LONGUS

PERONEUS BREVIS

Tibia

Fibula

Tendon of TIBIALIS POSTERIOR

Calcaneal (Achilles) tendon (cut)

(a)

(b)

**Figure 21-3** Muscles that move the foot and toes: (a) superficial posterior view; (b) deep posterior view.

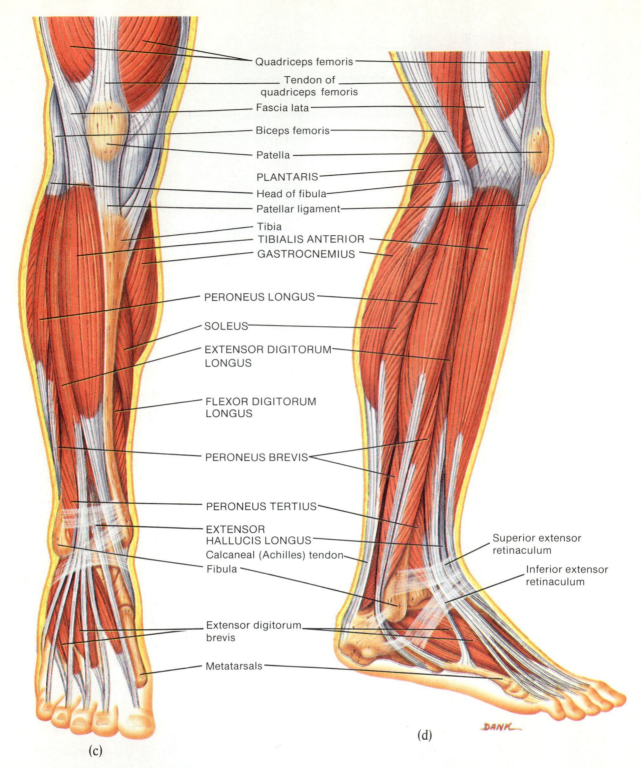

Quadriceps femoris
Tendon of quadriceps femoris
Fascia lata
Biceps femoris
Patella
PLANTARIS
Head of fibula
Patellar ligament
Tibia
TIBIALIS ANTERIOR
GASTROCNEMIUS
PERONEUS LONGUS
SOLEUS
EXTENSOR DIGITORUM LONGUS
FLEXOR DIGITORUM LONGUS
PERONEUS BREVIS
PERONEUS TERTIUS
EXTENSOR HALLUCIS LONGUS
Calcaneal (Achilles) tendon
Fibula
Extensor digitorum brevis
Metatarsals
Superior extensor retinaculum
Inferior extensor retinaculum

(c)

(d)

DANK

**Figure 21-3 *(Continued)*** (c) superficial anterior view; (d) right lateral superficial view. (From G. J. Tortora and S. R. Grabowski, *Principles of Anatomy and Physiology,* 7th ed., New York: HarperCollins, 1993. © Leonard Dank.)

## MUSCLES OF THE LOWER LEG

Some of the muscles that move the foot, ankle, and toes are the following:

1. *Gastrocnemius* (Figure 21-4)
   This is the largest muscle of the lower leg. It plantar flexes the foot and flexes the knee.
   a. To demonstrate this action, stand on your tip-toes. Feel the contraction.

**b.** In a sitting position, with both feet flat on the floor, raise your heels, keeping the ball of each foot on the floor. Feel the contraction of the entire muscle with your hand.

**Figure 21-4** Gastrocnemius (posterior view). (From P. J. Donnelly and G. A. Wistreich, *Exploring the Human Body,* New York: Harper & Row, 1989.)

2. *Peroneus longus* (Figure 21-5)
This muscle plantar flexes the foot and abducts and everts the foot.

To demonstrate this action, from a sitting position, raise your right foot, and turn it (evert) laterally. Place your fingers on the lateral surface of your leg, to feel the muscle contraction.

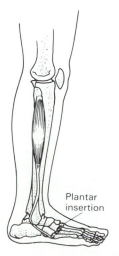

**Figure 21-5** Peroneus longus (lateral view). (From P. J. Donnelly and G. A. Wistreich, *Exploring the Human Body,* New York: Harper & Row, 1989.)

3. *Tibialis anterior* (Figure 21-6)
The tibialis dorsiflexes and inverts the foot.

To demonstrate this action, from a sitting position, raise your right leg slightly and point your toes upward. Then turn your foot medially (inversion). Place your fingers on the anterior surface of your leg to feel the contraction.

**Figure 21-6** Tibialis anterior (anterior view). (From P. J. Donnelly and G. A. Wistreich, *Exploring the Human Body,* New York: Harper & Row, 1989.)

4. *Flexor digitorum longus* (Figure 21-7)
This flexor muscle bends the toes two through five. It also plantar flexes the foot and inverts the foot.

To demonstrate this action, from a sitting position, raise your right foot, and flex the second through fifth toes. Then turn your foot inward (inversion) and plantar flex it.

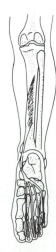

**Figure 21-7** Flexor digitorum longus (posterior view). (From P. J. Donnelly and G. A. Wistreich, *Exploring the Human Body,* New York: Harper & Row, 1989.)

5. *Extensor digitorum longus* (Figure 21-8)

This muscle extends the toes, dorsiflexes the foot, and everts the foot. The body of the muscle cannot be felt, but the tendons that insert on the dorsal surface of the phalanges of the second through fifth toes can be seen and felt.

To demonstrate this action, from a sitting position, raise your right foot, and straighten (extend) those toes. Then move the foot toward you (dorsiflexion), and turn it outward (eversion). If you cannot see the tendons, try to spread your toes.

**Figure 21-8** Extensor digitorum longus (anterior view). (From P. J. Donnelly and G. A. Wistreich, *Exploring the Human Body*, New York: Harper & Row, 1989.)

# LABORATORY REVIEW *21*

## *MUSCLES OF THE LOWER LEG*

1. Color the muscles of the lower leg in Figure Q21-1. Review the origin, insertion, and action of each muscle. Demonstrate as many of them as possible on your own leg.

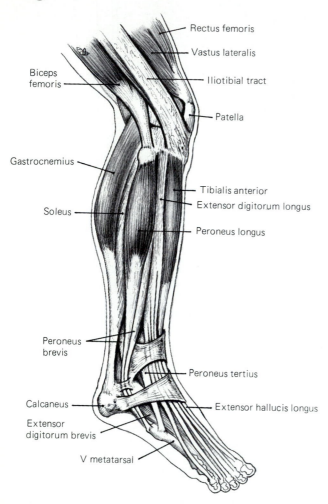

Rectus femoris

Vastus lateralis

Biceps femoris

Iliotibial tract

Patella

Gastrocnemius

Tibialis anterior

Extensor digitorum longus

Soleus

Peroneus longus

Peroneus brevis

Peroneus tertius

Calcaneus

Extensor hallucis longus

Extensor digitorum brevis

V metatarsal

**Figure Q21-1**

2. Figure Q21-2 reviews the muscles of the hip, thigh, and lower leg. Identify as many muscles as you can in the spaces provided. Color the drawings.

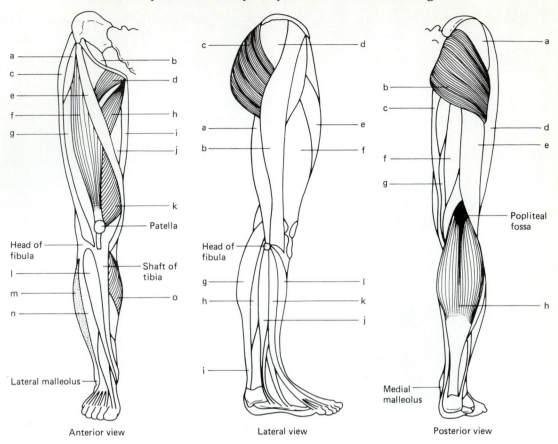

Anterior view          Lateral view          Posterior view

**Figure Q21-2** (From P. J. Donnelly, *Laboratory Manual for Human Anatomy,* revised ed., New York: Harper & Row, 1988.)

**ANTERIOR VIEW**

a. _____   f. _____   k. _____

b. _____   g. _____   l. _____

c. _____   h. _____   m. _____

d. _____   i. _____   n. _____

e. _____   j. _____   o. _____

**LATERAL VIEW**

a. _____   e. _____   i. _____

b. _____   f. _____   j. _____

c. _____   g. _____   k. _____

d. _____   h. _____   l. _____

**POSTERIOR VIEW**

a. _____   d. _____   g. _____

b. _____   e. _____   h. _____

c. _____   f. _____

# Cat Muscle Dissection
## Color Plates

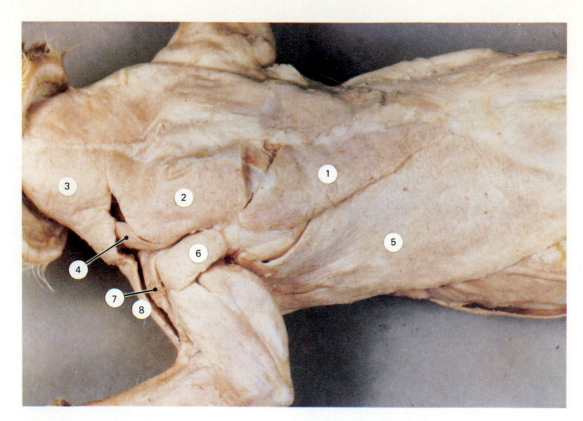

**Plate 16-1** Superficial back muscles.

1. Spinotrapezius
2. Acromiotrapezius
3. Clavotrapezius
4. Levator scapulae ventralis
5. Latissimus dorsi
6. Spinodeltoid
7. Acromiodeltoid
8. Clavodeltoid or clavobrachialis

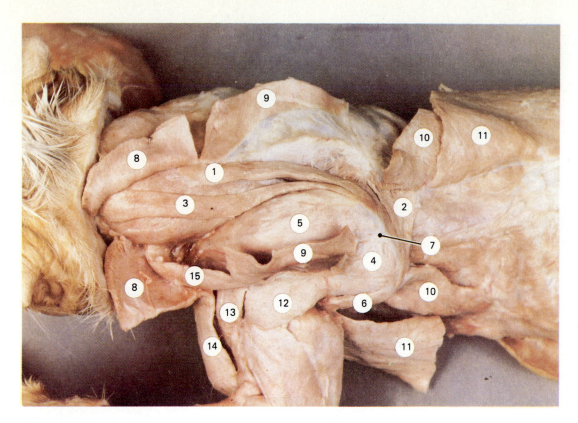

**Plate 16-2** Muscles associated with scapula.

1. Rhomboideus minor
2. Rhomboideus major
3. Rhomboideus capitis or occipitoscapularis
4. Infraspinatus
5. Supraspinatus
6. Teres major
7. Scapular spine (bone)
8. Clavotrapezius (reflected)

9. Acromiotrapezius (reflected)
10. Spinotrapezius (reflected)
11. Latissimus dorsi
12. Spinodeltoid
13. Acromiodeltoid
14. Clavodeltoid or clavobrachialis
15. Levator scapulae ventralis

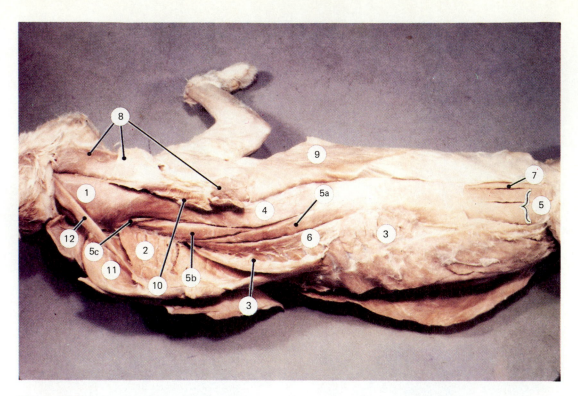

**Plate 16-3** Deep back muscles.

1. Splenius
2. Serratus ventralis
3. Serratus dorsalis (partially reflected)
4. Spinalis dorsi
5. Longissimus dorsi (posterior part)
   a. Longissimus thoracis
   b. Longissimus cervicis
   c. Longissimus capitis

6. Iliocostalis
7. Multifidus spinae
8. Trapezius group (reflected)
9. Latissimus dorsi (reflected)
10. Rhomboideus (cut end)
11. Subscapularis
12. Rhomboideus capitis

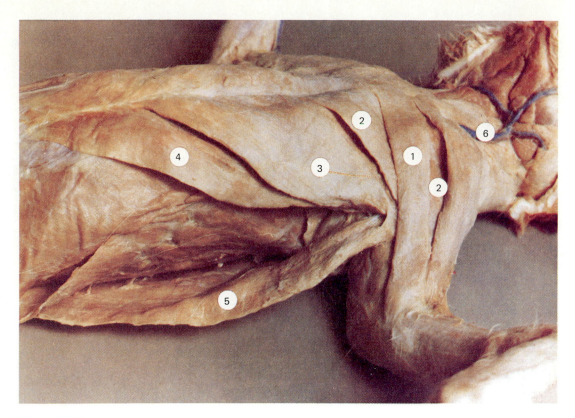

**Plate 17-1** Pectoral muscles.

1. Pectoantebrachialis
2. Pectoralis major
3. Pectoralis minor
4. Xiphihumeralis
5. Latissimus dorsi
6. Jugular vein and tributaries

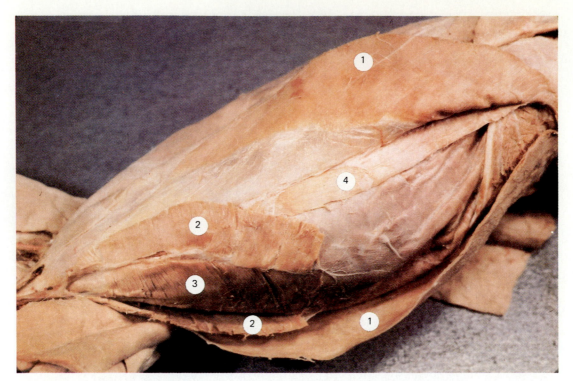

**Plate 17-2** Abdominal muscles.

1. External oblique (reflected)
2. Internal oblique (reflected)
3. Transversus abdominis
4. Rectus abdominis

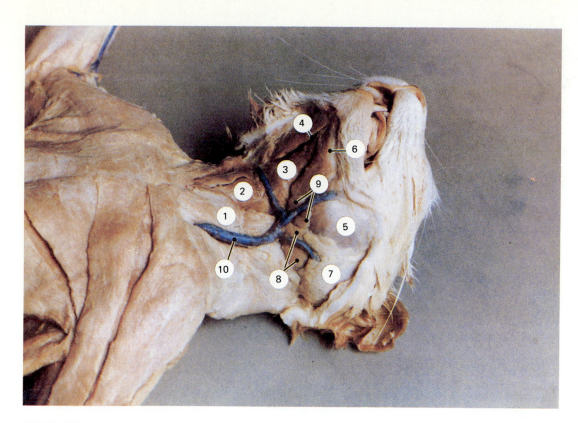

**Plate 18-1** Superficial neck and jaw muscles.

1. Sternomastoid
2. Sternohyoid
3. Mylohyoid
4. Digastric
5. Masseter

6. Mandible (bone)
7. Parotid gland
8. Submandibular gland
9. Lymph nodes
10. External jugular vein and tributaries

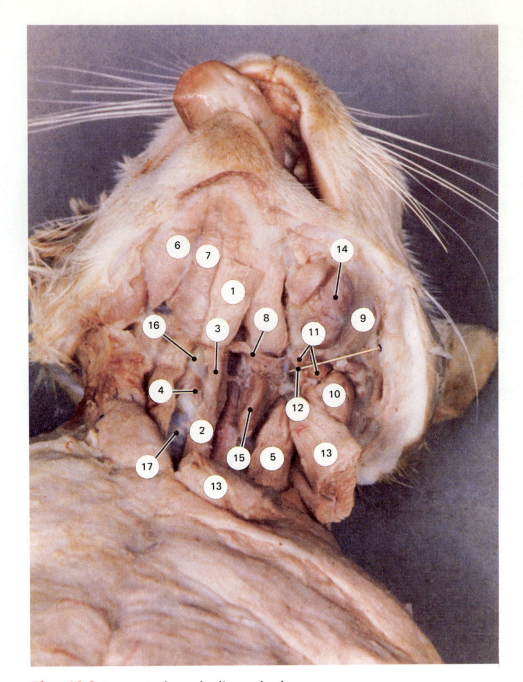

**Plate 18-2** Deep muscles and salivary glands.

1. Sternohyoid (reflected)
2. Sternothyroid
3. Thyrohyoid
4. Cricothyroid
5. Cleidomastoid
6. Digastric
7. Mylohyoid
8. Stylohyoid
9. Masseter

10. Submaxillary gland
11. Sublingual gland
12. Salivary duct
13. Sternomastoid (reflected)
14. Lymph nodes (reflected)
15. Lymph node
16. Larynx
17. Trachea

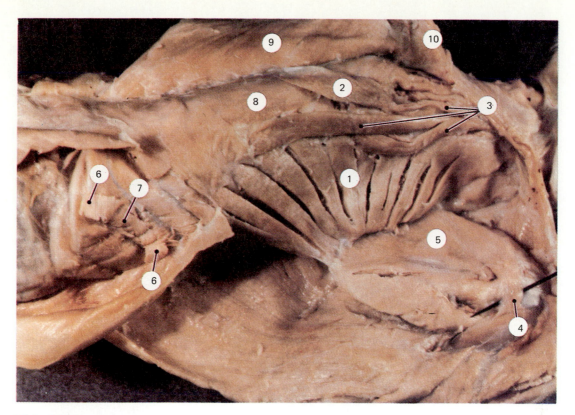

**Plate 19-1** Deep chest muscles.

1. Serratus ventralis
2. Transversus costarum
3. Scalenus
4. Coracobrachialis
5. Subscapularis

6. External intercostal (reflected)
7. Internal intercostal
8. Rectus abdominis
9. Pectoralis minor (reflected)
10. Pectoralis major (reflected)

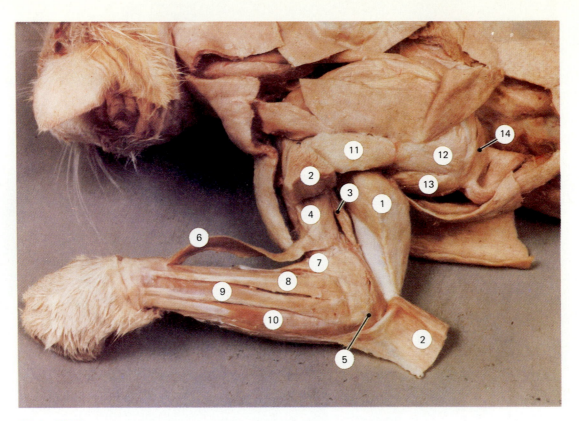

**Plate 19-2** Superficial arm muscles, lateral surface.

1. Triceps brachii, long head
2. Triceps brachii, lateral head (reflected)
3. Triceps brachii, medial head
4. Brachialis
5. Anconeus
6. Brachioradialis
7. Extensor carpi radialis
8. Extensor digitorum communis
9. Extensor digitorum lateralis
10. Extensor carpi ulnaris
11. Spinodeltoid
12. Infraspinatus
13. Teres major
14. Rhomboideus major

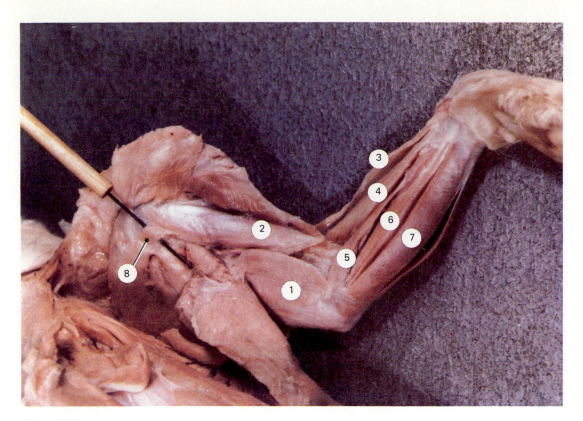

**Plate 19-3** Superficial arm muscles, medial view.

1. Epitrochlearis
2. Biceps brachii
3. Brachioradialis
4. Extensor carpi radialis
5. Pronator teres
6. Flexor carpi radialis
7. Palmaris longus
8. Coracobrachialis

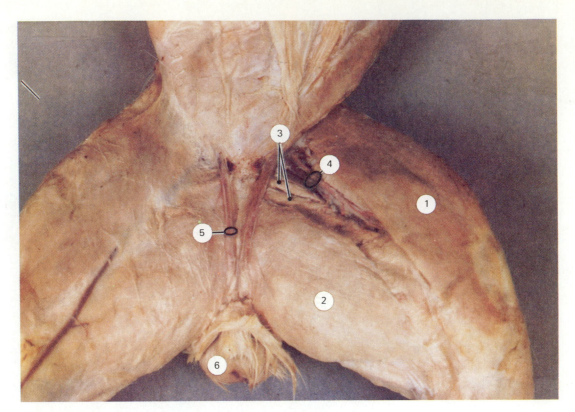

**Plate 20-1** Superficial muscles of thigh, medial view.

1. Sartorius
2. Gracilis
3. Adductor muscles
4. Blood Vessels
5. Spermatic cord
6. Scrotum

Femoral triangle

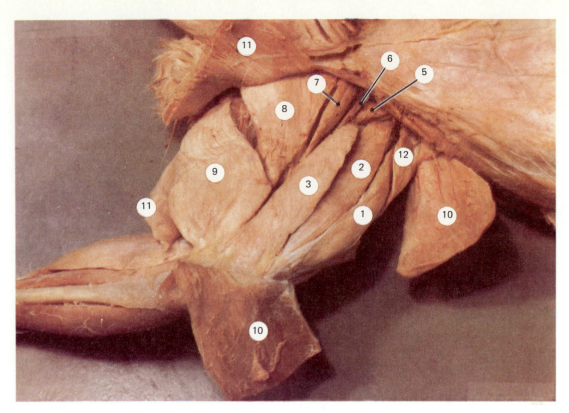

**Plate 20-2** Deep muscles of medial thigh.

1. Vastus lateralis
2. Rectus femoris
3. Vastus medialis
4. Vastus intermedius (not shown)

                                                  Quadriceps femoris

5. Iliopsoas
6. Pectineus
7. Adductor longus
8. Adductor femoris
9. Semimembranosus
10. Sartorius
11. Gracilis
12. Tensor fasciae latae

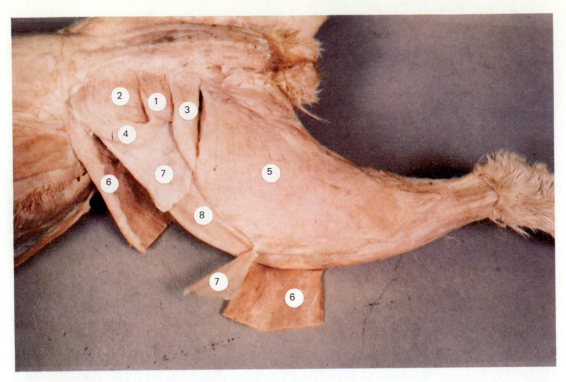

**Plate 20-3** Muscles of hip and thigh.

1. Gluteus maximus
2. Gluteus medius
3. Caudofemoralis
4. Tensor fasciae latae

5. Biceps femoris
6. Sartorius
7. Fascia lata (transected)
8. Vastus lateralis

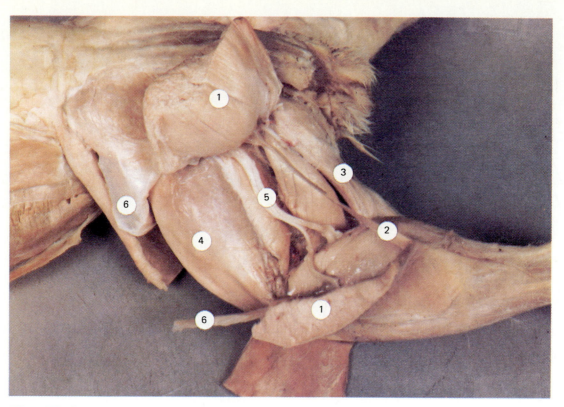

**Plate 20-4** Deep muscles of thigh, lateral view.

1. Biceps femoris (reflected)
2. Tenuissimus
3. Semitendinosus
4. Vastus lateralis
5. Sciatic nerve
6. Tensor fasciae latae (fascia transected)

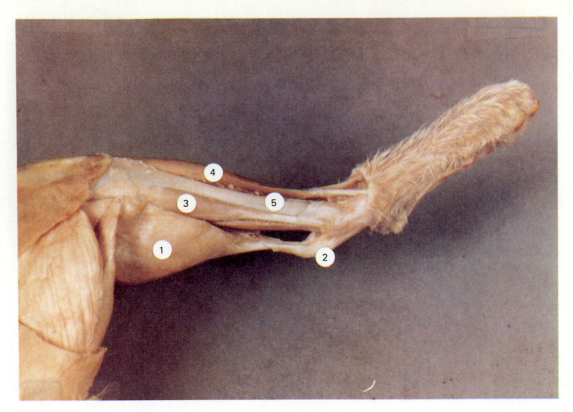

**Plate 21-1** Superficial muscles of lower leg, medial view.

1. Gastrocnemius
2. Tendon of Achilles
3. Flexor digitorum longus

4. Tibialis anterior
5. Tibia (bone)

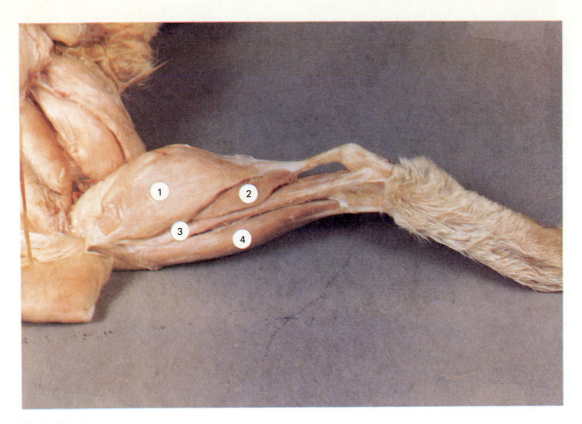

**Plate 21-2** Superficial muscles of lower leg, lateral view.

1. Gastrocnemius
2. Soleus

3. Peroneus
4. Extensor digitorum longus

# Muscle Ultrastructure and Chemistry

## After completing this exercise, you should be able to

1. Distinguish—on the basis of microscopic examination—among smooth, skeletal, and cardiac muscle
2. List the four protein components in muscle, and give the functions of each in muscle contraction
3. Interpret the striated pattern of skeletal muscle fibers in an electron micrograph
4. Describe the effects of temperature, adenosine triphosphate (ATP), potassium chloride (KCl), magnesium chloride ($MgCl_2$), and pH on muscular contraction

---

Skeletal muscle consists of thousands of elongated, cylindrical cells termed *myofibers* or muscle fibers. Each individual muscle fiber is covered by a plasma membrane (or *sarcolemma*) that surrounds the muscle cell's cytoplasm, known as *sarcoplasm*. Many nuclei are located within the sarcoplasm and close to the sarcolemma (Color Plates 20 and 21).

An electron microscopic view of skeletal muscle fibers reveals that they contain threadlike, cylindrical structures termed *myofibrils*. These myofibrils run longitudinally within the muscle fiber and consist of two types of smaller structures—thin and thick *myofilaments*. Alternating thick and thin myofilaments are stacked parallel to one another in units termed *sarcomeres*. (Figure 22-1).

The sarcomere, which measures 2–3 μm in length, is bounded at both ends by a narrow zone of dense material termed a *Z line*. Along the length of a muscle cell, there are other conspicuous bands, zones, or striations that are repeated regularly. These include the rel-atively clear, broad, *I* (iostropic) *band,* which is bisected by the Z line; and the large, dark, dense area termed the *A* (anisotropic) *band* in the center of the sarcomere (Figure 22-1). The muscle's striated appearance is due to the combination of the alternating dark A bands and the lighter I bands.

Thin myofilaments contain three proteins, the most abundant of which is *actin*. Actin gives thin myofilaments their characteristic shape. The two other proteins are *tropomyosin* and *troponin*. Together, these molecules are referred to as the *tropomyosin–troponin complex,* and they are involved in the regulation of muscle contractions.

Thick myofilaments overlap the free ends of the thin myofilaments. *Myosin* is the major protein found with thick myofilaments. In intact muscle, one end of the thin filaments is attached to the Z line. These thin filaments extend perpendicularly from both of its surfaces. Transmission electron micrographs show that sarcomeres consist of a set of thick filaments arranged in the spaces between the thin filaments projecting from the Z lines. The two types of filament are arranged in parallel groups and overlap for part of their lengths (Figures 22-1(c) and 22-2). The spacing of the filaments is such that the two sets can slide past one another without restriction. This sliding motion is responsible for the shortening reaction during muscle contraction. All events associated with the movement of myofilaments are known as the *sliding-filament theory* (Figure 22-2).

A stimulus must be applied to skeletal muscle for it to contract. Such a stimulus is delivered by a *neuron* or nerve cell, specifically known as a *motor neuron.* A long threadlike process of the neuron, termed an *axon,* branches into terminals that come into close association with a muscle fiber's sarcolemma (Color Plate 22). The axon terminal of the motor neuron, together with the portion of the sarcolemma involved, form a *motor end plate* (Figure 22-3).

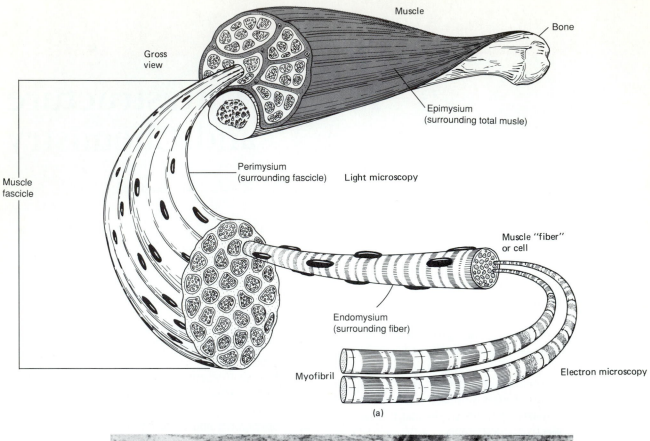

Gross view

Muscle

Bone

Epimysium (surrounding total musle)

Perimysium (surrounding fascicle)

Light microscopy

Muscle fascicle

Muscle "fiber" or cell

Endomysium (surrounding fiber)

Myofibril

Electron microscopy

(a)

A

H

Z

Z

Sarcomere

(b)

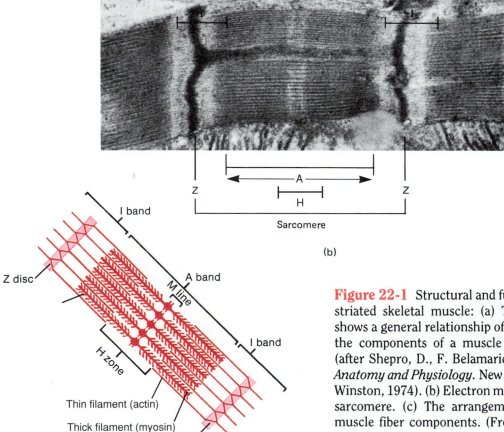

I band

Z disc

A band

M line

I band

H zone

Thin filament (actin)

Thick filament (myosin)

Z disc

(c)

**Figure 22-1** Structural and functional components of striated skeletal muscle: (a) This composite diagram shows a general relationship of skeletal muscle to bone, the components of a muscle fiber, and a sarcomere (after Shepro, D., F. Belamarich, and C. Levy. *Human Anatomy and Physiology.* New York: Holt, Rinehart and Winston, 1974). (b) Electron micrograph of a cylindrical sarcomere. (c) The arrangement and relationship of muscle fiber components. (From G. J. Tortora and S. R. Grabowski, *Principles of Anatomy and Physiology,* 7th ed., New York: HarperCollins, 1993.)

2 Sarcomeres

I band | A band

H zone

Thick filament

Z disc

Thin filament

Z disc

Z disc

(a)

(b)

(c)

**Figure 22-2** Selected positions of sarcomeres showing the sliding-filament theory of muscle contraction in action: (a) the relaxed state; (b) contracting; (c) fully or maximally contracted. Note that the lengths of both types of myofilaments do not change. (From G. J. Tortora and S. R. Grabowski, *Principles of Anatomy and Physiology,* 7th ed., New York: HarperCollins, 1993.)

The energy for muscle contraction is provided by the hydrolysis of ATP. In addition to the four proteins mentioned earlier, an ATP muscle contraction requires the presence of calcium ions. Calcium provides a means of regulation.

This exercise considers the microscopic properties of muscle and the particular effects that selected chemical and physical factors have on muscle contraction.

## MICROSCOPIC FEATURES OF SKELETAL MUSCLE

### Materials

1. The following prepared slides should be provided for class use:
   a. Teased skeletal muscle fibers
   b. Longitudinal section of skeletal muscle
2. Prepared slides stained to differentiate motor end plates should be provided for a class demonstration.

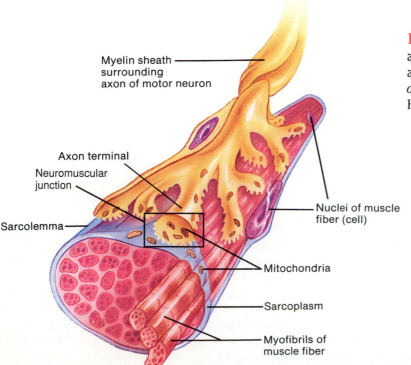

Myelin sheath surrounding axon of motor neuron

Axon terminal

Neuromuscular junction

Sarcolemma

Nuclei of muscle fiber (cell)

Mitochondria

Sarcoplasm

Myofibrils of muscle fiber

**Figure 22-3** A diagrammatic view of a motor end plate. (From G. J. Tortora and S. R. Grabowski, *Principles of Anatomy and Physiology,* 7th ed., New York: HarperCollins, 1993.)

## Procedure

This procedure is to be performed by students individually.

1. Examine a prepared slide of muscle under both low- and high-power objectives. (Refer to Color Plates 5 and 20).

2. Select a single fiber, and identify the following components: *muscle cell nuclei, sarcolemma* (a thin membrane that covers each fiber), *A bands* (repeating dark zones, perpendicular to the long axis of the muscle fiber), and *I bands* (repeating light zones, perpendicular to the long axis of the muscle fiber).

3. Sketch a representative field, showing the components listed in Step 2 in the subsequent results and observations section.

4. Examine a prepared slide of a longitudinal section of skeletal muscle. Identify the following components: *muscle cell nuclei, endomysium* (connective tissue layer surrounding each muscle fiber and external to the sarcolemma), and *local areas of contraction* (darker areas along the length of a muscle fiber).

5. Sketch a representative field showing the components listed in Step 4 in the results and observations section.

6. Examine the demonstration slide showing motor end plates. Locate the following: a *motor end plate, nerve fiber,* and *nerve fiber branches*. Note that each muscle fiber has only one end plate (refer to Color Plate 22).

7. Sketch and label a representative field in the results and observations section.

8. Answer the questions pertaining to this procedure in the results and observations section.

## THE CHEMISTRY OF MUSCLE CONTRACTION

### Materials

The following items should be provided for class use:

1. Rabbit psoas muscle strips in 50 percent glycerol (obtainable from Carolina Biological Supply Company)

2. The following prepared solutions in containers with droppers:
   a. Adenosine triphosphate (ATP)
   b. Magnesium chloride ($MgCl_2$)
   c. Potassium chloride (KCl)
   d. Three separate glycerol preparations maintained at 4°C, 25°C (room temperature), and 37°C
   e. Three separate Ringer's preparations, with pH of 5, 7, and 8, maintained at 37°C

3. Microscope slides
4. Forceps
5. Millimeter rulers
6. Microscopes
7. Water bath maintained at 37°C
8. Ice bath or refrigerator to maintain appropriate solutions at 4°C
9. Eye droppers
10. Stopwatch or other timing device

### Procedure 1: Effects of Temperature, Adenosine Triphosphate (ATP), Potassium Chloride (KCl), and Magnesium Chloride ($MgCl_2$) on Muscle Contraction

This procedure should be performed by students in groups of four.

1. On a glass slide, place two rabbit psoas muscle fibers in a small amount of the glycerol preparation kept at 25°C (room temperature). Examine them under low power. Are striations and nuclei evident?

2. Measure the length of each fiber in millimeters. Enter your findings in the results and observations section.

3. Remove the glycerol, and replace it with the glycerol preparation kept at 37°C. Wait 30 seconds. Measure the length of the fibers, and enter your findings. If the length is different from the room-temperature measurement at 25°C, calculate the percentage of contraction by dividing the contraction distance by the fiber length at room temperature. Enter your findings in the results and observations section.

4. Next, replace the glycerol with the preparation kept at 4°C. Wait 30 seconds. Measure the length of the fibers, and enter your findings. If this length is different from the length at 25°C calculate the percentage of contraction, as described in Step 3, and enter your findings.

5. Replace the glycerol with the preparation kept at 25°C. Wait 30 seconds, and repeat Steps 3 and 4, using the 4°C and 37°C glycerol solutions for 60 seconds.

6. Now replace the glycerol with the preparation kept at 25°C. Wait 30 seconds, and measure the length of the fibers again. This measurement will be used as a baseline value for calculations of the effects of

chemicals added to muscle preparation in the following steps.

7. Replace the glycerol with the 4°C preparation, and after 30 seconds cover the fibers with the KCl solution. Wait 30 seconds, and measure the length of the fibers. If there is a difference, calculate the percentage of contraction, as described in Step 3. Enter your findings in the results and observations section.

8. Repeat Step 7, using the following preparations separately, first with the 4°C and then with the 37°C glycerol preparations. Enter your findings in Table 22-1.
   a. $MgCl_2$
   b. ATP
   c. A mixture of KCl, $MgCl_2$, and ATP

9. Repeat Steps 7 and 8, exposing the psoas muscle to the different preparations for 60 seconds each.

## Procedure 2: Effects of pH on Muscle Contraction

This procedure should be performed by students in groups of four.

1. Place two rabbit psoas muscle fibers in a small amount of glycerol (37°C) on a glass slide.
2. Measure the length of each fiber in millimeters. Enter your findings in the results and observations section.
3. Replace the glycerol with the pH 7 Ringer's solution, wait 30 seconds, measure the length of the fibers, and enter your findings in Table 22-2.
4. Add to the fibers a mixture of KCl, $MgCl_2$, and ATP. Wait 30 seconds, measure the length of the fibers, and enter your findings.
5. Repeat Steps 3 and 4, using the Ringer's solution with pH 5. If there is a difference in contraction distance from that obtained with the pH 7 solution, calculate the percentage of contraction by dividing the contraction distance obtained at pH 5 by the fiber length at pH 7.
6. Repeat Step 5 with the Ringer's solution with pH 8.

## RESULTS AND OBSERVATIONS

### *Microscopic Features of Muscle*

1. Sketch and label representative microscopic fields of the muscle specimens provided.

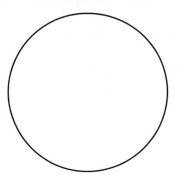

Teased skeletal
muscle fibers

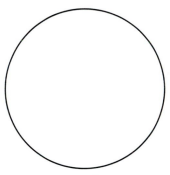

Longitudinal
section

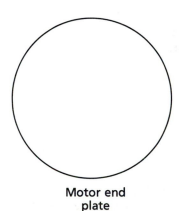

Motor end
plate

2. List three areas in the human body where smooth muscle characteristically is found.

   a. _____

   b. _____

   c. _____

3. Is striated muscle multinucleated? _____

### Effects of Temperature, ATP, KCl, MgCl₂, and pH on Muscle Contraction

1. Enter your findings in the appropriate column of Table 22-1. Note that this table combines the results obtained from Procedure 1 in this part of the exercise.

2. Enter your findings obtained from Procedure 2 in Table 22-2.

3. Was muscular contraction affected by temperature? In what way? _____ _____

**Table 22-1**

| Muscle Fiber Condition | Length at 30 Seconds (in mm) | Percentage of Contraction (if applicable) | Length at 60 Seconds (in mm) | Percentage of Contraction (if applicable) |
|---|---|---|---|---|
| Glycerol (25°C) | | | | |
| Glycerol (4°C) | | | | |
| ATP and glycerol (4°C) | | | | |
| KCl and glycerol (4°C) | | | | |
| MgCl₂ and glycerol (4°C) | | | | |
| ATP, KCl, MgCl₂, and glycerol (4°C) | | | | |
| Glycerol (37°C) | | | | |
| ATP and glycerol (37°C) | | | | |
| ATP and glycerol (4°C) | | | | |
| KCl and glycerol (37°C) | | | | |
| KCl and glycerol (4°C) | | | | |
| MgCl₂ and glycerol (37°C) | | | | |
| MgCl₂ and glycerol (4°C) | | | | |
| ATP, KCl, MgCl₂, and glycerol (37°C) | | | | |
| ATP, KCl, MgCl₂, and glycerol (4°C) | | | | |

**Table 22-2**

| Muscle Fiber Condition | Length at 30 Seconds (in mm) | Percentage of Contraction (if applicable) |
|---|---|---|
| Glycerol (37°C) | | |
| Ringer's solution (pH 7) | | |
| Ringer's solution (pH 7) and ATP, KCl, MgCl$_2$ | | |
| Ringer's solution (pH 5) | | |
| Ringer's solution (pH 5) and ATP, KCl, MgCl$_2$ | | |
| Ringer's solution (pH 8) | | |
| Ringer's solution (pH 8) and ATP, KCl, MgCl$_2$ | | |

4. Was muscular contraction affected by pH? In what way? _____

_____

5. Was muscular contraction affected by the absence of ATP? In what way? _____

_____

6. What is the role of calcium in muscle contraction?

_____

_____

# LABORATORY REVIEW 22

## MUSCLE ULTRASTRUCTURE AND CHEMISTRY

### COMPLETION AND DEFINITION

1. Draw a simple diagram to show the relationships among the components of a skeletal muscle, including muscle fibers, myofibrils, and sarcomere.

2. Draw a simple diagram to show the A bands, I bands, H zones, and Z lines within a sarcomere. (Include the overlapping of myosin and actin filaments.)

3. List four major proteins involved in muscle contraction.

   a. _____   c. _____

   b. _____   d. _____

## ANSWERS

4. Briefly describe the events taking place in a muscle contraction. Attach a separate sheet, if diagrams are used. _____

_____

_____

5. Define or explain:

    a. Sarcomere _____

    _____

    b. Myofibril _____

    _____

    c. Motor end plate _____

    _____

    d. Skeletal muscle _____

    _____

    e. Actin _____

    _____

    f. Myosin _____

    _____

## MULTIPLE CHOICE

Insert the correct answers in the spaces provided.

1. The ability of muscle tissue to receive and respond to stimuli is referred to as which of the following?
    a. tetanus
    b. contraction
    c. extension
    d. excitability
    e. relaxation
2. The sections of a muscle fiber separated by *two* lines are known as?
    a. sarcolemma
    b. sarcomere
    c. synapse
    d. T tubule
    e. myofilament
3. Which of the following is the plasma membrane surrounding a muscle fiber?
    a. sarcolemma
    b. sarcomere
    c. synapse
    d. T tubule
    e. myofilament
4. Which of the following tissues is striated and voluntary?
    a. smooth muscle
    b. cardiac muscle
    c. skeletal muscle
    d. visceral muscle
    e. heart muscle

1. _____

2. _____

3. _____

4. _____

5. _____

6. _____

7. _____

8. _____

9. _____

10. _____

5. The sarcomere portion consisting of thin myofilaments only is which of the following?
   a. H band
   b. Z line
   c. A band
   d. I band
   e. S zone

6. What is the connective tissue layer that surrounds muscle fiber bundles known as?
   a. myomysium
   b. Z band
   c. ectomysium
   d. perimysium
   e. endomysium

7. What muscle tissue is known to be nonstriated and involuntary?
   a. skeletal muscle
   b. cardiac muscle
   c. visceral muscle
   d. smooth muscle

8. Which of the following is known as the sarcomere portion where thin and thick myofilaments overlap?
   a. H band
   b. Z line
   c. A band
   b. I band
   e. S zone

9. Which of the following is known as the sarcomere portion in a sarcomere containing thick myofilaments only?
   a. H band
   b. Z line
   c. A band
   d. I band
   e. S zone

10. Which of the following is the muscle tissue known to contain intercalated discs?
    a. smooth muscle
    b. cardiac muscle
    c. skeletal muscle
    d. visceral muscle
    e. striated muscle

# Introduction to Skeletal Muscle Contraction, Using a Laboratory Animal

**After completing this exercise, you should be able to**

1. Define or explain *myofibril, sarcoplasm, stimulus,* and the *all-or-none response*
2. Describe the events associated with a single muscle twitch and tetany
3. Prepare a laboratory animal for the demonstration of skeletal muscle contraction
4. Use appropriate equipment and stimuli to demonstrate a single muscle twitch, summation and tetanic contraction, determination of work, and muscle fatigue

Muscle is the most common tissue in the body of most vertebrates. As described in Exercise 22, it consists of *elongated cells* or *fibers* specialized for contraction. The contractile components within the muscle fiber are the *myofibrils*. These fibrils run lengthwise through the unspecialized cytoplasm of muscles or the *sarcoplasm*. Muscle bundles, termed *fascicles,* are formed through the joining of fibers to other parallel muscle cells by intertwining connective tissue. The fascicles, in turn, are organized into larger units referred to as *muscle organs.*

The ability of muscle to respond to stimulus is termed *irritability.* The response takes the form of a contraction. The stimulus is normally an impulse transmitted to the muscle cell membrane by the axon, or nerve fiber, of a neuron. Each fiber supplies from a few up to more than a hundred individual muscle cells. As described in Exercise 22, the portion of the membrane directly under the end of the axon is termed a *myoneural* or *motor end plate* (Color Plate 22). The impulse causes changes within the myofibril and shortening of the sarcomere and the entire muscle cell.

In response to a single stimulus, individual muscle fibers either contract completely or do not contract at all (according to the *all-or-none-law*). Muscular contraction depends on how frequently nerve impulses are received by the muscle fibers and how many fibers are stimulated. The number of individual fibers contracting also determines the strength of a muscle contraction.

Skeletal muscle contracts only when it has been stimulated adequately. It does not contract spontaneously. In this exercise, isolated muscles from a frog are used to demonstrate how a whole muscle contracts and how changes in a stimulus can modify muscular contractions. Muscle used for such study is isolated and attached to appropriate equipment in such a way as to record contractions. An electrical stimulus either to the muscle itself or to an associated nerve produces a recorded response.

A single stimulus produces a single brief contraction known as a *muscle twitch*. An examination of a recorded response (Figure 23-1) reveals three principal events: (1) the *latent period*, which is the period between the stimulus and the beginning of contraction; (2) the *contraction period;* and (3) the *relaxation period*. The extent of the twitch depends on the stimulus applied. The stimulus must reach some *threshold level* before the muscle responds at all. Increased stimulus increases to the event of contraction until some maximum is obtained. As the stimulus is increased, more fibers are activated until all of them are contracting in response to the stimulus. Once a muscle is stimulated enough to contract, it briefly loses its ability to respond to additional stimuli. This is known as the *refractory period*.

If a second stimulus is applied after the refractory period, but before the contraction is over, the muscle will respond to both, and the additive effect of both stimuli will produce a stronger contraction. This is

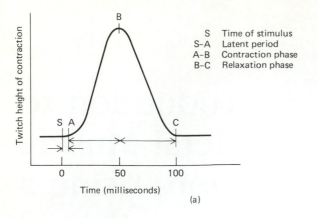

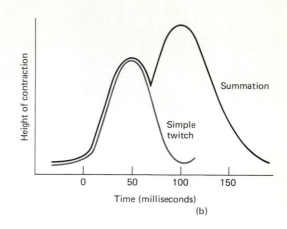

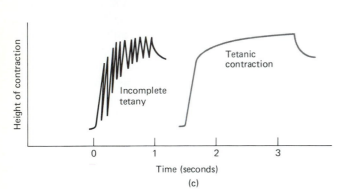

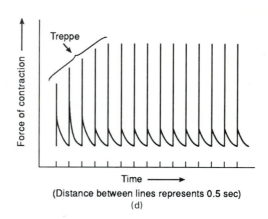

**Figure 23-1** Muscle contraction responses: (a) muscle twitch; (b) summation of contraction; (c) tetanic contraction; (d) treppe.

known as a *summation of twitches* or *wave summation* (Figure 23-1b).

The application of two successive stimuli of maximum intensity causes two separate maximal contractions. If successive stimuli are applied so quickly that the muscle can relax only partially between stimuli, it remains in a state of contraction. This summation and fusion of individual twitches is known as *incomplete tetanus* (Figure 23-1c). *Complete tetanus* is the condition in which stimuli are applied so quickly that the muscle does not relax at all. Tetany is an important action because most muscular contractions are short-term tetanic contractions, and are thus smooth sustained contractions. Skeletal muscle contractions are all tetanic in nature, even the blinking of an eye.

The condition known as *treppe* occurs when a skeletal muscle contracts more forcefully in response to the same strength of stimulus after it has contracted several times (Figure 23-1d). It can be demonstrated by stimulating an isolated muscle with a series of stimuli at the same intensity and frequency, but not at a rate fast enough to produce tetanus (Figure 23-1c).

## SPECIMEN PREPARATION

A laboratory animal is used in this exercise, to demonstrate some fundamental physiological principles. To ensure the success of the experiments, care must be exercised to avoid unnecessary injury, irritation, or pain to the laboratory animal. Any of these conditions will render the animal less capable of "normal" responses.

In this exercise, the living frog is prepared by a procedure known as *pithing*. Refer to Appendix B for the "pithing procedure."

### Materials

The following materials should be provided per two students or for instructor's demonstration:

1. One normal, healthy single-pithed frog
2. Dissecting tools (needle, scalpel, probe, scissors)
3. Three sterile gauze pads (2 × 2 inch)
4. Frog Ringer's solution
5. Thread

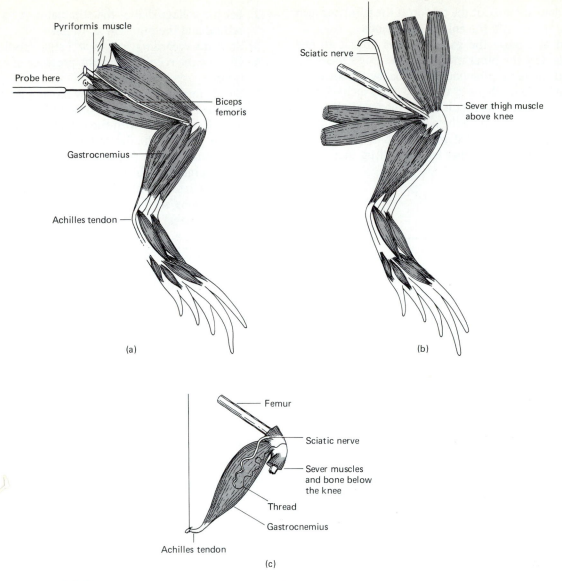

**Figure 23-2** Diagrammatic representation of the musculature of the dorsal surface of a frog's leg: (a) musculature of dorsal surface; (b) relationship of sciatic nerve and gastrocnemius muscle; (c) appearance of muscle preparation.

## Procedure: Preparation of Muscle

This procedure should be performed by students in pairs.

1. Remove the skin from one leg of the prepared frog by first cutting the skin as high as possible around the thigh. Roll the loosened skin back a short distance. Grasp it with one hand and the frog's thigh with the other. Peel the skin completely off the leg by pulling it quickly.

2. Use Figure 23-2a to locate the *gastrocnemius muscle*. Keep the frog's tissues moist with Ringer's solution during the remainder of the exercise.

3. Carefully insert a probe to separate the thigh muscles and to expose the sciatic nerve (Figure 23-2b).

4. Tie a piece of thread about 10 cm long (4 inches) around the sciatic nerve, and cut the nerve above the tie.

5. With the thread, gently ease the nerve from its location, and lay it on the surface of the gastrocnemius muscle.

6. Cut the thigh muscles first near their origins on the pelvic girdle and then just above the knee (Figure 23-2b).

7. Use the probe again to separate and free the gastrocnemius muscle from the other muscle of the frog's leg (Figure 23-2c).

8. Locate the Achilles tendon (Figure 23-2c). Tie a piece of thread about 15 cm long (6 inches) around

its lowest end. Cut the tendon at the point between the thread and the animal's heel bone.

9. Lift and move the gastrocnemius muscle to one side. Cut the other muscles of the frog's lower leg and the bone (tibiofibula) just below the knee.

10. Cut the femur bone to about 2.5 cm (1 inch) (Figure 23-2c). Your instructor will demonstrate the procedure for attaching the nerve–muscle preparation to the recording system. Be certain to keep this material and the other leg of the frog moist with Ringer's solution.

## MUSCLE ACTIVITY MEASUREMENTS

### Materials

The following materials should be provided for class use:

1. Stimulating apparatus
2. Physiograph or related recording system
3. Weight pans
4. Weights (5- and 10-g units)
5. Millimeter ruler
6. Litmus paper
7. Examples of muscle contraction tracings or records
8. Frog Ringer's solution
9. Sterile gauze pads (2 × 2 inch)

The procedures in this part of the exercise should be performed by students in groups of four. One individual should operate the equipment used. A sufficient number of tracings should be made for all members of the group.

### Procedure 1: The Single Muscle Twitch

To cause a muscle twitch, it is necessary to produce a *threshold stimulus*—that is, the lowest voltage of electrical stimulation that produces a contraction. As the voltage is increased, more and more motor units are activated. The lowest voltage that activates all such units and produces a maximum contraction is called the *maximal stimulus.*

A stimulator is used to regulate the strength of the stimulus (voltage) and the frequency of stimuli (number of pulses or stimuli per second). The duration of each stimulus is measured in microseconds ($\mu$sec). The muscle stimulated to contract raises the writing device so that a trace of the contraction is made.

1. Set up the stimulating apparatus and recording system, as directed by your instructor. Refer to Appendix A for details of physiograph operation. Examine the stimulator, and note the position of its components.

2. Set the voltage dial of the stimulator at its lowest setting and the duration at 10 $\mu$sec.

3. Start the recording system at a slow speed.

4. Observe the recording for the presence of contractions.

5. Slowly increase the voltage by moving the voltage dial until contractions are observed.

6. Mark the response with the voltage needed to produce it. This is the threshold stimulus. Make four recordings of this response so that each member of the group can have one. Enter this value, and attach a representative tracing to the results and observations section. Refer to Figure 23-1a as a guide for labeling purposes.

7. Turn off the system. Moisten the muscle preparation with Ringer's solution.

8. Set the voltage dial at the threshold value. Turn on the recording system and the stimulating system.

9. Increase the voltage by one increment (step value), and stimulate the muscle.

10. Continue to stimulate the muscle and increase the voltage by one voltage increment at a time until the muscle contractions do not increase further. The voltage at this point is the *maximal stimulus,* and the effect produced is known as the *maximal response.*

11. Mark the record, and enter the value for the maximal stimulus in the results and observations section.

### Procedure 2: Summation, Incomplete Tetanus, and Complete Tetanus

1. Use the same muscle preparation, stimulating apparatus, and recording system.

2. Set the voltage dial at the maximal stimulus value.

3. Stimulate the muscle at a frequency of 1 pulse per second (pps).

4. Record the effect for 15 seconds.

5. Increase the frequency to 2 pps for 15 seconds and then to 4 pps for 15 seconds.

6. Allow the muscle preparation to rest for 2 minutes. During this time, moisten the preparation again.

7. Continue to double the frequency, with a rest break following, until the muscle preparation goes into complete tetany. Make sufficient records for all members of the group.

8. Label the record, and attach it to the results and observations section. Be certain to indicate the following events on the record (refer to Figure 23-1).
   a. Point of maximal stimulus
   b. Wave summation
   c. Incomplete tetanus
   d. Complete tetanus

## Procedure 3: Determination of Work Performed

1. Use the same muscle preparation, stimulating apparatus, and recording system.
2. Attach a weight pan or comparable device to the muscle preparation. Follow the directions of your instructor, and make the necessary adjustments.
3. Determine the threshold stimulus. This is the value for unloaded contraction. Enter your findings in the results and observations section.
4. Add a 5-g load to the weight pan, and determine the threshold stimulus. Enter your findings in the results and observations section.
5. Increase the voltage substantially above the threshold stimulus. Obtain a single muscle twitch. Mark the starting point of the stimulus application. Make enough recordings for each member of the group. Attach a labeled record in the results and observations section.
6. Continue to add 5-g loads until the muscle preparation can no longer contract. Mark the record with the total weight lifted. Measure the height or distance the weight was lifted. Enter your findings, and attach a label record in the results and observations section.
7. Calculate the work performed by the muscle preparation, using the following formula:

$$\frac{\text{Work}}{\text{performed}} = \text{weight of load} \times \text{distance lifted}.$$

Work can be recorded in units of gram-millimeter.
8. Turn off the stimulating apparatus and the recording system, and moisten the muscle preparation with Ringer's solution.

## Procedure 4: Muscle Fatigue

Muscle fatigue represents a loss of *irritability* (ability to respond to stimuli) by muscle. It is caused by the accumulation of metabolic products such as acid phosphates, carbon dioxide, and lactic acid.

1. Use the same muscle preparation, stimulating apparatus, recording system, and weight pan.
2. Add a 10-g load to the weight pain, and determine the threshold stimulus. Record your finding in the results and observations section.
3. Increase the stimulus to 20 volts above the threshold value. Stimulate the muscle preparation about 15 times at a rate of 2 pps. Repeat this step until the recording line begins to slope downward toward the baseline. Mark this point. It represents the onset of fatigue. Continue to stimulate the muscle at the same rate. Determine the time it takes for the muscle preparation to reach the bottom of the record. This event is complete fatigue.

4. Turn off the stimulating apparatus and the recording system.
5. Remove the muscle tissue preparation and cut it into two parts. Press one cut end on a piece of red litmus paper and the other cut end on a piece of blue litmus paper. What happened? Enter and explain your findings in the results and observations section.

## RESULTS AND OBSERVATIONS

### *The Single Muscle Twitch*

1. Threshold voltage: _____
2. Maximal stimulus: _____
3. Attach a labeled representation of the single muscle twitch.

4. Distinguish between maximal stimulus and maximal response. _____
_____
_____

### Summation, Incomplete Tetanus, and Tetanus

1. Attach a labeled representation of summation, incomplete tetanus, and tetanus.

2. Do the heights of individual contractions differ in summation? _____

## *Determination of Work Performed (Gram-Millimeter)*

1. Threshold stimulus in unloaded contraction: ____
2. Threshold stimulus with first 5-g load: _____
3. Attach a labeled representation of a single twitch with first 5-g load and intense voltage.

4. Attach a labeled representation of maximum work performed.

5. Total weight (load) with which no further contraction of the muscle preparation occurred:

_____

6. Work performed by the muscle preparation:

$$\text{Work performed} = \text{load} \times \text{distance}$$

$$\text{\_\_\_\_} = \text{g} \times \text{\_\_\_ mm}$$

## *Muscle Fatigue*

1. Threshold stimulus with 10-g load: _____
2. Threshold stimulus with increased voltage: ____
3. Time for complete fatigue to occur: _____
4. What were the results of litmus test? Explain what happened. _____

_____

_____

# LABORATORY REVIEW *23*

NAME _____

LAB SECTION _____ DATE _____

## *INTRODUCTION TO SKELETAL MUSCLE CONTRACTION, USING A LABORATORY ANIMAL*

## COMPLETION AND DEFINITION

1. Draw and label a representation (in the form of a graph) of the events of a single muscle twitch.

Height of contraction

Time (milliseconds)

2. Explain or define:

   a. Tetanus (tetany) _____
   _____

   b. Threshold stimulus _____
   _____

   c. Summation _____
   _____

   d. Foramen magnum _____
   _____

   e. Maximal stimulus _____
   _____

3. Distinguish between single pithing and double pithing. _____
   _____

4. What substances are responsible for muscle fatigue? _____
   _____

## ANSWERS

1. _____
2. _____
3. _____
4. _____
5. _____
6. _____
7. _____
8. _____
9. _____
10. _____

## MULTIPLE CHOICE

Provide the correct answers in the spaces.

1. Which of the following is the weakest stimulus from a neuron that can still initiate a contraction?
   a. threshold stimulus
   b. subliminal stimulus
   c. tetanus
   d. twitch
   e. muscle tone
2. What is the rapid, jerky response to a single stimulus called?
   a. threshold stimulus
   b. subliminal stimulus
   c. tetanus
   d. twitch
   e. muscle tone
3. Which of the following is a stimulus that cannot initiate a stimulus?
   a. threshold stimulus
   b. subliminal stimulus
   c. tetanus
   d. twitch
   e. muscle tone
4. Which of the following factors are known to decrease the strength of muscle contractions?
   a. fatigue
   b. lack of nutrients
   c. lack of oxygen
   d. choices *a, b,* and *c*
   e. choices *b* and *c* only
5. Once a threshold stimulus is applied, individual muscle fibers of a motor unit will contract to their fullest extent or will not contract at all, provided conditions remain constant. This characteristic best describes which of the following?
   a. treppe
   b. wave summation
   c. sliding-filament theory
   d. tetanus
   e. all-or-none principle
6. What is the condition in which a skeletal muscle contracts more forcefully to the same strength of stimulus after it has contracted several times?
   a. treppe
   b. wave summation
   c. sliding filament theory
   d. tetanus
   e. all-or-none principle
7. What is the brief period between the application of a stimulus and the beginning of a contraction called?
   a. refractory period
   b. wave summation
   c. latent period
   d. relaxation period
   e. contraction period
8. Which of the following best describes the period during which two stimuli are applied, one immediately after the other, causing the muscle to respond to the first stimulus but not to the second?

  a. refractory period
  b. wave-summation period
  c. latent period
  d. relaxation period
  e. contraction period

9. Which of the following best describes the situation when two stimuli are applied, and the second is delayed until the refractory period is over, causing the skeletal muscle to respond to both stimuli?
  a. refractory period
  b. wave summation
  c. latent period
  d. relaxation period
  e. contraction period

10. A state of sustained muscle contraction is known as which of the following?
  a. relaxation period
  b. tetanus
  c. treppe
  d. threshold
  e. latent period

# Human Muscle Contraction

## After completing this exercise, you should be able to

1. Distinguish between *muscle tone* and *muscle contractions* that are associated with body movements
2. Differentiate between *isotonic* and *isometric* muscle contraction
3. Perform simple experiments to demonstrate muscle fatigue and to measure work done

The only action a muscle is capable of is contraction. Some other force, including gravity, must be exerted to return the muscle to its uncontracted or relaxed state. For example, the biceps (a flexor) bends the arm, and the triceps (an extensor) must contract to pull the forearm back into place and also to stretch the biceps muscle (Figure 24-1). The force involved in this case is generated by the sliding filaments (Exercise 22) and is exerted parallel to the muscle fibers.

When an individual uses the biceps muscle—for example, to lift a heavy object—the force exerted by the contracting muscle on the object is referred to as *muscle tension*. The force exerted on the muscle by the object is termed the *load*. In order for a heavy object or load to be moved, muscle tension must be greater than the load, because these two factors are opposing forces. With *isotonic* (equal tension) contractions, the muscle fibers shorten, but the tension remains the same (Figure 24-2a). A contraction of this type takes place when an individual uses the biceps to lift heavy weights or other objects. With *isometric* (equal length) contraction, another type of muscle activity, a muscle's length remains essentially the same (does not shorten), but the tension changes (Figure 24-2b). Contractions of this type occur when a muscle supports a load in a fixed position; for example, contraction of the quadriceps muscles of the leg when an individual is in a standing position.

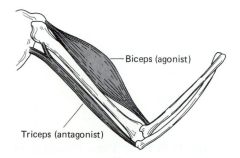 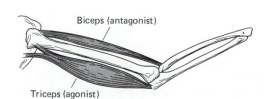

**Figure 24-1** Muscle contraction and the relaxed state. The movement of flexing the forearm involves contraction of the biceps and simultaneous relaxation of the triceps. To return the arm to the straight position the triceps contracts, extending the forearm, while the biceps relaxes and yields to the action of the triceps. The muscle that performs the action is the *agonist;* the muscle that is in opposition is the *antagonist.*

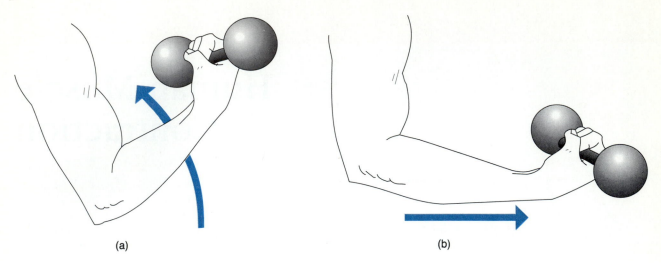

(a)                                                     (b)

**Figure 24-2** Muscle contractions: (a) isotonic; (b) isometric.

*Muscle tone (tonus)* is a reflex stretch response to motor-nerve impulses originating in the spinal cord. These impulses stimulate small numbers of different muscle fibers to contract at different times, resulting in a partial but sustained contraction of the entire muscle. Several muscles involved with body posture are associated with this muscle property. When muscles are not being stretched, as in the case of postural muscles during sleep or unconsciousness, the body collapses. The loss of muscle tone is due to an interference with the conduction of nerve impulses. It may result from certain disease states, such as poliomyelitis, or from accidents involving destruction to motor-nerve fibers. Muscles with less than normal tone are referred to as *flaccid.* (FLAK-sid).

Some interesting aspects of muscle contractions can be revealed by *electromyography,* the procedure used to record the action potentials (electric discharges) from muscles. The records are termed *electromyograms.* When a nerve impulse arrives at the muscle fibers of a *motor unit* (a neuron and the group of muscle cells it stimulates), a reversal of the electrical charge on the muscle fiber membranes occurs. Such a reversal is the action potential that spreads throughout the muscle fibers to cause contraction. Electromyograms are recorded with the aid either of a needle or wire electrodes inserted directly into muscles or of surface electrodes placed on the skin. The action potentials from individual motor units are picked up by the needle or wire electrodes, while surface electrodes detect integrated action potentials from whole muscles. In some situations, electromyograms can be used to demonstrate which muscles are responsible for or control particular movements. For example, by placing surface electrodes over the muscles of the shin and calf of a test subject and then having the individual respond to

a command such as "lean backward," "sit," or "stand," an electromyogram would show which muscles contract during specific movements and to what extent they contract (Figure 24-3).

This exercise deals with various aspects of muscle activities, including isometrics.

## Materials

The following materials should be provided per four students to perform Procedures 1–4:

1. One 5- or 10-lb barbell
2. One cloth tape measure (1 meter length)
3. One relatively soft tennis or rubber ball
4. Stopwatch or other timing device
5. An ergograph or similarly constructed device to measure work performed
6. Weight pan or similar container to hold gram weights
7. Gram weights
   a. 500 g (2)
   b. 250 g (2)
   c. 100 g (5)
   d. 50 g (2)
   e. 10 g (10)
8. 1-meter stick

## Procedure 1: Muscle Tone (Biceps, Triceps, Gastrocnemius, and Masseter Muscles)

This procedure should be performed by students in pairs.

1. Take turns in carrying out the following tasks. Note the degree of contraction of the biceps and triceps in each case by feeling them for *tightening* and/or

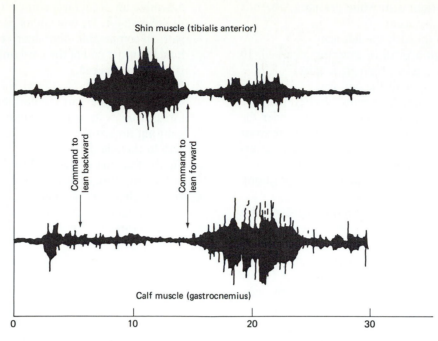

**Figure 24-3** A diagrammatic representation and comparison of electromyograms obtained from human muscles of the shin and calf, during the execution of specific verbal commands. The test subject first was told to lean backward, and then to lean forward. The involvement of a specific muscle and the extent of muscle contractions following the command can be seen. (Adapted from Witherspoon, J. D., R. Flowers, M. Isaacson, and S. Keslensky. "Electromyograms." *Muscle Contraction* (audiovisual tape). Houston: Nacro Bio-Systems, Inc., 1973.)

*hardening.* Enter your findings, and answer the questions in the results and observations section.

a. Sit in a comfortable position, with your forearms supported on a level surface and with palms down.

b. Sit in the same position, but with palms up, and note any differences.

c. Sit in a comfortable position. Grasp the barbell, and lift it with palms up. This is known as a weight curl. (A book may be substituted for the weight.)

d. Stand erect, and allow both arms to hang freely at the sides of your body.

e. Stand erect and hold the barbell with one hand at the side of your body.

2. Take turns in carrying out the following tasks. Note the degree of contraction of the gastrocnemius muscle. Enter your findings, and answer the questions in the results and observations section.

a. Lie down on your abdomen (prone position).

b. Sit on the edge of a chair so that both feet touch the floor.

c. Sit on the edge of a table so that both feet are dangling and not touching the floor.

d. Stand erect, and lock your knees for 5 seconds.

e. Stand erect with legs relaxed.

f. Stand on your toes with your arms extended overhead, and reach toward the ceiling.

3. Locate the masseter muscle in your partner's face. Place your hands on it. Have your partner first relax and then tighten this muscle. Describe what happens in each case in the results and observations section.

## Procedure 2: Distinguishing Between Isotonic and Isometric Muscle Contractions

This procedure should be performed by students in pairs.

1. Use a tape measure to measure the distance around the midline (circumference) of the biceps and triceps muscles of both arms while performing each of the following tasks. Enter your findings, and answer the questions in the results and observations section.

a. Extend both arms directly to the front of the body.

b. Flex both arms with palms facing upward, much like a weight-lifting exercise, 10 times in succession, then remeasure.

c. Extend the right arm while grasping a barbell or weight substitute.

d. Repeat Step (c) with the left arm.

e. Flex the right arm while grasping a barbell, 10 times in succession, then remeasure.

f. Repeat Step (e) with the left arm.

2. With a tape measure, determine the circumference of the midline calf region of both legs, under the condition of each of the following tasks. Enter your findings, and answer the questions in the results and observations section.

a. Lie down on your abdomen (prone position), with gastrocnemius relaxed.

b. Repeat Step (a), but contract the gastrocnemius muscles.

c. Stand erect.

d. Stand on your toes, and extend your arms upward.

e. Stand on your toes, with your arms at the side.

f. Sit on a table, so that your legs are dangling and do not touch the floor.

g. Repeat Step (f), but try to move your feet forward and upward with someone pushing against them (try to prevent any movement backward or downward).

3. Change places with your partner, and repeat the steps and tasks of this procedure.

## Procedure 3: Determining Muscle Fatigue

This procedure should be performed by students in groups of four.

1. Hold a tennis ball (or soft rubber ball) in your right hand. As you squeeze the ball as rapidly as possible, have your laboratory partners count the number of complete contractions during each 10-second interval for a total period of 3 minutes nonstop. Enter the count in the results and observations section.

2. Continue to time while you squeeze the ball until fatigue is complete. Record the time. This is the end of Trial 1.

3. Wait 1 minute, and repeat Steps 1 and 2. This is the end of Trial 2.

4. Wait 5 minutes and repeat Steps 1 and 2. This is the end of Trial 3.

5. Repeat all steps with your left hand.

6. Now have each of your laboratory partners repeat Steps 1–4. Enter your findings, and answer the questions in the results and observations section.

## Procedure 4: Determination of Work Performed

This procedure should be performed by students in groups of four.

1. Assemble an ergograph similar to the one shown in Figure 24-4. Tie one end of the cord to a weight pan or comparable container. Have one partner hold the other end of the cord until the experiment is ready to proceed.

2. Put the ergograph on a level surface, and sit next to it so that you can comfortably place your left forearm upon its surface. Your fingers should be pointing toward the pulley system. Strap the forearm to the board.

3. Tie the free end of the cord to the middle phalanx of the index finger of the hand.

4. Have a laboratory partner hold a meter stick next to the weight pan and note the level at which the pan is located.

5. Have a laboratory partner add weights, in 10-g increments, to the pan. Lift the weight pan by raising your index finger. Continue to add weights until the weight pan cannot be lifted more than 5 mm. In carrying out this task, your hand must not be lifted from the ergograph's surface, nor should the wrist be flexed. Enter the maximum number of grams lifted and the distance (in cm) the weight was lifted in the results and observations section.

6. Repeat Steps 3–5 with the middle finger and again with four fingers together, not including the thumb.

7. Repeat this entire procedure with your laboratory partners.

8. Calculate the work performed (weight lifted in grams × distance lifted in centimeters). Enter your calculations, and answer the questions in the results and observations section.

## Materials

The following materials should be provided for class use, to produce electromyograms (Procedure 5):

1. Skin electrodes (8 mm or 12.5 mm diameters)

2. Electrode washers (8 mm or 12.5 mm diameters) or micropore tape for attaching electrodes to the skin

3. Skin cleanser to prepare the skin

4. Electrode gel

5. Kimwipes® or paper towels

6. Sterile, disposable alcohol prep-pads

7. Scissors

8. Physiograph

## Procedure 5: Electromyograms from Leg Muscles

This procedure should be performed by students in pairs. Your instructor will have a physiograph or similar equipment ready for class use. Refer to Appendix A for a general review of the component parts of a physiograph.

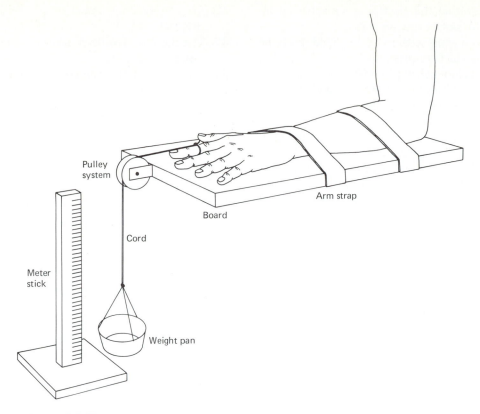

**Figure 24-4** An ergograph: This device can be used to determine work performed by skeletal muscle; it consists of a board, pulley, cord, weight pan, and restraining straps.

1. Obtain four of the skin electrodes, and—using the individual alcohol prep-pads—wipe them and allow them to dry on a clean surface.
2. Turn on the power switch of the physiograph, and allow the instrument to warm up for at least 10 minutes.
3. Using a paper towel and the skin cleaner provided, wipe the skin over the tibialis anterior (shin) and gastrocnemius (calf) muscles (See Figure 24-5.)
4. Apply electrode gel onto the sterilized electrode surfaces. Use a Kimwipe® to smooth the gel evenly so that it fills the well (depression) between the metal surfaces and surrounding plastic or similar type of ring of the electrode.
5. Cut four 1-inch micropore tape strips, or remove the paper coverings over the adhesive surfaces of the electrode washers.
6. Use the adhesive strips to apply two skin electrodes to the tibialis anterior muscle and two skin electrodes to the gastrocnemius muscle in the pattern shown in Figure 24-5. Make certain that the electrodes are placed parallel to one another and that one of the recording electrodes is placed over the proximal portion of the muscle and the second electrode is positioned over the distal portion.

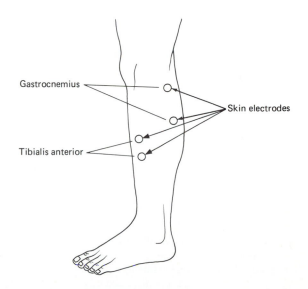

**Figure 24-5** Placement of biopotential skin electrodes for electromyograms (recordings) from the tibialis anterior muscle of the shin and the gastrocnemius muscle of the calf. The two muscles may contract simultaneously, as when standing. At other times, one muscle contracts while the other relaxes. This occurs when the toe end of the foot is raised.

7. Check to see that the electrodes are adhering securely to the skin. Press on them.
8. Make the following settings on two modules to be used for the electromyographs.
   a. Time constant setting to 0.03
   b. Sensitivity setting to between 20 and 100
   c. Gain knob setting to ×100
   d. Chart speed to 1.0 cm/sec
9. Plug the electrode into the appropriate modules, and turn on the physiograph to record.
10. Establish a baseline recording by having your partner relax in the standing position.
11. Next, issue the following commands to your partner to execute in the order given. Mark the electromyogram with each command. Refer to Figure 24-3.
    a. Stand straight with both legs together
    b. Sit (on the chair provided)
    c. Stand again
    d. Lean forward from the standing position
    e. Bend the knees slightly from the standing position.
    Note the degree of contraction for each muscle by examining the electromyogram. Enter your findings, and answer the questions in the results and observations section.

12. Change the chart speed to 0.5 cm/sec, and repeat Steps 10 and 11.
13. Change places with your laboratory partner, and repeat Steps 1–12.
14. Remove your electromyogram, and attach it to the space provided in the results and observations section.

## RESULTS AND OBSERVATIONS

### Muscle Tone

1. Enter your findings on the biceps and triceps and those of your laboratory partner in Table 24-1.
2. Enter your findings on the gastrocnemius and those of your laboratory partner in Table 24-2.
3. Describe the effects observed with the masseter muscle. _____
4. In which of the tasks performed was muscle tonus evident? _____

**Table 24-1**

| Task | Degree of Contraction* of Biceps and Triceps | |
|---|---|---|
| | Your Results | Partner's Results |
| Forearms supported with palms down | | |
| Forearms supported with palms up | | |
| Weight curl | | |
| Standing with arms at side | | |
| Standing with one hand grasping weight | | |

*Determine by increases in size or firmness.

**Table 24-2**

| Task | Degree of Contraction* of Gastrocnemius | |
|---|---|---|
| | Your Results | Partner's Results |
| Body in prone position | | |
| Sitting with feet touching the floor | | |
| Sitting with legs dangling over an edge | | |
| Standing with knees locked | | |
| Standing in a relaxed manner | | |
| Standing erect on tiptoes | | |

*Determine by increases in size or firmness.

## Distinguishing Between Isotonic and Isometric Muscle Contractions

1. In Table 24-3, enter both your and your laboratory partner's findings on the biceps.
2. In Table 24-4, enter both your and your laboratory partner's findings on the gastrocnemius.
3. Were any differences found between circumference measurements of the right and left arms? If so, explain these differences. _____

_____

4. Were any differences found between circumference measurements of the right and left gastrocnemius muscles? If so, explain. _____

_____

5. What variables influence the reading of measurements? _____

_____

**Table 24-3**

| Task | Measurements | | | |
|---|---|---|---|---|
| | Your Results | | Partner's Results | |
| | *Right Arm* | *Left Arm* | *Right Arm* | *Left Arm* |
| Body arms extended | | | | |
| Both arms flexed with palms up | | | | |
| Right arm extended with weight | | | | |
| Left arm extended with weight | | | | |
| Right arm flexed with weight | | | | |
| Left arm flexed with weight | | | | |

**Table 24-4**

| Task | Measurements | | | |
|---|---|---|---|---|
| | Your Results | | Partner's Results | |
| | *Right Leg* | *Left Leg* | *Right Leg* | *Left Leg* |
| Body in prone position with gastrocnemius relaxed | | | | |
| Body in prone position with gastrocnemius contracted | | | | |
| Standing erect | | | | |
| Standing on toes with arms extended | | | | |
| Standing on toes with arms at side | | | | |
| Sitting with legs dangling | | | | |
| Sitting with force applied to dangling legs | | | | |

## Determining Muscle Fatigue

1. Enter your findings in Table 24-5.
2. Indicate the time required for complete fatigue to occur.

   Student 1: _____

   Student 2: _____

   Student 3: _____

   Student 4: _____

3. Were all subjects able to squeeze the ball for 3 minutes? If not, why not? _____

   _____

4. Explain any differences noted in the results obtained with the right hand or the left hand. _____

   _____

   _____

## Determination of Work Performed

1. Enter both your measurements on the ergograph and those of your laboratory partners in Table 24-6.
2. Which of the fingers performed the most work?

   _____

3. Did your results differ from those of your partners?

   _____

4. How can you account for any differences in the work performed? Explain your answer by considering the muscles involved. _____

**Table 24-5**

| Student | Trial | Number of Contractions During 10-sec Intervals | | | | | | | | | | | | | | | |
|---|---|---|---|---|---|---|---|---|---|---|---|---|---|---|---|---|---|
| | | 10 | | 20 | | 30 | | 40 | | 50 | | 60 | | 120 | | 180 | |
| | | R* | L | R | L | R | L | R | L | R | L | R | L | R | L | R | L |
| | 1 | | | | | | | | | | | | | | | | |
| | 2 | | | | | | | | | | | | | | | | |
| | 3 | | | | | | | | | | | | | | | | |
| | 1 | | | | | | | | | | | | | | | | |
| | 2 | | | | | | | | | | | | | | | | |
| | 3 | | | | | | | | | | | | | | | | |
| | 1 | | | | | | | | | | | | | | | | |
| | 2 | | | | | | | | | | | | | | | | |
| | 3 | | | | | | | | | | | | | | | | |
| | 1 | | | | | | | | | | | | | | | | |
| | 2 | | | | | | | | | | | | | | | | |
| | 3 | | | | | | | | | | | | | | | | |

*R = right hand; L = left hand.

**Table 24-6**

| Student | Weight Lifted (in g) | | | Distance Lifted (in cm) | | | Work Performed* | | |
|---|---|---|---|---|---|---|---|---|---|
| | IF† | MF | FF | IF | MF | FF | IF | MF | FF |
| | | | | | | | | | |
| | | | | | | | | | |
| | | | | | | | | | |

*Work performed = weight lifted × distance lifted.
†IF = index finger; MF = middle finger; FF = four fingers.

## Electromyograms from Leg Muscles

1. Place the appropriate portion of the electromyogram, or sketch the results in the space provided here.

2. Enter your findings in Table 24-7. Indicate which muscle(s) contracted during the execution of the commands listed. Use either a " + " or a " − ."

**Table 24-7**

| Command | Muscles Involved | |
|---|---|---|
| | Tibialis Anterior (shin) | Gastrocnemius (calf) |
| Sit | | |
| Stand | | |
| Lean forward | | |
| Knee bend | | |
| Extend toe | | |

# LABORATORY REVIEW 24

NAME _____

LAB SECTION _____ DATE _____

## HUMAN MUSCLE CONTRACTION

**ANSWERS**

## COMPLETION AND DEFINITION

1. Distinguish between isotonic and isometric contractions.

   _____

   _____

2. What is muscle tension? _____

   _____

3. Lifting a weight exerts a force on a muscle. What is this force called? _____

4. Define the following:

   a. Flaccid _____

   b. Muscle fatigue _____

   c. Muscle tone _____

   d. Electromyogram _____

   _____

5. Distinguish between flexor and tensor muscles. _____

   _____

## MULTIPLE CHOICE

Provide the correct answers in the spaces.

1. What is the force exerted by a contracting muscle used to lift a heavy object?
   a. load
   b. muscle tension
   c. tonus
   d. flaccid
   e. potential
2. What is the force exerted by the object being lifted on the muscle in Question 1?
   a. load
   b. muscle tension
   c. tonus
   d. flaccid
   e. potential
3. In order for a heavy object or load to be moved, muscle tension must have what quantitative relationship to the load?
   a. less than
   b. greater than
   c. same as

1. _____

2. _____

3. _____

4. _____

5. _____

6. _____

7. _____

8. _____

9. _____

10. _____

4. What type of contraction takes place when an individual uses the biceps to lift a heavy weight?
   a. isometric
   b. equal length
   c. isotonic
   d. impulse
   e. flaccid

5. What is a reflex stretch response to motor-nerve impulses originating in the spinal cord?
   a. a full contraction
   b. muscle tension
   c. muscle tone
   d. relaxed state
   e. flaccid

6. How do you describe a muscle with less than normal tone?
   a. in a full contraction
   b. showing muscle tension
   c. relaxed
   d. tonus
   e. flaccid

7. What is the actual record of action potentials from muscles called?
   a. an electrode
   b. a contraction
   c. electromyograph
   d. an electromyogram
   e. isometrics

8. In the case of an isometric contraction, what happens to a muscle's length?
   a. shortens
   b. remains the same
   c. lengthens

9. In the case of an isometric contraction, what happens to the tension on the muscle?
   a. changes
   b. does not change

10. With isotonic contractions, what happens to muscle fibers?
    a. shorten
    b. lengthen
    c. do not change

# Nervous System Coordination and Control

*T*he nervous and endocrine systems regulate the activities and functions of the body. These two systems, however, operate at different rates. The nervous system acts rapidly, while the endocrine system operates more slowly. This section considers the nervous system.

The nervous system is divided into the central nervous system, the peripheral nervous system, and the autonomic nervous system.

The *central nervous system (CNS)* includes the brain and the spinal cord. It contains the centers where connections are made between nerve fibers bringing information from receptor (sense) organs and nerve fibers carrying information to the effector organs (muscles and glands).

The *peripheral nervous system (PNS)* consists of the nerves connecting the peripheral parts of the body with the central nervous system. Twelve pairs of nerves arise from the brain and are called the cranial nerves; 31 pairs of nerves arise from the spinal cord and are called the spinal nerves.

The *autonomic nervous system (ANS)* is composed of nerves, nerve trunks with ganglia, and special ganglia that govern the operation of smooth muscle, cardiac muscle, and glands, which—for the most part—are not under voluntary control.

The exercises in this section consider the general structure and organization of the nervous system, as well as selected activities and functions associated with the components of the system.

### GENERAL REFERENCES

Brown, A. G. *Nerve Cells and Nervous Systems: An Introduction to Neuroscience.* New York: Springer-Verlag, 1991.

Geschwind, N. "Specializations of the Human Brain." *Scientific American,* 241: 180–199, 1979.

Guyton, A. C. *Basic Neuroscience: Anatomy and Physiology,* 2nd ed. Philadelphia: Saunders, 1991.

Iversen, L. L. "The Chemistry of the Brain." *Scientific American,* 241: 134–149, 1979.

Netter, F. H. *The CIBA Collection of Medical Illustrations: Vol. 1, The Nervous System.* Summit, NJ: CIBA Pharmaceutical Products, 1953.

Noback, C. R., N. L. Strominger, and R. J. Damarst. *The Human Nervous System: Introduction and Review.* Philadelphia: Lea & Febiger, 1991.

Peters, A., S. L. Palay, and H. de F. Webster. *The Fine Structure of the Nervous System: Neurons and Their Supporting Cells,* 3rd ed. New York: Oxford University Press, 1991.

Roberts, P. A. *Neuroanatomy (Oklahoma Notes),* 2nd ed. New York: Springer-Verlag, 1991.

Tortora, G. J., and S. R. Grabowski. *Principles of Anatomy and Physiology,* 6th ed. New York: HarperCollins, 1993.

## *SPECIFIC REFERENCES*

Aoki, C., and P. Siekevitz. "Plasticity in Brain Development." *Scientific American,* 259: 56–64, 1988.

Bennett, D. A., and D. A. Evans. "Alzheimer's Disease." *Disease-a-Month.* 38: 1–64, 1992.

Dunant, Y., and M. Israël. "The Release of Acetylcholine." *Scientific American,* 252: 58–66, 1985.

Houston, J. P. *Fundamentals of Learning and Memory,* 2nd ed. New York: Academic Press, 1981.

Stevens, C. F. "The Neuron." *Scientific American,* 241: 54–65, 1979.

Williams, M. *Brain Damage, Behavior and the Mind.* New York: Wiley, 1979.

Winter, P. M., and J. N. Miller. "Anesthesiology." *Scientific American,* 252: 124–131, 1985.

# Anatomy of the Nervous System

**After completing this exercise,
you should be able to**

1. Name the basic parts of the mammalian brain
2. Locate the cranial and spinal nerves
3. Dissect the mammalian brain
4. Identify the basic parts of the autonomic nervous system on a dissected cat specimen

## Materials

The following materials should be provided for class use:

1. Sheep brains (preserved)
2. Bone saw
3. Bone forceps
4. Model of human nervous system
5. Charts of human nervous system
6. Human brain (preserved)
7. Autonomic trunk and ganglia of a dissected cat specimen
8. Cat-dissection specimen

## SHEEP BRAIN, DORSAL SURFACE

For detailed study, it is best to use a sheep's brain (either fresh or preserved). It is large enough to see the parts clearly and is easily obtained. If a sheep's brain is not available and you wish to use the cat's brain, there are instructions for its dissection at the end of this exercise.

Covering the brain are three membranes or *meninges*. Usually, only the outermost and toughest membrane, the *dura mater*, is visible. The middle membrane is the *arachnoid*, and the innermost pigmented membrane is the *pia mater*.

In a dorsal view (Figure 25-1), the brain reveals a large anterior portion, the *cerebrum or cerebral hemispheres*. Posterior to the cerebrum is the *cerebellum*. The *medulla oblongata* (Figure 25-2) is located anterior to the spinal cord. If the *cerebellum* is gently drawn back (Figure 25-1), some of the midsection of the dorsal brain is revealed. One can see four distinct spherical bodies, the *corpora quadrigemina,* which make up the dorsal midbrain, and which will be seen more clearly when the brain is cut in half. Much of the dorsal surface of the brain that is visible in lower animals is obscured by the drastic folding of both cerebral and cerebellular tissue. Observe both dorsal and ventral views, and locate the following:

**IN A DORSAL VIEW (Figure 25-1)**

1. Cerebral hemispheres
2. Midbrain (corpora quadrigemina)
   a. Superior colliculi
   b. Inferior colliculi
3. Cerebellum
4. Medulla oblongata

**IN A VENTRAL VIEW (Figure 25-2)**

1. Olfactory bulb
2. Cerebral hemispheres
3. Infundibulum and pituitary gland (if present)
4. Mammillary bodies
5. Cerebral peduncle
6. Pons varolii
7. Medulla oblongata

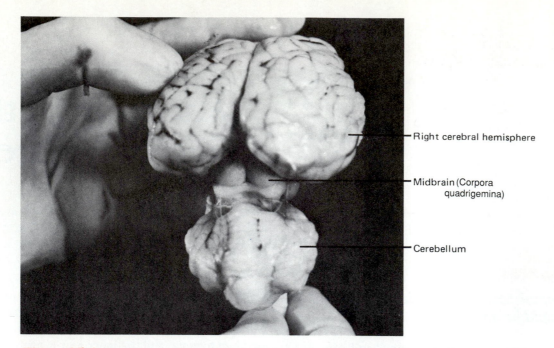

**Figure 25-1** Sheep brain (dorsal view). Separate the cerebrum and cerebellum to see the midbrain.

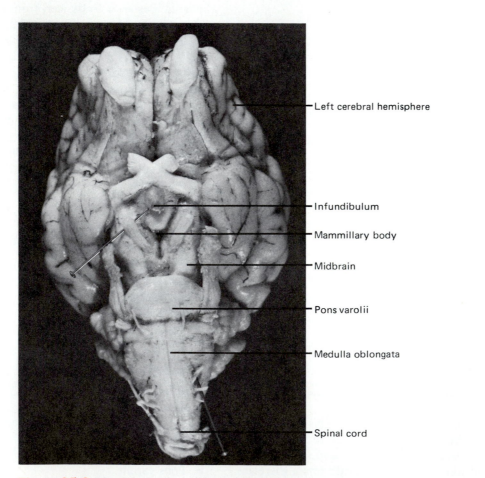

**Figure 25-2** Sheep brain (ventral view) (with pituitary removed). The pins underline right cranial nerve III and left cranial nerves IX, X, and XI.

## SHEEP BRAIN, VENTRAL SURFACE

Using Figures 25-2, 25-3, and your own specimen, compare the sheep brain with the human brain (Figure 25-4).

Once again, locate the cerebral hemispheres, cerebellum, and medulla oblongata. In addition, identify the following:

1. *Olfactory nerve tracts* cross the ventral surface of the hemispheres, ending in two enlarged areas— the *olfactory bulbs.*
2. *Optic nerves* cross over to form *optic tracts;* the crossover point is the *optic chiasma.*
3. The *pituitary gland,* an endocrine gland, is located behind the optic chiasma. If it is present, it is about the size of a pea. If it has been removed, it leaves a small tissue mass called the *infundibulum,* which connects the pituitary to the brain.

4. *Mamillary bodies* are small U-shaped regions in the midline, posterior to the pituitary gland.
5. *Cerebral peduncles* are the masses surrounding the mamillary bodies and extending back about one inch. This area gives rise to the third pair of cranial nerves.
6. The *pons varolii* is a bridgelike area in the cerebellar region.
7. The *medulla oblongata* has already been identified on the dorsal surface and can also be seen clearly on the ventral surface. Posterior to it is the spinal cord.
8. The *trapezium* is the right or left side of the medulla of the sheep brain.

Also note the large vascular network on the ventral surface. Blood flows toward the brain in the *basilar artery,* which divides in the region of the cerebral peduncles to form circular vessels around the pituitary

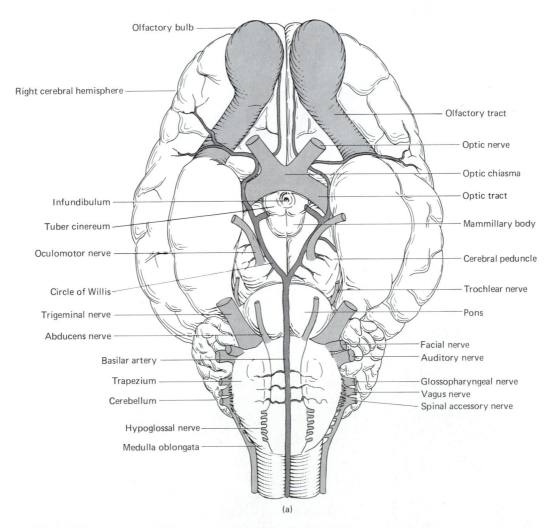

(a)

**Figure 25-3** Sheep brain (ventral view): (a) Diagram shows pituitary removed and cranial nerves identified by name.

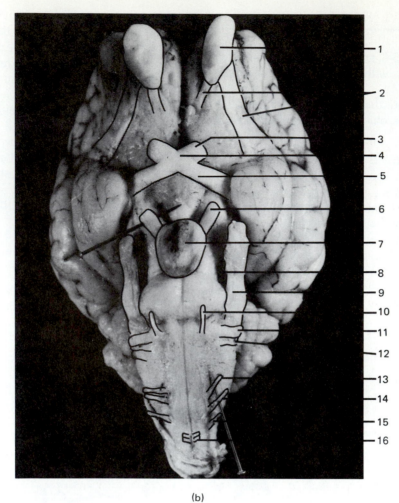

(b)

1. Olfactory bulb
2. Olfactory tracts
3. Optic nerve (II)
4. Optic chiasma
5. Optic tract
6. Oculomotor nerve (III)
7. Pituitary gland
8. Trochlear nerve (IV)
9. Trigeminal nerve (V)
10. Abducens nerve (VI)
11. Facial nerve (VII)
12. Acoustic nerve (VIII)
13. Glossopharyngeal nerve (IX)
14. Vagus nerve (X)
15. Spinal accessory nerve (XI)
16. Hypoglossal nerve (XII)

**Figure 25-3 *(Continued)*** (b) Photograph with highlighting and labels shows pituitary in place and cranial nerves identified by number.

ANTERIOR

CEREBRUM

Cranial nerves

Olfactory bulb

Olfactory tract

Pituitary gland

Optic tract

Tuber cinereum

Mammillary body

Cerebral peduncle of midbrain

PONS

Middle cerebellar peduncle

MEDULLA OBLONGATA

Pyramids

Olive

Decussation of pyramids

Spinal cord

CEREBELLUM

C1

Olfactory (I) nerve fibers

Optic (II) nerve

Oculomotor (III) nerve

Trochlear (IV) nerve

Trigeminal (V) nerve

Abducens (VI) nerve

Facial (VII) nerve

Vestibulocochlear (VIII) nerve

Glossopharyngeal (IX) nerve

Vagus (X) nerve

Accessory (XI) nerve

Hypoglossal (XII) nerve

POSTERIOR

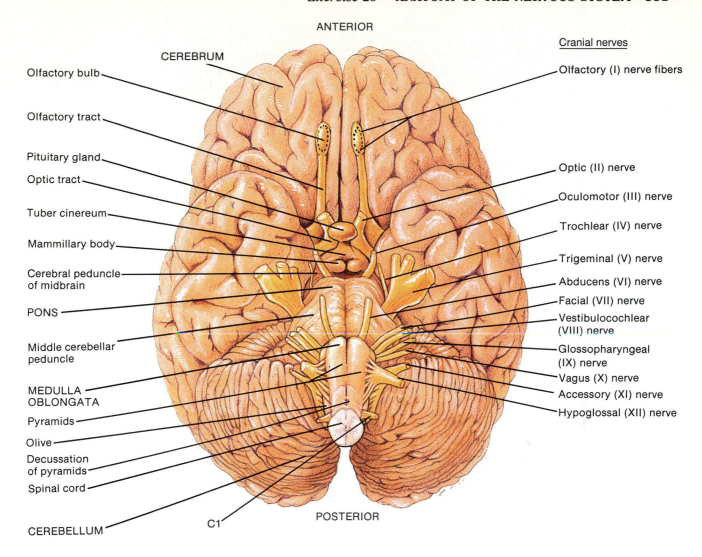

**Figure 25-4** Human brainstem: Ventral surface of the brain, showing the structure of the brainstem in relation to the cranial nerves and associated structures. (From G. J. Tortora and S. R. Grabowski, *Principles of Anatomy and Physiology,* 7th ed., New York: HarperCollins, 1993.)

and the optic chiasma. This circuit, which has other contributing sources, is termed the *circle of Willis*. It is discussed in more detail in a later numbered exercise.

## CRANIAL NERVES

The cranial nerves exit from the brain on the ventral surface. They can be identified with the aid of Figures 25-2 and 25-3. Compare the cranial nerves with their counterparts in the human brain (Figure 25-4).

I. *Olfactory nerves* contain nerve fibers from the nasal membranes and cannot be seen in this preparation. The nerve fibers synapse in the olfactory bulb. The second set of neurons continue in the

*olfactory tract* and terminate in the olfactory areas of the cerebral hemispheres.

II. *Optic nerves* arise in the retina of the eye and cross over anterior to the pituitary gland to form the *optic chiasma*. They enter the brain as the *optic tracts*. Some fibers within the nerve cross over at the chiasma; others do not. The optic nerves are the only cranial nerves that cross over.

III. *Oculomotor nerves* are medium-sized nerves posterior to the mamillary bodies. They supply four of the six pairs of extrinsic eyeball muscles with nerve fibers.

IV. *Trochlear nerves* are small and arise from the midbrain. They project from the cleft between the posterior end of the cerebrum and the pons varolii. They supply nerve fibers to one pair of the extrinsic eyeball muscles.

V. *Trigeminal nerves* are large and arise laterally from the pons. Together with cranial nerve VII, they supply nerves to much of the head and face.

VI. *Abducens nerves* are small and arise medial to the trigeminal nerves from the anterior end of the medulla oblongata, at the border of the trapezium. They supply nerves to one pair of extrinsic eyeball muscles.

VII. *Facial nerves* are medium-sized and arise directly behind the trigeminal nerves from the trapezium area. They supply nerves to the head and face.

VIII. *Acoustic (auditory, statoacoustic, or vestibulocochlear) nerves* are large, short nerves that arise from the lateral trapezium area, posterior to cranial nerve VII. They are associated with the cochlea (hearing) and the semicircular canals (equilibrium).

XI. *Glossopharyngeal nerves* are small nerves that arise laterally from the medulla oblongata. They are associated primarily with the tongue and pharynx regions.

X. *Vagus nerves* are small and arise from the lateral medulla posterior to the glossopharyngeal nerves. They supply nerves to the pharynx and larynx and extend posteriorly into the thoracic and abdominal cavities to supply visceral organs with parasympathetic fibers.

XI. *Accessory nerves* have several roots of origin in the most posterior part of the lateral medulla. They innervate several neck and shoulder muscles.

XII. *Hypoglossal nerves* also arise from several roots in the medulla oblongata, medial to cranial nerves XI. They are associated with the tongue muscles.

## SHEEP BRAIN IN MIDSAGITTAL SECTION

### Brain Ventricles or Cavities

Cut the sheep brain in half, using a sharp, long-bladed knife. Examine the cut surface, and identify as many parts as possible, referring to Figures 25-5 and 25-6. Compare the midsagittal section of the sheep brain with similar views of the human brain (Figures 25-7 and 25-8).

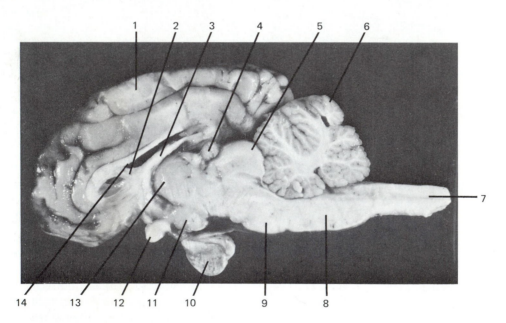

1. Cerebrum
2. Septum pellucidum
3. Fornix
4. Pineal gland
5. Midbrain
6. Cerebellum
7. Spinal cord
8. Medulla oblongata
9. Pons
10. Pituitary
11. Mammillary body
12. Optic chiasma
13. Massa intermedia
14. Corpus callosum

**Figure 25-5** Sheep brain (midsagittal section).

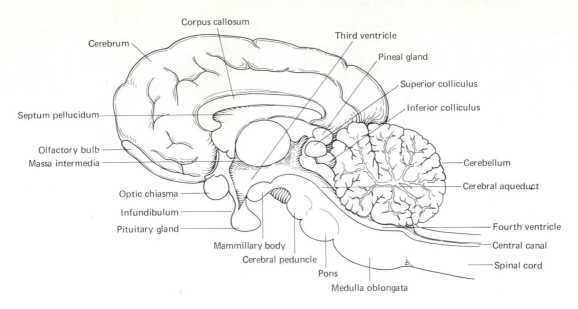

**Figure 25-6** Sheep brain (midsagittal section).

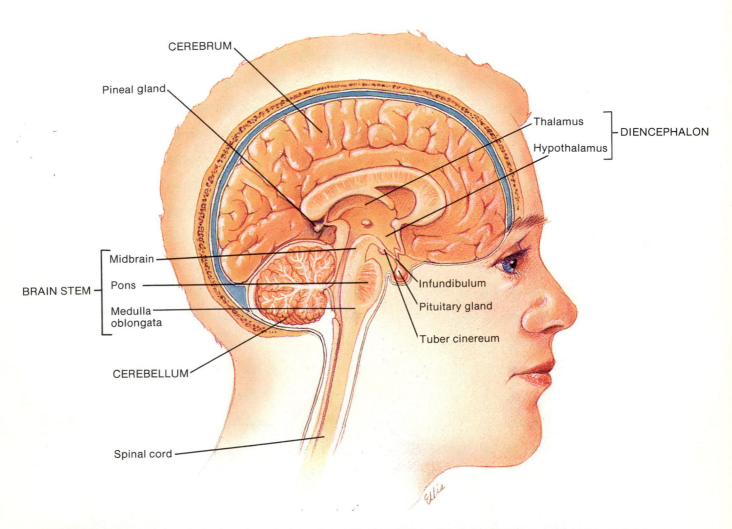

**Figure 25-7** Human brain: principal parts (seen in sagittal section). The infundibulum and pituitary gland are discussed, in conjunction with the endocrine system, in Exercise 28. (From G. J. Tortora and S. R. Grabowski, *Principles of Anatomy and Physiology,* 7th ed., New York: HarperCollins, 1993.)

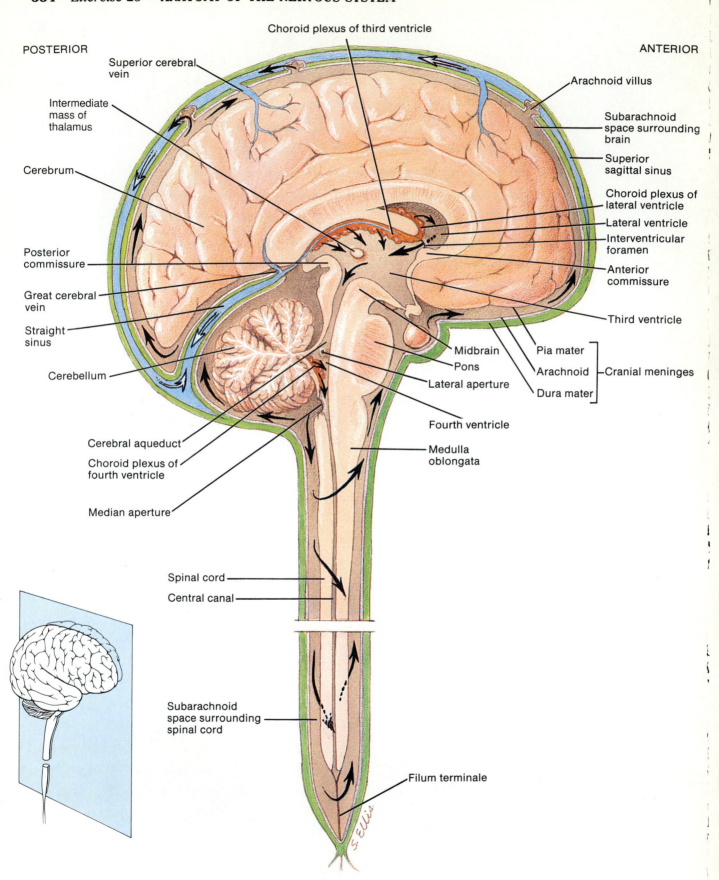

**Figure 25-8** Meninges and ventricles of the brain: brain, ventricles, spinal cord, and meninges (seen in sagittal section). Arrows indicate the direction of flow of cerebrospinal fluid. (From G. J. Tortora and S. R. Grabowski, *Principles of Anatomy and Physiology*, 7th ed., New York: HarperCollins, 1993.)

Within the brain are cavities filled with cerebrospinal fluid. These are *brain ventricles* and are numbered or named as follows:

I and II (lateral ventricles)—located in the cerebral hemispheres

III (third ventricle)—located in the diencephalon region

IV (fourth ventricle)—located in the medulla oblongata

The lateral ventricles connect with the third ventricle by channels termed the *interventricular foramina* or *foramina of Munro*. Ventricle III is connected to ventricle IV by the *cerebral aqueduct* or *aqueduct of Sylvius*. Figures 25-8 and 25-9 show the location of each ventricle in the human brain. The locations and shapes of the ventricles in the sheep and cat brains are similar.

## Brain Parts

### Telencephalon—Cerebrum

This region becomes highly developed in all mammals and particularly so in the human. It is composed of the right and left *cerebral hemispheres* and contains the lateral ventricles. (See "cerebrum" on Figures 25-7, 25-8, and 25-9.) Note the extent of folding of the *cerebral hemispheres* back over the brainstem. The large mass of cerebral tissue growing back over other parts of the brain is an evolutionary phenomenon. It has reached its greatest extent in the human brain. Near the floor of the cerebrum is a tough white area called the *corpus callosum,* the fibers of which connect the right and left hemispheres (see Figure 25-8).

### Diencephalon

The *diencephalon* (Figure 25-7) is a less distinct region than the cerebrum. It consists of several structures:

1. *Pineal gland*—a small, dorsal projection of endocrine nature
2. *Infundibulum*—a ventral downgrowth attached to the hypophysis of the pituitary gland
3. *Anterior choroid plexus*—a network of blood vessels roofing the third brain ventricle (see Figure 25-8). (Together with other choroid plexuses, it contributes to the formation of cerebrospinal fluid.)
4. *Fornix*—part of the olfactory pathways of the brain
5. *Massa intermedia* or *intermediate mass*—gray matter that connects the right and left thalamus regions (centrally located)
6. *Thalamus*—represents the lateral walls of the diencephalon, which are difficult to see in a midsagittal section
7. *Hypothalamus*—the floor of the diencephalon, which gives rise to the infundibulum
8. *Epithalamus*—rooflike portion made up of a vascular network termed the *anterior choroid plexus*

### Mesencephalon

The *mesencephalon* or *midbrain* is part of the brainstem and consists of the following parts (featured in Figure 25-6, for sheep brain):

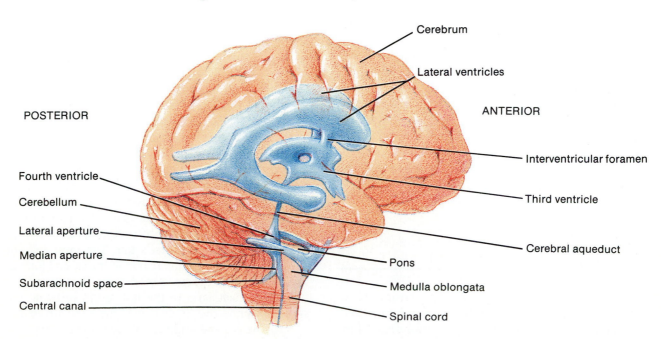

POSTERIOR

ANTERIOR

Cerebrum

Lateral ventricles

Interventricular foramen

Third ventricle

Fourth ventricle

Cerebellum

Lateral aperture

Median aperture

Subarachnoid space

Central canal

Cerebral aqueduct

Pons

Medulla oblongata

Spinal cord

**Figure 25-9** Human brain, ventricles. (From G. J. Tortora and S. R. Grabowski, *Principles of Anatomy and Physiology*, 7th ed., New York: HarperCollins, 1993.)

1. *Corpora quadrigemina*—located dorsally and composed of (a) the superior colliculi, which are visual reflex centers; and (b) the inferior colliculi, which are auditory reflex centers.
2. *Cerebral peduncles*—located ventrally

## Metencephalon

The embryonic metencephalon develops into the following adult parts.

1. *Cerebellum*—The cerebellum is the second largest part of the brain. It is dorsally located and highly convoluted, and its gray cortex and inner white matter are distinct. The branching patterns of the white matter within the cerebellum are termed the *arbor vitae* because they resemble the branches of a tree. The middle region of the cerebellum is the *vermis.*
2. *Pons*—This is an important relay center in the adult brain. It is located ventrally (Figure 25-4) and consists primarily of fiber tracts.
3. *Trapezium*—The trapezium is a fiber-tract area, which is considered a part of the medulla oblongata in the cat and sheep (Figure 25-3a) but is a part of the human pons.

## Myelencephalon

The *medulla oblongata* is the most posterior part of the sheep brain (Figure 25-6). It merges with the spinal cord. Within it is the fourth ventricle. Roofing ventricle IV is another blood vessel plexus, the *posterior choroid plexus* (see Figure 25-8, showing human brain). Large motor tracts descending from the cerebrum cross over in the medulla so that skeletal muscle fibers on the right side of the body are controlled by the left side of brain. Thus, left-side muscle function is controlled by the right cerebral hemisphere.

Notice that the brain ventricles connect with the central canal of the spinal cord. The first and second ventricles can be seen by making a cross-section of the cerebral hemispheres. Refer to Figure 25-8.

The *pons varolii* is a bridgelike region of the same embryonic area. It consists primarily of fiber tracts.

## THE HUMAN BRAIN: SURFACE ANATOMY OF THE CEREBRUM

Study the external anatomy of the cerebral hemispheres, using Figure 25-10. If a human brain is not available, use the sheep brain. The convolutions of the brain form grooves and prominences. The grooves may be either shallow *(sulci-*singular, *sulcus)* or deep *(fissures)* The *central sulcus* forms a key landmark on the brain, separating the lobes. The rounded prominences on either side of the grooves are termed *gyri.* Two prominent gyri are as follows:

> *Precentral gyrus*—located anterior to the central sulcus; represents the location of an important motor area of the cortex
>
> *Postcentral gyrus*—located posterior to the central sulcus; represents the location of important sensory areas of the cortex

Each hemisphere of the brain may be divided into four basic areas, termed *lobes,* as follows:

1. *Frontal lobes*—parts of the brain that constitute the area beneath the frontal bone, located anterior to the central sulcus
2. *Parietal lobes*—paired areas located posterior to the central sulcus and the frontal lobes
3. *Temporal lobes*—located laterally and separated from the frontal lobes by the *lateral cerebral sulcus*
4. *Occipital lobes*—occupying the area beneath the occipital bone. They are separated from the parietal lobe by the *parietooccipital sulcus*

Each hemisphere is separated from the other by the *longitudinal fissure.* The cerebrum is separated from the cerebellum by the *transverse fissure.*

## HUMAN SPINAL CORD AND SPINAL NERVES

The human spinal cord and its 31 pairs of spinal nerves should be studied from models and charts (see Figure 25-11). The cord begins posterior to the medulla oblongata and continues down the middle of the back for about 16 to 18 inches. There are two enlargements, one in the neck region (*cervical enlargement*) and the other in the midback region (*lumbar enlargement*). During embryonic development, the spinal cord does not keep pace with body growth in length, and it terminates at the level of the second lumbar vertebra. The last seven or eight pairs of nerves run posteriad beyond the spinal cord until they reach their designated exits, creating what looks like a bundle of hairs, appropriately named the *cauda equina,* or horse's tail. The most inferior portion of the cord, in the region of the second lumbar vertebra, is called the *conus medullaris.* Extending from the conus medullaris to the last sacral vertebra is a strand of connective tissue named the *filum terminale.*

The names and numbers of each spinal nerve correspond to the names and numbers of the vertebra at

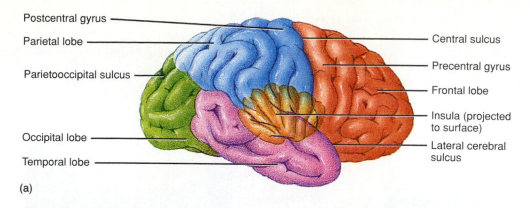

Postcentral gyrus

Parietal lobe

Parietooccipital sulcus

Occipital lobe

Temporal lobe

Central sulcus

Precentral gyrus

Frontal lobe

Insula (projected to surface)

Lateral cerebral sulcus

(a)

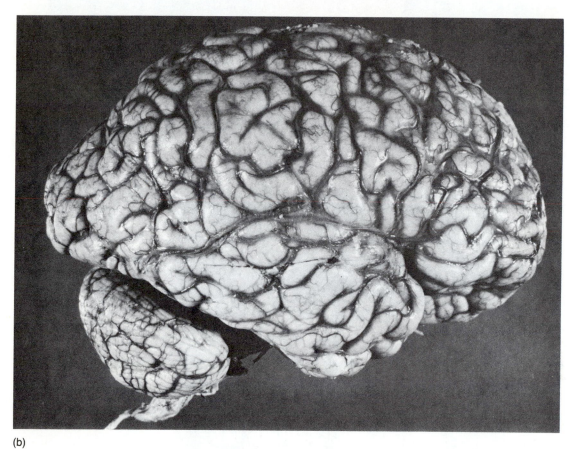

(b)

**Figure 25-10** Lobes and fissures of the human cerebrum: (a) Diagram of right lateral view; because the insula cannot be seen externally, it has been projected to the surface. (b) Photograph of right lateral view (courtesy of Martin Rotker, Taurus Photos).

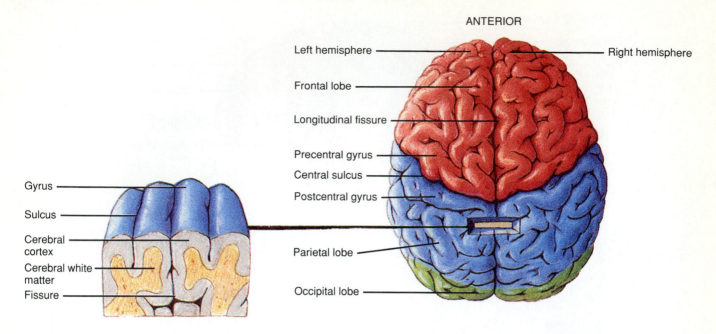

ANTERIOR

Left hemisphere

Right hemisphere

Frontal lobe

Longitudinal fissure

Precentral gyrus

Central sulcus

Postcentral gyrus

Gyrus

Sulcus

Cerebral cortex

Cerebral white matter

Fissure

Parietal lobe

Occipital lobe

(c)

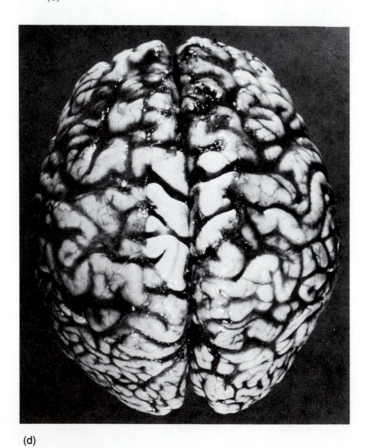

(d)

**Figure 25-10** *(Continued)* Lobes and fissures of the human cerebrum: (c) Diagram of superior view; the insert to the left indicates the relative differences among a gyrus, a sulcus, and a fissure. (d) Photograph of superior view (courtesy of Martin Rotker, Taurus Photos).

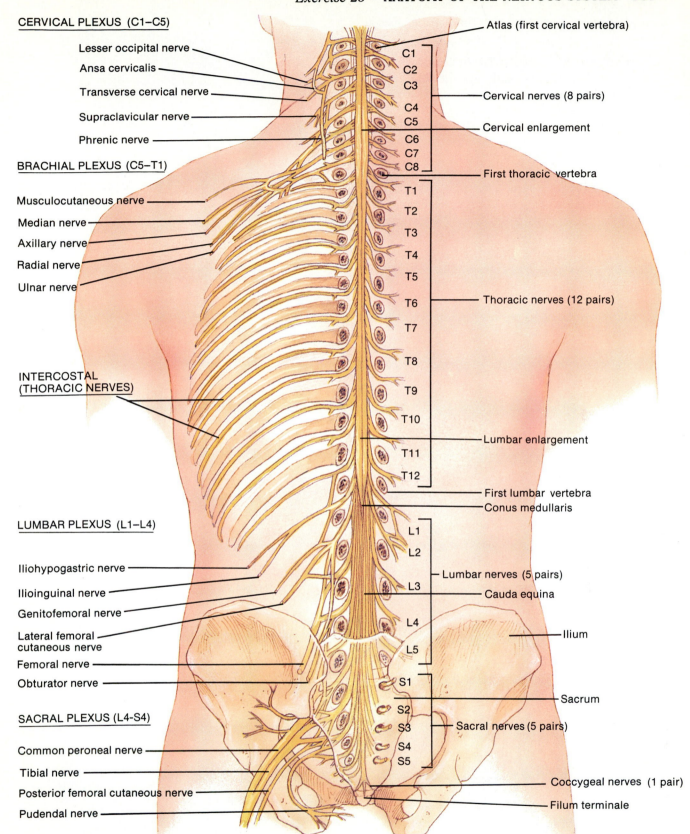

CERVICAL PLEXUS (C1–C5)

Lesser occipital nerve

Ansa cervicalis

Transverse cervical nerve

Supraclavicular nerve

Phrenic nerve

BRACHIAL PLEXUS (C5–T1)

Musculocutaneous nerve

Median nerve

Axillary nerve

Radial nerve

Ulnar nerve

INTERCOSTAL (THORACIC NERVES)

LUMBAR PLEXUS (L1–L4)

Iliohypogastric nerve

Ilioinguinal nerve

Genitofemoral nerve

Lateral femoral cutaneous nerve

Femoral nerve

Obturator nerve

SACRAL PLEXUS (L4-S4)

Common peroneal nerve

Tibial nerve

Posterior femoral cutaneous nerve

Pudendal nerve

Atlas (first cervical vertebra)

C1
C2
C3
C4
C5
C6
C7
C8

Cervical nerves (8 pairs)

Cervical enlargement

First thoracic vertebra

T1
T2
T3
T4
T5
T6
T7
T8
T9
T10
T11
T12

Thoracic nerves (12 pairs)

Lumbar enlargement

First lumbar vertebra

Conus medullaris

L1
L2
L3
L4
L5

Lumbar nerves (5 pairs)

Cauda equina

Ilium

S1
S2
S3
S4
S5

Sacrum

Sacral nerves (5 pairs)

Coccygeal nerves (1 pair)

Filum terminale

**Figure 25-11** Spinal cord and spinal nerves (seen in posterior view). (From G. J. Tortora and S. R. Grabowski, *Principles of Anatomy and Physiology,* 7th ed., New York: HarperCollins, 1993.)

their point of exit (Figure 25-11). They are classified as follows:

1. *Cervical nerves* (8 pairs)—The first cervical nerves ($C_1$) exit between the skull and the first cervical vertebra. The eighth cervical nerves ($C_8$) exit from the intervertebral foramina between the seventh cervical vertebra and the first thoracic vertebra.
2. *Thoracic nerves* (12 pairs)—The first thoracic nerves ($T_1$) exit from the intervertebral foramina between the first and second thoracic vertebrae. The twelfth thoracic ($T_{12}$) nerves exit between the twelfth thoracic vertebra and the first lumbar vertebra.
3. *Lumbar nerves* (5 pairs)—The first lumbar nerves ($L_1$) exit from the intervertebral foramina between the first and second lumbar vertebrae. The fifth lumbar nerves ($L_5$) exit between the fifth lumbar vertebra and the sacrum.
4. *Sacral nerves* (5 pairs)—The first sacral nerves ($S_1$) exit between fused first and second sacral vertebrae. The fifth sacral nerves ($S_5$) exit inferior to the sacrum.
5. *Coccygeal nerves* (1 pair)—The coccygeal nerves ($Co_1$) exit at the coccyx.

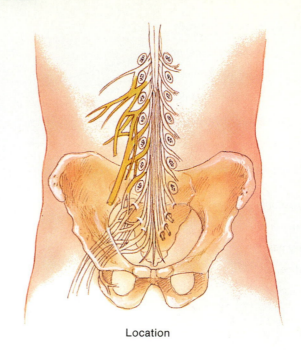

Location

## NERVE PLEXUSES

A nerve *plexus* is a place where two or more nerves join and exchange nerve fibers. Figure 25-12 shows an example of the type of network formed by adjacent spinal nerves as they leave the spinal column. Such networks are formed on both sides of the body. Figure 25-12 shows the lumbar plexus in detail.

There are several important nerve plexuses. Using your text and charts, identify the following (Figures 25-11 and 25-12):

1. *Cervical plexus*—involves nerves $C_1$ through $C_5$
2. *Brachial plexus*—involves nerves $C_4$ through $C_8$ and $T_1$ and $T_2$
3. *Lumbar plexus*—involves nerves $L_1$ through $L_4$
4. *Sacral plexus*—involves nerves $L_4$, $L_5$, and $S_1$ through $S_4$

## THE AUTONOMIC NERVOUS SYSTEM

Using the diagram of the autonomic nervous system (Figure 25-13), locate all its parts on a demonstration cat. You can expose all of these nerves and ganglia on your laboratory cat more easily at the end of the course, when it will be possible to remove most abdominal and thoracic organs.

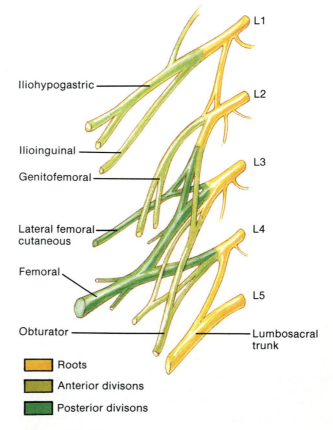

Iliohypogastric
Ilioinguinal
Genitofemoral
Lateral femoral cutaneous
Femoral
Obturator
L1
L2
L3
L4
L5
Lumbosacral trunk

■ Roots
■ Anterior divisons
■ Posterior divisons

**Figure 25-12** Human lumbar plexus (anterior view). (From G. J. Tortora, *Principles of Human Anatomy*, 6th ed., New York: HarperCollins, 1992.)

**Figure 25-13** Autonomic nervous system of the cat.

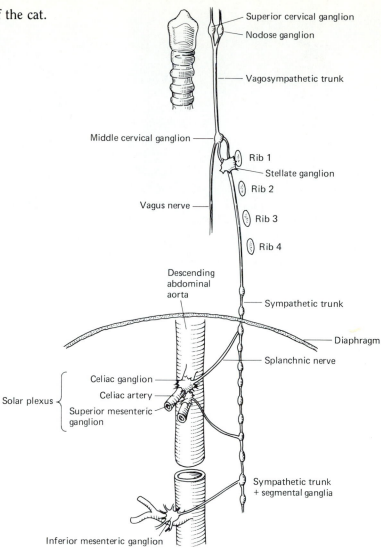

Superior cervical ganglion
Nodose ganglion
Vagosympathetic trunk
Middle cervical ganglion
Rib 1
Stellate ganglion
Rib 2
Vagus nerve
Rib 3
Rib 4
Descending abdominal aorta
Sympathetic trunk
Diaphragm
Splanchnic nerve
Celiac ganglion
Celiac artery
Solar plexus
Superior mesenteric ganglion
Sympathetic trunk + segmental ganglia
Inferior mesenteric ganglion

## REMOVAL OF THE CAT BRAIN (OPTIONAL)

This dissection may be done, even though the sheep's brain has already been studied. Use extreme care in this procedure. The bone is hard and the nervous tissue is soft. When removing bone tissue, be careful not to damage the soft brain beneath it. Also guard against cutting your fingers with sharp dissecting tools. Do not use the razor-blade-type scalpel because it is not durable enough for this exercise.

Begin by removing all the skin and muscle from the dorsal surface of the head (Figure 25-14a). The area to be skinned extends from the tip of the nose to the second cervical vertebra, between the ears.

When the skull is cleanly exposed, begin the removal of the bone. The ideal tool for this purpose is a

bone saw. If this is not available, take a hammer, and crack the bone as much as possible over the entire exposed surface, being careful not to injure the soft brain below. The thickest bone is in the posterior part of the skull. Remove this first with large bone forceps. Expose about ½ inch of the spinal cord. Proceed to remove the top of the skull until the entire dorsal surface of the brain is exposed (Figure 25-14b).

Cut through the spinal cord, and lift the brain out slowly by cutting the cranial nerves that hold it to the base of the skull. Leave as much of the nerves as possible so that they can be identified.

Study the dorsal and ventral surfaces and the midsagittal section, following instructions for the sheep brain and using Figures 25-15 and 25-16, to identify parts of the cat brain.

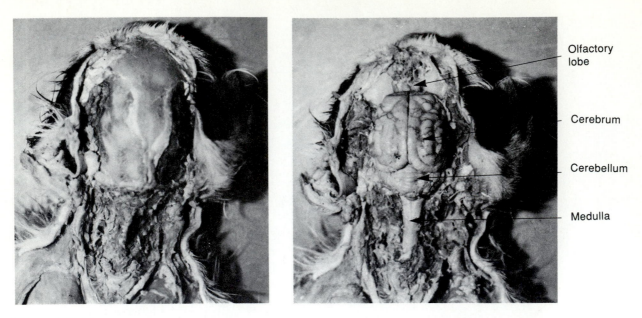

Olfactory lobe

Cerebrum

Cerebellum

Medulla

**Figure 25-14** (a) Cat skull, skin, and muscle removed; (b) cat skull, bone removed.

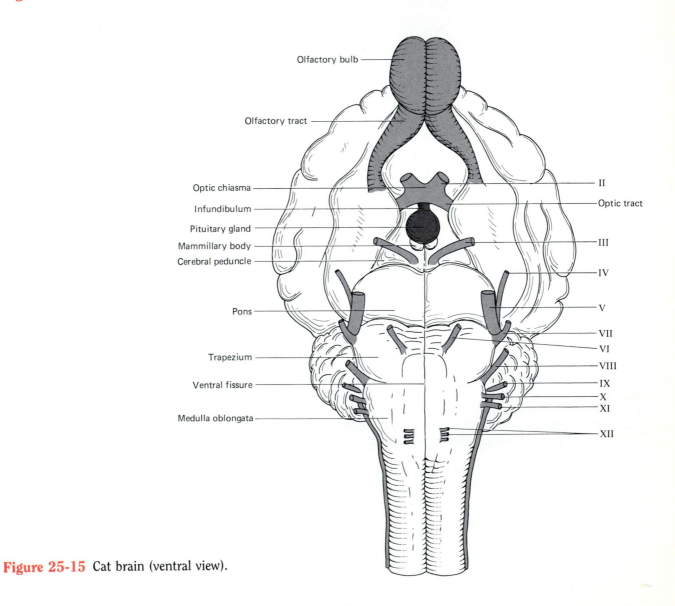

Olfactory bulb

Olfactory tract

Optic chiasma

Infundibulum

Pituitary gland

Mammillary body

Cerebral peduncle

Pons

Trapezium

Ventral fissure

Medulla oblongata

II

Optic tract

III

IV

V

VII

VI

VIII

IX

X

XI

XII

**Figure 25-15** Cat brain (ventral view).

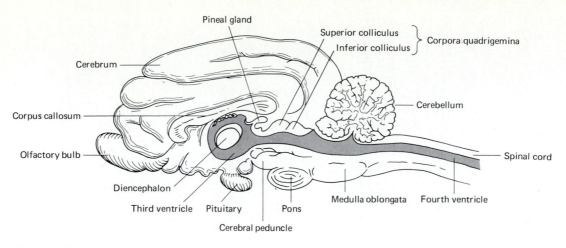

**Figure 25-16** Cat brain (midsagittal section).

# LABORATORY REVIEW *25*

NAME _____

LAB SECTION _____ DATE _____

## ANATOMY OF THE NERVOUS SYSTEM

### COMPLETION, FILL-IN, AND DEFINITION

1. Fill in the names of the cranial nerves.

   I. _____     VII. _____

   II. _____     VIII. _____

   III. _____     IX. _____

   IV. _____     X. _____

   V. _____     XI. _____

   VI. _____     XII. _____

2. Define:
   a. Nerve plexus _____

   _____

   b. Ganglion _____

   _____

   c. Nerve _____

   _____

   d. Neuron _____

   _____

   e. Nerve fiber _____

   _____

   f. Brain ventricle _____

   _____

   g. Meninges _____

   _____

3. Name the lobes of the cerebral hemispheres:

   a. _____

   b. _____

   c. _____

   d. _____

4. How does a nerve differ from a nerve tract? _____

   _____

   _____

5. Name four nerve plexuses:

    a. _____

    b. _____

    c. _____

    d. _____

6. What is a choroid plexus?

_____

_____

7. What is the difference between a gyrus and a sulcus?

_____

_____

_____

_____

8. What is the difference between a sulcus and a fissure?

_____

_____

_____

_____

# Demonstration of Selected Cranial Nerve and Cerebellar and Cerebral Functions in the Human

**After completing this exercise, you should be able to**

1. Explain the events taking place in a reflex arc
2. List the components of the reflex arc
3. List the functions of the cerebellum
4. Name and list the functions of each cranial nerve
5. Perform simple tests to demonstrate the functions of the cerebellum and selected cranial nerves
6. Briefly describe the essential processes of an overall memory system
7. Describe basic differences between short-term memory and long-term memory
8. Use selected techniques to demonstrate the influence and involvement of organizational processes with memory

*Reflex* actions are rapid automatic responses to stimuli, over which an individual appears to have little or no control. Such responses attempt to restore the body to *homeostasis*. An example of a well-known reflex is the knee-tendon reflex, which occurs when the knee is hit by a small rubber mallet. The lower leg jerks in a characteristic way, which appears to be uncontrolled. For a reflex to occur, a nerve impulse must travel along a neural pathway termed a *reflex arc* (Figure 26-1), the functional unit of the body's nervous system. Figure 26-1 shows the direction or path of a nerve impulse, from its origin in one part of the body (*receptor*) to its termination elsewhere in the body (*effector*). The basic components of this *conduction pathway* in order of involvement are as follows:

1. *Receptor*—the distal end of a dendrite or sensory neuron that responds to a stimulus by initiating a nerve impulse in the sensory neuron
2. *Sensory neuron*—a nerve cell that passes the nerve impulse from the receptor to the central nervous system.
3. *Center*—a region within the central nervous system where an incoming sensory impulse generates an outgoing motor impulse. This event occurs when the sensory neuron synapses with an association or a motor neuron.
4. *Motor neuron*—a nerve cell that transmits the impulse generated by the sensory or association neuron in the center to the effector body part that will respond, such as a muscle or gland.
5. *Effector*—the effector body part that responds to the motor-nerve impulse.(An impulse traveling to the nerve fiber in an involved muscle causes it to contract.)

Reflexes can be categorized according to the effector organ involved. Thus, reflexes that result in the contraction of skeletal muscles are termed *somatic reflexes*. Those that involve brain centers and cranial nerves are known as *cranial reflexes*. Finally, reflexes that either cause the contraction of cardiac or smooth muscle or cause gland secretions are *visceral* (autonomic) reflexes.

The cerebellum plays an important role in balance, posture, and coordination. It receives a steady flow of impulses from receptors embedded in joints and muscles, as well as from the eyes and portions of the ear. It registers the body's position, as well as its rate of movement and the coordination of muscular activities. This laboratory exercise includes a number of tests to demonstrate these functions of the cerebellum and of selected cranial nerves.

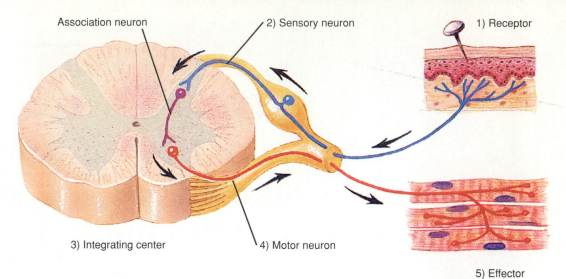

**Figure 26-1** A generalized reflex arc. The components of the reflex arc and the direction of the conduction pathway (arrows) are shown. (From G. J. Tortora and S. R. Grabowski, *Principles of Anatomy and Physiology,* 7th ed., New York: HarperCollins, 1993.)

## Table 26-1   A General Summary of the Cranial Nerves

| Nerve Designation | Composition | Associated Functions |
|---|---|---|
| I. Olfactory | Sensory | Smell |
| II. Optic | Sensory | Sight |
| III. Oculomotor | Motor and sensory | Eye movement and focusing |
| IV. Trochlear | Motor and sensory | Nerve supply to eye muscle |
| V. Trigeminal | Motor and sensory | Chewing, sensation from the head region, between eyes and teeth |
| VI. Abducens | Motor and sensory | Eye movement |
| VII. Facial | Motor and sensory | Facial expression and taste |
| VIII. Statoacoustic | Sensory | Posture and hearing |
| IX. Glossopharyngeal | Motor and sensory | Swallowing, taste, and general sensation |
| X. Vagus | Motor and sensory | Visceral muscle movement and sensation |
| XI. Accessory | Motor and sensory | Swallowing and head movement |
| XII. Hypoglossal | Motor and sensory | Speech and swallowing |

The brain gives rise to 12 pairs of cranial nerves that supply motor and sensory fibers to the head, neck, and some visceral organs. These nerves form part of the peripheral nervous system that connects the brain with receptors, muscles, and glands. Table 26-1 lists the cranial nerves and their composition and function.

This exercise also considers a fascinating integrative function of the cerebrum, namely, *memory.* The

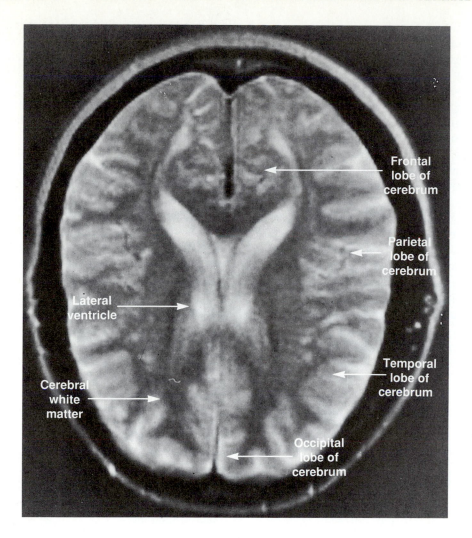

**Figure 26-2** A magnetic resonance image (MRI) of the human brain, showing portions of the cerebrum. The frontal lobe forms the anterior portion of each cerebral hemisphere.

frontal lobe (Figure 26-2), as well as other brain portions—such as the parietal, occipital, and temporal lobes, and the diencephalon—are associated with memory.

Memory is the ability of an individual to recall thoughts. For an experience to become part of memory, it must cause changes in the CNS that represent the experience.

The essential processes of an overall memory system are encoding, storage, and retrieval. Together, these three components form the foundation of the major information-processing sequence in the body. While these components have not been defined to everyone's satisfaction, the following definitions should be sufficient for purposes of this exercise. *Encoding* refers to processes by which a memorial representation of physical objects and events is developed. *Storage* may be defined as the persistence of information over time, while *retrieval* refers to the search for and utilization of stored information. It should be noted that even information that is stored may not be easily located and used. For example, an individual may know all of the states in the United States, but at any given moment, may not be able to retrieve all of them.

Two types of memory are generally recognized: short-term (*primary memory*) and long-term (*secondary memory*). With *short-term memory,* every item (stimulus) that is perceived is kept temporarily in the conscious mind. Such items may be lost or forgotten unless they are rehearsed. If an item (such as a phone number, name, etc.) is rehearsed, it remains in the short-term memory, and may enter the subconscious mind or long-term memory. Information stored in *long-term memory* is relatively permanent and is subject to a very slow decay process.

Both long-term and short-term memory are very flexible systems. Each is capable of encoding materials in a number of ways. Short-term memory is kept in mind by the processes of reverberation and synaptic facilitation. With *reverberation,* thoughts continue to cycle in closed chains of neurons until the threshold for excitation is no longer reached. In *synaptic facilitation,* neurotransmitting chemicals temporarily are produced in larger amounts, facilitating the transfer of impulses between particular neurons. The growth of synaptic contacts and the formation of memory molecules (specific protein and ribonucleic acid molecules, either individually or together) retain long-term memory.

From the results of various investigations, it appears that the overall memory system of the human is governed by principles of organization and the grouping of perceptions. Information is not passively stored in exactly the same form in which it is received. It is structured, interrelated with other items, and organized so that both the storage and the retrieval of the information will be efficient. Several techniques may be used to demonstrate and measure such activities and/or processes. These include category *clustering* (grouping related information) *for recall,* and *visual coding* (encoding information in terms of visual images). Several approaches are used in order to demonstrate the influence of selected organizational processes of memory.

## SELECTED CRANIAL NERVE AND CEREBELLAR FUNCTIONS

### Materials

The following materials should be provided for class use:

1. Containers of
   a. Garlic
   b. Ground coffee
2. Tuning forks
3. Sterile cotton swabs and tongue depressors
4. Timers or stopwatches that have an audible ticking sound
5. Flashlights or penlights

### Procedure 1: Cranial Nerve Function Tests

This procedure should be performed by students in pairs.

1. Take turns in performing the tasks listed for each cranial nerve. Describe your responses and those of your laboratory partner in the results and observations section.

2. Read each of the tasks first before performing them.
   a. *Olfactory nerve.* Close your eyes while your partner passes containers of garlic and ground coffee by your nose. Can you distinguish the aromas?
   b. *Optic nerve.* Determine how many words on one page of a textbook or other publication you can read in 1 minute.
   c. *Oculomotor nerve.* Keep your head still while your partner slowly moves a pencil up and down in front of you. Can you follow the movement with your eyes?
   d. *Trochlear and abducens nerves.* Look straight ahead while your partner holds a pencil about 1 meter from the side of your head and moves it forward and backward. Indicate at what point you cannot see the pencil. Have your partner determine whether the pupil opens or closes more during this task.
   e. *Trigeminal nerve.* Close your mouth and clench your teeth. Have your partner hold one hand firmly under your chin. Try to open your mouth. (This is a test for a motor response.)
   f. *Facial nerve.* Carry out the following facial movements: smile and show your teeth, inflate your cheeks, wrinkle your forehead, and raise either one or both eyebrows. Can you raise each eyebrow separately?
   g. *Vestibulocochlear (acoustic) nerve.* Close your eyes. Have your partner hold a timer or stopwatch about 1 meter away from the left or right side of your head. Can you determine the direction from which the sound is coming? Is this a test of the cochlear or the vestibular portion of the nerve?
   h. *Glossopharyngeal and vagus nerves.*
      i. Open your mouth wide. Have your partner carefully touch the back of your throat. What is the response?
      ii. Have your partner hold down your tongue with a sterile tongue depressor and direct a penlight into your mouth, while you say "ah." What happens to the uvula (the fleshy mass near the back of the mouth)?
   i. *Spinal accessory nerve.*
      i. Sit down, and have your partner push firmly on your shoulders. Try to raise your shoulders. (Note that this does not mean to stand up.) What muscles are involved?
      ii. Have your partner place his or her hands firmly on the sides of your face in front of the ears. Try to turn your head. What muscles are involved?
   j. *Hypoglossal nerve.* Stick your tongue out and try to keep it as straight as you can.

## Procedure 2: Demonstration of Cerebellar Function

This procedure should be performed by students in pairs.

1. Take turns in performing the following tasks. Describe your responses and those of your laboratory partner in the results and observations section.
2. Have your laboratory partner stand approximately 2 meters in front of you. You will be the examiner. Read each of the tasks out loud, and determine the ease with which they are performed.
   a. Stretch both arms to the side and move fingers rapidly.
   b. Stretch both arms to the side and snap the fingers of both hands.
   c. Turn to the side and, looking straight ahead, walk a straight line placing the heel of one foot directly in front of the toe of the other foot.
   d. Close your eyes and stand erect for 1 minute.
   e. Close your eyes and touch your nose with the index finger of your left hand.
   f. Close your eyes and touch your nose with the index finger of your right hand.
   g. Touch the index finger on my right hand. (Examiner should move the right hand accordingly.)
   h. Stand erect and move the heel of your right foot first up and then down the shin of your left leg.

## RESULTS AND OBSERVATIONS

### Cranial Nerve Function Tests

1. Enter your responses and those of your partner in Table 26-2.
2. How would you test for the sensory responses of the trigeminal nerve? _____

   _____

3. How could you distinguish between the sensory responses of the vestibular and the cochlear portions of the vestibulocochlear nerve? _____

   _____

   _____

### Demonstration of Cerebellar Function

1. Enter your responses and those of your partner in Table 26-3. Indicate whether the task was performed and whether any difficulties were observed.
2. Suggest two other possible tasks to test cerebellar function. _____

   _____

**Table 26-2**

| Cranial Nerve Tested | Your Response | Partner's Response |
|---|---|---|
| Olfactory | | |
| Optic | | |
| Oculomotor | | |
| Trochlear and abducens | | |
| Trigeminal | | |
| Facial | | |
| Vestibulocochlear (acoustic) | | |
| Glossopharyngeal and vagus | | |
| Spinal accessory | | |
| Hypoglossal | | |

**Table 26-3**

| Task | Your Response | Partner's Response |
|---|---|---|
| Move fingers quickly | | |
| Snap fingers | | |
| Walk a straight line in tandem without loss of balance | | |
| Stand erect with feet together and eyes closed without loss of balance | | |
| Touch index finger of each hand to the nose while eyes closed | | |
| Touch examiner's index finger on command | | |
| Move heel of one foot up and down shin of opposite leg | | |

## CATEGORY CLUSTERING (GROUPING PHENOMENON)

### Materials

List of words supplied in Table 26-4.

### Procedure: This procedure should be performed by students individually.

1. Read the list of 30 words in Table 26-4 only once.
2. Turn immediately to the results and observations section, and write down as many words as you can recall. Use any order you wish.
3. After you cannot recall any additional terms, count the correct responses, and enter the number in the results and observations section.
4. Answer the questions concerning this experiment in the results and observations section.

## VISUAL VERSUS VERBAL CODING

### Materials

The following materials should be provided for general class use.

1. Two sets of 20 anatomical diagrams or photographs of body structures (each item should be numbered)
2. Two sets of 20 cards or sheets, each containing one term for an anatomical structure

### Procedure

This procedure is to be performed by students in pairs.

1. Have your laboratory partner sit comfortably, with his or her laboratory manual opened to the results and observation section for this exercise.

**Table 26-4   Grouping Phenomenon**

| | | | | | |
|---|---|---|---|---|---|
| Snake | Sycamore | Red | Pumpkin | blue | Elm |
| London | Brown | New York | Maple | Fox | Lettuce |
| Yellow | Tomato | Hemlock | Redwood | Broccoli | Black |
| Orange | Pine | Peas | Parrot | Moose | Paris |
| Los Angeles | Cow | Zebra | Moscow | Seattle | Carrot |

2. Obtain one set of 20 anatomical photographs or diagrams. Show the illustrations to your partner one at a time and only once.

3. After you have shown your partner all the illustrations, ask your partner to write the names of all the structures he or she can recall in the results and observations section.

4. Obtain one set of 20 cards or sheets containing individual terms for anatomical structures. Show the cards to your partner one at a time and only once.

5. When you have shown your partner all the cards, ask your partner to write all the terms he or she can recall in the results and observations section.

6. Change places with your partner, and repeat Steps 1 through 5 with a second and different set of illustrations and word cards.

## RESULTS AND OBSERVATIONS

### *Category Clustering*

1. Write the words recalled from Table 26-4.

| | |
|---|---|
| 1. _____ | 16. _____ |
| 2. _____ | 17. _____ |
| 3. _____ | 18. _____ |
| 4. _____ | 19. _____ |
| 5. _____ | 20. _____ |
| 6. _____ | 21. _____ |
| 7. _____ | 22. _____ |
| 8. _____ | 23. _____ |
| 9. _____ | 24. _____ |
| 10. _____ | 25. _____ |
| 11. _____ | 26. _____ |
| 12. _____ | 27. _____ |
| 13. _____ | 28. _____ |
| 14. _____ | 29. _____ |
| 15. _____ | 30. _____ |

Correct number recalled: _____

2. a. Were the words toward the end of the list recalled better than earlier ones? _____

   b. What type of memory accounts for your response: short-term or long-term? _____

3. Were the items in the middle of the list difficult to recall? _____

4. What categories or groups were formed in the list?

   _____

   _____

5. Were you consciously aware of grouping during the listing process? _____

6. Compare your results with those of a laboratory partner. Were they comparable as to groupings?

   _____

### *Visual Versus Verbal Coding*

1. Write the names of the anatomical structures recalled from the illustrations shown.

| | |
|---|---|
| a. _____ | k. _____ |
| b. _____ | l. _____ |
| c. _____ | m. _____ |
| d. _____ | n. _____ |
| e. _____ | o. _____ |
| f. _____ | p. _____ |
| g. _____ | q. _____ |
| h. _____ | r. _____ |
| i. _____ | s. _____ |
| j. _____ | t. _____ |

Correct number recalled: _____

2. Write the anatomical terms recalled from the cards shown.

| | |
|---|---|
| a. _____ | k. _____ |
| b. _____ | l. _____ |
| c. _____ | m. _____ |
| d. _____ | n. _____ |
| e. _____ | o. _____ |
| f. _____ | p. _____ |
| g. _____ | q. _____ |
| h. _____ | r. _____ |
| i. _____ | s. _____ |
| j. _____ | t. _____ |

Correct number recalled: _____

3. In which situation was the larger number of correct recall responses found? _____

4. Were the recall responses of your partner similar?

_____

5. What conclusion can you draw from this experiment? _____

_____

_____

# LABORATORY REVIEW *26*

NAME _____

LAB SECTION _____ DATE _____

## DEMONSTRATION OF SELECTED CRANIAL NERVE AND CEREBELLAR AND CEREBRAL FUNCTIONS IN THE HUMAN

### COMPLETION AND DEFINITION

1. Where is the cerebellum located in the human with respect to other parts of the brain? _____

   _____

2. List two functions of the cerebellum.
   a. _____
   b. _____

3. What are the cranial nerves? _____

   _____

4. How many cranial nerves are there in the human? _____

5. Give the functions of the cranial nerves listed.

   a. Trochlear _____

   b. Facial _____

   c. Glossopharyngeal _____

   d. Vagus _____

   e. Hypoglossal _____

6. For each item listed, indicate which specific cranial nerve in the human could be tested by using the item.

   a. Odor of fresh brewed coffee _____

   b. Reading letters on an eye chart _____

   c. Having a penlight flashed into an eye _____

   d. Moving a timer or stopwatch with an audible ticking sound on either side of an individual whose eyes are closed _____

   e. Touching the individual's uvula _____

7. List and briefly explain the components of the overall memory system.

   a. _____
   b. _____
   c. _____
   d. _____

8. Distinguish between short-term memory and long-term memory. _____

   _____

   _____

## ANSWERS

9. What is category clustering? _____

_____

10. Is the memory for pictures better than the memory for names of the same pictures?

_____

## MULTIPLE CHOICE

Insert the correct answers in the spaces provided.

1. _____
2. _____
3. _____
4. _____
5. _____
6. _____
7. _____
8. _____
9. _____
10. _____
11. _____
12. _____
13. _____
14. _____

1. Which of the following is the specific part of a reflex that responds to a motor impulse?
   a. effector
   b. center
   c. sensory neuron
   d. receptor
   e. dendrite

2. Which of the following is the part of a neuron that conducts nerve impulses away from the cell body?
   a. effector
   b. center
   c. sensory neuron
   d. receptor
   e. dendrite

3. What is the type of neuron that conducts nerve impulses toward the central nervous system?
   a. efferent
   b. internuncial
   c. association
   d. central
   e. afferent

4. What is the distal end of a sensory neuron that responds to a stimulus by starting a nerve impulse?
   a. efferent
   b. internuncial
   c. association
   d. central
   e. afferent

5. What is the region within the central nervous system where an incoming sensory impulse generates an outgoing motor impulse?
   a. motor neuron
   b. receptor
   c. sensory neuron
   d. center
   e. effector

6. What kind of reflex results in the contraction of skeletal muscles?
   a. somatic
   b. visceral
   c. cranial
   d. cardiac

7. What kind of reflex involves brain centers and cranial nerves?
   a. somatic
   b. visceral
   c. cranial
   d. cardiac

8. What are reflexes causing smooth muscle contraction called?
   a. somatic
   b. visceral
   c. cranial
   d. cardiac

9. What are reflexes that cause heart muscles to contract considered to be?
   a. somatic
   b. visceral
   c. cranial
   d. cardiac

10. Which of the following cerebral components is involved with responses related to memory, reasoning, and judgment?
    a. temporal lobe
    b. parietal lobe
    c. occipital lobe
    d. all of these
    e. only choices *a* and *b*.

11. In addition to *encoding* and *storage,* what is an essential process of an overall memory system?
    a. short-term memory
    b. primary memory
    c. retrieval
    d. organization
    e. long-term memory

12. What is the process by which a memory representation can be developed for physical objects and events?
    a. encoding
    b. storage
    c. retrieval
    d. short-term memory
    e. long-term memory

13. Which of the following is the search for and use of stored information?
    a. encoding
    b. storage
    c. retrieval
    d. short-term memory
    e. long-term memory

14. Short-term memory is kept in mind primarily by which of the following?
    a. reverberation
    b. memory molecule formation
    c. synaptic facilitation
    d. RNA molecules
    e. choices *a* and *c* only.

# Demonstration of Reflexes in the Human

**After completing this exercise, you should be able to**

1. List at least five commonly encountered reflex actions in the human
2. Perform simple procedures to identify and to demonstrate common reflex actions

As indicated in Exercise 26, discrete nervous pathways to the spinal cord or brain are involved in the performance of a reflex action. The knee-jerk response is a well-known example. Another situation involving a reflex action familiar to most is the response to touching a hot object. A stimulus received by the heat- or pain-sensitive endings in the finger is transmitted as an impulse to the spinal cord, where it relays the signal to a motor neuron. The motor neuron completes the arc by stimulating muscle contraction in the finger, moving it away from the unpleasant stimulus. Reflexes carried out by the spinal cord are *spinal reflexes.*

Because normal reflexes depend on normal neuron functions, reflexes often are utilized to obtain information concerning the condition of the nervous system. An anesthesiologist, for example, may use a reflex in a patient being anesthetized to determine how a particular drug is affecting the nervous system. Some specific reflexes also are of importance in diagnosing nervous-system disorders and in locating injured tissue. Lesions of the central nervous system may occur at all levels and may result from various causes, including physical injuries, infections, and biochemical changes. In some situations, exaggerated spinal reflexes result, while in others, underactive reflexes occur. Examples of reflexes of diagnostic significance include the abdominal reflex, Archilles reflex (ankle jerk), Babinski sign, biceps reflex,

patellar reflex (knee jerk), and the triceps reflex. Table 27-1 briefly describes both the normal and the abnormal responses for these reflexes and some of the conditions responsible for abnormal reactions. The specific spinal segments involved also are indicated in this table.

This exercise considers several types of reflex actions. These include *deep* and *superficial* reflexes. *Deep reflexes,* also termed *somatic reflexes,* result in the contraction of skeletal muscles and are triggered by a sharp tap on an appropriate muscle or tendon. The associated receptors for these reflexes are situated in the muscle and not in the tendon. Tapping the tendon causes the muscle to be stretched; this stretching, in turn, activates the muscle spindle, thus triggering the response. Somatic reflexes are named according to the response they produce and include the *stretch reflex* and the *withdrawal* (or *flexor*) *reflex.*

A stretch reflex involves only two neurons and one synapse in the pathway, and is therefore termed a *monosynaptic reflex arc.* Because the receptor and effector organs of the stretch reflex involve structures on the same side of the spinal cord, the reflex arc is referred to as an *ipsilateral reflex.* The patellar (knee-jerk) reflex (Figure 27-1) and all other monosynaptic reflex arcs are examples of an ipsilateral reflex.

A *withdrawal* or *flexor reflex* consists of a *polysynaptic reflex arc.* These reflexes involve association neurons (termed *interneurons*), in addition to sensory and motor neurons. Visceral reflexes, which also are polysynaptic because at least two motor-neuron synapses are always involved, are included in this exercise.

*Superficial reflexes,* also referred to as *cutaneous reflexes,* are caused by irritating or tactile stimulation. Stroking or scratching the skin surface triggers these reflexes. Figure 27-2 shows the features of clinically important reflexes.

Abnormal responses to stimuli may be *reduced (hypoflexia), exaggerated (hyperreflexia),* or *pathological.*

**Table 27-1   Normal and Abnormal Reflex Actions**

| Reflex | Spinal Segment[a] | Normal Response | Abnormal Response |
|---|---|---|---|
| Abdominal | $T_8$, $T_9$, $T_{10}$ (above umbilicus) $T_{10}$, $T_{11}$, $T_{12}$ (below umbilicus) | Abdominal muscles underlying the skin contract in response to stroking the side of the abdomen. Movement of the umbilicus toward the region stimulated also occurs. | Absence of reflex. Conditions causing abnormal reactions include lesions of peripheral nerves, lesions of reflex centers in the chest region of the spinal cord, and multiple sclerosis. |
| Achilles | $S_1$, $S_2$ | Extension of the foot in response to tapping the Achilles tendon. | Absence of reflex or exaggerated reflex. Conditions causing first of abnormal reactions include damage to nerves supplying posterior leg muscles, damage to nerve cell in spinal cord's lumbosacral region, chronic diabetes mellitus, neurosyphilis, alcoholism, and subarachnoid bleeding. Exaggerated responses may be due to injuries of the cord's sacral region. |
| Babinski sign (positive) | $L_4$, $L_5$, $S_1$, $S_2$ | Extension of the great (big) toe in response to a light stimulation of the outer margin of the sole of the foot. Other toes may or may not respond. Children under 1½ years of age exhibit the sign because of an incomplete development of the nervous system (corticospinal tract) at this time. | Presence of reflex after 1½ years of age. Conditions causing abnormal reaction include interruption of corticospinal tract development and lesion in the upper corticospinal tract. |
| Biceps reflex | C5, C6 | Flexion of the forearm occurs in response to tapping the biceps tendon. | Absence of reflex or exaggerated response. Conditions causing abnormal reactions include injury to the cervical segments of the spinal cord. |
| Patellar | $L_2$, $L_3$, $L_4$ | Extension of the leg in response to tapping the patellar ligament. | Absence of reflex or exaggerated response. Conditions causing abnormal reactions include injury to nerves in the lumbar segments of the spinal cord, chronic diabetes, and neurosyphilis. Exaggerated responses may result from injury or disease in the corticospinal tract. |
| Triceps reflex | C7, C8 | Extension of the forearm occurs in response to tapping the triceps tendon. | Absence of reflex or exaggerated response. Conditions causing abnormal reactions include injury to the cervical segments of the spinal cord. |

[a]C = cervical; L = lumbar; S = sacral; and T = thoracic.

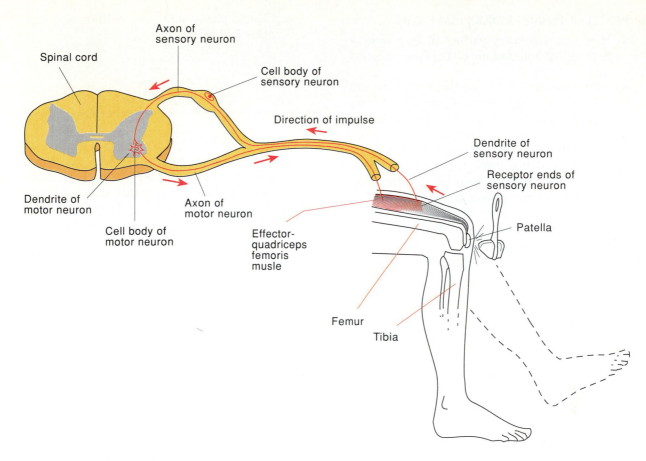

**Figure 27-1** Simple reflex arc involved in the knee-jerk (patellar) response.

Aging, injury to neurons, and malnutrition are among the causes of hypoflexia. Hyperreflexia may result from the loss of inhibitory control by the motor cortex and from some types of poisonings. Pathological reflex responses involve one or more muscles other than the muscle where the stimulus originates.

## Materials

The following materials should be provided for class use:

1. Rubber mallets (percussion hammers)
2. Sterile cotton swabs
3. Plastic disposable tubes (15 ml capacity)
4. Lemon juice or a safe substitute
5. 5 percent glucose solution
6. Small pieces of ice
7. Container of disinfectant
8. Marking pencils or pens

## Procedure: Demonstrating Common Reflex Actions

The following tests should be performed by students in pairs (except Test 6, which requires three students). Enter brief descriptions of the reactions produced in Table 27-2, and refer to Figure 27-2, as needed.

## SOMATIC REFLEXES OF THE LEG

**1. ACHILLES TENDON REFLEX (ANKLE JERK)**

a. Have your laboratory partner kneel on a chair or table so that both feet hang over the edge.

b. Locate the Achilles tendon of the right foot, and tap it with a rubber hammer or the lateral edge of your palm. Describe the response in the results and observations section of this exercise.

c. Change places with your partner, and repeat Steps (a) and (b). (Refer to Figure 27-2a.)

## 2. PATELLAR TENDON REFLEX (KNEE JERK)

a. Have your laboratory partner sit on a table or desk so that both feet are off the floor and hang freely.

b. Locate the patellar tendon (the depression just below the kneecap; refer to Figure 27-1d) of the right leg, and tap the tendon with the rubber mallet or the edge of your hand. Note the response in the results and observations section.

c. Have your partner count backward from 99 to 32. Tap the patellar tendon of the right leg again. Was the response greater or less than the first time?

d. Repeat Steps (b) and (c) with the left leg. Note the extent of the response.

e. Change places with your partner, and repeat Steps (a)–(d).

## SOMATIC REFLEXES OF THE ARM

### 3. BICEPS REFLEX

a. Have your laboratory partner rest his or her arm on a table or desk.

b. Place your thumb on the biceps tendon, as shown in Figure 27-2b.

c. Tap the first digit of your thumb with the reflex hammer. Note the extent of the response.

d. Change places with your partner, and repeat Steps (a)–(c).

### 4. TRICEPS REFLEX

a. Have your laboratory partner flex one arm at the elbow.

b. Hold the wrist of that arm, as shown in Figure 27-1e.

c. Tap the triceps tendon above the elbow, using the pointed end of the reflex mallet, and note the response.

d. Test each arm in a flexed, relaxed, state.

e. Change places with your partner, and repeat Steps (a)–(d).

## SUPERFICIAL REFLEX OF THE FOOT

### 5. BABINSKI SIGN

a. Have your partner remove the shoe and sock from the right foot.

b. Take the tip of a rubber mallet or the blunt end of a pen, and run it on the sole of the foot firmly in one upward movement from the heel to the transverse arch. (Refer to Figure 27-1c.) Record your findings in the results and observations section.

c. Change places with your partner, and repeat Steps (a) and (b).

## VISCERAL REFLEXES

### 6. CILIOSPINAL REFLEX

a. A total of three persons are needed for this activity. Have one laboratory partner sit comfortably in a chair. Observe this partner's pupils.

b. Have another laboratory member unexpectedly touch a small piece of ice on the back of the seated individual. Note the response of the pupils. Record your findings in the results and observations section.

c. Change places with your partners, and repeat Steps (a) and (b).

### 7. SALIVARY REFLEX

a. Have your partner rinse his or her mouth with water.

b. Collect all saliva secreted during a 3-minute period in a plastic tube. Mark the top level of fluid in the tube. Dispose of the tube, as indicated by your instructor.

c. Dip a cotton swab in the glucose solution and place the swab under your partner's tongue. Collect the saliva secreted during a 3-minute period and mark the fluid level.

d. Dip another swab in lemon juice, and repeat Step (c). In which of the three 3-minute intervals was more saliva produced? Enter your findings in the results and observations section.

e. Change places with your partner, and repeat Steps (a)–(d).

## RESULTS AND OBSERVATIONS

1. Enter in Table 27-2 brief descriptions of your reactions and those of your partner in the reflex-testing procedures.

2. Were differences in patellar tendon responses observed? Explain. _____

_____

3. Were there any differences in the flexed versus the relaxed state of biceps tendon noted? Explain.

_____

_____

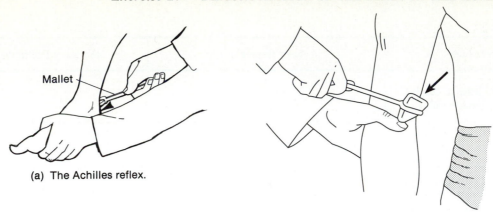

(a) The Achilles reflex.

(b) Biceps reflex

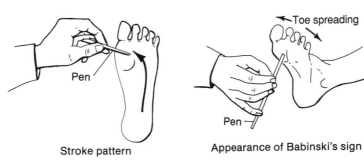

(c) Babinski's sign

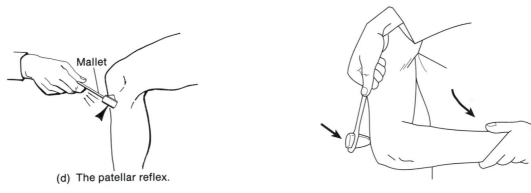

(d) The patellar reflex.

(e) Triceps reflex

**Figure 27-2** Selected clinically important reflexes.

4. What is the medical significance of the Babinski response? _____
_____

5. What type of nerve is stimulated in the ciliospinal reflex? _____
_____

6. Can the salivary reflex be controlled? Explain.
_____
_____
_____

**Table 27-2**

| Reflex Procedure | Your Response | Laboratory Partner's Response |
|---|---|---|
| Somatic reflexes of the leg<br>   Achilles tendon reflex | | |
| Patellar tendon reflex<br>   Right leg (initial) | | |
| Right leg (after counting) | | |
| Left leg (initial) | | |
| Left leg (after counting) | | |
| Somatic reflexes of the arm | | |
| Biceps reflex | | |
| Triceps reflex | | |
| Flexed arm | | |
| Relaxed arm | | |
| Superficial reflex of the foot:<br>   Babinski sign | | |
| Visceral reflexes | | |
| Ciliospinal reflex | | |
| Salivary reflex<br>   Normal | | |
| With glucose solution | | |
| With lemon juice | | |

# LABORATORY REVIEW 27

NAME _____

LAB SECTION _____ DATE _____

## DEMONSTRATION OF REFLEXES IN THE HUMAN

### COMPLETION AND DEFINITION

1. What is a reflex? _____

_____

2. Diagram a reflex arc.

3. In the following tests for reflex action, what body structure is stimulated?

a. Knee jerk _____

b. Babinski sign _____

c. Ankle jerk _____

4. With the aid of your text and other references, define or explain the following.

a. Tonus _____

_____

b. Flaccid _____

c. Ipsilateral _____

d. Reciprocal innervation _____

_____

5. What is a flexion or a withdrawal response? _____

_____

6. Distinguish between deep and superficial reflexes. Give one example of each type.

a. _____

b. _____

## ANSWERS

1. _____
2. _____
3. _____
4. _____
5. _____
6. _____
7. _____
8. _____
9. _____
10. _____

## MULTIPLE CHOICE

Insert the correct answers in the spaces provided.

1. The extension of the forearm by tapping the tendon of the triceps indicates the proper functioning of which of the following spinal-cord segments?
   a. sacral 1 and 2
   b. cervical 7 and 8
   c. cervical 5 and 6
   d. lumbar 2, 3, and 4
   e. lumbar 4 and 5, and sacral 1 and 2

2. A normal extension of the leg by tapping the patellar ligament shows the proper functioning of which of the following spinal-cord segments?
   a. sacral 1 and 2
   b. cervical 7 and 8
   c. cervical 5 and 6
   d. lumbar 2, 3, and 4
   e. lumbar 4 and 5, and sacral 1 and 2

3. Which of the following spinal-cord segments is responsible for the Achilles reflex?
   a. sacral 1 and 2
   b. cervical 7 and 8
   c. cervical 5 and 6
   d. lumbar 2, 3, and 4
   e. lumbar 4 and 5, and sacral 1 and 2

4. Which of the following spinal-cord segments is responsible for the biceps reflex?
   a. sacral 1 and 2
   b. cervical 7 and 8
   c. cervical 5 and 6
   d. lumbar 2, 3, and 4
   e. lumbar 4 and 5, and sacral 1 and 2

5. What is the term for reflexes that result from skeletal-muscle contraction, which are triggered by a sharp tap on an appropriate area?
   a. superficial
   b. hyperreflexia
   c. cutaneous
   d. hypoflexia
   e. somatic

6. What do abnormal reduced reflex responses exemplify?
   a. superficial reflexes
   b. hyperreflexia
   c. cutaneous responses
   d. hypoflexia
   e. withdrawal reflexes

7. What do abnormal exaggerated reflex responses exemplify?
   a. superficial reflexes
   b. hyperreflexia
   c. cutaneous responses
   d. hypoflexia
   e. withdrawal reflexes

8. Stretch reflexes involve (how many?) neurons and (how many?) synapses.
   a. 2, 1
   b. 1, 1
   c. 2, 2
   d. 3, 1
   e. 3, 2

9. Which of the following do somatic reflexes include?
    a. withdrawal reflexes
    b. flexor reflexes
    c. stretch reflexes
    d. choices *a* and *b* only
    e. choices *a, b,* and *c*
10. Of what is the patellar reflex an example?
    a. ipsilateral reflex
    b. polysynaptic reflex
    c. withdrawal reflex
    d. flexor reflex
    e. superficial reflex

# The Endocrine System

*I*n addition to the nervous system, the endocrine glands provide a communication and regulatory system. While the two often supplement one another and frequently perform separately, the interplay between endocrine secretions and nerve impulses often functions to control total human behavior.

The endocrine secretions, or *hormones,* do not belong to any one category of chemical compounds. They have varying chemical structures. Some are proteins, while others are amines (amino acid derivatives) or steroids (types of lipids).

The concentration of a hormone in the blood is controlled by a negative-feedback mechanism that regulates production, storage, and elimination processes. Sensors read the status of the body's internal environment and then direct the appropriate responses. The presence of a greater than normal concentration of a hormone will initiate a response to inhibit the gland's production. Glands may be under direct neural control or under the control of other hormones or both.

The exercises in this section point out the location of the human endocrine glands and demonstrate selected properties of the human endocrine system.

### GENERAL REFERENCES

Gorbman, A., W. W. Dickhoff, S. R. Vigna, N. B. Clark, and C. L. Ralph. *Comparative Endocrinology.* New York: Wiley, 1983.

Greenspan, F. S., and P. H. Forsham (Eds.). *Basic and Clinical Psychology.* Los Altos, CA: Lange Medical Publications, 1983.

Hardy, R. N. *Endocrine Physiology.* Baltimore: University Park Press, 1982.

Tortora, G. J., and S. R. Grabowski. *Principles of Anatomy and Physiology,* 7th ed. New York: HarperCollins, 1993.

### SPECIFIC REFERENCES

Carmichael, S. W., and H. Winkler. "The Adrenal Chromaffin Cell." *Scientific American,* 253: 40–49, 1985.

DeVos, A. M., M. Ultsch, and A. A. Kossiakoff. "Human Growth Hormone and Extracellular Domain of Its Receptor: Crystal Structure of the Complex." *Science,* 255: 306–312, 1992.

Dunn, J. T. "Iodine Deficiency—The Next Target for Elimination." *The New England Journal of Medicine,* 326: 267–268, 1992.

Dupre, J. "Insulin Therapy: Progress and Prospects." *Hospital Practice,* 18: 171–179, 1983.

Golde, D. W., and J. C. Gasson. "Hormones That Stimulate the Growth of Blood Cells." *Scientific American,* 259: 62–70, 1988.

Reichlin, S. "Somatostatin." *The New England Journal of Medicine,* 309: 1495–1501, 1983.

Reiter, R. J. (Ed.). *The Pineal Gland.* New York: Raven Press, 1984.

Volpe, R. "Autoimmune Thyroid Disease." *Hospital Practice,* 19: 141–158, 1984.

# Location and Histology of Human Endocrine Glands

## After completing this exercise, you should be able to

1. Name the major endocrine glands and state their locations
2. Associate glands with the hormones that they produce
3. Recognize microscopic features of selected endocrine glands

Many endocrine glands are *organs* that secrete chemical regulators, or *hormones,* into the blood. The approximate locations of endocrine glands are shown in Figure 28-1. Hormones are carried by the vascular system to all the organs of the body. Some hormones, however, only affect the metabolism of specific organs; these organs are termed the *target organs* of the hormone (Figure 28-1). Other hormones influence every body cell. Hormones are not enzymes, but they affect the metabolism and activities of their target organs by regulating the activity of specific enzymes in these and related structures. In this way, the endocrine system plays an important role in the regulation of (1) growth and development, (2) metabolic homeostasis, and (3) reproduction. Because hormones are powerful chemicals, their circulating levels must be carefully controlled. When the level of a hormone is adequate to meet existing body needs, its further release is stopped. This form of control is termed *negative feedback.*

Both overactivity and underactivity of an endocrine gland are examples of malfunction. Conditions of this type cause endocrine diseases such as those listed in Figure 28-2. When a gland secretes an excessive amount of a specific hormone, it is *hyperactive.* This type of activity may be caused by various conditions, including an enlarged gland or a tumor. A gland that fails to secrete its hormones or that simply secretes inadequate quantities is considered to be *hypoactive.* Conditions that may cause such limited activity include infection, tumors, or the unfavorable effects of surgical or radioactive treatments. Glands that are decreased in size and consequently secrete inadequate amounts of hormones are referred to as being *atrophied.*

This exercise presents the location and the microscopic features of normal endocrine glands, and some endocrine disorders and diseases commonly found in humans.

## ENDOCRINE GLANDS—LOCATION AND BRIEF DISCUSSION

### Materials

The following materials should be provided for class use:

1. Human torso models
2. Head models
3. Human skull models
4. Charts showing endocrine gland locations

Endocrine glands secrete into the blood and do not have ducts. They therefore are termed the *ductless glands.* All other large secretory masses produce substances that are carried to their specific destinations by one or more ducts. These are termed *ducted glands* or *exocrine glands.* The products of endocrine glands are termed *hormones,* while products of exocrine glands are given such specific names as saliva, bile, and pancreatic juice. The advantage of an endocrine gland is obvious. Once in the blood, the secretions may be carried to all cells of the body.

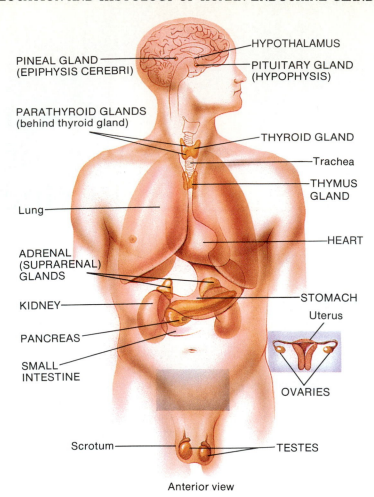

**Figure 28-1** Location of human (male and female) endocrine glands. (From G. J. Tortora and S. R. Grabowski, *Principles of Anatomy and Physiology,* 7th ed., New York: HarperCollins, 1993.)

## Pituitary Gland

The pituitary gland (Figure 28-3) is located at the base of the brain, within the hypophysial fossa of the sella turcica of the sphenoid bone. It is suspended by a stalk termed the *infundibulum* and is composed of two major parts: the anterior and posterior lobes.

   The pituitary gland arises embryonically from two different sites. The *anterior lobe (adenohypophysis)* is a derivative of part of the embryonic mouth epithelium, termed *Rathke's pocket.* The *posterior lobe (neurohypophysis)* is a derivative of the diencephalon region of the embryonic brain. There is an *intermediate lobe* in the pituitary's early development, but it fuses with the anterior lobe. Its embryonic origin is the same as that of the anterior lobe.

   Both the anterior and the posterior lobes are attached to the inferior surface of the brain by the *infundibulum.* The hormones found in the posterior lobe of the pituitary are synthesized in the *hypothalamus* of the brain. They are merely stored in the pituitary, but they are still thought of as posterior-lobe hormones. Some of the hormones produced by the pituitary gland and the actions of these hormones are listed in Table 28-1.

1. **ANTERIOR LOBE (ADENOHYPOPHYSIS) HORMONES**

   a. *Somatotropin or human growth hormone (HGH).* This hormone regulates growth of body tissue but specifically acts on the epiphysial cartilages of the long bones (refer to Exercise 13) to effect growth in height. Hypersecretion of this hormone in childhood may result in *gigantism,* and hyposecretion may result in a *midget* or *pituitary dwarf.* A hypersecretion in the adult may cause *acromegaly,* a disfiguring condition in which some growth areas in the skeleton still respond and grow absurdly out of

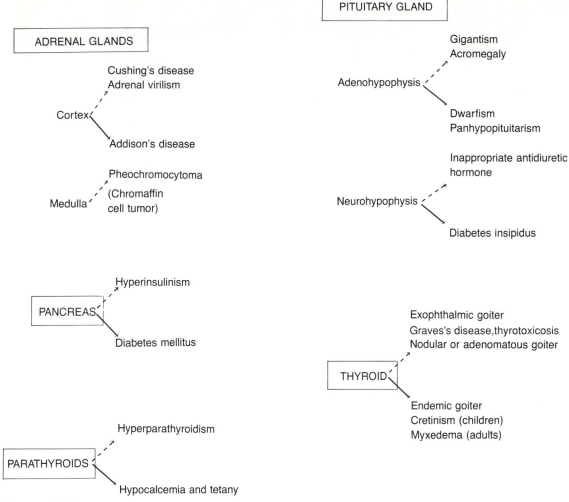

**Figure 28-2** Endocrine diseases. The hyperactivity of a gland is indicated by a dotted line; the hypoactivity is indicated by a solid line.

shape, and others, such as those that affect height, do not continue to grow.

b. *Prolactin.* This hormone acts on the mammary glands, causing them to produce milk before and after childbirth.

The following tropic hormones are those that act as stimuli on other endocrine tissue:

c. *Thyrotropic hormone (TH).* This hormone regulates the growth, development, and activity of the *thyroid gland.*

d. *Adrenocorticotropic hormone (ACTH).* This hormone regulates the growth and development of the *adrenal cortex.*

e. *Follicle-stimulating hormone (FSH).* This is one of two *gonadotropic hormones.* It stimulates and regulates the early developmental stages of the ovarian (graafian) follicle in the female and

of the spermatozoa in the seminiferous tubules of the male testis. It has other functions as well.

f. *Luteinizing hormone (LH).* This hormone finishes what FSH has begun in the ovary or the testis. The follicle containing the ovum (or egg) completes its development and ovulates. The spermatozoa are anatomically completed and released from the seminiferous tubules. In the male, the luteinizing hormone often is termed the *interstitial-cell-stimulating hormone (ICSH).*

2. **POSTERIOR LOBE (NEUROHYPOPHYSIS) HORMONES**

a. *Oxytocin.* This hormone is found only in the female and stimulates uterine contractions. It is one of the major factors in initiating labor at the time of birth.

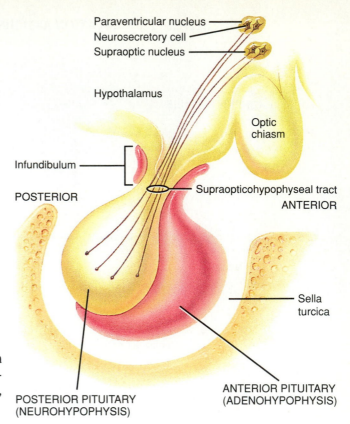

**Figure 28-3** Hypothalamic-hypophysial tract. (From G. J. Tortora and S. R. Grabowski, *Principles of Anatomy and Physiology,* 7th ed., New York: HarperCollins, 1993.)

**Table 28-1   Hormones of the Pituitary Gland**

| Gland | Hormones Produced | Action |
|-------|-------------------|--------|
| Anterior lobe | Somatotropic hormone or human growth hormone (HGH) | Regulates normal body growth, particularly that of the skeleton |
| | Thyroid-stimulating hormone (TSH) | Regulates growth, development, and normal function of thyroid gland |
| | Adrenocorticotropic hormone (ACTH) | Regulates growth, development, and normal function of adrenal cortex |
| | Follicle-stimulating hormone (FSH) | Initiates development of ova in the female and of sperm in the male |
| | Luteinizing hormone, termed luteotropin (LH) in the female and interstitial cell stimulating hormone (ICSH) in the male | Stimulates ovulation and development of corpus luteum in the female and development of interstitial cells to produce testosterone in the male |
| | Prolactin (PR) | Stimulates milk production by mammary glands |
| | Melanocyte-stimulating hormone (MSH) | Increases skin pigmentation |
| Posterior lobe | Oxytocin | Stimulates contraction of uterine muscle |
| | Antidiuretic hormone (ADH) or vasopressin | Causes kidneys to remove water from urine from distal convoluted tubules<br>Causes constriction of arterioles, to produce rise in blood pressure |

b. *Antidiuretic hormone (ADH)* (also termed *vasopressin*). This hormone causes reabsorption of water from the distal convoluted tubules in the kidney, thereby preventing severe water loss. The hormone also causes a rise in blood pressure. Hyposecretion leads to *diabetes insipidus*.

## Thyroid Gland

The thyroid gland (Figure 28-4) is a large mass of tissue on the anterior surface of the trachea just inferior to the larynx. It is composed of right and left lateral lobes, connected by a slim *isthmus* on the anterior surface of the trachea. The thyroid gland is an endodermal derivative of the embryonic gut.

The gland contains a large quantity of iodine, which it uses to produce three hormones:

1. Thyroxin (T4)
2. Triiodothyronine (T3)
3. Thyrocalcitonin

Thyroxin and triiodothyronine control the metabolic rate of all cells in the body, while thyrocalcitonin regulates blood calcium.

*Hypothyroidism* results in a slow-moving, sleepy, obese individual, whereas *hyperthyroidism* produces the opposite. A severe lack of the thyroid hormones in the fetus or in the child results in *cretinism*. In the adult, a similar physiological need produces an individual with *myxedema*. Lack of iodine in the diet also produces an abnormal gland that increases in size, due to *compensatory hypertrophy*, creating a *goiter,* which is an enlarged thyroid gland. A severe hyperthyroid condition causes another type of goiter, sometimes termed *exophthalmic goiter*. The two types of goiter are not the same because one is *endemic* and caused by iodine lack in certain geographic areas, whereas the other is a failure of the thyroid gland to function properly, even in the presence of iodine-rich soil and water.

## Parathyroid Glands

The parathyroid glands (Figure 28-5) are usually four in number (two pairs embedded in the posterior medial surface of each thyroid lobe). The number is not constant, however. Some people have only one pair, and others have three pairs of the glandular masses. They arise from embryonic visceral arch tissue and produce the hormone *parathormone (PTH),* which controls calcium and phosphate metabolism in the body. *Hypoparathyroidism* results in brittle bones and low blood calcium and *hyperparathyroidism* results in soft bones, loss of teeth, and high blood calcium.

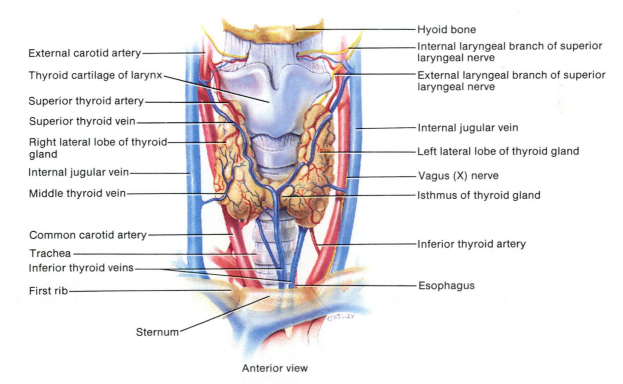

External carotid artery
Thyroid cartilage of larynx
Superior thyroid artery
Superior thyroid vein
Right lateral lobe of thyroid gland
Internal jugular vein
Middle thyroid vein
Common carotid artery
Trachea
Inferior thyroid veins
First rib
Sternum

Hyoid bone
Internal laryngeal branch of superior laryngeal nerve
External laryngeal branch of superior laryngeal nerve
Internal jugular vein
Left lateral lobe of thyroid gland
Vagus (X) nerve
Isthmus of thyroid gland
Inferior thyroid artery
Esophagus

Anterior view

**Figure 28-4** Location, blood supply, and innervation of the thyroid gland in anterior view. The pyramidal lobe of the thyroid, when present, may be attached to the hyoid bone by muscle. (From G. J. Tortora, *Principles of Human Anatomy,* 6th ed., New York: HarperCollins, 1992.)

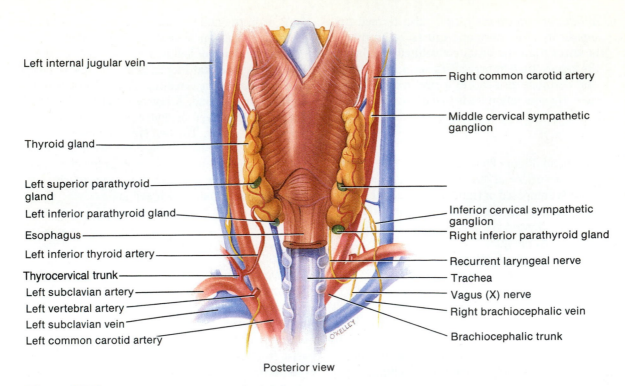

Left internal jugular vein

Thyroid gland

Left superior parathyroid gland

Left inferior parathyroid gland

Esophagus

Left inferior thyroid artery

Thyrocervical trunk

Left subclavian artery

Left vertebral artery

Left subclavian vein

Left common carotid artery

Right common carotid artery

Middle cervical sympathetic ganglion

Inferior cervical sympathetic ganglion

Right inferior parathyroid gland

Recurrent laryngeal nerve

Trachea

Vagus (X) nerve

Right brachiocephalic vein

Brachiocephalic trunk

O'KELLEY

Posterior view

**Figure 28-5** Location, blood supply, and innervation of the parathyroid glands (posterior view). (From G. J. Tortora, *Principles of Human Anatomy,* 6th ed., New York: HarperCollins, 1992.)

## Suprarenal or Adrenal Glands

The suprarenal glands (Figure 28-6) are located above the kidneys in the human. In most other animals, they are found lying beside the kidneys—hence the names *suprarenal* and *adrenal*. The gland has two distinct regions that have different embryonic origins:

1. The *suprarenal medulla* is the core of the gland, and it originates from embryonic neural crest cells. The hormones produced in this region are termed *epinephrine* (adrenaline) and *norepinephrine*. They have a function similar to that of the sympathetic division of the autonomic nervous system: They prepare the body for an emergency when secreted in large amounts.

2. The *suprarenal cortex* secretes many hormones, which are divided into categories:
   a. *Glucocorticoids—Cortisone* is an example. These hormones act as anti-inflammatory agents, function in protein digestion and storage, and provide resistance to stress.
   b. *Mineralocorticoids—Aldosterone* is an example. These hormones regulate sodium and potassium metabolism.
   c. *Adrenal sex hormones* or *gonadocorticoids*— These estrogens and androgens are secreted in minute quantities and have little effect in the normal body. However, they can cause sexual

anomalies when secreted in large amounts, as the result of a tumor. Hypersecretion leads to Cushing's syndrome. On the other hand, hyposecretion of the adrenocorticoids may result in a debilitating and progressive condition termed Addison's disease.

## Islets of Langerhans of the Pancreas

The pancreas is a large exocrine gland (see Figure 28-7) that produces powerful digestive enzymes that are carried to the small intestine by ducts. Between the enzyme-producing tubules, there are islands of cells termed *alpha cells* and *beta cells* (the islets of Langerhans; see Color Plate 31). The beta cells produce *insulin,* which enables the body cells to obtain blood sugar needed for physiological oxidation. A hyposecretion of this hormone causes *diabetes mellitus*. The alpha cells secrete the hormone *glucagon,* which, by converting stored glycogen to glucose, increases the blood sugar level.

## Gonads (Ovaries and Testes)

The paired ovaries (Figure 28-8) are the primary reproductive organs of the female and are located in the pelvic region of the body cavity (Figure 28-1). They have two types of hormone-producing tissue. The hormone *estrogen* is secreted by the cells of the *graafian follicle*

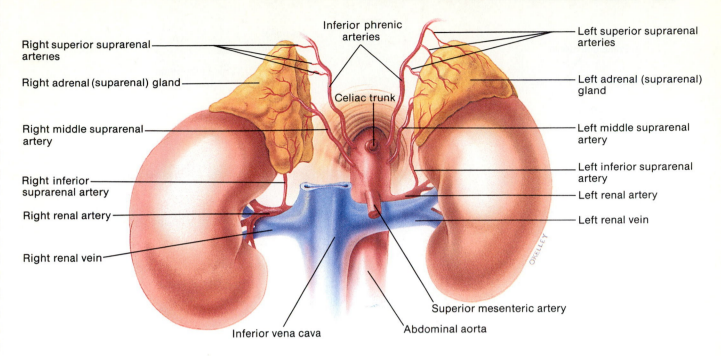

**Figure 28-6** Location and blood supply of the adrenal (suprarenal) glands (anterior view). (From G. J. Tortora and S. R. Grabowski, *Principles of Anatomy and Physiology,* 7th ed., New York: HarperCollins, 1993.)

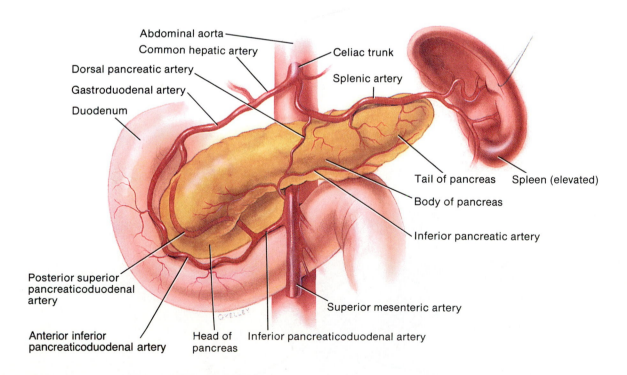

**Figure 28-7** Location and blood supply of the pancreas (anterior view). (From G. J. Tortora and S. R. Grabowski, *Principles of Anatomy and Physiology,* 7th ed., New York: HarperCollins, 1993.)

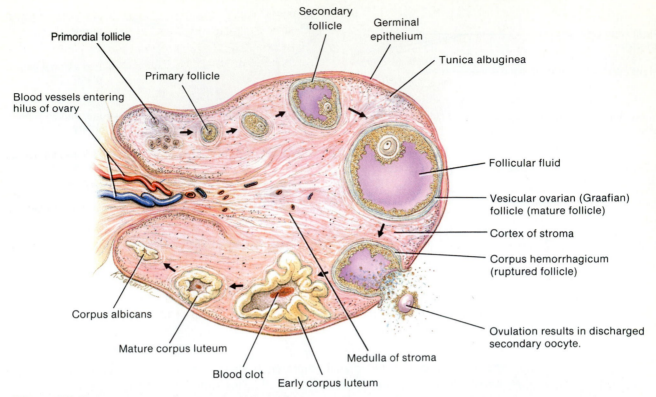

**Figure 28-8** Human ovary (sectional view). (From G. J. Tortora and S. R. Grabowski, *Principles of Anatomy and Physiology,* 7th ed., New York: HarperCollins, 1993.)

and *progesterone* by the cells of the *corpus luteum.* There is a cyclic relationship between the ovarian hormones and the gonadotropic hormones of the pituitary gland. Estrogen is responsible for the development and maintenance of the female secondary sexual characteristics.

The paired testes (Figure 28-9) are the primary reproductive organs of the male. They lie outside the body cavity (see Figure 28-1), within a pouch of skin termed the *scrotum.* The testes develop in the embryo's pelvic region and descend into the scrotum before birth. The lower temperature (facilitated by external location) in the scrotum facilitates sperm development. In addition to producing and releasing spermatozoa, the testis contains interstitial cells that produce the male sex hormone *testosterone.* This hormone controls the growth and development of the male secondary sexual characteristics.

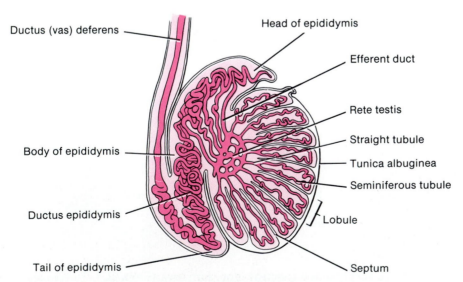

**Figure 28-9** Human testis (sectional view). (From G. J. Tortora and S. R. Grabowski, *Principles of Anatomy and Physiology,* 7th ed., New York: HarperCollins, 1993.)

## Pineal Gland

This small mass of endocrine tissue is a dorsal outgrowth of the diencephalon region of the brain (see Figures 25-5 and 25-6).

The products of the pineal gland are *melatonin,* which plays a part in ovarian function; *adrenoglomerulotropin,* which stimulates the adrenal cortex; and *serotonin,* which is found in normal brain tissue and acts as a neurotransmitter. The physiology of this gland has not been specifically defined.

## Thymus Gland

The thymus gland is a dark, irregularly shaped, two-lobed structure. It is located behind the sternum and between the lungs (Figure 28-10). The two lobes are held together by connective tissue, and each lobe is enclosed by a connective tissue *capsule.* Microscopic examination of the gland shows that each lobe is divided further into smaller units termed *lobules.* The outer *cortex* of lobules is filled with lymphocytes (germinal centers), while the inner *medulla* regions hold fewer lymphocytes. (Color Plate 23 shows the structure of the thymus gland.)

The thymus gland is quite obvious in an infant. It reaches its maximum size at the time of puberty. After this period, much of the functional areas of thymic tissue are replaced by fat and connective tissue. Despite such replacement, the gland continues to be functional. The major role of the thymus gland is to help produce and distribute specialized cells, termed *T lymphocytes,* to other lymphoid organs (lymph nodes, spleen, etc.). Such cells are associated with the immune responses that result in the direct or indirect destruction of disease-causing microorganisms, cancers, and other foreign factors. Hormones produced by the thymus gland, such as *thymosin* and *thymopoietin,* promote the production and maturation of T lymphocytes.

## HISTOLOGY OF SELECTED ENDOCRINE GLANDS

### Materials

The following prepared slides should be provided for class use:

1. Adrenal gland
2. Pancreas
3. Thymus gland (prepuberty)
4. Thymus gland (adult)
5. Thyroid glands (normal and goiter specimens)
6. Parathyroid gland

## Thyroid Gland

1. Examine the thyroid-gland slide with the low-power objective.
2. Identify the circular, single-celled layers termed *follicles* within the specimen. (Refer to Figure 28-11 and Color Plate 24.) Such follicles usually enclose solid-appearing materials known as *colloids,* which contain several proteins, including thyroxin. Some follicles will be empty.
3. Sketch and label a representative portion of the prepared slide in the results and observations section.

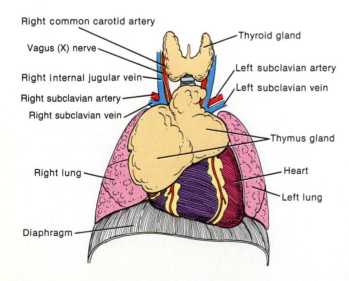

Right common carotid artery
Vagus (X) nerve
Right internal jugular vein
Right subclavian artery
Right subclavian vein
Right lung
Diaphragm
Thyroid gland
Left subclavian artery
Left subclavian vein
Thymus gland
Heart
Left lung

**Figure 28-10** Location of the thymus gland in a young child. (From G. J. Tortora and S. R. Grabowski, *Principles of Anatomy and Physiology,* 7th ed., New York: HarperCollins, 1993.)

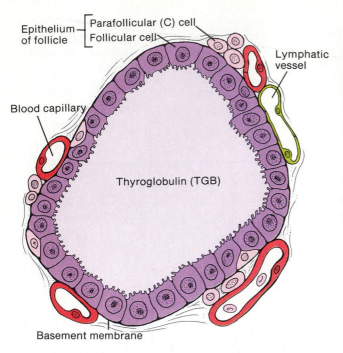

**Figure 28-11** Histology of the thyroid gland, showing one follicle. (From G. J. Tortora and S. R. Grabowski, *Principles of Anatomy and Physiology,* 7th ed., New York: HarperCollins, 1993.)

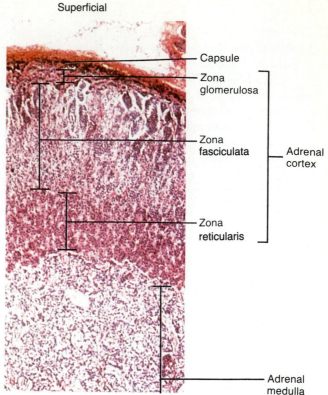

**Figure 28-12** Histology of the adrenal (suprarenal) glands: photomicrograph of a section of the adrenal gland, showing its subdivisions and zones at a magnification of 45×. (Courtesy of Andrew Kuntzman.)

## Thyroid Gland and Goiter

1. Examine the thyroid gland preparation from a goiter case, with the low-power objective (Refer to Color Plate 25.)
2. Compare this specimen with the normal thyroid gland preparation. Sketch any obvious differences for a representative portion in the results and observations section.

## Parathyroid

1. Examine the prepared slide of the parathyroid gland, with the low-power objective first and then with the high-power objective.
2. Identify the principal and oxyntic cells of the gland. (Refer to Color Plates 26 and 27.)
3. Sketch and label a representative portion of the gland in the results and observations section.

## Adrenal Glands

1. Examine a prepared slide of an adrenal gland, with the low-power or scanning objective. Refer to your textbook, Figure 28-12 and Color Plates 28, 29, and 30 to identify the following parts.
   a. *Capsule*—thin, outer covering
   b. *Cortex*—outer section
   c. *Medulla*—central region

2. Examine the cortex more closely and locate its three zones.
   a. *Zona glomerulosa*—region next to the capsule containing thin layer of cells in clumps or even arcs
   b. *Zona fasciculata*—thicker central zone containing cells arranged in long cords
   c. *Zona reticularis*—zone next to the medulla containing cords of cells arranged in a netlike fashion.
3. Examine the medulla. Its cells are arranged in irregular groups, with several blood vessels separating them.
4. Sketch and label a general representation of the different parts of the adrenal gland in the results and observations section.

## Pancreas

1. Examine the prepared slide of the pancreas, with the low-power objective. (Refer to Figure 28-13 and Color Plate 31.)

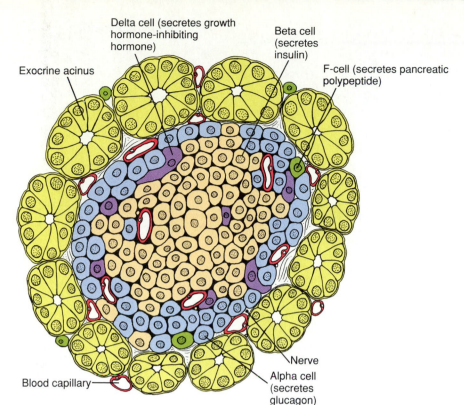

Delta cell (secretes growth hormone-inhibiting hormone)

Beta cell (secretes insulin)

Exocrine acinus

F-cell (secretes pancreatic polypeptide)

Nerve

Blood capillary

Alpha cell (secretes glucagon)

**Figure 28-13** Islet of Langerhans. (From G. J. Tortora and S. R. Grabowski, *Principles of Anatomy and Physiology,* 7th ed., New York: HarperCollins, 1993.)

2. Identify the following portions of the gland.
   a. *Acini*—small tubules that consist of clusters of cells and a lumen (the exocrine part of the pancreas)
   b. *Islets of Langerhans*—clusters of alpha and beta cells, surrounded by capillaries and exocrine structures; small granules in alpha cells are darker than those found in beta cells
3. Sketch and label a general representation of the different parts of the pancreas in the results and observations section.

### *Thymus Gland*

1. Examine the two forms of thymus gland preparation with the low-power and high-power objectives.
2. Note the cortex, the medulla, and the cell types in the preparation. (Refer to Color Plate 23.)
3. Answer the questions in the results and observations section.

## RESULTS AND OBSERVATIONS

1. Thyroid gland

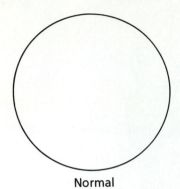

Normal

Goiter

a. Were there any differences between the normal thyroid and goiter specimens? _____
_____

b. What is a goiter? _____
_____

2. Parathyroid gland

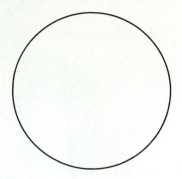

3. Adrenal gland

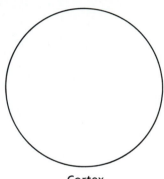

Cortex

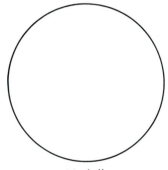

Medulla

4. Pancreas

5. Thymus

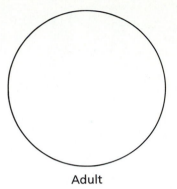

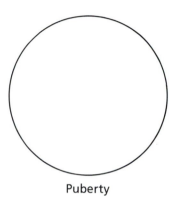

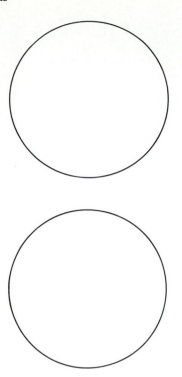

Adult

Puberty

a. Which cell types are more numerous in the adult thymus gland? _____

b. Which cell types are more numerous in the pre-puberty thymus gland? _____

c. Why are there such differences between the two preparations? _____

_____

_____

d. What are T lymphocytes? _____

_____

# LABORATORY REVIEW 28

## LOCATION AND HISTOLOGY OF HUMAN ENDOCRINE GLANDS

### MATCHING AND FILL-IN

1. Match the hormones on the left with the endocrine glands that produce them.

   a. _____ Thyroxin

   b. _____ Parathormone

   c. _____ Aldosterone

   d. _____ Insulin

   e. _____ Epinephrine

   f. _____ Human growth hormone

   g. _____ Cortisone

   h. _____ Glucagon

   i. _____ Adrenocorticotropic hormone

   j. _____ Thyrocalcitonin

   k. _____ Estrogen

   l. _____ Prolactin

   m. _____ Testosterone

   n. _____ Melatonin

### ANSWERS

1. Pituitary, anterior lobe
2. Pituitary, posterior lobe
3. Thyroid
4. Parathyroid
5. Adrenal cortex
6. Adrenal medulla
7. Pancreas
8. Ovary
9. Testis
10. Thymus
11. Pineal

2. Complete Table Q28-1 by indicating at least one specific abnormality that results from the hypo- or hypersecretion of the hormones indicated.

**Table Q28-1**

| Hormone | Hormonal Level | Abnormality |
|---|---|---|
| Glucocorticoids | Hyposecretion | a. |
| | Hypersecretion | b. |
| Parathyroid hormone | Hyposecretion | c. |
| | Hypersecretion | d. |
| Thyroxin | Hyposecretion (child) | e. |
| | Hypersecretion (adult) | f. |
| Antidiuretic hormone | Hyposecretion | g. |
| Human growth hormone | Hypersecretion (childhood) | h. |
| | Hypersecretion (adult) | i. |

3. Complete Table Q28-2 from the one clue provided in each case.

**Table Q28-2**

| Endocrine Gland | Hormone | Associated Activity |
|---|---|---|
| a. | b. | Regulates growth in height |
| c. | Follicle-stimulating hormone | d. |
| e. | f. | Stimulates contraction of the uterus |
| g. | h. | Stimulates water reabsorbtion by the kidneys |
| i. | Parathyroid hormone | j. |
| Thymus | k. | l. |
| m. | Prolactin | n. |
| o. | Insulin | p. |
| Alpha cells of the pancreas | q. | r. |
| s. | Serotonin | t. |
| u. | Thyroxin | v. |

4. Distinguish between an *exocrine* and an *endocrine* gland. _____

_____

_____

5. Identify the parts (a)–(d) of the thyroid gland indicated in Figure Q28-1.

    a. _____

    b. _____

    c. _____

    d. _____

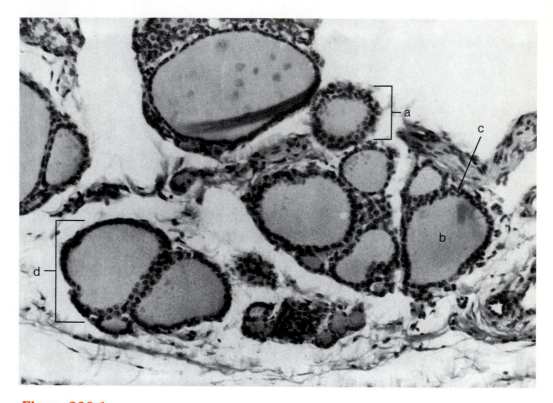

**Figure Q28-1**

## MULTIPLE CHOICE

Enter the correct answers in the spaces provided.

1. Once hormones have reached levels that are adequate to meet existing body needs, further release is controlled by which of the following?
   a. hyperactivity of endocrine glands
   b. hypoactivity of endocrine glands
   c. hypoactivity of target organs
   d. hyperactivity of target organs
   e. negative feedback mechanism
2. What is the term for overactivity of an endocrine gland, in relation to hormone production?
   a. hypoactivity
   b. hyperactivity
   c. homeostasis
   d. negative feedback
   e. either choices *a* or *c*
3. Hyposecretion of HGH in childhood may result in which of the following conditions?
   a. gigantism
   b. acromegaly
   c. midget state
   d. exophthalmic goiter
   e. diabetes insipidus
4. Hypersecretion of HGH in childhood may result in which of the following states?
   a. gigantism
   b. acromegaly
   c. midget state
   d. exophthalmic goiter
   e. diabetes insipidus
5. Hypersecretion of HGH in an adult may produce which of the following conditions?
   a. gigantism
   b. acromegaly
   c. midget state
   d. exophthalmic goiter
   e. diabetes insipidus
6. Which of the following hormones is not produced by the posterior lobe of the pituitary?
   a. prolactin
   b. oxytocin
   c. FSH
   d. ACTH
   e. thyrotropic hormone
7. Diabetes insipidus results from the hyposecretion of which of the following hormones?
   a. oxytocin
   b. ADH
   c. FSH
   d. ACTH
   e. prolactin

| ANSWERS | |
|---|---|
| 1. | _____ |
| 2. | _____ |
| 3. | _____ |
| 4. | _____ |
| 5. | _____ |
| 6. | _____ |
| 7. | _____ |
| 8. | _____ |
| 9. | _____ |
| 10. | _____ |
| 11. | _____ |
| 12. | _____ |
| 13. | _____ |
| 14. | _____ |
| 15. | _____ |

8. Which of the hormones is a major factor in initiating labor at the time of birth?
   a. oxytocin
   b. ADH
   c. FSH
   d. ACTH
   e. prolactin

9. Which of the following is known to produce estrogen?
   a. corpus luteum
   b. follicle cells associated with a graafian follicle
   c. islets of Langerhans
   d. thyroid glands
   e. thymus gland

10. Cretinism is associated with a lack of hormones produced by which of the following glands?
    a. corpus luteum
    b. follicle cells associated with a graafian follicle
    c. islets of Langerhans
    d. thyroid glands
    e. thymus gland

11. A goiter results from a deficiency or disorder in which of the following endocrine glands?
    a. thymus gland
    b. thyroid gland
    c. parathyroid glands
    d. adrenal gland
    e. pineal gland

12. Diabetes mellitus results from the hyposecretion of which of the following hormones?
    a. glucagon
    b. ADH
    c. cortisone
    d. testosterone
    e. insulin

13. What is a hormone known for its anti-inflammatory activity?
    a. glucagon
    b. ADH
    c. cortisone
    d. testosterone
    e. insulin

14. Which of the following hormones control calcium and phosphate metabolism in the body?
    a. insulin
    b. parathormone
    c. ADH
    d. aldosterone
    e. adrenocorticoids

15. Hypersecretion of which of the following hormones can lead to Cushing's syndrome?
    a. insulin
    b. parathormone
    c. ADH
    d. aldosterone
    e. adrenocorticoids

# The Senses

$S$pecialized sensory receptors pick up environmental stimuli and transmit them to the central nervous system. Mechanical, chemical, and visual receptors each receive different types of stimuli. The nature of the sensation provoked by such stimuli is determined by the brain. Table VIII-1 describes several sensory receptors.

The exercises in this section consider the general structure of the eye and several procedures to test its normal functioning, the structure and functioning of the ear, the sense of smell and its relationship to taste, and various sensory properties of the skin. Attention also is given to several types of optical illusions, including those involving the interpretation of spatial relationships.

**Table VIII-1  The Sensory Receptors**

| Sensory Receptor | Location | Type of Receptor and Type of Stimulus |
|---|---|---|
| Rods and cones | Retina of each eye | Visual—light |
| Hair cells | | |
| For converting sound waves to nerve impulses | Organ of Corti in cochlear duct of inner ear | Mechanical—vibrations or movements of fluid in inner ear |
| For positioning of head and body | Utricle of inner ear | Mechanical—movements of $CaCO_3$ (calcium carbonate) particles in direction of gravity |
| For changes in movement of body | Ampulla of semicircular ducts of inner ear | Mechanical—rapid movement of $CaCO_3$ particles |
| Taste buds (Gustatory cells) | Papillae of tongue | Chemical—specific chemicals in solution |
| Olfactory cells | Olfactory epithelium in nasal cavity | Chemical—dissolved gaseous substances |
| Touch and pressure cell endings | Dermis of skin, smooth muscles of visceral organs | Mechanical—touch |

**Table VIII-1** *(Continued)*

| Sensory Receptor | Location | Type of Receptor and Type of Stimulus |
|---|---|---|
| Vibration | Dermis of skin | Mechanical—touch |
| Temperature (heat and cold) | Dermis of skin, smooth muscles of visceral organs | Mechanical—hot and cold |
| Pain cell endings | Dermis of skin, smooth muscles of visceral organs | Mechanical—all intense stimuli |
| Proprioceptors | Muscles, tendons, joints | Mechanical—movement |

## GENERAL REFERENCES

Barlow, H. B., and J. D. Mollon (Eds.). *The Senses.* New York: Cambridge University Press, 1982.

Pickles, J. O. *An Introduction to the Physiology of Hearing.* New York: Academic Press, 1982.

Rodieck, R. W. *The Vertebrate Retina, Principles of Structure and Function.* San Francisco: Freeman, 1976.

Tortora, C. J., and S. R. Grabowski. *Principles of Anatomy and Physiology,* 7th ed. New York: HarperCollins, 1993.

## SPECIFIC REFERENCES

Borg, E., and S. A. Counter. "The Middle Ear Muscles." *Scientific American.* 261: 74–81, 1989.

Cytowic, R. E. "Tasting Colors, Smelling Sounds." *Sciences,* September/October, 32–37, 1988.

Doty, R. L. "Olfactory Communication in Humans." *Chemistry of the Senses,* 6: 351–376, 1981.

Loeb, G. E. "The Functional Replacement of the Ear." *Scientific American,* 252: 104–111, 1985.

Koretz, J. F., and G. H. Handelman. "How the Human Eye Focuses." *Scientific American,* 259: 92–99, 1988.

Krachmer, J. H., and D. A. Palay. "Corneal Disease." *The New England Journal of Medicine,* 325: 1804–1806, 1991.

Mahowald, M. A., and C. Mead. "The Silicon Retina." *Scientific American,* 264: 76–82, 1991.

Nathans, J. "The Genes for Color Vision." *Scientific American,* 260: 42–49, 1989.

Wolfe, J. M. "Hidden Visual Processes." *Scientific American,* 248: 94–103, 1983.

# Anatomy of the Eye and Ear

### After completing this exercise, you should be able to

1. Identify the parts of the mammalian eye
2. Examine the muscles that move the eyeball
3. List and identify all parts of the ear
4. Trace the path of sound waves through the parts of the ear
5. Use an otoscope and become familiar with its applications

## Dissection of a Mammalian Eye

### Materials

The following materials should be provided for class use:

1. Fresh eyes of sheep or cow
2. Preserved eyes of sheep or cow
3. Model of human eye
4. Chart of human eye

## OUTER SURFACE OF EYEBALL AND EXTRINSIC EYEBALL MUSCLES

Because the cat eye is too small to be of much use in studying the anatomy of the organ, it is recommended that fresh cow or sheep eyes be substituted.

Examine the surface of the eyeball. It is almost round, with a clear surface in front and with a muscular area surrounded by fat in the rear. Examine the back of the eyeball carefully before removing the fat. Locate the *optic nerve,* and isolate it, so that it will not be removed accidentally with the fat.

Remove as much fat as possible, so that the *extrinsic eyeball muscles* can be studied before the eyeball is opened. Most mammalian eyes have six extrinsic muscles per eye, plus a set of *retractor muscles (retractor bulbi)* not present in the human. Using Figure 29-1, locate the following muscles:

1. Superior rectus
2. Inferior rectus
3. Lateral rectus
4. Medial rectus
5. Superior oblique
6. Inferior oblique
7. Retractor bulbi (not found in the human eye)

## Opening the Eyeball

Hold the sheep or cow eye between thumb and forefinger, with the thumb on the posterior surface and the forefinger on the cornea (Figure 29-2). Make an incision about halfway between poles, dividing the eye into anterior and posterior halves. Continue this incision around the equator with your scissors. Separate the two halves on a dissecting tray or in a pan of water. Compare the parts of the eye of your dissected specimen with those shown in Figure 29-3.

The anterior portion of the eyeball contains the thick cornea and the circular ciliary muscles (body) and iris. The latter two are pigmented and appear black, in contrast to the translucency of the cornea. The aqueous fluid, or aqueous humor, in this part of the eye has probably drained away and will not be seen. The aqueous fluid in the living specimen is found in two chambers of the eye: the *anterior chamber* between the cornea and the iris and the *posterior chamber* between the iris and the lens. It is secreted by the cornea and drained

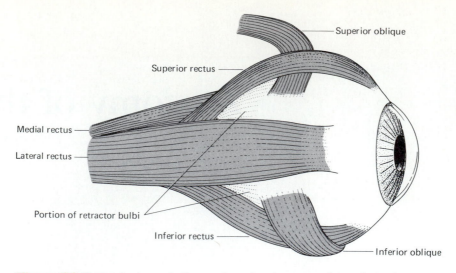

Superior oblique

Superior rectus

Medial rectus

Lateral rectus

Portion of retractor bulbi

Inferior rectus

Inferior oblique

**Figure 29-1** Extrinsic eyeball muscles of a sheep eye (lateral view).

**Figure 29-2** Using scissors to open the eye around its circumference.

into the canal of Schlemm, a venous sinus. Refer to your textbook for a more complete discussion.

The portion of the eyeball from the lens to the retina contains the jellylike *vitreous body,* which creates pressure within the eyeball to keep it from collapsing. Embedded in it is the large, transparent *lens.* The lens usually remains attached to the vitreous jelly by black strands, the *suspensory ligaments.*

Remove the lens and vitreous body and find the *retina,* a thin, transparent membrane forming the inner coat of the eyeball. It contains the sensory receptors (*rods* and *cones*).

Below the retina is the *choroid coat* (an extension of the ciliary body). In sheep or beef eyes, it has a shiny, greenish-blue layer called the *tapetum,* which improves night vision. The tapetum is not found in the human eye.

The region where the retinal fibers turn into the *optic nerve* is termed the *optic disc* or *blind spot.* Close to the optic disc on the retina is the yellowish (in the

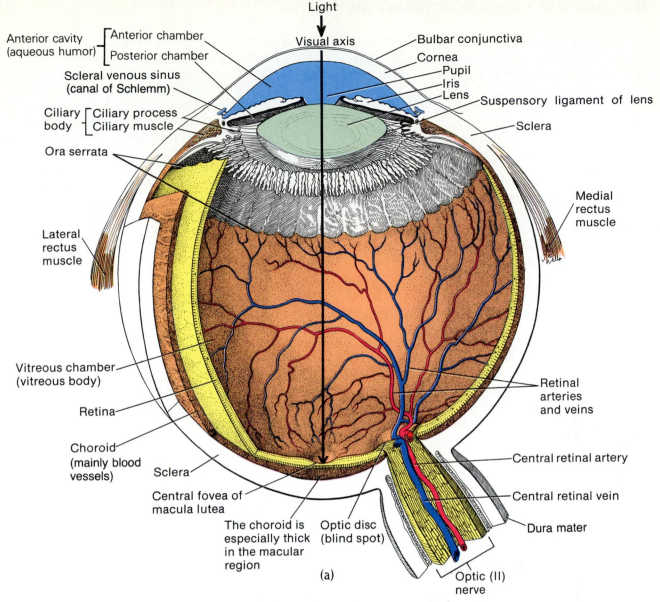

Light

Visual axis

Anterior cavity (aqueous humor) — [Anterior chamber / Posterior chamber]

Scleral venous sinus (canal of Schlemm)

Ciliary body — [Ciliary process / Ciliary muscle]

Ora serrata

Lateral rectus muscle

Vitreous chamber (vitreous body)

Retina

Choroid (mainly blood vessels)

Sclera

Central fovea of macula lutea

The choroid is especially thick in the macular region

Optic disc (blind spot)

(a)

Bulbar conjunctiva

Cornea

Pupil

Iris

Lens

Suspensory ligament of lens

Sclera

Medial rectus muscle

Retinal arteries and veins

Central retinal artery

Central retinal vein

Dura mater

Optic (II) nerve

Transverse section of left eyeball (superior view)

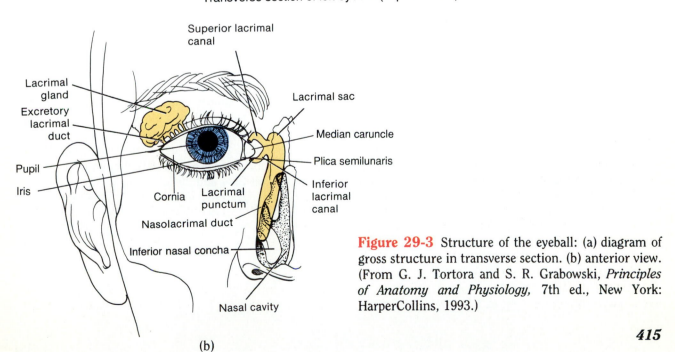

Superior lacrimal canal

Lacrimal gland

Excretory lacrimal duct

Pupil

Iris

Cornia

Lacrimal punctum

Nasolacrimal duct

Inferior nasal concha

Nasal cavity

Lacrimal sac

Median caruncle

Plica semilunaris

Inferior lacrimal canal

(b)

**Figure 29-3** Structure of the eyeball: (a) diagram of gross structure in transverse section. (b) anterior view. (From G. J. Tortora and S. R. Grabowski, *Principles of Anatomy and Physiology*, 7th ed., New York: HarperCollins, 1993.)

**415**

living animal) *macula lutea.* In the center of the macula lutea is the *fovea centralis,* the area of sharpest vision, with a high concentration of cone cells.

## *Observations of the Cat Eye and Human Eye*

The layer outside the choroid coat is the *sclerotic coat* or *sclera.* It is tough and gives much support to the eyeball. The extrinsic eyeball muscles insert on it, to move the eyeball.

Compare your specimen with Figure 29-3, and with the human eye model.

Examine the cat's eye without removing it. Observe the eyelids. There are upper eyelids, lower eyelids, and a third eyelid, or *nictitating membrane,* located in the nasal area of the eyeball. The nictitating membrane is a fold of tissue that can be released to cover the entire corneal surface. Only a remnant of the third eyelid is found in the human; it is termed the *plica semilunaris.* Locate the plica semilunaris on your own or someone else's eye. Because it no longer has a function, it is said to be "vestigial." Covering most of the corneal surface is a thin epithelial extension of the eyelids termed the *conjunctiva.*

## ANATOMY OF THE HUMAN EAR

### Materials

The following materials should be provided for class use.
1. Model of human ear
2. Chart of human ear
3. Otoscope
4. Alcohol-treated pads

Structurally the human ear consists of three regions: the *external* or *outer ear,* the *middle ear,* and the *internal* or *inner ear* (Figure 29-4).

The *outer ear* is made up of the *auricle,* or *pinna,* and the *external auditory canal.* The auricle is the large, soft flap on the side of the head that is supported by cartilage. Within it is the opening to the external auditory canal, a tube about 2.5 cm (1 inch) long, which terminates at the tympanic membrane, or eardrum. The human auricle is vestigial, but the external auditory canal functions as a funnel that traps high-frequency vibrations and directs them to the middle ear. The skin lining this canal contains hair follicles and *cerumen* or *wax glands.*

The *middle ear* is an air-filled cavity in the temporal bone. It extends from the eardrum to the cochlea. The *tympanic membrane* arises from all three embryonic germ layers but is generally considered to be part of the middle ear. Attached to the tympanic membrane are three ear ossicles that have developed from the embryonic visceral arches and are therefore a part of the visceral skeleton. The ossicles, termed *malleus* (hammer), *incus* (anvil), and *stapes* (stirrup), bridge the span of the middle ear and terminate at the *cochlea.* Also located in the middle ear is the opening to the *eustachian tube,* which leads to the pharynx.

The lower region of the *inner ear* is known as the *cochlea.* It contains the sensory apparatus for hearing and is shaped like a snail shell, coiled 2¾ times. The entire structure is enclosed in a bony labyrinth within the temporal bone.

The stapes of the middle ear lies up against the *oval window* of the cochlea. Sound waves passing over the ear ossicles cause the membrane to vibrate, setting into motion the fluid in the cochlear channels, called the *scala vestibuli, scala tympani,* and *scala media* (cochlear duct) (see Figure 29-5). The membranes separating these fluid-filled canals are the *vestibular membrane* and the *basilar membrane.* The basilar membrane supports the *organ of Corti* in the *scala media* (cochlear duct), which is the sensory part of the ear in contact with cranial nerve VIII. At the end of the scala tympani is another membranous window, the *round window,* which serves as an outlet for sound-wave vibrations in the cochlea. Just above the cochlea of the inner ear are three small semicircular canals. These function together as an organ of balance.

The parts of the ear are too small to make dissection feasible and are almost inaccessible, lying deep within the petrous region of the temporal bone. The location of the parts of the ear may be studied using models and diagrams. Trace the path of a sound wave, beginning with the pinna and ending with the auditory nerve, using Figures 29-4 and 29-5.

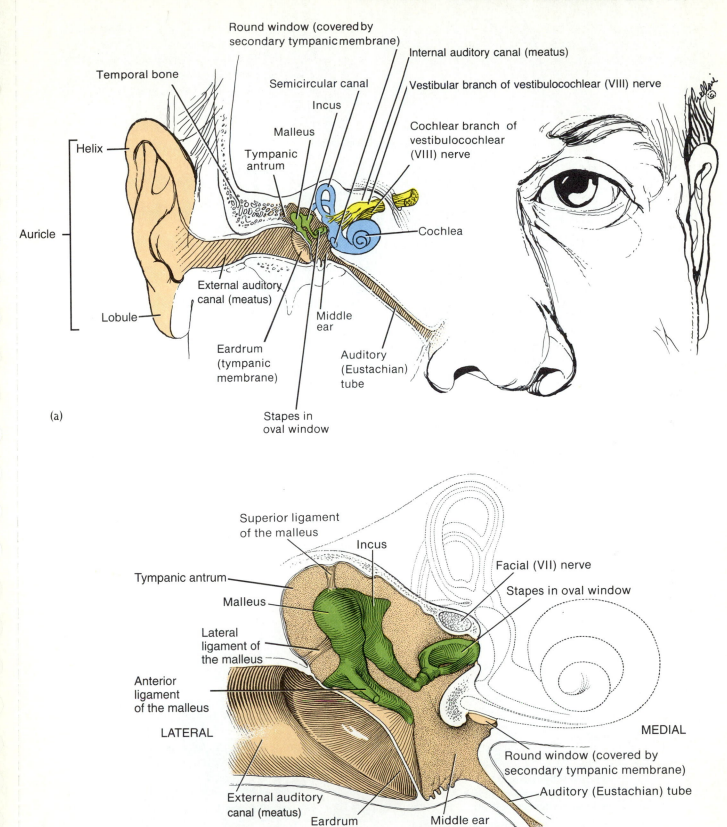

**Figure 29-4** Structure of the auditory apparatus: (a) divisions of the right ear into external, middle, and internal portions (seen in a frontal section through the right side of the skull); (b) details of the middle ear and body labyrinth of the internal ear. (From G. J. Tortora and S. R. Grabowski, *Principles of Anatomy and Physiology,* 7th ed., New York: HarperCollins, 1993.)

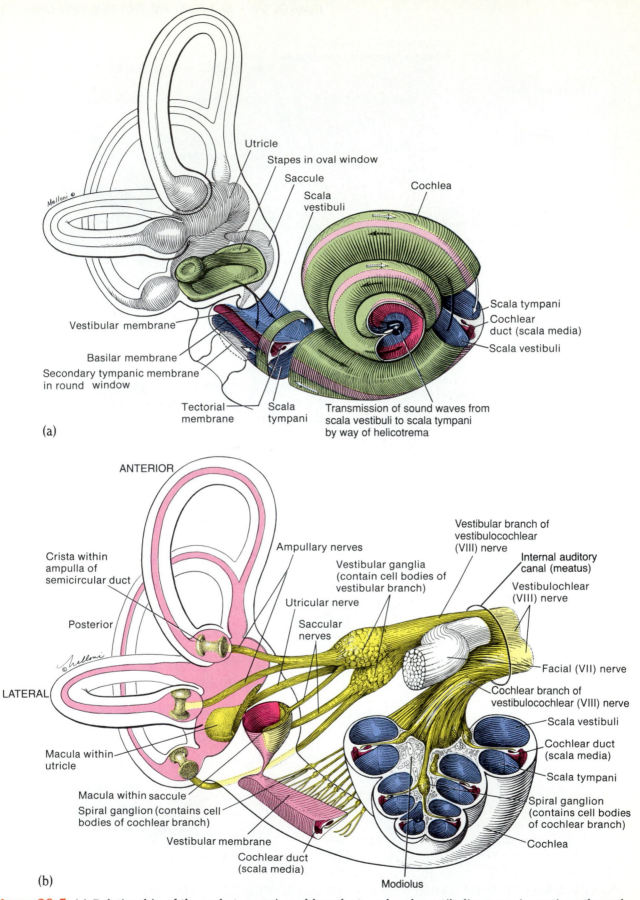

Utricle
Stapes in oval window
Saccule
Scala vestibuli
Cochlea
Vestibular membrane
Basilar membrane
Secondary tympanic membrane in round window
Tectorial membrane
Scala tympani
Scala tympani
Cochlear duct (scala media)
Scala vestibuli
Transmission of sound waves from scala vestibuli to scala tympani by way of helicotrema

(a)

ANTERIOR
Crista within ampulla of semicircular duct
Posterior
LATERAL
Macula within utricle
Macula within saccule
Spiral ganglion (contains cell bodies of cochlear branch)
Vestibular membrane
Cochlear duct (scala media)
Ampullary nerves
Vestibular ganglia (contain cell bodies of vestibular branch)
Utricular nerve
Saccular nerves
Vestibular branch of vestibulocochlear (VIII) nerve
Internal auditory canal (meatus)
Vestibulochlear (VIII) nerve
Facial (VII) nerve
Cochlear branch of vestibulocochlear (VIII) nerve
Scala vestibuli
Cochlear duct (scala media)
Scala tympani
Spiral ganglion (contains cell bodies of cochlear branch)
Cochlea
Modiolus

(b)

**Figure 29-5** (a) Relationship of the scala tympani, cochlear duct, and scala vestibuli as seen in sections through the cochlea. The arrows indicate the transmission of sound waves, which are discussed in this exercise. (b) The origins of the vestibular and cochlear branches of the vestibulocochlear (VIII) nerve. (From G. J. Tortora and S. R. Grabowski, *Principles of Anatomy and Physiology,* 7th ed., New York: HarperCollins, 1993.)

(a)

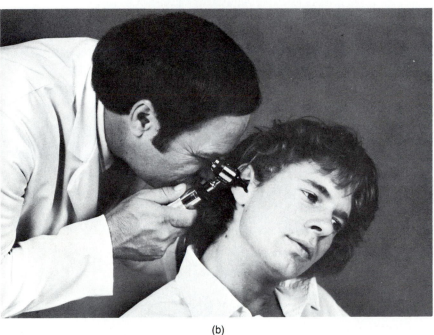

(b)

**Figure 29-6** The otoscope: (a) the instrument; (b) use of the instrument.

## EXAMINING THE EAR

If an otoscope (Figure 29-6) is available, make an examination of the external auditory canal and the tympanic membrane. Your instructor will explain the use and cleaning of the instrument. Carefully wipe the tip of the scope with a cotton square containing disinfectant. Let it air dry. Turn on the light, and look into the auditory canal of your lab partner. Record here what you see:

## RESULTS AND OBSERVATIONS: EXAMINING THE EAR

Description of view through otoscope. _____
_____
_____

# LABORATORY REVIEW *29*

## ANATOMY OF THE EYE AND EAR

1. Label the parts of the mammalian eye in Figure Q29-1. Use the following key.
   - **a.** Sclerotic coat
   - **b.** Choroid coat
   - **c.** Retina
   - **d.** Vitreous chamber
   - **e.** Posterior chamber
   - **f.** Anterior chamber
   - **g.** Fovea
   - **h.** Optic disc (blind spot)
   - **i.** Cornea
   - **j.** Lens
   - **k.** Iris
   - **l.** Ciliary body
   - **m.** Suspensory ligaments
   - **n.** Pupil
   - **o.** Eyelid
   - **p.** Plica semilunaris
   - **q.** Eyelash
   - **r.** Optic nerve

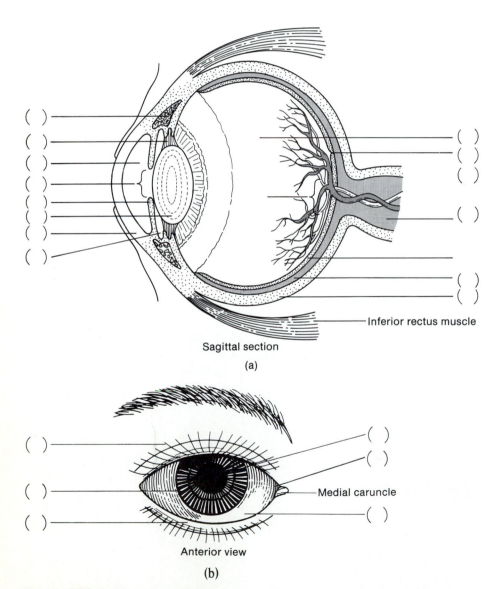

Inferior rectus muscle

Sagittal section

(a)

Medial caruncle

Anterior view

(b)

**Figure Q29-1** The mammalian eye.

2. Trace the pathway of light from outside the eye to the brain. Some of the steps are provided. You fill in the rest.

a. Conjunctiva _____

b. _____

c. _____

d. _____

e. _____

f. Lens _____

g. _____

h. _____

i. _____

j. Brain _____

3. Label the parts of the ear in Figure Q29-2. Use the key below. (Refer to the "Physiology of Hearing" exercise.)

a. External auditory canal
b. Tympanic membrane
c. Cochlea
d. Eustachian tube opening
e. Oval window
f. Round window
g. Scala media
h. Basilar membrane

i. Organ of Corti
j. Scala tympani
k. Scala vestibuli
l. Vestibular membrane
m. Malleus
n. Incus
o. Stapes
p. Tectorial membrane

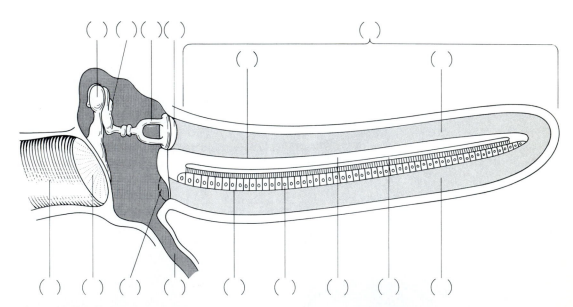

**Figure Q29-2** Parts of the ear.

4. Trace the pathway of sound from a source outside the ear to the brain. Some of the steps are provided. You fill in the rest.

a. External auditory canal

b. _____

c. _____

d. Incus

e. _____

f. _____

g. _____

h. Scala tympani

i. _____

j. _____

k. _____

l. Brain

# The Physiology of Vision

## After completing this exercise, you should be able to

1. Describe the focusing of light rays on the retina
2. Perform and interpret tests used to determine visual acuity
3. Perform two procedures to test pupillary reflexes
4. Perform a simple test to detect color blindness
5. Define or explain visual acuity, astigmatism, color blindness, and optical illusions

Light rays reflected from objects pass into the eye through the cornea and on through the anterior chamber and the lens. As light passes through these structures, it is focused on a light-sensitive region, the retina (Figure 30-1). The iris regulates the amount of light entering the eye, and the position and shape of the lens can be shifted slightly, to allow the eye to focus on objects at varying distances. This exercise includes a physical examination of the eye with the aid of an ophthalmoscope (Figure 30-3a).

Tiny inverted images are focused on the retina, and the varying intensities and wavelengths of the light energy stimulate combinations of the photoreceptors—the cones and rods—to initiate a series of nerve impulses. These impulses are coordinated by other neurons in the retina and are transmitted by the optic nerves to the optic centers of the brain, for final interpretation and integration. The normal human eye can sufficiently deflect light rays centering on an object 20 meters away to focus a clear object on the retina. An eye with a normal ability to refract light is referred to as being *emmetropic*. With a normal human lens, objects far in the distance, up to as close as 20 meters will form images on the light-sensitive area of the ret-

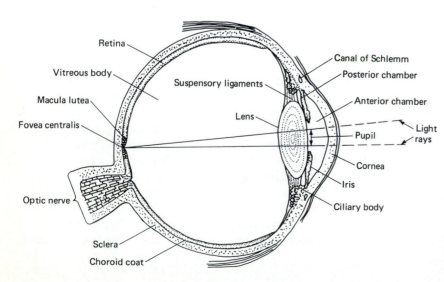

**Figure 30-1** The components of the human eye (shown in sagittal—lengthwise—section): The focusing of light rays onto the fovea centralis, the area of sharpest vision is shown.

Retina
Vitreous body
Macula lutea
Fovea centralis
Optic nerve
Sclera
Choroid coat
Suspensory ligaments
Lens
Canal of Schlemm
Posterior chamber
Anterior chamber
Light rays
Pupil
Cornea
Iris
Ciliary body

ina. With objects closer than 20 meters, the lens has to sharpen its focus, by means of ciliary muscles.

Various types of abnormalities related to improper *refraction* (deflecting light from a straight path) are known. These include *nearsightedness (myopia), far-sightedness (hypermetropia),* and *astigmatism. Astigmatism* refers to irregularities in the surface of the cornea or lens, causing an inability of the eye to focus a ray of light sharply on the retina. Procedures commonly used to study the functioning of the eye related to the series of events described here are included in this exercise. They are a test of *visual acuity* (the degree of detail the eye can distinguish); a test of *astigmatism,* using the well-known astigmatism test chart (Figure 30-2), and the determination of the individual's location of the *blind spot.*

The opening in the center of the iris is the pupil, which controls light transmission. In the darkness, the diameter of the pupil is wide, but while in bright light, it decreases. The dimensions of the pupil are controlled by the muscles of the iris, which are under reflex control. Procedures to test pupillary reflexes are included in this exercise.

Sensation of any color can be obtained by the appropriate mixture of three colors of light: red, green, and blue. In color blindness, a sex-linked genetic disorder, the number of colors that may be distinguished by an individual is greatly reduced. With humans, the most common type is red–green color blindness. A less common type is an inability to distinguish between blues and yellows. In the most severe forms of the genetic disorder, all objects appear to be some shade of gray. Approximately 8.0 percent of males and 0.5 percent of females exhibit some degree of color blindness. A test for color blindness is included in this exercise.

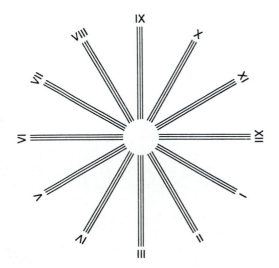

**Figure 30-2** The astigmatism test chart.

Various drugs cause changes in the diameter of the pupil. For example, homatropine, which is a drug used in ophthalmic examinations, makes the pupil larger than normal. Drugs such as cocaine and epinephrine cause a similar reaction. In contrast, morphine brings about a constriction of the pupil size.

Individuals constantly make judgments related to their perception of the size and distance of the objects they see. Such estimations depend on several factors, including the relative positions of an individual's eyes, the size of the retinal images, the amount of detail seen, and the color and clarity of the outlines of objects. A false interpretation of a visual sensation results in an optical illusion. Examples of such illusions include objects appearing farther away or larger than they really are. A portion of this exercise considers depth perception measurement and demonstrates some optical illusions.

The retina is the location of the photoreceptor cells and is essentially an extension of the brain. Before light can reach the rods and cones, it must first penetrate the many nerve cell bodies and processes that form the remaining layers of the retina. These nerve fibers all come together in a thick bundle near the back of the eye, to form the *optic nerve,* which completely interrupts the retina. Because there are no receptors in this area, no sensation of vision is produced when light falls on it. This area is termed the *blind spot.* The existence of the blind spot is demonstrated in this exercise.

Most individuals are normally quite unaware of the large gap in their visual field because the movements of their eyes continuously place images on that part of the retina where the sharpest vision occurs, the central fovea in the macula lutea. The macula lutea lies to the side of the blind spot, and it forms a disk, the outer rim of which contains both rods and cones (see Figures 30-1 and 30-3c). Toward the center, the rods disappear, so that the central foveal area contains only the cones. This is the area transmitting the sharpest image to the brain.

The photoreceptive rods and cones contain light-sensitive pigments that decompose when they absorb light energy. Rods contain a purple pigment known as *rhodopsin.* In the presence of light, rhodopsin breaks down into the yellowish substance *retinene,* which is made from vitamin A, and a colorless protein known as *scotopsin.* This photochemical dissociation of rhodopsin is termed the *bleaching reaction.* In bright light, the sensitivity of retinal rods is greatly reduced.

Individual cones are composed of retinene, combined with one of three different types of visual proteins: red, blue, and green. The particular color experienced by an individual depends on the particular set or combination of cones stimulated by a light source. If all three types of cones are stimulated, a person will see white; if none of the cones is stimulated, a person will see black. Cones are less sensitive to light than rods are. Cones provide color vision and greater visual acuity.

In situations when an individual becomes adapted to a bright light in the form of a light bulb and then quickly views a blank surface such as a wall, the image of the light bulb will appear. This effect is termed a *positive afterimage* and results from the continued activity (firing) of the photoreceptors. Usually after a short period of time, the *negative afterimage,* a dark image of the bulb, will appear against a lighter background. This effect is caused by the bleaching of the visual pigment of the involved receptors. This exercise also considers a visual pigment activity.

## EXAMINING THE EYE

### Materials

The following materials should be provided for class use:

1. Model of the eye
2. Chart of the anatomy of the eye
3. Ophthalmoscope
4. A darkened room
5. Timers
6. Disinfectant and cotton

### Procedure

This procedure is to be performed by students in pairs.

1. Review the locations of the components of the eye (Figure 30-1) and of the retina (Figures 30-3b and 30-3c, and Color Plate 32.).
2. Examine the parts of the ophthalmoscope. Locate all of the parts indicated in Figure 30-3a. Pay particular attention to the diopter window, lens-selection disk, rheostat control, and the viewing aperture.
3. Take the ophthalmoscope in one hand, and turn it on by depressing the rheostat lock button and then rotating the rheostat control. Hold the instrument about 5 cm (2 inches) above the desk, and rotate the rheostat control over its entire range of settings. Note how the intensity of the light changes from very bright to very dim with the movement of the control.
4. Set the diopter window on "0" and the light source on the green spot of the instrument. The ophthalmoscope is now ready for the eye-examination procedure.

5. Next, using Figures 30-4a through 30-4d, practice holding the ophthalmoscope and using the proper positions for examining both the right and left eyes. (Note that the ophthalmoscope is held in the right hand for the examination of the right eye and in the left hand for the left eye.)
6. In a darkened room, have your partner seated on your right and facing you.
7. Grasp the ophthalmoscope, resting your index finger on the lens-selection disk (see Figure 30-4c), and check to see that the "0" setting is in the diopter window and that the light is on.
8. Instruct your partner to set the timer for 1 minute and to look straight ahead at a fixed object or spot at his or her eye level. *(A safe time period for light exposure is 1 minute.)*
9. Begin viewing your partner's right eye at a distance of about 12 inches, as shown in Figure 30-4a. Shine the beam of light onto the pupil, and examine the lens and vitreous body of the eye. A red reflex (red-light reflection) will appear as you look through the pupil. If the image seen is not a sharp one, adjust the focus with the lens-selection disk of the ophthalmoscope.
10. Move to within 4 cm (2 inches) of your partner, while still keeping the pupil in focus (Figure 30-4b). Try to direct the ophthalmoscope's beam of light toward the pupil's edge rather than its center. (The red reflection should be more obvious.)
11. Look for the *optic disk* of the retina, and adjust the focus to obtain a sharp image. The blood vessels radiating from the optic disk should be clearly in view. Follow the path of at least two blood vessels to the edge of the retina.
12. Locate and identify the *macula* by having your partner look directly into the light beam for *1 second only.* Do not increase this exposure time. Can you see any blood vessels in the macula?
13. Examine the entire surface of the retina. Pay particular attention to surface depressions, protrusions (bumps) or any other irregularity. Bring any unusual condition to the attention of your instructor.
14. Keeping in mind the 1-minute beam-exposure time, examine the outer limits (periphery) of the retina. Ask your partner to look up, so that you can see the superior portion of the retina, and to look down so that the inferior portion of the retina can be observed.
15. Shut off the ophthalmoscope, and answer the questions in the results and observations section.
16. Change places with your partner, and repeat Steps 1 through 15.

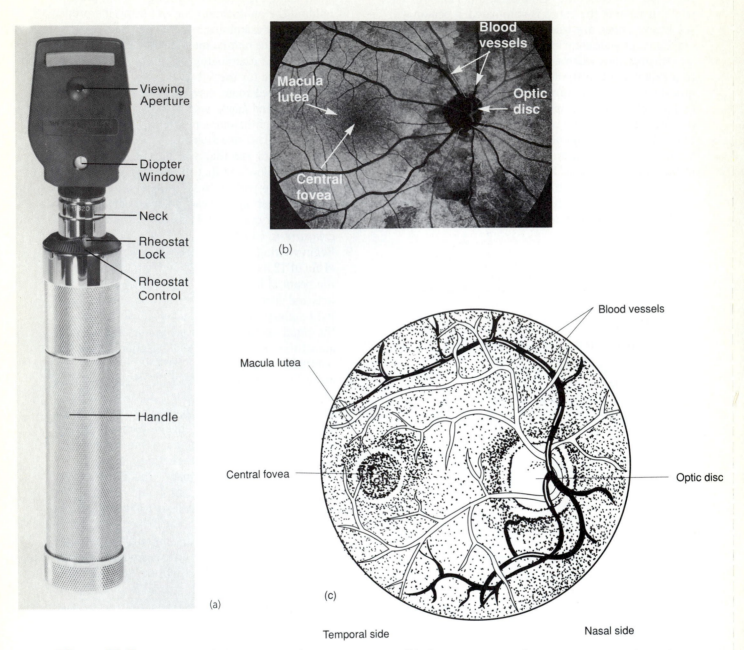

**Figure 30-3** (a) The ophthalmoscope and its components; (b) the appearance of a retina, as seen through an ophthalmoscope; (c) diagram of the fundus of the right retina, as seen through an ophthalmoscope.

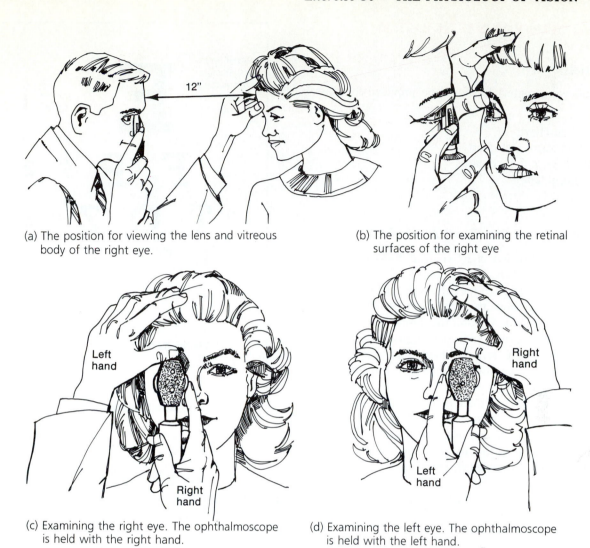

(a) The position for viewing the lens and vitreous body of the right eye.

(b) The position for examining the retinal surfaces of the right eye

Left hand

Right hand

Right hand

Left hand

(c) Examining the right eye. The ophthalmoscope is held with the right hand.

(d) Examining the left eye. The ophthalmoscope is held with the left hand.

**Figure 30-4** Proper positions during eye examinations.

## VISUAL ACUITY AND ASTIGMATISM TESTING

### Materials

The following materials should be provided for class use:

1. Snellen chart
2. Astigmatism test chart
3. Blank 3- × 5-inch cards
4. Masking tape
5. Yardstick

### Procedure 1: Snellen Visual Acuity Test

1. Place the Snellen chart (Figure 30-5) in a well-lighted area of the laboratory.
2. Use the yardstick to measure a distance of 20 feet on the floor from the Snellen chart. Put a short piece of masking tape on the 20-foot site.
3. If you are wearing corrective lenses (eyeglasses or contact lenses), remove them. Face the Snellen chart, and cover your left eye with a blank 3- × 5-inch card.
4. Read the letters of the fourth line from the top of the chart. Next, select and read the smallest line of letters that you can see clearly. As you call out the letters, have your laboratory partner write them in the space provided in the results and observations section.
5. Repeat Steps 3 and 4 with the right eye, reading from right to left.
6. If you normally wear corrective lenses, repeat Steps 3–5 with the lenses on.
7. Determine the visual acuity for each eye without and with glasses. The measure of visual acuity states the distance at which the chart is read (20 feet) and the distance indicated on the chart for the smallest line the individual can read. For example, an individual standing 20 feet from the chart, who can read line 2, but nothing smaller, is said to have a visual acuity of 20/100.
8. Compare the letters read with the actual letters on the Snellen chart. Enter the correct Snellen letter in the spaces provided in the results and observations section, in the correct left-to-right or right-to-left sequence.

### Procedure 2: Astigmatism Testing

This procedure should be performed by students in pairs.

1. Place the astigmatism test chart (see Figure 30-2) in a well-lighted area of the laboratory.
2. Use the yardstick to measure a distance of 10 feet from the chart. Put a short piece of masking tape on the 10-foot site.

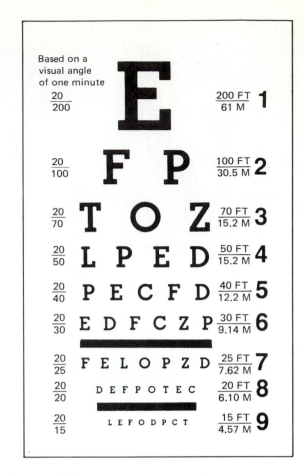

**Figure 30-5** The Snellen chart: This chart is used to determine whether an individual has normal or abnormal vision; the standard of normal vision (20/20) is the ability to stand 20 feet from the chart and be able to read Line 8.

3. If you are wearing corrective lenses, remove them. Test each eye separately. Look at the chart, and determine which, if any, of the line groups appear exceptionally thicker and darker than others. Indicate such line groups by number, in the space provided in the results and observations section.
4. If you wear corrective lenses, repeat Step 3 with them on.

## DEMONSTRATION OF BLIND SPOT

### Materials

The following materials should be provided per each two students:

1. Five coins
2. Four blank 3- × 5-inch cards
3. A millimeter ruler

## Procedure

This procedure should be performed by students in pairs.

1. Place the five coins in a straight line about 8 millimeters apart.
2. Look at the coins, and concentrate your view on the middle coin.
3. Close your right eye and hold a blank card over it. How many coins can you see? Have your partner record your findings in the results and observations section.
4. Repeat Step 3 with your left eye closed.
5. Assist your laboratory partner with this procedure, and record the findings in the results and observations section.

## PUPILLARY REFLEXES

### Materials

The following materials should be provided per each two students:

1. One small flashlight (or penlight)
2. A millimeter ruler

### Procedure 1: Light Reflex

This procedure should be performed by students in pairs.

1. Hold the millimeter ruler under one eye of your laboratory partner, and determine the diameter of the pupil.
2. With the ruler in the same position, quickly and briefly flash a light into the eye. Quickly measure the diameter of the pupil.
3. Remove the light source, and record your findings in the results and observations section.
4. Change places with your partner, and repeat Steps 1–3.

### Procedure 2: Accommodation Reflex

This procedure should be performed by students in pairs.

1. Hold the millimeter ruler under one eye of your laboratory partner, and determine the diameter of the pupil.
2. With the ruler in the same position, have your partner look first at an object located at the far end of the room and second at your shoulder. Measure the diameter of the pupil in each situation.
3. Record your findings in the results and observations section.
4. Change places with your partner, and repeat Steps 1–3.

## DEPTH PERCEPTION TESTING

### Materials

The following materials should be provided for class use:

1. Depth-perception testing system, with the following items in place.
   b. Calibrated base with adjustment string attached
   b. Two slider bases, equipped with arrow holders and arrows
2. Sunglasses
3. Commercially prepared disposable alcohol pads
4. Container for used alcohol pads

### Procedure

This procedure should be performed by students in pairs.

1. Examine the depth-perception testing system provided. Make certain that all appropriate parts are operational. Note that one of the slider bases is attached to the adjustment string. Refer to Figure 30-6.
2. Place the testing system flat on a table or desk in a well-lighted area.
3. Have your laboratory partner sit comfortably on a chair situated 8 feet away from the depth-perception testing system. The zero end of the scale should be closest to the chair. If your partner wears glasses or contact lenses, have them removed.
4. Ask your partner to close his or her eyes, while the unattached slider base is moved to some position along the testing system's millimeter scale (Figure 30-6). Make certain that the arrows are facing one another and that they are at the eye level of your partner.
5. Give the adjustment string to your partner, to align the two arrows by moving the slider base attached to the string.
6. When your partner is satisfied with the alignment, note the position of the string-attached (movable) slider base relative to the unattached one. Record any difference shown on the depth-perception testing system scale in millimeters in Table 30-6 in the results and observation section.
7. Repeat Steps 4–6, using another position for the attached slider base on the testing system's scale.
8. Have your partner repeat Steps 4–7, first with the right eye closed. Record all findings in Table 30-6, in the results and observations section.
9. Repeat Steps 4–8, with your partner wearing a pair of sunglasses. Record all findings in Table 30-7.
10. Change places with your partner, and perform Steps 1–9.
11. Answer questions listed in the results and observations section.

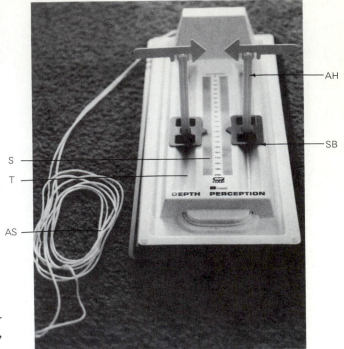

**Figure 30-6** A depth-perception tester. The components shown include arrow holders (AH), troughs (T), slider bases (SB), scale (S) and adjustment string (AS).

## OPTICAL ILLUSIONS

### Materials

The following materials should be provided for class use:

1. Millimeter rulers
2. Magnifying glasses (optional)

### Procedure

This procedure should be performed by students individually.

1. Examine the three squares in Figure 30-7.
2. Determine visually which of the squares is smaller and which of the squares appears taller. Record your findings in the results and observations section.
3. Measure each square with a ruler. Record your findings in the results and observations section.
4. Examine the diagrams in Figure 30-8.
5. Determine visually whether the horizontal lines are of equal length. Record your findings in the results and observations section.
6. Measure the lines with a ruler. Record your findings in the results and observations section.
7. In the space provided in the results and observations section, construct an optical illusion diagram of your own making. Test it with your laboratory partner.

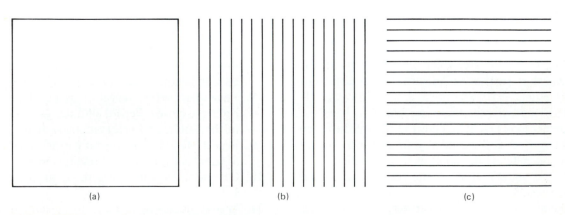

(a)  (b)  (c)

**Figure 30-7** Optical illusion: Which square appears to be the smallest?

(a)                                                    (b)

**Figure 30-8** Optical illusion: Which horizontal line is longer?

## TEST FOR COLOR-BLINDNESS

### Materials

The following materials should be provided for students in groups of four:

1. One Holmgren Color Vision Test
2. One Ichikawa or Ishihara Color Blindness Test
3. Correct explanation keys for color-blindness tests

### Procedure

This procedure should be performed by students in groups of four.

1. Each member of the group should take the Holmgren Color Vision Test.
   a. Select and match 12 different wool strands with those mounted in the binder. List your responses.
   b. Consult the key, and list the key's colors. Circle incorrect responses in the results and observations section.
2. Each member of the group should take the Ichikawa or Ishihara Color Blindness Test.
   a. Examine each of the plates in the test booklet.
   b. Enter your impressions as to the symbols on these plates and the key explanation for your responses in the results and observations section.
3. Compare the findings obtained with both color-blindness tests, and answer the questions in the results and observations section.
4. The instructor should compile test results for the whole class. Determine the number of individuals who are color-blind. Answer the questions in the results and observations section.

## VISUAL PIGMENT ACTIVITY

### Materials

The following materials should be provided for class use:

1. Laboratory lamps (40-watt bulbs) without covers or flashlights
2. Squares of bright, glossy red, yellow, green, and blue paper (4- × 4-inch) mounted in the center of the individual solid, black poster boards or papers (8½ × 11 inch)
3. Squares of bright, glossy, red, yellow, green, and blue paper (4- × 4-inch), with center-cut holes, mounted in the center of individual white poster boards or papers (8½ × 11 inch)
4. White poster board or paper sheets (8½ × 11 inch)

### Procedure

This procedure should be performed by students in pairs.

1. Have your laboratory partner sit comfortably approximately 6 feet in front of a blank wall and a table with a lamp without a shade or cover.
2. Turn on the lamp, and direct your partner to stare at the bright light for 20 seconds and then immediately to look away toward the blank wall.
3. Ask what afterimages (positive and negative) appeared. Record the findings in the results and observations section.
4. Stare at the bright light once again for 20 seconds. Now have your partner immediately hold in front of you a white poster board with a red paper square. Record a description of the types of afterimages seen in the results and observations section.
5. Did you see the colored square?

6. Next, have your partner first hold a black poster board with a red paper square in front of you. Stare at the square for 60 seconds. Quickly have your partner hold a white poster board in front of you. Note the color of the afterimages. Did you see a colored square? Answer the questions in the results and observations section.
7. Repeat Step 6 with the yellow, green, and blue paper squares.
8. Change places with your partner, and repeat Steps 1–7.

## RESULTS AND OBSERVATIONS

### Examining the Eye

1. Description of the right eye, as seen through the ophthalmoscope. _____

_____

2. Description of the left eye, as seen through the ophthalmoscope. _____

_____

3. Were you able to find all the parts of the eye shown in Figures 30-3b and 30-3c? List the parts that you were able to observe. _____

_____

4. Were any unusual structures observed? _____

_____

### Visual Acuity and Astigmatism Testing

1. Enter the results of the Snellen test in Table 30-1.
2. Enter the results of the astigmatism test in Table 30-2.

### Demonstration of Blind Spot

1. Enter your findings and those of your laboratory partner in Table 30-3.
2. Would you expect to see more or fewer coins if the distance between coins were increased? _____

_____

3. What is the blind spot? _____

_____

### Pupillary Reflexes

1. Enter your findings on the light reflex in Table 30-4.
2. Enter your findings on the accommodation reflex in Table 30-5.

**Table 30-1**

| Eye Condition | Letters Read | Actual Sequence on Chart | Visual Acuity |
|---|---|---|---|
| Left eye (without glasses) | | | |
| Left eye (with glasses, if applicable) | | | |
| Right eye (without glasses) | | | |
| Right eye (with glasses, if applicable) | | | |

**Table 30-2**

| **Eye Condition** | **Thicker and Darker Line Groups** |
|---|---|
| Without glasses | |
| With glasses | |

**Table 30-3**

| **Student** | **Number of Coins Seen** | |
|---|---|---|
| | Right Eye Closed | Left Eye Closed |
| | | |
| | | |
| | | |
| | | |

**Table 30-4**

| **Student** | **Pupil Diameter** | |
|---|---|---|
| | Without Light | With Light |
| | | |
| | | |

**Table 30-5**

| **Student** | **Pupil Diameter** | | |
|---|---|---|---|
| | At Start | Distant Object | Near Object |
| | | | |
| | | | |

## *Depth Perception Testing*

1. Enter your findings and those of your laboratory partner, obtained with conditions of normal lighting in Table 30-6.
2. Was alignment achieved in one or both attempts, with having both eyes open? _____

   _____

3. Was alignment achieved with either the right eye or the left eye closed? _____

4. a. If alignment was not achieved with both eyes open, or with either eye closed, which situation produced a greater millimeter distance away from the fixed slider base? _____

   _____

   b. Offer an explanation for any differences that occurred. _____

   _____

5. Enter your findings and those of your partner, obtained with the use of sunglasses, in Table 30-7.

**Table 30-6**

| Student | Depth Perception Measurement Difference in Millimeters | | | | | |
|---|---|---|---|---|---|---|
| | Both Eyes Open | | Right Eye Closed | | Left Eye Closed | |
| | *1* | *2* | *1* | *2* | *1* | *2* |
| _____ (without glasses) | | | | | | |
| _____ (with glasses, if applicable) | | | | | | |
| _____ (without glasses) | | | | | | |
| _____ (with glasses, if applicable) | | | | | | |

**Table 30-7**

| Student | Depth Perception Measurement Differences in Millimeters | | | | | |
|---|---|---|---|---|---|---|
| | Both Eyes Open | | Right Eye Closed | | Left Eye Closed | |
| | *1* | *2* | *1* | *2* | *1* | *2* |
| _____ (without glasses) | | | | | | |
| _____ (with glasses, if applicable) | | | | | | |
| _____ (without glasses) | | | | | | |
| _____ (with glasses, if applicable) | | | | | | |

6. a. Was alignment more difficult with the sunglasses? _____

   _____

   b. Explain your findings. _____

   _____

   _____

## *Optical Illusions*

1. Which of the squares appears to be smallest?

   _____ Which of the squares appears to

   be tallest? _____

2. Record the dimensions of the squares in Table 30-8.
3. Which of the horizontal lines appears to be longer?

   _____

4. Record the length of the horizontal lines in Table 30-9.
5. Construct your own optical-illusion diagram in the space provided.

**Table 30-8**

| Square | Height (in mm) | Length (in mm) |
|--------|----------------|----------------|
| a. | | |
| b. | | |
| c. | | |

**Table 30-9**

| Line | Length |
|------|--------|
| a. | |
| b. | |

2. Optical-illusion diagram.

## Color-Blindness Tests

1. Enter the responses to the Holmgren Color Vision Test in Table 30-10.
2. Enter the responses to the Ichikawa or Ishihara Color Blindness Test in Table 30-11.

3. Were the findings obtained with the two color-blindness tests similar? If not, explain the results obtained. _____

_____

**Table 30-10**

| Thread Color | Student Response | | | | Key Color |
|---|---|---|---|---|---|
| | 1 | 2 | 3 | 4 | |
| | | | | | |
| | | | | | |
| | | | | | |
| | | | | | |
| | | | | | |
| | | | | | |
| | | | | | |
| | | | | | |
| | | | | | |
| | | | | | |
| | | | | | |

**Table 30-11**

| Plate | Student Response | | | | Key Explanation |
|---|---|---|---|---|---|
| | 1 | 2 | 3 | 4 | |
| 1 | | | | | |
| 2 | | | | | |
| 3 | | | | | |
| 4 | | | | | |
| 5 | | | | | |
| 6 | | | | | |
| 7 | | | | | |
| 8 | | | | | |
| 9 | | | | | |
| 10 | | | | | |
| 11 | | | | | |
| 12 | | | | | |
| 13 | | | | | |
| 14 | | | | | |
| 15 | | | | | |
| 16 | | | | | |
| 17 | | | | | |
| 18 | | | | | |
| 19 | | | | | |
| 20 | | | | | |
| 21 | | | | | |
| 22 | | | | | |
| 23 | | | | | |
| 24 | | | | | |

4. Summarize class results in Table 30-12.

5. How many color-blind individuals were unaware of their color-blindness? _____

## Visual Pigment Activity

1. Did you see all color squares as afterimages in the situations of the experiment? If not, which colors were not seen? _____

_____

2. Were the center holes in any of the colored squares observed in the afterimages? If so, which ones?

_____

_____

3. What causes an afterimage? _____

_____

## Table 30-12

| Students | Number Color-Blind | Type of Color-Blindness | Percentage Color-Blind |
|----------|--------------------|-----------------------|----------------------|
| Female | | | |
| Male | | | |
| All | | | |

# LABORATORY REVIEW 30

NAME _____

LAB SECTION _____ DATE _____

## THE PHYSIOLOGY OF VISION

### COMPLETION AND DEFINITION

1. Define or explain the following:
   a. Visual acuity _____

   _____

   b. Astigmatism _____

   _____

   c. Color-blindness _____

   _____

2. What is an optical illusion? _____

   _____

   _____

3. List five factors that can affect judgment of size and distance.

   a. _____

   b. _____

   c. _____

   d. _____

   e. _____

4. List two pupillary responses and the normal result for each.

   a. _____

   _____

   b. _____

   _____

5. Are the following Snellen chart test results considered normal, better than normal, or worse than normal?

   a. 20/20 _____

   b. 20/15 _____

   c. 20/100 _____

   d. 70/100 _____

# ANSWERS

1. _____
2. _____
3. _____
4. _____
5. _____
6. _____
7. _____
8. _____
9. _____
10. _____
11. _____
12. _____

# MULTIPLE CHOICE

Insert the correct answers in the spaces provided.

1. Which of the following parts of the eye holds the aqueous humor?
   a. macula lutea
   b. retina
   c. anterior chamber
   d. posterior chamber
   e. pupil
2. Which of the following findings of an optical examination would indicate the best visual acuity?
   a. 20/20
   b. 20/30
   c. 20/40
   d. 20/50
   e. 20/60
3. The condition of nearsightedness also can be referred to as which of the following?
   a. hyperopia
   b. astigmatism
   c. emmetropia
   d. myopia
   e. hypoopia
4. Which of the following is the small area of the retina where no image is formed?
   a. conjuctiva
   b. blind spot
   c. iris
   d. sclera
   e. cornea
5. Which portion of the eye regulates the amount of light entering the eye?
   a. conjuctiva
   b. blind spot
   c. iris
   d. sclera
   e. cornea
6. What is the term for the degree of detail the eye can distinguish?
   a. reflex control
   b. astigmatism
   c. macula lutea
   d. visual acuity
   e. myopia
7. In a dark environment, what does the diameter of the pupil do?
   a. increases
   b. decreases
   c. does not change
8. In bright light, what happens to the sensitivity of retinal rods?
   a. greatly increased
   b. greatly reduced
   c. is not changed
9. If all three types of retinal cones are stimulated, what will an individual see?
   a. black
   b. white
   c. red
   d. blue
   e. green

10. What does the central fovea contain?
    a. only rods
    b. only cones
    c. both rods and cones
11. In determining visual acuity with a Snellen chart, when reading the results, higher numbers following the "20/" (e.g., 20/50) indicate the relative size of the letter on the chart, for the individual being tested to see it clearly. Does a higher number indicate a larger or a smaller relative size?
    a. smaller
    b. larger
12. What is the term for an eye with a normal ability to refract light?
    a. myopic
    b. astigmatic
    c. hypermetropic
    d. emmetropic
    e. hyperopic

# The Physiology of Hearing

1. Trace the path of sound waves through the parts of the ear
2. Perform experiments to demonstrate both normal and abnormal physiological aspects of hearing

The ear has two distinct functions: (1) The auditory component is concerned with the sensation of hearing, while (2) the vestibular portion provides information to the brain about the movement and position of the head. The sensory elements of both components are hair cells in the inner ear (see Figure 29-5). These are excited by the bending of hairs and are referred to as *mechano-receptors*.

The ear responds to mechanical disturbances produced by a vibrating body. For example, if the arms of a tuning fork are struck, they will swing back and forth, and a mechanical disturbance of molecules will be created through the surrounding air in all directions. Such vibrations are perceived by the mechanoreceptors as sound. The human ear can distinguish among the properties of sound, which include intensity, pitch, and timbre. Intensity or loudness is measured in units termed *decibels*. *Pitch* is determined by the frequency of vibrations, with higher frequencies producing higher-pitched sounds. The range of frequencies to which the human ear is sensitive is from 20 to 20,000 hertz (vibration cycles per second). *Timbre* refers to the quality of sound and depends on the presence of several frequencies.

Sound is transmitted through the ear by first passing through the external auditory meatus (see Figure 31-1) and causing the tympanic membrane to vibrate.

These vibrations are transferred to the bones of the middle ear and are amplified as they reach the oval window. Vibrations of the oval window cause the fluid of the scala vestibuli to vibrate. This disturbance causes certain portions of the basilar membrane to bulge into the cochlear duct. Basilar membrane movement stimulates the hair cells of the organ of Corti, which results in the conduction of impulses along the cochlear nerve. Vibrations of the fluid of the scala vestibuli are also transferred to the scala tympani and are spread as the round window bulges toward the middle ear. Interference with or damage to this conduction system or to the nerve fibers or receptor cells results in hearing loss or deafness. *Deafness* is a severe form of hearing loss that makes everyday auditory communication difficult or impossible. Based on the part of the hearing mechanism that fails to function adequately, hearing loss or deafness can be classified into three main types: conductive, sensorineural, and central hearing loss.

*Conductive hearing loss* may be caused by simple situations such as the accumulation of wax in the ear. However, it also may be due to an inflammation and infection of the middle ear, known as *otitis media,* or to *otosclerosis,* a condition in which new bone growth may prevent the normal vibration of the stapes in the oval window.

*Sensorineural hearing loss* results from injury or disease involving the organ of Corti or the cochlear nerve. Causes of this form of hearing loss include intense noise, ear-damaging drugs, bacterial and viral infections of the inner ear, and aging.

*Central hearing loss* results from defects in the auditory tracts or centers in the brain. Several causes are associated with this condition and include traumatic injury, tumors, infections, blockage of blood vessels in the brain, and a progressive hearing loss that occurs with age, known as *presbyacusis*.

The vestibular apparatus, or *nonauditory laby-rinth,* is the component of the inner ear that detects

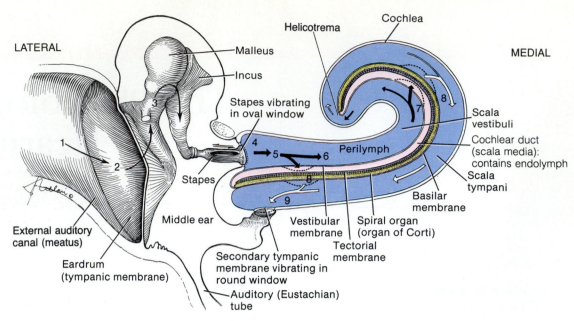

**Figure 31-1** The mechanism of normal hearing: The arrows show the direction in which sound travels. (From G. J. Tortora and S. R. Grabowski, *Principles of Anatomy and Physiology,* 7th ed., New York: HarperCollins, 1993.)

specific movements and positions of the head. Compensatory responses of the muscles of the neck, trunk, and limbs are initiated by the vestibular apparatus and help to keep the body in balance and in a proper orientation with respect to gravity. Motion sickness (a temporary disturbance) and Meniere's disease are known to interfere with the functioning of the organs of Corti. Motion sickness results from repetitive changes in the rate and direction of movement and from conflicting visual and vestibular signals. Meniere's disease may be caused by allergies, infections, or trauma. It results in an irritation or injury to the inner ear, leading to the accumulation of an excess of endolymph (see cochlear duct on Figure 31-1).

Procedures in this exercise consider certain properties of sound, and an individual's sensitivity to sound, static equilibrium response, and spatial orientation to sound.

## HEARING TESTS

### Materials

The following materials should be provided for class use:

1. Watch, clock, or other small device with audible ticking sound
2. Tuning fork
3. Rubber reflex hammer
4. Individual cotton plugs
5. Yardstick
6. Masking tape

### Procedure 1: Auditory Acuity

This procedure should be performed by students in pairs.

1. Have your laboratory partner sit down with eyes closed. Carefully insert a cotton plug into the left ear.
2. Hold a ticking watch or clock in line with your partner's right ear approximately 12 feet away.
3. As you move slowly toward your partner, have your partner indicate exactly when the ticking sound is heard.
4. Stop and measure the distance between the sound source and your partner's ear.
5. Repeat this procedure with the other ear, using a fresh cotton plug. Enter your findings in the results and observations section.
6. Change places with your partner, and repeat Steps 1–5.
7. Answer the questions in the results and observations section.

### Procedure 2: Sound Localization

This procedure is to be performed by students in pairs.

1. Have your laboratory partner sit down with eyes closed.

2. Bring a ticking watch into the hearing range of your partner, from the following locations:
   a. Front of head
   b. Rear of head
   c. Left side of head
   d. Right side of head
   e. One arm's length above head on the right side
   f. One arm's length above head on the left side
3. Have your partner point to the position from which the sound is coming.
4. In the results and observations section, indicate whether the sound localization was correctly identified.
5. Change places and repeat Steps 1–4, but changing the order of the locations in Step 2.

### Procedure 3: Rinne Test (Test for Middle-Ear Deafness)

This procedure is to be performed by students in pairs.

1. Have your partner sit down.
2. Locate the mastoid process of the temporal bone behind the left ear.
3. Strike a tuning fork and place the handle of the vibrating fork against the mastoid. (The sensation felt is bone conduction.)
4. Have your partner indicate when the vibrations are no longer heard, and immediately hold the tuning fork just outside the external ear. (In middle-ear deafness, no sound is heard by air conduction.)
5. Describe your partner's responses in the results and observations section.
6. Change places with your partner, and repeat Steps 1–5.

### Procedure 4: Weber Test (Test for Middle-Ear Deafness and Nerve Defects)

This procedure is to be performed by students in pairs.

1. Have your partner sit down in a relatively quiet room.
2. Strike the tuning fork, and place the handle of the vibrating fork against the middle of your partner's forehead.
3. Have your partner indicate the direction from which the sound appears to be coming. (With normal hearing, the sound appears to be referred to the middle of the head. With middle-ear deafness, the sound is detected in the defective ear. With nerve defects, vibrations are received better with the normal ear.)
4. Describe your partner's responses in the results and observations section.
5. Change places with your partner, and repeat Steps 1–4.

## RESULTS AND OBSERVATIONS

### *Hearing Tests*

1. Enter your findings in the auditory acuity test in Table 31-1.
2. Was the auditory acuity of both ears the same for you? _____ For your partner? _____
3. Enter your findings in the sound-localization test in Table 31-2. Indicate correct sound localization by a "+" and an incorrect identification by a "−."

**Table 31-1**

| Student | Distance at Which Sound Is Detected (Right Ear) | Distance at Which Sound Is Detected (Left Ear) |
|---|---|---|
| | | |
| | | |

**Table 31-2**

| Student | Front | Rear | Left Side | Right Side | One Arm's Length to Left | One Arm's Length to Right |
|---|---|---|---|---|---|---|
| | | | | | | |
| | | | | | | |

4. a. Describe your response to the Rinne test. ____

_____

b. Describe your partner's response. _____

_____

5. a. Describe your response to the Weber test.

_____

b. Describe your partner's response. _____

_____

## STATIC EQUILIBRIUM AND SPATIAL ORIENTATION TESTS

### Materials

The following materials should be provided for class use:

Rotating stools or swivel chairs without casters

### Procedure 1: Static Equilibrium Response

The following tests should be performed by students in pairs (except Procedure 2, which will require three students):

1. Have one laboratory partner stand erect and still, with both feet as close together as possible. This position should be maintained for 2 minutes. Note the kinds and amount of movement necessary to keep the body in this position. Record your findings in the results and observations section.

2. Have your laboratory partner repeat Step 1, but with eyes closed. (*Note:* Two students should stand by to prevent a fall.) Enter your findings in the results and observations section.

3. Change places with your laboratory partner, and repeat steps 1 and 2.

### Procedure 2: Spatial Orientation

1. Have your laboratory partner stand erect, with both arms extended to the sides and then bring the tips of the index fingers together.

2. Repeat Step 1 with eyes closed until the index fingers are brought together successfully at least three times.

3. Have your partner sit in a swivel chair with eyes closed. Twirl your partner around rapidly 10 times. Repeat Step 2. Record your findings in the results and observations section.

4. Change places with your partner, and repeat Steps 1–3.

## RESULTS AND OBSERVATIONS

1. In Table 31-3, enter brief descriptions of your reactions and those of your partner in the reflex-testing procedures.

2. What structures in the human body contribute to an individual's static equilibrium? _____

_____

### Table 31-3

| Reflex Procedure | Your Response | Laboratory Partner's Response |
|---|---|---|
| Static equilibrium Eyes open | | |
| Eyes closed | | |
| Spatial orientation Eyes open | | |
| Eyes closed | | |
| After twirling | | |

# LABORATORY REVIEW 31

## THE PHYSIOLOGY OF HEARING

### COMPLETION, DEFINITION, AND DESCRIPTION

1. Give the two distinct functions of the ear and the specific portions of the ear concerned with each.

   _____

   _____

   _____

2. What is sound? _____

   _____

3. List three specific properties of sound, which the human ear can distinguish.

   a. _____

   b. _____

   c. _____

4. List the three main types of hearing loss and the particular portion of the hearing mechanism involved.

   a. _____

   b. _____

   c. _____

5. Briefly describe two tests for the detection of middle-ear deafness.

   a. _____

   _____

   b. _____

   _____

6. What parts of the ear are associated with balance? _____

   _____

7. List and briefly describe two disturbances involved with the functions of the organs of balance.

   a. _____

   b. _____

8. Define or explain the following:

   a. Otitis media _____

   b. Otosclerosis _____

   c. Presbyacusis _____

   d. Decibel _____

## ANSWERS

1. _____
2. _____
3. _____
4. _____
5. _____
6. _____
7. _____
8. _____
9. _____
10. _____

## MULTIPLE CHOICE

Insert the correct answers in the spaces provided.

1. What is the organ of hearing located within the inner ear?
   a. semicircular ducts
   b. oval window
   c. membranous labyrinths
   d. organ of Corti
   e. round window

2. Name the openings between the middle and inner ears:
   a. semicircular ducts
   b. oval window
   c. round window
   d. organ of Corti
   e. choices *b* and *c* only

3. What is the specific unit used to measure the loudness, or intensity, of sound?
   a. hertz
   b. decibel
   c. vibration
   d. pitch
   e. pitched sound unit

4. What are the sense organs of static equilibrium?
   a. semicircular ducts
   b. oval window
   c. membranous labyrinths
   d. organ of Corti
   e. round window

5. What is the unit used to express vibration cycles per second?
   a. hertz
   b. decibel
   c. vibration
   d. pitch
   e. pitched sound unit

6. Which of the following forms of hearing loss can be caused by the accumulation of cerumen?
   a. conductive
   b. sensorineural
   c. central hearing
   d. presbyacusis

7. Which of the following forms of hearing loss can be caused by otitis media?
   a. conductive
   b. sensorineural
   c. central hearing
   d. presbyacusis

8. What is the hearing loss that occurs with age called?
   a. otitis media
   b. myopia
   c. hypermetropia
   d. presbyacusis
   e. astigmatism

9. What is a medical term for a middle-ear infection?
   a. otitis media
   b. myopia
   c. hypermetropia
   d. presbyacusis
   e. astigmatism
10. Which of the following hearing losses would an intense noise over a period of time probably cause?
   a. conductive
   b. sensorineural
   c. central hearing
   d. presbyacusis
   e. otitis media

# The Physiology of Smell and Taste

## After completing this exercise, you should be able to

1. Indicate the location of the olfactory receptors
2. List the factors on which sensations of smell depend
3. Describe two roles of the sense of smell
4. Locate the areas of specific taste reception on the tongue

Chemoreceptors that respond to chemical changes in the external environment are referred to as *exteroceptors*. These include the sense of smell, or *olfactory receptors* that respond to gaseous molecules in the air, and the sense of taste, or *gustatory receptors* that respond to chemicals dissolved in food or drink.

The receptors for the sense of smell are located in the olfactory cells in the mucous membrane lining of the upper portion of the nasal cavity. Only volatile substances in solution can stimulate these receptors. The gaseous particles enter the nasal cavity by diffusion from the circulating air in the external environment. Glands that surround olfactory cells secrete fluid into which the particles dissolve, and this solution acts as a chemical stimulus to a nerve impulse (Figure 32-1).

In humans, the sense of smell is less important than in other animals. However, it does play a role in starting the flow of certain digestive juices at mealtime. What have you experienced when you are quite hungry and smelled the aroma of food? The sensation of smell also can play a role in reviving a person who has fainted. Smelling salts stimulate the endings of the trigeminal cranial nerve and produce pain. This effect initiates respiratory reflexes and eventually results in reviving individuals who have fainted.

The epithelial surface of the tongue contains taste receptors, which are specialized cells grouped into barrel-shaped structures termed taste buds. (See Figure 32-2a and Color Plate 33.) Each taste bud consists of a taste (*gustatory*) pore, which communicates with the external environment, supporting cells, and receptor cells, which have little gustatory hairlike processes that project through the taste pore (see Figure 32-2a). Molecules dissolved in saliva at the tongue's surface interact with the hairlike processes, causing receptor cells to release a neurotransmitting chemical. This response, in turn, triggers sensory nerve endings in the taste buds.

Despite the fact that individuals experience a variety and range of taste sensations, there are only four basic responses: sweet, sour, salty, and bitter. Figure 32-2b shows the general locations of the taste sensations.

Several experiments demonstrating the features of the sensation of smell and taste are considered in this exercise.

## Materials

The following materials should be provided for class use:

1. Containers (with tops or plugs) of
   a. Freshly ground coffee
   b. Garlic powder
   c. Oil of cloves
   d. Oil of wintergreen
   e. Oregano
   f. Vanilla extract
   g. Vinegar
   h. Orange or lemon peels
   i. Perfume
   j. Distilled water
   k. 5 percent NaCl solution
   l. 5 percent sucrose solution
   m. 0.5 percent quinine-sulfate solution

453

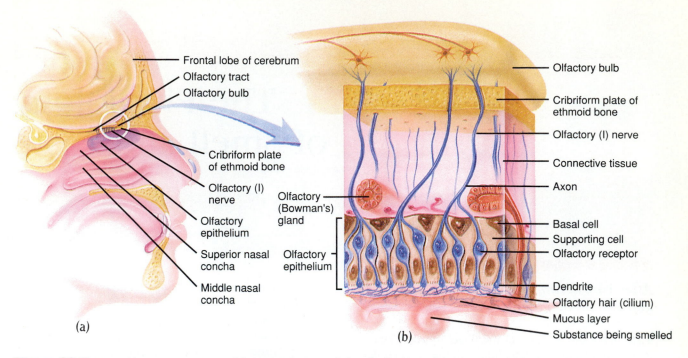

**Figure 32-1** The olfactory receptors: (a) Sagittal view of the skull shows location of receptor in the naval cavity (circled area); (b) an enlarged view of the receptors in action while a substance is being smelled—small arrows show the direction of the sensation. (From G. J. Tortora and S. R. Grabowski, *Principles of Anatomy and Physiology,* 7th ed., New York: HarperCollins, 1993.)

2. Equally cut slices of apple, onion, and potato in separate and covered containers
3. Cotton swabs
4. Tongue depressors
5. Container for the disposal of swabs and tongue depressors
6. Distilled water for drinking
7. Paper towels

## Procedure 1: Demonstration of Olfactory Stimulus

This procedure is to be performed by students in groups of four.

1. Dip a cotton swab into the perfume, and gently squeeze out the excess fluid on the inner side of the container.
2. Apply the perfume to the forearm of one of your laboratory partners.
3. Note the time elapsed before each of your other laboratory partners first detects the aroma of the perfume after its application.
4. Note the time elapsed from application until the aroma is no longer detectable for each partner.
5. Record your findings and answer the questions in the results and observations section.

## Procedure 2: Stimulus Identification

This procedure is to be performed by students in pairs.

1. Smell the contents of each of the containers.
2. Close all containers. Wait 5 minutes.
3. Close your eyes. Have a laboratory partner open each container, and pass it slowly about 3 inches from your nose. Your exposure should be no longer than 3 seconds for each container.
4. Identify each stimulus by name. Have your partner record whether each response is correct or incorrect in the results and observations section.
5. Immediately after the first round of testing, repeat Steps 3 and 4 two more times.
6. Repeat Steps 1–5 with each member of your group.
7. Answer the questions in the results and observations section.

## Procedure 3: Taste Receptor Distribution

This procedure is to be performed by students in pairs.

1. Have your partner rinse his or her mouth with distilled water and then use a paper towel to dry the surface of the tongue. Dispose of the towel in an appropriate container.

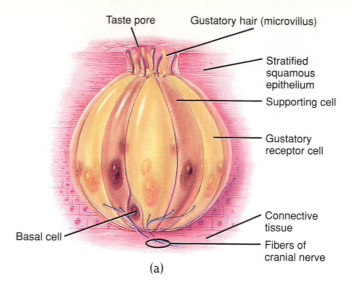

Taste pore  Gustatory hair (microvillus)

Stratified squamous epithelium

Supporting cell

Gustatory receptor cell

Connective tissue

Fibers of cranial nerve

Basal cell

(a)

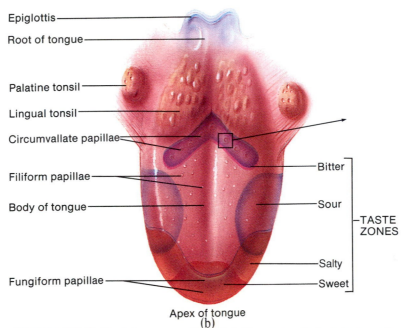

Epiglottis

Root of tongue

Palatine tonsil

Lingual tonsil

Circumvallate papillae

Filiform papillae

Body of tongue

Fungiform papillae

Apex of tongue

Bitter

Sour

Salty

Sweet

TASTE ZONES

(b)

**Figure 32-2** Gustatory reception: (a) the general structure and organization of a taste bud; (b) general structure of the tongue—areas of specific taste reception on the tongue are indicated. (From G. J. Tortora and S. R. Grabowski, *Principles of Anatomy and Physiology,* 7th ed., New York: HarperCollins, 1993.)

2. Moisten a clean cotton swab with the 5 percent sucrose solution, and remove any excess fluid by pressing the swab against the inside surface of the container.

3. Touch several regions on your partner's tongue with the moistened swab, and ask your partner to indicate the type of sensation and where it is detected. Indicate the site(s) on the sketch of the tongue in the results and observations section.

4. Have your partner rinse her or his mouth with distilled water again, and repeat Steps 2 and 3 with the 5 percent NaCl, vinegar, and 0.5 percent quinine-sulfate solutions.

5. Change places with your partner, and repeat Steps 1–4.

6. Compare your responses.

7. Answer the questions in the results and observations section and those pertaining to this portion of the exercise in the laboratory review section.

## Procedure 4: The Relationship Between Smell and Taste

This procedure is to be performed by students in pairs.

1. Have your laboratory partner sit comfortably with eyes closed and nostrils held shut.
2. Place a small piece of apple or potato on your partner's tongue. Instruct your partner not to swallow and to identify the material placed on the tongue. Note whether each identification is correct or incorrect in the results and observations section.
3. Have your partner rinse his or her mouth with water.
4. Repeat Step 2 with the other food, either apple or potato. Record your findings in the results and observations section.
5. Repeat Steps 2–4 with nostrils open.
6. Place a piece of onion on your partner's tongue. Remove it, and repeat Steps 1–5. Are the results obtained after the use of the onion similar to those obtained without the onion?
7. Change places with your partner, and repeat Steps 1–6.

## RESULTS AND OBSERVATIONS

### Demonstration of Olfactory Stimulus

1. Enter your findings on Table 32-1.
2. Did the laboratory partner on whom the perfume was applied detect the aroma earlier than the other laboratory partners? _____ Did the laboratory partner on whom the perfume was applied detect the aroma after the others could no

longer smell it? _____

Explain. _____
_____

### Stimulus Identification

1. Enter all findings in Table 32-2. Indicate correct responses by a "+" and incorrect responses by "−."
2. Did the number of correct responses increase or decrease with each repetition of the test?

_____

### Taste Receptor Distribution

1. Circle and label the tongue outline in Figure 32-3, to indicate the location where a particular taste sensation was detected.

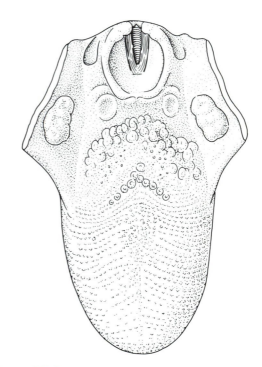

**Figure 32-3** Taste receptor distribution (to be labeled during Procedure 3).

## Table 32-1

| Student | Time Elapsed Until Odor Detected (in min) | Time Elapsed Until Odor No Longer Detected (in min) |
|---|---|---|
| | | |
| | | |
| | | |
| | | |

**Table 32-2**

| Stimulus | Student | | | Partner | | |
|---|---|---|---|---|---|---|
| | Test 1 | Test 2 | Test 3 | Test 1 | Test 2 | Test 3 |
| | | | | | | |
| | | | | | | |
| | | | | | | |
| | | | | | | |
| | | | | | | |
| | | | | | | |
| | | | | | | |
| | | | | | | |
| | | | | | | |

2. a. Were all taste sensations present on your tongue? _____

   b. Were any taste sensations particularly more distinct? If so, which one(s)? _____
   _____

3. a. Did any of the taste receptor locations overlap? If so, which one(s)? _____
   _____

b. How can this situation be explained? _____

_____

### The Relationship Between Smell and Taste

Enter your findings in Table 32-3. Indicate correct identifications by a " + " and incorrect identifications by a " − ."

**Table 32-3**

| Student | Before Onion | | | | After Onion | | | |
|---|---|---|---|---|---|---|---|---|
| | Nostrils Closed | | Nostrils Open | | Nostrils Closed | | Nostrils Open | |
| | *Potato* | *Apple* | *Potato* | *Apple* | *Potato* | *Apple* | *Potato* | *Apple* |
| | | | | | | | | |
| | | | | | | | | |

# LABORATORY REVIEW 32

NAME _____

LAB SECTION _____ DATE _____

## THE PHYSIOLOGY OF SMELL AND TASTE

### COMPLETION AND IDENTIFICATION

1. Where are the receptors for the sense of smell located?

   _____

2. What types of substances stimulate the receptors of smell?

   _____

3. On what physical process does the sensation of smell depend? _____

4. Identify the location of specific taste reception on the tongue in Figure Q32-1.

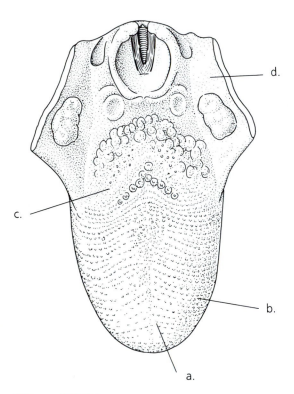

**Figure Q32-1**

a. _____     c. _____

b. _____     d. _____

5. Explain the relationship between the sensation of smell and the use of smelling salts
   (ammonium salts) to revive an individual who has fainted. _____

   _____

   _____

6. How would a nasal cold affect an individual's sense of taste?

   _____

   _____

## ANSWERS

1. _____
2. _____
3. _____
4. _____
5. _____

## MULTIPLE CHOICE

Insert the correct answers in the spaces provided.

1. Which part of the tongue reacts strongest to sweet tastes?
   a. tip
   b. center
   c. back
   d. sides
   e. choices *c* and *d* only

2. Where are the receptors for gustatory sensations?
   a. olfactory bulbs
   b. olfactory nerves
   c. taste pores
   d. taste buds
   e. supporting cells

3. Where are the taste buds most numerous?
   a. soft palate
   b. throat
   c. salivary glands
   d. tongue
   e. choices *a* and *c* only

4. Which part of the tongue reacts most strongly to sour flavors?
   a. tip
   c. center
   c. back
   d. sides
   e. choices a *and* c only

5. Which of the following is an example of a chemoreceptor that is an exteroceptor?
   a. taste bud
   b. Meissner's corpuscle
   c. pacinian corpuscle
   d. lamellated corpuscle
   e. touch corpuscle

# Cutaneous Sensory Systems

**After completing this exercise, you should be able to**

1. List and give the locations of touch receptors
2. Identify a pacinian corpuscle in a prepared slide of human skin
3. Perform simple experiments to demonstrate the receptors of heat, cold, touch, pain, and pressure
4. Explain the importance of touch receptors

The sensations of touch include those of pressure, heat, cold, and pain. In humans and many other animals many of the receptors of these sensations are located in the skin (Figure 33-1). Touch receptors enable many animals to monitor their environments.

Structure and different functions serve as the basis on which sensory organs can be categorized. From a structural standpoint, the sense organ can be the dendrites of sensory neurons, which are either free, as in the case of those found in the skin mediating pain and temperature, or enclosed within nonneural structures, such as the pressure receptors in the skin. Other sense receptors, including the photoreceptors in the eye (Exercise 30) and taste buds (Exercise 32) synapse with sensory dendrites.

As indicated earlier, the structures of certain skin receptors for the different sensations are quite similar. The types of impulses they send to the brain are also very much alike. The impulse that registers heat, for example, is exactly the same as that for cold or for touch. Humans experience them as different sensations because they are transmitted to different centers in the brain. If the pathways were changed so that the cold receptors sent impulses to the heat center in the brain, one would probably feel a sensation of heat every time a piece of ice was touched.

Pain receptors are nerve endings not covered by a myelin sheath. Upon injury of the nerve ending, an impulse is sent to the brain, which is interpreted as pain. Even if the nerve ending itself is not injured, it can pick up signals from those that are and can transmit the same message. In a condition termed *referred pain*, the sensation is not projected back to the point of stimulation. Pain receptors are more numerous than receptors for heat, cold, or touch, especially in humans.

Other cutaneous receptors are quite complex, with their dendritic endings encapsulated in other types of cells. *Meissner's corpuscles* (Figure 33-1 and Color Plate 34), commonly referred to as the *touch receptors* because they respond to light pressure, are located just beneath the epidermis in the dermal papillary layer of the skin. *Krause's end bulbs* and *Ruffini's corpuscles* are considered to be *thermoreceptors* that respond to cold and heat, respectively. All of these receptor types are quite similar, with the possible exception of the *pacinian corpuscles (deep pressure receptors)*, which are anatomically more distinctive (Color Plate 35).

The application of cold at times has been used to reduce or eliminate pain in some clinical situations, such as times of tooth or mouth pain, bone marrow aspiration (removal), and postoperative recovery. The two forms of cold therapy currently in use are in ice massage (the direct application of ice with a circular or back and forth motion) and cold packs. The areas in which cold is applied include the skin site over or surrounding pain, a region between the pain and the brain, the opposite side of the body corresponding to the pain site, acupuncture points, and apparently unrelated areas. This exercise contains an experiment to test the effect of cold applications.

In this exercise, consideration is given to sensory receptors that are (1) *thermoreceptors*, which respond

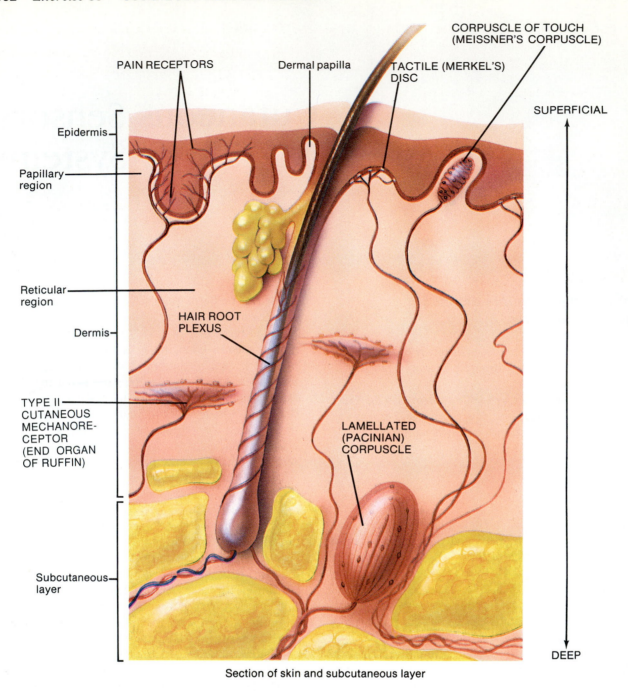

Section of skin and subcutaneous layer

**Figure 33-1** The structure and location of cutaneous receptors. (From G. J. Tortora, *Principles of Human Anatomy,* 6th ed., New York: HarperCollins, 1992.)

to heat and cold, (2) *mechanoreceptors,* such as those for touch and pressure, which are stimulated by deformation of the sensory dendrite cell membrane; and (3) *nociceptors,* or pain receptors, which are stimulated by chemicals released from damaged cells.

## MICROSCOPIC IDENTIFICATION OF THE SKIN AND SKIN RECEPTORS

### Materials

The following materials should be provided for class use:

1. Prepared slides of human skin, showing both epidermis and dermis
2. Prepared slides of human skin, showing skin receptors

### Procedure

This procedure is to be performed by students individually.

1. Examine the prepared slides of human skin.
2. With the aid of Figure 33-1 and your textbook, identify the specific layers of the epidermis and the various glands typically found in the dermis.
3. Examine the preparation of human skin and identify the skin receptors present. Refer to Color Plates 34 and 35.
4. Answer the questions for review at the end of the exercise.

## SKIN RECEPTORS

### Materials

The following materials should be provided for class use:

1. Pens or markers with washable ink
2. Millimeter rulers
3. Metal compass or dividers with points blunted
4. Large beakers (500 ml) with water maintained at the following temperatures—
   a. 4°C (cold)
   b. 25°C (room temperature)
   c. 45°C (hot)
5. Metal probes or teasing needles
6. Ice with or without plastic cups
7. Paper or cloth towels
8. Timers

### Procedure 1: Heat and Cold Receptors

This procedure is to be performed by students in pairs.

1. On the underside of your laboratory partner's forearm, use a fine-tip marking pen to draw a 4 × 14 mm rectangle, divided into 2-mm squares (total of fourteen 2-mm squares). Allow the grid to dry.
2. Place a metal probe in each of the temperature-controlled beakers of water. Allow probes to remain in the beakers for 5 minutes before use.
3. Touch one square with a probe from one of the beakers. Touch the same square with the probes from the other beakers, one at a time. Ask your partner to identify each touch as cold, room, or hot, and record in the results and observations section whether each response is correct or incorrect.
4. To maintain the temperature of the probes, return them to the appropriate beaker frequently during the experiment.
5. Repeat Step 3 on each of the remaining squares. Vary the order in which the probes for the three beakers are used.
6. Change places with your partner, and repeat Steps 1–5.

### Procedure 2: Touch Discrimination (Two-Point Sensibility)

This procedure is to be performed by students in pairs.

1. Have your laboratory partner sit with eyes closed and one arm resting comfortably on a flat surface.
2. Adjust the compass so that the points are about 3 cm apart. Press it down gently on the resting arm. Can your partner feel *two* points?
3. Move the points of the compass closer together, and again press it gently on the arm. Continue to bring the points closer together until your partner reports feeling only one point. Then move the points just far enough apart to be detected as two when pressed on your partner's arm. This is the *two-point threshold.* Measure the distance between the two points of the compass in millimeters, and record it in the results and observations section.
4. Determine the two-point threshold for the tip of the index finger and thumb, the left and right palms, the back surface of both hands, and the back of the neck. Record your findings in the results and observations section.
5. Change places with your partner, and repeat Steps 1–4.

## Procedure 3: Pressure Detection

This procedure is to be performed by students in pairs.

1. Place one hand palm down on a flat surface.
2. Observe the fine hairs on the back of the hand.
3. Have your partner move one hair in several directions with a metal probe. Can you feel this movement?
4. Repeat this step with a hair on the forearm.
5. Have your partner flick a hair rapidly several times. Can you feel a sensation?
6. Describe your sensations in the results and observations section.
7. Change places with your partner, and repeat Steps 1–6.

## Procedure 4: The Effect of Cold or Pain

This procedure is to be performed by students in pairs.

1. Obtain a plastic cup with ice and a supply of paper towels.
2. Have your laboratory partner sit comfortably and place his or her right hand on a flat surface.
3. Pinch the web region between the thumb and index finger with your fingers until your partner indicates that a fairly intense level of pain is felt. *Do not inflict unnecessary pain.*
4. Immediately, start timing and ask your partner to indicate when the pain has disappeared completely. Record the time in the results and observations section. This reaction will serve as the baseline for comparisons with the steps that follow.
5. Repeat Step 3 with the left hand. As soon as the intense pain is felt, have your partner apply ice with a circular motion to the site. Use the paper towels to absorb the water runoff from the ice.
6. Immediately, start timing and ask your partner to indicate when the pain has disappeared completely. Use the paper towels to absorb the water runoff from the ice.
7. Record the time in the appropriate space in the results and observations section.
8. Repeat Steps 3 and 4, again using the opposite hand, and then apply the ice with a circular motion directly below the elbow of the same side.
9. Change places with your laboratory partner, and repeat Steps 1–8.

## RESULTS AND OBSERVATIONS

### Heat and Cold Receptors

1. Enter your findings in Table 33-1. Indicate correct temperature responses with a "+" and incorrect with a "−."
2. Were all temperatures identified correctly? _____
3. Were there any areas on which only one or two temperatures were not identified correctly? Explain. _____

   _____

4. Were your laboratory partner's responses significantly different from yours? Explain. _____

   _____

### Touch Discrimination

1. Enter your findings in Table 33-2.
2. Which of the areas tested appeared to be the most sensitive? _____

3. Were the two-point threshold sensitivities of your partner similar to yours? Explain. _____

   _____

   _____

### Pressure Detection

1. Describe the sensations felt with the movement of hairs. _____

   _____

   _____

2. Were there differences in sensations felt on the back of the hand and on the forearm? Explain. _____

   _____

   _____

**Table 33-1**

| Square Number | Temperature Tested | | |
|---|---|---|---|
| | Room | Cold | Hot |
| 1 | | | |
| 2 | | | |
| 3 | | | |
| 4 | | | |
| 5 | | | |
| 6 | | | |
| 7 | | | |
| 8 | | | |
| 9 | | | |
| 10 | | | |
| 11 | | | |
| 12 | | | |
| 13 | | | |
| 14 | | | |

**Table 33-2**

| Skin Area Tested | Laboratory Partner's Two-Point Threshold (in mm) | Your Two-Point Threshold (in mm) |
|---|---|---|
| Forearm | | |
| Tip of index finger | | |
| Tip of thumb | | |
| Palm of left hand | | |
| Palm of right hand | | |
| Back surface of left hand | | |
| Back surface of right hand | | |
| Back of neck | | |

## *The Effect of Cold on Pain*

1. Enter your findings in the first column and those of your laboratory partner in the second column of Table 33-3.
2. Did the cold application reduce the time the pain was felt, when compared to the situation without a cold application? _____
3. Were your responses similar to those of your laboratory partner? _____ If not, offer possible explanations of this finding. _____ _____ _____

**Table 33-3**

| Site With or Without Cold Application | Reduction or Elimination of Pain Sensation (in seconds) | |
|---|---|---|
| | 1 | 2 |
| Pain site without cold | | |
| Cold application over pain site | | |
| Cold application between site and brain | | |

# LABORATORY REVIEW *33*

## *CUTANEOUS SENSORY SYSTEMS*

### COMPLETION, DEFINITION, AND IDENTIFICATION

1. Identify the parts of the skin in Figure Q33-1.

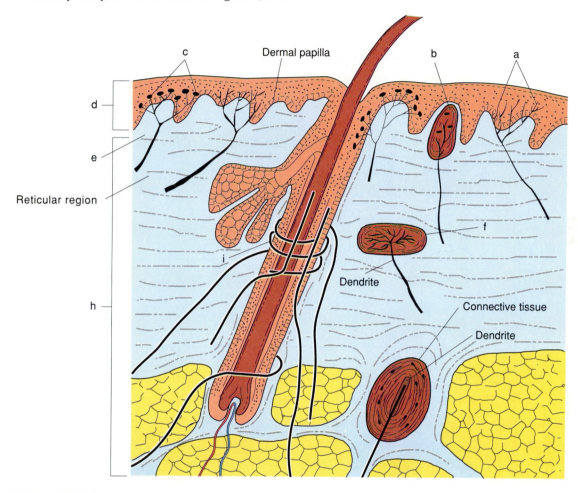

**Figure Q33-1** Diagrammatic view of the skin.

a. _____

b. _____

c. _____

d. _____

e. _____

f. _____

g. _____

h. _____

i. _____

**ANSWERS**

2. What is a receptor? _____

_____

3. List the specific cutaneous sensations

   a. _____

   b. _____

   c. _____

   d. _____

   e. _____

4. What function do pain receptors serve? _____

_____

5. List two sense receptors in the skin that are free:

   a. _____

   b. _____

6. List a sense receptor in the skin that is enclosed within a non-neural structure.

   _____

7. Briefly describe the functions and/or activities of the following sense receptors in skin.

   a. Thermoreceptor _____

   b. Mechanoreceptor _____

   c. Nociceptor _____

## MULTIPLE CHOICE

Insert the correct answers in the spaces provided.

1. What may be considered tactile sensations, in addition to pain and pressure?
   a. sound
   b. heat
   c. cold
   d. choices *a* and *c*
   e. choices *b* and *c*

2. Which of the following sensations are detected by mechanoreceptors?
   a. smell
   b. taste
   c. gustation
   d. heat
   e. pressure

3. What would be an example of a touch receptor?
   a. vibration corpuscle
   b. pacinian corpuscle
   c. Meissner's corpuscle
   d. thermoreceptive corpuscle
   e. nociceptor

1. _____

2. _____

3. _____

4. _____

5. _____

4. What would be an example of a pain receptor?
   a. vibration corpuscle
   b. pacinian corpuscle
   c. Meissner's corpuscle
   d. thermoreceptive corpuscle
   e. nociceptor
5. What would be an example of a pressure receptor?
   a. vibration corpuscle
   b. pacinian corpuscle
   c. Meissner's corpuscle
   d. thermoreceptive corpuscle
   e. nociceptor

# Cat Dissection:
# An Introduction
# to the Visceral
# Organs

*I*t is not always possible to dissect human cadavers. The cat is a good substitute because its anatomy is similar to that of the human, and it is relatively inexpensive and easily available. The perfect situation is for each student to dissect his own specimen, but if finances do not permit this, it is possible to assign two or three students to each specimen. Throughout these exercises, references are made to human anatomy, especially where it varies in important ways from the anatomy of the cat. Frequent comparisons with photos and line drawings of the human make it possible to discuss human organ systems while working on the cat. Never forget that the ultimate goal in the class is to learn the human anatomy.

This exercise can be used to acquaint the student with the internal organs. Major organs can readily be identified, regions delineated, and organ systems pointed out. Gently move the exposed organs about, so that you can appreciate their depth. Do not cut anything out unless specifically told to do so by your instructor. Carefully separate one part from another without tearing membranes, and try to find as many of the organs mentioned in Exercise 34 as possible before going on to detailed discussion of the anatomy and physiology of each system in other exercises.

# Opening the Body Cavity of the Cat

**After completing this exercise, you should be able to**

1. Open the abdominal and thoracic cavities
2. Examine the contents of the body cavities, and recognize many of the more prominent organs before making more comprehensive dissections

## Materials

The following materials should be provided for class use:

1. Cat-dissection specimens
2. Dissecting tools
3. Bone forceps

## OPENING THE ABDOMINAL CAVITY

To open the abdominal cavity of the cat, place the animal on its back, and begin a midventral incision line from a point between the legs. Continue this line anteriorly until you reach the diaphragm. The diaphragm can be seen or felt as you move forward. Do not cut through the diaphragm at this time because its anatomy is better appreciated if it is in one piece. When you reach the diaphragm level, cut laterally both to the right and to the left, following the contour of the diaphragm, which is a curved line. Also cut laterally on both sides near the beginning of your first incision line. Now you may fold the two flaps of muscle to the side and expose the entire abdominal cavity (Figure 34-1). Gently move the abdominal organs so that the diaphragm can be observed.

Begin your study with the muscular diaphragm, which is divided into two parts: the *vertebral portion* and the *sternocostal portion*. The origins of the slips of the sternocostal portion are on the *xiphoid process* of the sternum and the last five ribs. The muscle fibers interdigitate with those of the *transverse abdominis* muscles. The origin of the vertebral portion is by a strong tendon from the ventral surface of the second, third, and fourth lumbar centra. The origin divides into two parts, with muscle fibers that surround the openings of the aorta and the esophagus.

The muscles of the diaphragm contract in such a way as to lower the dome-shaped structure during inhalation, thereby expanding the thoracic cavity above it. During exhalation, the muscles relax and the diaphragm rises, decreasing the size of the thoracic cavity and forcing air out of the lungs.

Make a brief examination of the contents of the body cavity. All organs are studied in the next few exercises, but for now, just try to become familiar with the gross appearance and location of the major organs, membranes, and cavities.

## OPENING THE THORACIC CAVITY

After the diaphragm has been studied, continue to cut forward to the neck region. Begin above the diaphragm about ½ inch to the left of the midline. The ribs attach to the sternum at the midline by *costal cartilages* (observe them on the cat skeleton), and it will be possible to cut through these cartilages with your scissors. If you encounter too much resistance, use the bone forceps. When you have cut through the first costal cartilage, continue cutting forward until you reach the neck region. Cut transversely if necessary. Open the

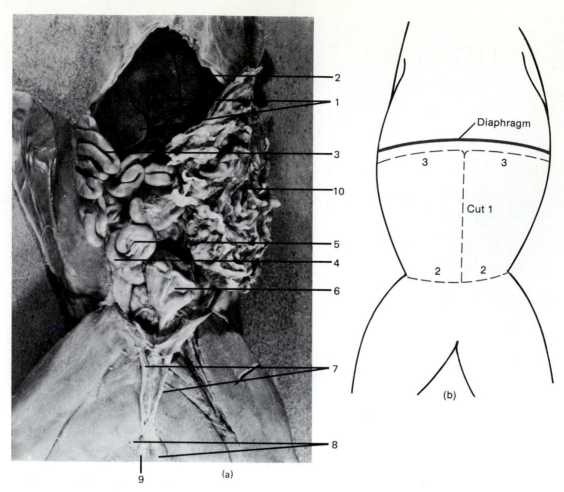

1. Liver, right medial lobe
2. Liver, left medial lobe
3. Jejunum
4. Ileum
5. Cecum
6. Urinary bladder
7. spermatic cords
8. Testes (in male)
9. Penis (in male)
10. Greater omentum

**Figure 34-1** Opening the abdominal cavity:
(a) Observable visceral organs—Photo
(b) Diagram indicating dissection strategy (Line drawing).

thoracic cavity on both the right and the left sides (Figure 34-2). In the midline there are some important blood vessels that must be severed as close to the chest plate as possible.

Another method of opening the thoracic cavity is shown in Figure 34-3. The chest plate is removed as a triangular piece and finally severed at the neck region. It may be saved and used to cover the thoracic viscera between laboratory sessions.

You now have exposed both the abdominal and thoracic cavities and are ready to begin a brief examination of the organs.

## ORGANS IN THE ABDOMINAL CAVITY

Examine the organs in the abdominal cavity. Covering a large part of the ventral body cavity region in the cat is a broad, fatty membrane, the *greater omentum*. If you wish to preserve this mesentery, it can be loosened, freed, and folded to the left side of the animal. If, after observation, you want to cut it off, free it carefully from all organs to which it is attached. It is usually discarded when heavily infiltrated with fat. Figure 34-1 shows the superficial abdominal viscera.

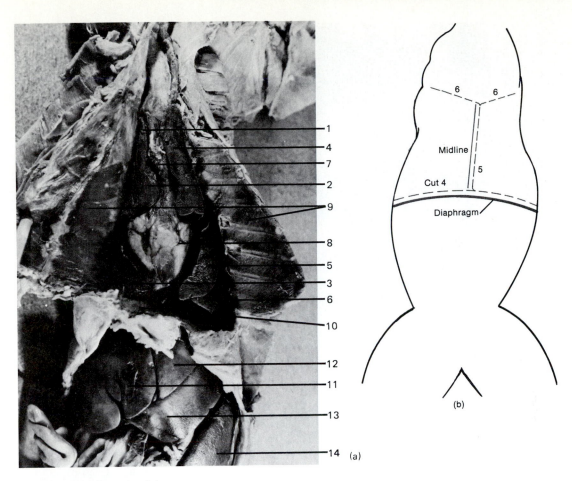

1. Lung, right anterior lobe
2. Lung, right medial lobe
3. Lung, right posterior lobe
4. Lung, left anterior lobe
5. Lung, left medial lobe
6. Lung, left posterior lobe
7. Thymus gland
8. Heart
9. Chest plate
10. Diaphragm
11. Liver, right medial lobe
12. Liver, left medial lobe
13. Liver, left lateral lobe
14. Spleen

**Figure 34-2** Opening the thoracic cavity:
(a) Observable visceral organs—Photo
(b) Diagram indicating dissection strategy (Line drawing).

## The Liver

The liver is the largest organ in the abdominal cavity. It is usually dark brown in color, due to the large amount of dried blood in it, and it becomes very hard and somewhat brittle in the preserved animal. Gently loosen its lobes to identify the following five parts.

1. *Right medial lobe*—largest of the lobes; the greenish *gallbladder* is located in a cleft within it

2. *Right lateral lobe (quadrate lobe)*—lateral and dorsal to the right medial lobe; it also has a deep cleft
3. *Right caudate lobe*—small, taillike lobe, which lies deep and is centrally located; it is covered by a mesentery, which may be removed to expose it
4. *Left medial lobe*
5. *Left lateral lobe*

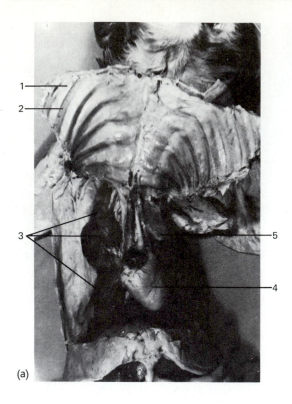

(a)

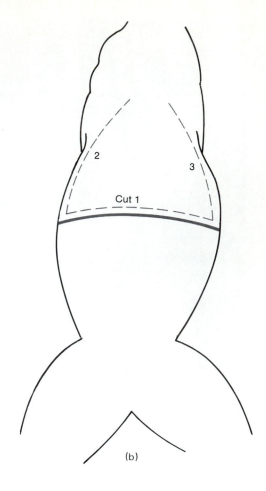

(b)

1. Chest plate
2. Rib
3. Lung lobes
4. Heart
5. Thymus gland (reflected)

**Figure 34-3** Alternative method of opening the thoracic cavity:
(a) Observable visceral organs—Photo
(b) Diagram indicating dissection strategy (Line drawing).

## The Digestive Tract

Under the left side of the liver is the J-shaped *stomach*. It begins at the *esophagus*, a small portion of which can be seen below the diaphragm, and it ends at the *pylorus*, which merges into the small intestine.

Follow the small intestine along its entire length. The *duodenum* is about 14 to 16 cm long and consists of two limbs. The *descending limb* runs 8 to 10 cm caudad of the pylorus. It then makes a U-shaped bend, extending cranially for 4 to 5 cm (the *ascending limb*). From that point, the small intestine turns posteriorly and is called the *jejunum*.

The jejunum and the *ileum* make up the rest of the small intestine. Connecting all parts of the small intestine is a large mesentery, the *mesentery proper*, and within it is a large lymph node (not found in the human), the *pancreas of Aselli*.

The ileum enters the *large intestine* or *colon*, which has a much greater diameter. It consists of the *ascending colon, transverse colon, descending colon,* and *rectum*. The rectum opens to the outside through the *anus*.

At the beginning of the ascending colon, just beyond the ileum, there is a small blind sac, the cecum. The cat has no *appendix*.

Examine the area between the duodenum and the stomach. You see a glandular organ, the *pancreas*, which has a longitudinal portion running alongside the duodenum, called the *head of the pancreas* and a transverse portion called the *tail of the pancreas*. A portion of the pancreas (the islets of Langerhans) is an endocrine gland and is discussed in Exercise 28.

On the cat's left side is a large, dark organ, the *spleen*.

## Urogenital Organs

At the posterior end of the abdominal cavity is the ventrally located *urinary bladder.* The size will vary with the amount of urine it contained at the time of death, which determined its distention.

Located deep toward the dorsal surface are two large, bean-shaped organs, the right and left kidneys. Above and medial to each kidney is a small hard *adrenal gland;* the adrenal gland is an endocrine gland that is discussed in more detail in Exercise 28. Extending from the kidney is the long, yellowish *ureter,* which terminates at the *urinary bladder.* The kidneys and ureters lie beneath the *parietal peritoneum* (retroperitoneal in position).

If you have a female cat, you will find *uterine horns* in the same area as the ureter. Capping each is a small oval *ovary* surrounded by the *ostium* of the short oviduct. In the male cat, the *testes* are located outside the body cavity in the *scrotal sac,* which was pointed out at the time the cat was skinned. One of the sperm-carrying ducts may be noted at this time. It is the *ductus deferens,* which loops over the posterior end of the ureter and enters the *urethra* posterior to the urinary bladder. The skin will be removed from the *scrotum* when the reproductive system is studied in detail, and at that time, the ductus deferens will be seen in its full length.

## Organs in the Thoracic Cavity

The most conspicuous organs in the thoracic cavity are the *lungs.* Lung tissue is soft, flexible, and dark brown in color. The lungs are composed of six lobes:

1. Right anterior lobe
2. Right medial lobe
3. Right posterior lobe
4. Left anterior lobe
5. Left medial lobe
6. Left posterior lobe

A seventh lobe, the *mediastinal lobe* on the right side, is recognized by some.

Lying in the midline among the lobes of the lung is the *heart.* It is surrounded by a tough sac, the *pericardium,* and will be dissected in a later exercise. Above the heart in a young cat is glandular tissue, which will vary in mass in different cats. It is the *thymus gland* (Figures 34-2 and 34-3). The younger the cat, the more thymus tissue it has. An old cat (usually with yellow fat) may have only a trace of the gland or none at all. The thymus gland has a similar fate in humans, reaching its full size in childhood. (See Exercise 28 for more details.)

Along the dorsal wall is the flattened *esophagus.* Follow it to the *diaphragm,* and see where it expands to form the *stomach* below the diaphragm. Lying ventral to the esophagus is the *trachea,* supported by cartilaginous rings. At the anterior end of the trachea is the *larynx,* which contains the *vocal cords.* Situated just posterior to the larynx are the dark lobes of the *thyroid gland,* another endocrine gland described in Exercise 28.

Major blood vessels in both the abdominal and thoracic cavities have been injected with a colored latex so that they can be traced. With few exceptions, the arteries are injected red and the veins blue. The blood vessels will be dissected in more detail in a later exercise.

# LABORATORY REVIEW *34*

NAME _____

LAB SECTION _____ DATE _____

## OPENING THE BODY CAVITY OF THE CAT

1. In the right-hand column, arrange the organs listed on the left, in order, according to the passage of food through the digestive tract from mouth to anus.
   - a. ileum
   - b. Ascending colon
   - c. Pharynx
   - d. Descending colon (rectum)
   - e. Esophagus
   - f. Jejunum
   - g. Mouth
   - h. Transverse colon
   - i. Stomach
   - j. Cecum
   - k. Duodenum
   - l. Anus

2. Name six organs visible in the thoracic cavity of the cat.

   a. _____

   b. _____

   c. _____

   d. _____

   e. _____

   f. _____

## ANSWERS

Mouth _____

_____

_____

_____

_____

_____

_____

_____

_____

_____

Anus _____

# The Cardiovascular System

*T*he human circulatory system performs a number of essential functions that include transporting various substances from one part of the body to another, defending the body against a wide variety of disease-causing microorganisms, assisting in the repair of damaged tissue, playing an important role in acid–base balance, and regulating body temperature. The cells of the body depend on the circulatory system for oxygen and essential nutrients and the removal of wastes. If the circulation to any region of the body is stopped, all the cells within that region soon die.

The cardiovascular system is the portion of the circulatory system that includes the heart and blood vessels. This system serves the body by moving blood between the cells of the body and the organs of the digestive, endocrine, integumentary, respiratory, and urinary systems, which communicate with the external environment. In performing this function, the heart works as a pump, forcing blood through blood vessels. These blood vessels, which form a closed system of channels, transport blood and allow for the exchange of gases, nutrients, and wastes between the blood and cells of the body. Although slight variations in heart activity occur normally, marked changes may be indications of cardiac abnormalities or disease. Specific exercises demonstrate the structure and organization of the components of the cardiovascular system. Other exercises in this section include the procedures used to demonstrate the normal and abnormal activities and processes associated with the cardiovascular system.

## GENERAL REFERENCES

Cohn, P. F. *Clinical Cardiovascular Physiology*. Philadelphia: Saunders, 1984.

Netter, F. H. *The CIBA Collection of Medical Illustrations, Vol. 5, The Heart*. Summit, NJ: CIBA Pharmaceutical Products, 1969.

Opie, L. H. *The Heart: Physiology and Metabolism*. New York: Raven Press, 1991.

Tortora, G. J., and S. R. Grabowski. *Principles of Anatomy and Physiology*, 7th ed. New York: HarperCollins, 1993.

## *SPECIFIC REFERENCES*

Brown, A. M. "Receptors Under Pressure: An Update on Baroreceptors." *Circulation Research,* 46: 1–100, 1980.

Farley, D. "High Blood Pressure, Controlling the Silent Killer." *FDA Consumer* 25: 28–33, 1991.

Fuster, V., L. Badimon, J. J. Badimon, and J. H. Chesebro. "Mechanisms of Disease and the Acute Coronary Syndromes." *The New England Journal of Medicine,* 326: 242–250, 1992.

Laks, M. M., and R. H. S. Selvester. "Computerized Electrocardiography—An Adjunct to the Physician." *The New England Journal of Medicine.* 325: 1803–1804, 1991.

Murphy, K. R. *ECG Essentials: A Pocket Reference for Systematic Interpretation.* Chicago: Quintessence Books, 1991.

Oliver, M. F. "Diet and Coronary Heart Disease." *British Medical Bulletin,* 37: 49–58, 1981.

Pickering, J. G., et al. "Blood Pressure During Normal Daily Activities, Sleep and Exercise." *Journal of the American Medical Association,* 247: 992–996, 1982.

Ross, J., Jr. "The Failing Heart and the Circulation." *Hospital Practice,* 18: 151–169, 1983.

# Mammalian Heart and Associated Blood Vessels

***After completing this exercise, you should be able to***

1. Identify all parts of the external surface of the mammalian heart
2. Relate all major blood vessels supplying and draining the heart to their specific heart chambers
3. Trace blood through the heart to the lungs and back again

## Materials

The following materials should be provided for class use:

1. Cat dissection specimens
2. Fresh sheep or beef hearts
3. Dissecting instruments
4. Model of the human heart
5. Charts of the human heart

## CAT HEART

Locate the heart in the thoracic cavity of the cat. It is surrounded by a tough membrane called the *pericardium,* or *pericardial sac.* Examine the sac carefully. Note that a pair of nerves attach to it laterally. These are the *phrenic nerves,* which supply the muscle of the diaphragm. Free the phrenic nerves (or at least the phrenic nerve on the left side of the cat) from the pericardial sac. Slit the pericardial sac, and fold it back out of the way—or remove it entirely.

Identify all parts of the heart that can be seen without removing it from the body. Locate the *left* and *right* ventricles, the *left* and *right atria* (seen externally as the flaplike auricles), and the *coronary artery* and its branches on the surface of the heart.

By lifting the ventricles slightly and pulling the heart toward the cat's left side you can see the large veins entering the right atrium. Identify the *superior vena cava,* which returns blood to the heart from the head and shoulders, and the *inferior vena cava,* which returns blood from the organs below the diaphragm.[1] The *pulmonary artery,* which carries blood from the right ventricle to the lung, can be seen prominently in ventral view. Remove enough connective tissue from the anterior region to expose the large *aorta,* which arches to the left just anterior to the heart. The *pulmonary veins,* which empty into the left atrium, can be seen if the ventricles are lifted and pulled to the cat's right side.

## SHEEP HEART

For a more complete study of the anatomy of the mammalian heart (which is remarkably similar in all species) use the freshly thawed or preserved sheep heart. As in the cat, the terms *superior* and *inferior* are used. Orient the heart so that you are looking at its ventral side. Find the following parts, using Figures 35-1 and 35-2: *left ventricle, right ventricle, left atrium, right atrium, pulmonary artery, aorta, superior vena cava, inferior vena cava,* and the *pulmonary veins.* There are four pulmonary veins, two from each lung, but they may be cut off in your specimen. Insert a probe into all major

---

[1]The student should be aware that the blood vessels in the cat were named to correspond with those in the human, so that it is common to find the adjectives *superior* and *inferior* replacing the more correct terms *anterior* and *posterior* for the four-legged animal. To be consistent with terminology used in most textbooks on the cat, the terms *superior* and *inferior* are used in these exercises.

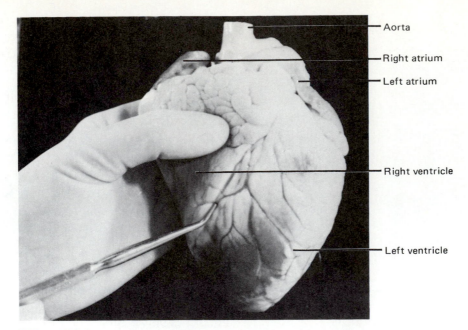

**Figure 35-1** Sheep heart, ventral view: The probe is pointing to the coronary artery at the septum between the right ventricle and the left ventricle.

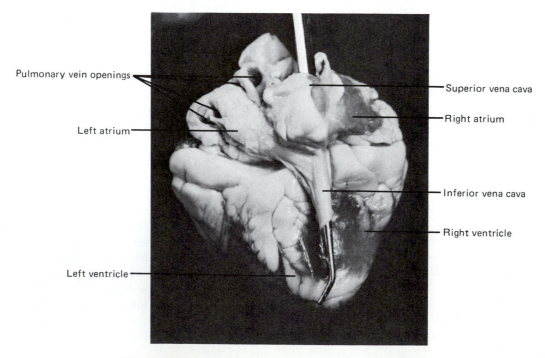

**Figure 35-2** Sheep heart, dorsal view: The probe passes through the right atrium from the superior vena cava to the inferior vena cava.

blood vessels to determine the chambers with which they connect.

To open the heart, begin at the end of the *pulmonary artery,* and cut back into the right ventricle (Figure 35-3a). The valve at the exit of the pulmonary artery from the right ventricle is the *pulmonary semilunar* valve.

Open the right ventricle carefully, and observe the delicate membranelike *tricuspid valve,* which separates the right ventricle from the right atrium (Figure 35-3b). The valve is held to the *papillary muscle* of the ventricle by connective tissue strands called *chordae tendineae,* which prevent the valve from being forced into the atrium when the ventricle contracts.

Now cut through the tricuspid valve into the right atrium (Figure 35-3c). The muscle layer or *myocardium* of the atrium is thinner than that of the ventricle, and when not filled with blood, the atrial wall collapses. Open the atrium wide, and locate the openings from the three veins that empty into it: *superior vena cava, inferior vena cava,* and *coronary sinus.* The coronary sinus, located between the *venae cavae,* is the expanded end of the *coronary vein,* which drains the heart tissue. It may be difficult to locate if the specimen has been preserved. The walls between right and left sides of the chambers are called *septa.* The *interatrial septum* sep-

arates right and left atria, and the *interventricular septum* separates the two ventricles. The muscle of the atrium is called the *pectinate muscle* due to its comblike appearance.

Turn to the left side of the heart. With your scalpel, cut through the *left atrium* and the *left ventricle,* through the *mitral* or *bicuspid valve,* to the apex of the heart. Like the tricuspid valve, the mitral valve is held to the ventricular wall by *chordae tendineae.* Opening into the left atrium are the four pulmonary veins, which return oxygenated blood to the heart from the lungs. (This is an example of a vein that carries oxygenated blood. Another example is the umbilical vein of the fetus, which returns blood from the *placenta.*)

Examine the left ventricle. Notice the thickness of the muscle. The left ventricle pumps blood through the aorta to distant parts of the body and must contract with greater force than the right ventricle. Locate the *aortic semilunar valve,* which leads to the aorta. Cut into the aorta and open it to see the two large openings near its base (Figure 35-3d). These are the openings into the *right* and *left coronary arteries,* which carry oxygenated blood to the heart muscle.

Examine the area where the *arch of the aorta* passes close to the *pulmonary artery.* Find a small ligament connecting the two (the *ligamentum arteriosum*). This

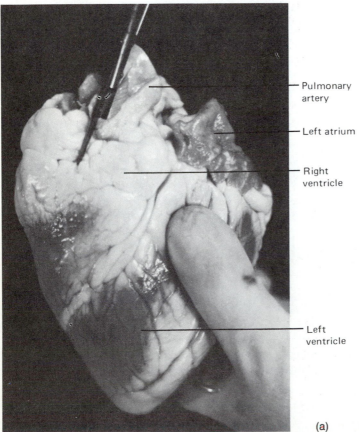

(a)

— Pulmonary artery

— Left atrium

— Right ventricle

— Left ventricle

**Figure 35-3** (a) Use scissors to cut through the pulmonary artery into the right ventricle.

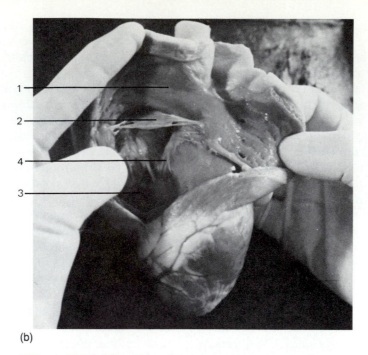

(b)

**Figure 35-3** *(Continued)* (b) Right ventricle opened to show tricuspid valve.
1. Right atrium
2. Tricuspid valve
3. Right ventricle
4. Chordae tendineae

(c)

(c) Right ventricle and right atrium opened to show the entire tricuspid valve and chordae tendineae. Probe runs beneath the tricuspid valve.
1. Right atrium
2. Tricuspid valve
3. Right ventricle

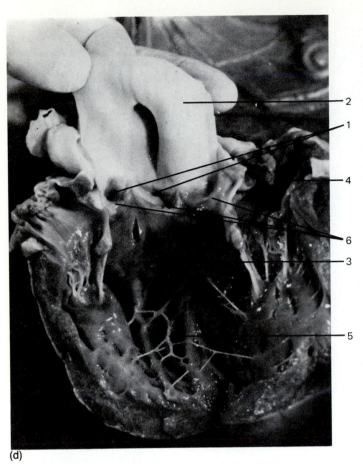

(d)

(d) Aorta cut lengthwise and held open to show the coronary artery openings.

1. Openings to coronary artery
2. Aorta
3. Left ventricle
4. Left atrium
5. Chordae tendineae
6. Aortic semilunar valves

is a remnant of the *ductus arteriosus,* an important fetal lung bypass, which closes and becomes ligamentous in the adult heart. Consult your text for a discussion of fetal circulation.

## THE HUMAN HEART

Diagrams of the human heart are shown in Figure 35-4. Its size and position in the thoracic cavity is much like that of the sheep. It is located in the *mediastinal cavity* and is surrounded by a tough pericardial sac, which is composed of two layers. The inner layer corresponds to the visceral peritoneum, and because it adheres tightly to the heart muscle, it is called the *epicardium.* The outer pericardial layer corresponds to the parietal peritoneum and lies very close to the inner layer. A potential cavity exists between the two layers of the pericardium. This cavity is filled with fluid, to prevent friction and is called the *pericardial space.* Figure 35-4 shows the relationship of the heart to major blood vessels associated with it.

## HEART STRUCTURE AND BLOOD PATHWAY IN THE HUMAN HEART

The human heart is a four-chambered muscle. The two superior chambers, the *atria,* are rather thin-walled, but the inferior chambers, the *ventricles,* are thick-walled. Separating the two atria is a muscular *interatrial septum.* Similarly, the ventricles are separated by a muscular *interventricular septum.* Valves between atria and ventricles (atrioventricular valves) are named for their anatomical configuration (Figure 35-5). The *tricuspid valve,* on the right side, has three flaps, and the *bicuspid valve,* on the left side, has two flaps. The latter is often called the *mitral valve.* Associated with the right atrium are three main veins: the *superior vena cava,* the *inferior vena cava,* and the *coronary sinus.* Blood returning to the heart through these veins passes through the tricuspid valve into the right ventricle. When the right ventricle contracts, and the tricuspid valve closes, the blood leaves the heart by way of the *pulmonary artery,* which terminates in pulmonary capillaries. Here, the gases oxygen and carbon dioxide are

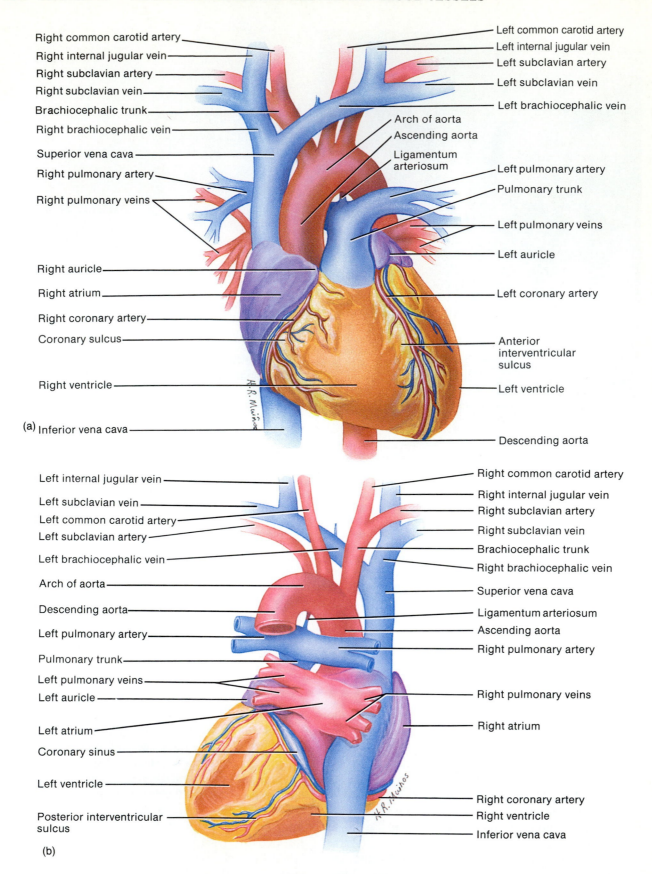

**Figure 35-4** Structure of the heart: (a) anterior external view; (b) posterior view; (c) anterior internal view; (d) path of blood through the heart. (From G. J. Tortora and S. R. Grabowski, *Principles of Anatomy and Physiology*, 7th ed., New York: HarperCollins, 1993.)

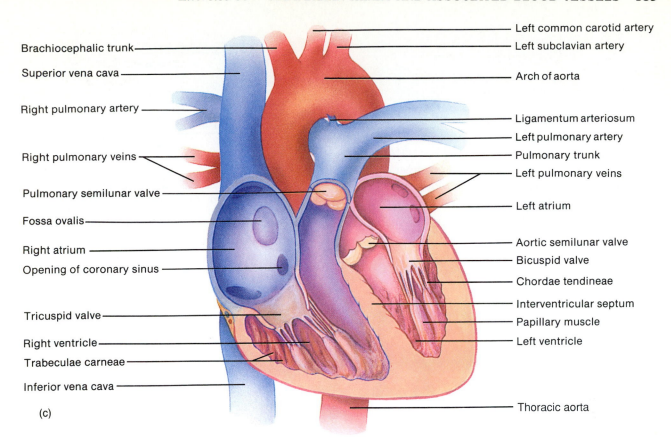

Brachiocephalic trunk

Superior vena cava

Right pulmonary artery

Right pulmonary veins

Pulmonary semilunar valve

Fossa ovalis

Right atrium

Opening of coronary sinus

Tricuspid valve

Right ventricle

Trabeculae carneae

Inferior vena cava

(c)

Left common carotid artery

Left subclavian artery

Arch of aorta

Ligamentum arteriosum

Left pulmonary artery

Pulmonary trunk

Left pulmonary veins

Left atrium

Aortic semilunar valve

Bicuspid valve

Chordae tendineae

Interventricular septum

Papillary muscle

Left ventricle

Thoracic aorta

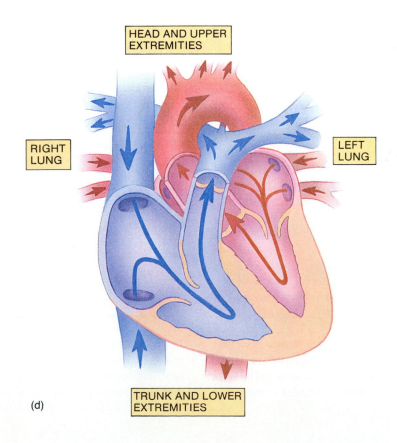

HEAD AND UPPER EXTREMITIES

RIGHT LUNG

LEFT LUNG

TRUNK AND LOWER EXTREMITIES

(d)

**Figure 35-5** Atrioventricular (AV) valves: (a) bicuspid valve open; (b) bicuspid valve closed. (The tricuspid valve operates in a similar manner.) (From G. J. Tortora and S. R. Grabowski, *Principles of Anatomy and Physiology*, 7th ed., New York: HarperCollins, 1993.)

exchanged, and the newly oxygenated blood returns to the left atrium in the *pulmonary vein*. When the left atrium contracts, blood flows through the mitral valves into the left ventricle. The mitral valve closes when the left ventricle contracts, and blood leaves the heart through the aorta to be distributed to all parts of the body. This tracing of a single drop of blood through the pulmonary circuit can be noted in the diagram at the end of this exercise (Figure Q35-1). Of course, it must be remembered that in the living, beating heart *both* atria contract simultaneously, and then *both* ventricles contract.

The tricuspid and mitral valves are made of fibrous tissue that is lined by the endothelial inner lining of the heart. They are not muscular. Both heart valves are held down to the ventricular muscle by connective tissue cords, the *chordae tendineae*. The large vessels that leave the ventricle (the pulmonary artery and the aorta) also possess valves at their bases, to prevent backflow. These valves are called the *semilunar valves*. Individually, they usually are referred to as the *pulmonary valve* and the *aortic valve*. All four valves are designed so that they permit blood flow in only one direction and do not allow for backflow. Damaged valves will, however, result in backflow.

The microscopic nature of the myocardium of the heart was studied in Exercise 6. It is striated, branched, and has centrally located nuclei and intercalated discs. The histology of cardiac muscle should be reviewed at this time. On opening the heart chambers, one finds a difference in the appearance of the muscle. In the atria, the muscles appear to be ridged, like the teeth of a comb, and are called *pectinate muscles*. In the ventricles, the muscles project as small cones and are called *papillary muscles*. Irregular ridges and folds in the ventricular musculature are called *trabeculae carnae*.

## CONDUCTION SYSTEM OF THE HEART

In addition to the extrinsic nerves from the autonomic nervous system, which regulate the beat of the heart, there is an *intrinsic* nervous system within the heart muscle, which originates from embryonic cardiac muscle cells and which consists of the following parts (Figure 35-6):

*Sinoatrial node* (SA node)—the pacemaker in the right atrium, which initiates the beat

*Atrioventricular node* (AV node)—in the right atrium, near the ventricles

*Atrioventricular bundle* (bundle of His)

*Right and left bundle branches*

*Purkinje cells*—extending into the muscle of each ventricle

A chick embryo heart can be seen to beat before any nerve fibers have formed. If the beating heart of a pithed frog is removed from the animal's body and placed in a dish of water, it will continue to beat. This indicates that nerve tissue is incorporated into the muscular tissue of the heart. This tissue cannot be seen with the naked eye. Use a stained slide of such tissue, if it is available.

## THE FETAL HEART

The heart is fully developed before birth, with only two major differences, which are called *lung bypasses*. These two lung bypasses enable the fetal heart to pump blood into the general circulation of the fetus without going

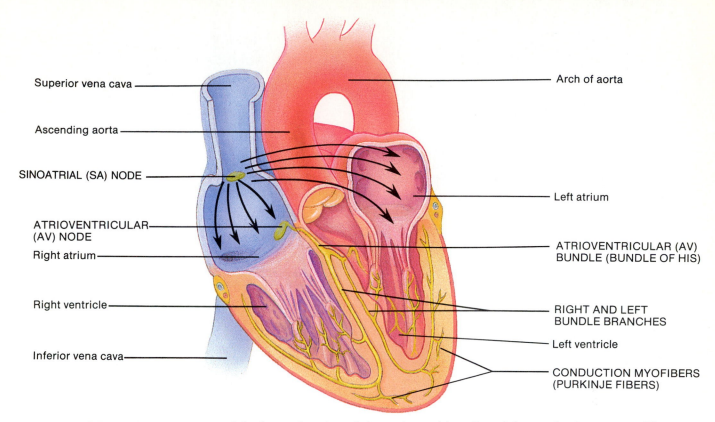

Superior vena cava

Ascending aorta

SINOATRIAL (SA) NODE

ATRIOVENTRICULAR (AV) NODE

Right atrium

Right ventricle

Inferior vena cava

Arch of aorta

Left atrium

ATRIOVENTRICULAR (AV) BUNDLE (BUNDLE OF HIS)

RIGHT AND LEFT BUNDLE BRANCHES

Left ventricle

CONDUCTION MYOFIBERS (PURKINJE FIBERS)

**Figure 35-6** Conduction system of the heart: location of the nodes and bundles of the conduction system. The arrows indicate the flow of action potentials through the atria. Exercise 37 presents the normal cardiac cycle, using an electrocardiogram. (From G. J. Tortora and S. R. Grabowski, *Principles of Anatomy and Physiology*, 7th ed., New York: HarperCollins, 1993.)

first to the lung. The pulmonary circuit is ready to go but is not used by the fetus, which has no access to air. The fetus obtains its oxygen and food from the maternal circulation and passes its wastes to the mother, whose organs will eliminate it. The arteries supplying the placenta are the right and left umbilical arteries, which carry unoxygenated blood, and the vein (single) from the placenta is the umbilical vein, which carries oxygenated blood. Figure 35-7 is a schematic illustration of blood flow before birth.

The two lung bypasses are

1. *Foramen ovale*—a hole in the interatrial septum, which allows blood to pass directly from right atrium to left atrium and thus bypass the lung. About 50% of the blood entering the right atrium passes through the foramen ovale.

2. *Ductus arteriosus*—a short artery, which connects the pulmonary artery with the aorta. The blood that enters the right ventricle and then the pulmonary artery is diverted directly into the aorta, again bypassing the lung.

Study Figure 35-7, and trace blood in the fetus from the right side of the heart to the umbilical arteries (branches of the internal iliac arteries), which take it to the placenta, where exchange is made. Wastes pass into the maternal circulation, and nutrients and oxygen enter the fetal circulation from the mother. Blood from the maternal circulation enters the single umbilical vein, which returns it to the fetal circulation, laden with the mother's oxygen and nutrients.

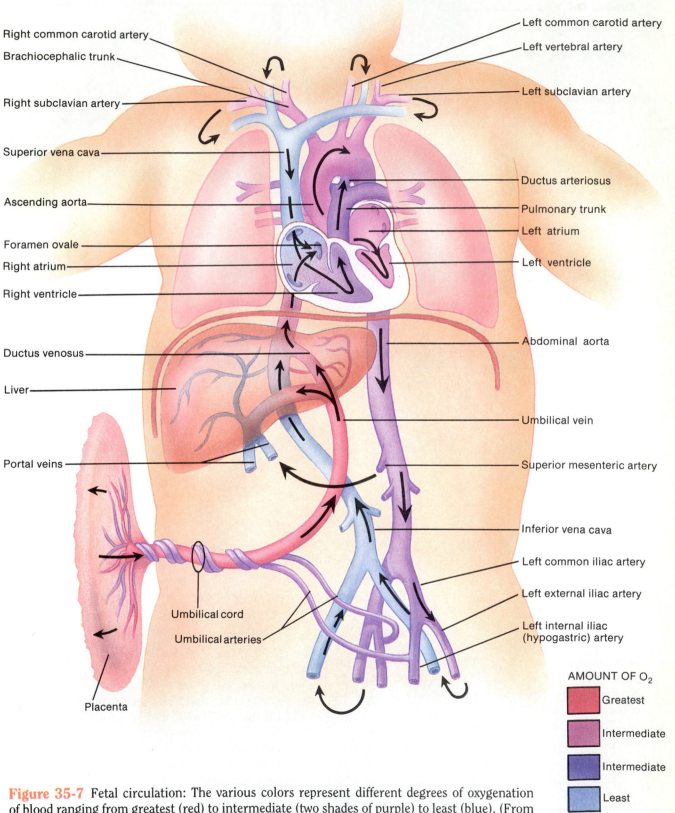

Right common carotid artery

Brachiocephalic trunk

Right subclavian artery

Superior vena cava

Ascending aorta

Foramen ovale

Right atrium

Right ventricle

Ductus venosus

Liver

Portal veins

Umbilical cord

Umbilical arteries

Placenta

Left common carotid artery

Left vertebral artery

Left subclavian artery

Ductus arteriosus

Pulmonary trunk

Left atrium

Left ventricle

Abdominal aorta

Umbilical vein

Superior mesenteric artery

Inferior vena cava

Left common iliac artery

Left external iliac artery

Left internal iliac (hypogastric) artery

AMOUNT OF O₂

Greatest

Intermediate

Intermediate

Least

**Figure 35-7** Fetal circulation: The various colors represent different degrees of oxygenation of blood ranging from greatest (red) to intermediate (two shades of purple) to least (blue). (From G. J. Tortora, *Principles of Human Anatomy,* 6th ed., New York: HarperCollins, 1992.)

# LABORATORY REVIEW 35

## MAMMALIAN HEART AND ASSOCIATED BLOOD VESSELS

1. Identify the parts of the mammalian heart in Figure Q35-1.

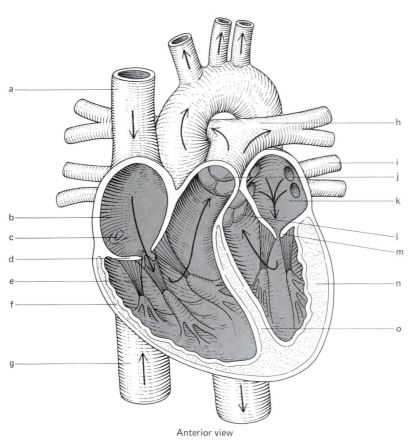

Anterior view

**Figure Q35-1** The mammalian heart. (From G. J. Tortora and N. P. Anagnostakos, *Principles of Anatomy and Physiology,* 3rd, 4th, 5th, 6th eds., New York: HarperCollins, 1981, 1984, 1987, 1990.)

2. Trace the pathway of a drop of blood through parts of the heart, cavities, and blood vessels, from the superior vena cava to the aorta. Some steps are provided. You fill in the rest.

   a. <u>Superior vena cava</u>

   b. _____

   c. _____

   d. <u>Right ventricle</u>

   e. _____

   f. _____

   g. _____

   h. _____

   i. <u>Left atrium</u>

   j. _____

   k. _____

   l. _____

   m. <u>Aorta</u>

# Blood Vessels and Lymphatics

## After completing this exercise, you should be able to

1. Locate and identify the major arteries and veins of the cat, as well as comparable human blood vessels
2. Trace blood and other substances through the vascular system from one point to another
3. Indicate the general locations of lymph nodes and vessels in the cat and the human
4. Locate lymph ducts on a cat specimen

## Materials

The following materials should be provided for class use:

1. Cat-dissection specimens
2. Dissecting tools
3. Charts of human blood vessels
4. Demonstration of the *portal system* in the cat

## MAJOR ARTERIES OF CAT THORAX, NECK, AND SHOULDER

As you locate the arteries discussed in this exercise, compare them with Figure 36-1. Dissect on the left side of the cat whenever possible.

Return to the heart. Locate again the *pulmonary artery,* which leaves the right ventricle and carries blood to the lung. Leaving the left ventricle is the *aorta.* The first part of the aorta, which gives rise to the *coronary arteries,* is the *ascending aorta.* The *arch of the aorta*

makes a sharp turn to the left and extends posteriorly as the *descending thoracic aorta.*

Arising from the region of the aortic arch are large arteries that supply the head and shoulder regions. On the right side of the arch, find a large straight artery that runs anteriorly for about an inch before branching. This is the *brachiocephalic artery.* Its branches are the *right subclavian artery* to the arm and the *right* and *left common carotid arteries,* which pass anteriorly alongside the trachea and give rise to a number of smaller branches in the neck and head region (labeled in Figure 36-1).

The subclavian artery reaches the underarm region, where it is called the *axillary artery,* and then passes into the arm, where it is known as the *brachial artery.* Some important branches of the subclavian, axillary, and brachial arteries are shown in Figure 36-1.

Examine the left side of the aortic arch. Notice that there is no brachiocephalic artery here. The *left subclavian artery* arises directly from the arch of the aorta and has the same branches as those on the right side of the body.

## MAJOR ARTERIES OF CAT ABDOMINAL CAVITY

Follow the *descending thoracic aorta* through the diaphragm into the abdominal cavity, where it is known as the *descending abdominal aorta* (Figure 36-2). Just below the diaphragm, the aorta gives off a short unpaired vessel, the *celiac artery,* which has three important branches: the *hepatic artery* to the liver, the *left gastric artery* to the stomach, and the *splenic artery,* to the spleen, stomach, and pancreas.

Return to the aorta. A short distance posterior to the celiac artery is the *superior mesenteric artery,* which

1. Ascending aorta
2. Arch of the aorta
3. Brachiocephalic
4. Common carotid
5. Subclavian
6. Vertebral
7. Costocervical
8. Internal mammary (thoracic)
9. Thyrocervical
10. Subscapular
11. Ventral (anterior) thoracic
12. Long thoracic
13. Axillary

14. Brachial
15. Radial collateral
16. Ulnar collateral
17. Superior thyroid
18. Laryngeal
19. Occipital
20. Internal carotid
21. External carotid
22. External maxillary
23. Posterior auricular
24. Superficial temporal
25. Internal maxillary

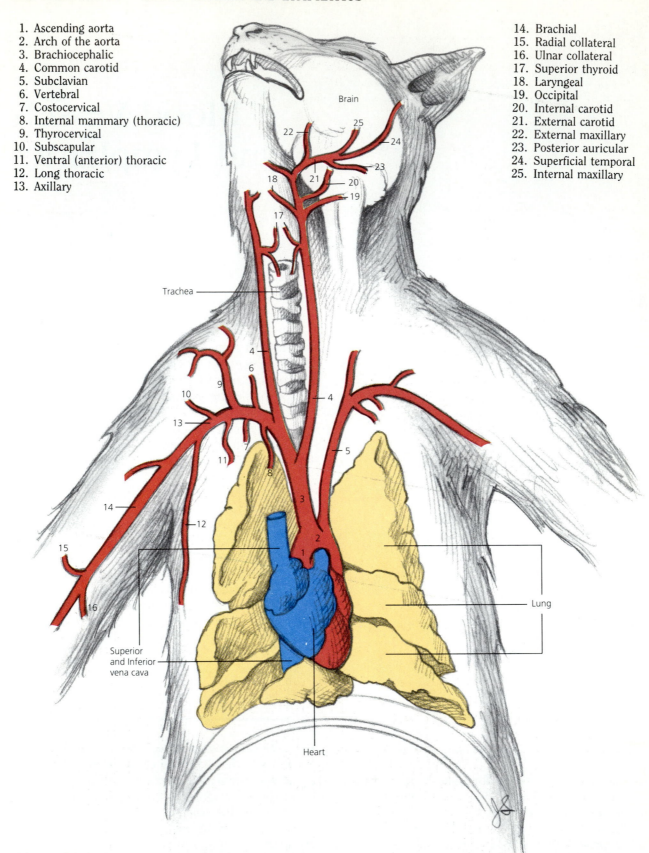

**Figure 36-1** Major arteries in the thoracic cavity, neck, and shoulder in the cat. (From P. J. Donnelly, *Laboratory Manual for Anatomy,* 2nd ed., New York: HarperCollins, 1993.)

1. Aorta
2. Celiac
3. Left gastric
4. Hepatic
5. Splenic
6. Superior mesenteric
7. Middle colic
8. Inferior pancreatico–duodenal
9. Ileocolic
10. Adrenolumbar
11. Phrenic
12. Renal
13. Ovarian (in females)
14. Spermatic (testicular) (in males)
15. Inferior mesenteric
16. Iliolumbar
17. Umbilical
18. External iliac
19. Internal iliac
20. Middle hemorrhoidal
21. Superior gluteal
22. Inferior gluteal
23. Caudal (medial sacral)
24. Femoral
25. Deep femoral

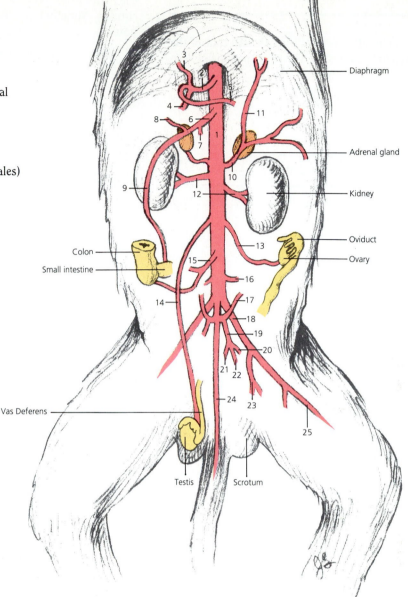

**Figure 36-2** Major arteries of the abdominal cavity in the cat. (From P. J. Donnelly, *Laboratory Manual for Anatomy,* 2nd ed., New York: HarperCollins, 1993.)

supplies the small intestine, pancreas, and part of the large intestine. It has several branches, labeled on Figure 36-2.

The next major arteries along the aorta are the small, paired *adrenolumbar arteries,* which supply the adrenal gland and muscles of the back and the diaphragm. Posterior to the adrenolumbars are the large, paired *renal arteries* to the kidneys. Posterior to the renal arteries are the paired *genital arteries.* If you have a male cat, this artery passes caudad to the testis and is called the *spermatic* (testicular) *artery.* If you have a female cat, it is called the *ovarian artery* and supplies the ovary. The genital arteries are small in sexually immature cats, but large in older, mature cats.

Caudad to the genital arteries is an unpaired vessel, the *inferior mesenteric artery,* which supplies the large

intestine. Trace its branches. The next major branches of the aorta are the paired *iliolumbar arteries* to the back muscles. Seven pairs of *lumbar arteries* to the muscles of the back branch off the aorta dorsally. Lift the aorta to see them.

Posterior to the iliolumbar arteries, the aorta gives off large *external iliac arteries,* which pass out of the body cavity into the legs. The aorta continues for a short distance below the external iliacs and gives rise to paired *internal iliac arteries* laterally and a single *caudal artery* (medial sacral), which continues into the tail. Branches of the internal iliacs are labeled on Figure 36-2.

The artery that penetrates the thigh from the abdominal cavity becomes known as the *femoral artery.* Find its branches.

## COMPARABLE HUMAN ARTERIES

Compare the arterial system of the cat with that of the human, shown in Figure 36-3. The relationship between the major arteries of the cat and the human is strikingly similar. Nonetheless, some differences in pattern are noted:

1. The left common carotid artery in the human

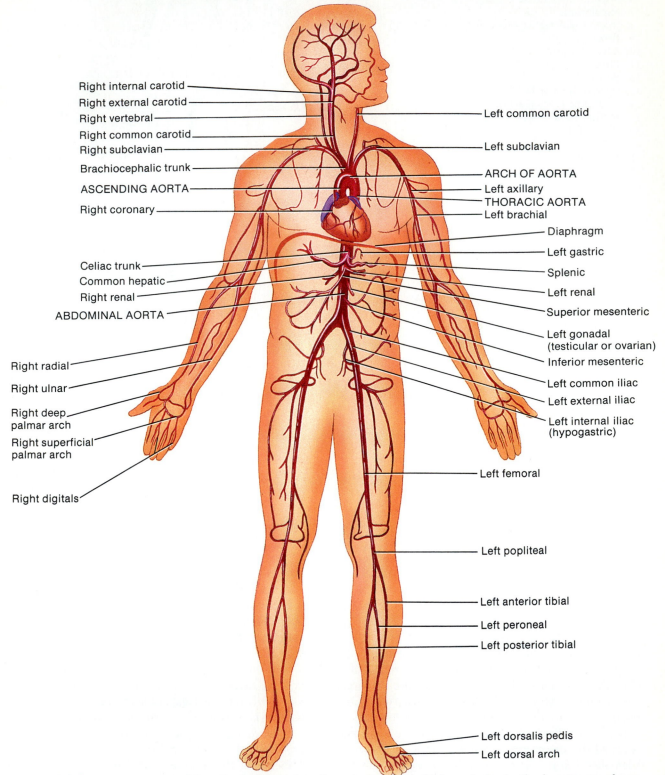

Right internal carotid
Right external carotid
Right vertebral
Right common carotid
Right subclavian
Brachiocephalic trunk
ASCENDING AORTA
Right coronary

Celiac trunk
Common hepatic
Right renal
ABDOMINAL AORTA

Right radial

Right ulnar

Right deep palmar arch

Right superficial palmar arch

Right digitals

Left common carotid
Left subclavian
ARCH OF AORTA
Left axillary
THORACIC AORTA
Left brachial
Diaphragm
Left gastric
Splenic
Left renal
Superior mesenteric
Left gonadal (testicular or ovarian)
Inferior mesenteric
Left common iliac
Left external iliac
Left internal iliac (hypogastric)

Left femoral

Left popliteal

Left anterior tibial

Left peroneal

Left posterior tibial

Left dorsalis pedis
Left dorsal arch

**Figure 36-3** Human aorta and its principal branches (in anterior view): Major arteries in the human are shown. (From G. J. Tortora and S. R. Grabowski, *Principles of Anatomy and Physiology*, 7th ed., New York: HarperCollins, 1993.)

branches directly from the aortic arch and not from the brachiocephalic artery. However, this same condition is found from time to time in the cat. The length of the brachiocephalic artery will vary in cat specimens. Students should compare dissections in this area to determine how much variation occurs among animals of the same species.

2. The human has a *common iliac artery,* but the cat does not. In the cat, the *external iliac artery* and the *internal iliac artery* are branches of the aorta.
3. The *internal carotid,* at its point of origin from the *common carotid* in the cat, is very small and often difficult to find. It is more prominent in the human.
4. The *left gastric* and the *hepatic arteries* of the cat sometimes have a common origin from the celiac. There are other minor variations of the celiac branches in the human.
5. The *medial sacral artery* of the human is very small, corresponding to the absence of a functional tail.

## MAJOR VEINS OF CAT THORAX, NECK, AND SHOULDER

Veins return blood to the heart. Most of them transport unoxygenated blood, which will be pumped to the lung, where its carbon dioxide will be exchanged for oxygen. Two veins that carry oxygenated blood toward the heart are the *pulmonary veins* and the *umbilical vein* of the fetus. The *pulmonary veins* return oxygenated blood from the lungs directly to the left atrium of the heart, and the *umbilical vein* returns oxygenated blood from the placental blood of the fetus to the maternal venous system.

The logical place to begin the tracing of the veins is at the heart, but it must be kept in mind that as you move away from the heart you are working back to tributaries of major veins and that the blood flow is always toward the heart. Locate the *superior vena cava* again, and follow it away from the heart (Figure 36-4). The small *internal mammary vein* from the chest wall empties into it on the ventral surface. This vein was cut when the thorax was opened. Lift the right side of the heart slightly, and you will see a large vein emptying into the dorsal side of the *superior vena cava* from the abdominal cavity. This is the *azygos vein* from the muscles of the back and the intercostal muscles. Flowing into the superior vena cava about an inch above the heart are the paired *brachiocephalic veins.*

Working back from the left or right brachiocephalic vein, find the jugular vein, which soon branches into the *internal jugular vein* and *external jugular vein,* from the head and neck regions. Trace the external jugulars to the neck region, where they are connected by the *transverse jugular vein* observed when muscles

in that region were dissected. Veins draining the head are shown in Figure 36-4.

The *subclavian vein* from the arm also joins the brachiocephalic vein in this area. The subclavian vein is known as the *axillary vein* in the underarm region and the *brachial vein* in the arm. Another prominent vein that empties into the venous channel at this point is the *subscapular vein* from the shoulder.

Locate once again the *pulmonary veins* from the lungs that empty into the left atrium.

## MAJOR VEINS OF THE CAT ABDOMINAL CAVITY

Trace the *inferior vena cava* into the abdominal cavity through the diaphragm (Figure 36-5). Just below the diaphragm, it receives a pair of veins from the liver, the *hepatic veins.* Follow the inferior vena cava posteriorly and locate the *adrenolumbar veins,* the *renal veins,* the *spermatic* (testicular) or *ovarian veins,* and the *iliolumbar veins.* These veins drain the same organs supplied by their arterial counterparts.

The venous drainage channel from the abdominal viscera is a large *portal vein,* which receives tributaries from the abdominal organs. A *portal system* of veins is one that begins in capillaries and ends in capillaries. In this case, the terminal capillaries are in the liver. This has great significance for the study of the physiology of these organs. Refer to Figure 36-6 for a diagram of the portal system of the cat. (Because it is not injected with colored latex in your cats, it is difficult to study in detail unless you have a special preparation in which the veins of this system are injected with a yellow dye.)

In the posterior abdominal region, the inferior vena cava is formed by the confluence of paired *common iliac veins* (see Figure 36-5). Near the point of juncture, find the single *caudal vein* or medial sacral from the tail and, somewhat more posteriorly, the *internal iliac vein* and the *external iliac vein.* Trace the external iliac vein to the leg, where it is called the *femoral vein.*

The venous system of the cat is notorious for its variability. Be alert for all deviations from the normal pattern. Compare your dissection with others in the class.

## COMPARABLE HUMAN VEINS

A diagram of the human venous system (Figure 36-7) is shown, for comparison with that of the cat. Study the similarities and the differences.

Some of the major differences in the areas we studied are

1. Inferior vena cava
2. Superior vena cava
3. Internal mammary (thoracic)
4. Brachiocephalic
5. Internal jugular
6. External jugular
7. Subclavian
8. Transverse scapular
9. Posterior facial
10. Anterior facial
11. Transverse jugular
12. Subscapular
13. Axillary
14. Long thoracic
15. Brachial
16. Posterior humeral circumflex

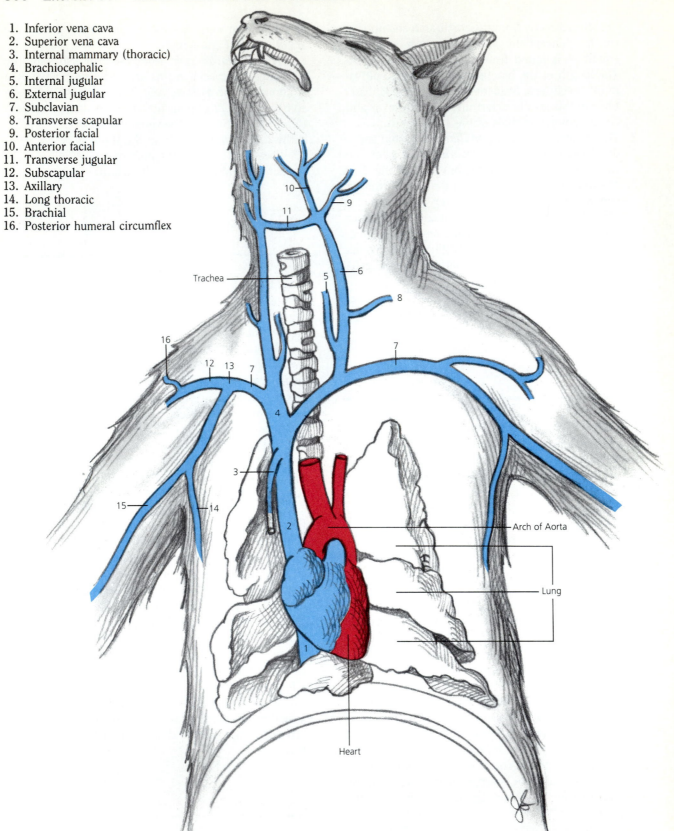

**Figure 36-4** Major veins of the thorax and shoulder in the cat. (From P. J. Donnelly, *Laboratory Manual for Anatomy,* 2nd ed., New York: HarperCollins, 1993.)

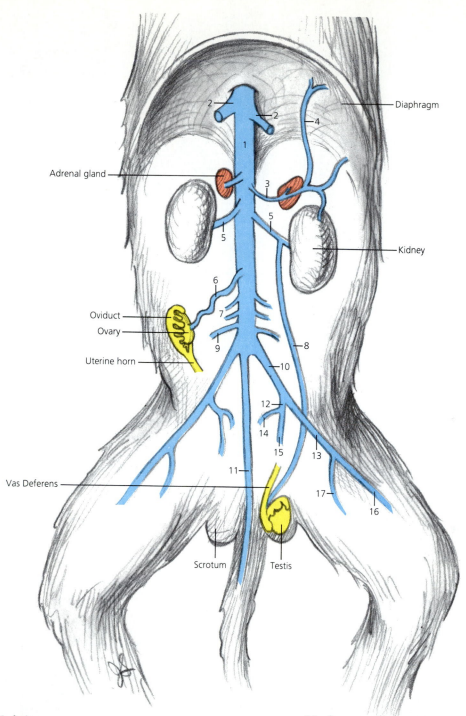

1. Inferior vena cava
2. Hepatic
3. Adrenolumbar
4. Phrenic
5. Renal
6. Ovarian (in females)
7. Lumbar
8. Spermatic (testicular) (in males)
9. Iliolumbar
10. Common iliac
11. Caudal (medial sacral)
12. Internal iliac
13. External iliac
14. Inferior gluteal
15. Middle hemorrhoidal
16. Femoral
17. Deep femoral

**Figure 36-5** Major veins of the abdomen and leg in the cat. (From P. J. Donnelly, *Laboratory Manual for Anatomy,* 2nd ed., New York: HarperCollins, 1993.)

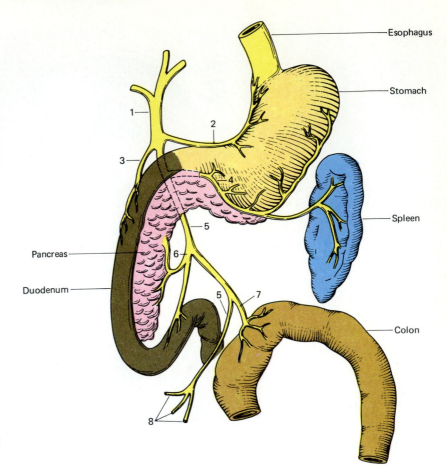

1. Portal vein
2. Coronary
3. Superior pancreaticoduodenal
4. Gastrosplenic
5. Superior mesenteric
6. Inferior pancreaticoduodenal
7. Inferior mesenteric
8. Intestinal

**Figure 36-6** Major tributaries of the portal system in the cat.

1. The *external* and *internal jugular* veins of the human flow separately into the *brachiocephalic,* whereas the cat usually has a short *common jugular* vein. Check your cat dissection to determine whether it differs from the human in this minor way.
2. In the cat, the *external jugular* is larger than the *internal jugular.* This is reversed in the human.
3. There is no *transverse jugular* in the human.
4. The human has an *anterior jugular* vein. It branches from the *external jugular* and is not present in the cat.
5. There are four pulmonary veins that open into the left atrium of the human. In the cat, there are three main pulmonary veins that open into the atrium by expanded veins called *dorsal, sinistral,* and *dextral sinuses.* Each of the three main pulmonary veins is formed by two veins, for a total of six (one from each major lung lobe).
6. The cat has a small *ulnar* vein joining the *brachial* midway down the forearm. In the human both *ulnar* and *radial* veins join the *brachial* at the elbow and are about equal in size.
7. An important arm vein in the human is the *basilic,* which is not found in the cat.

## OVERALL CIRCULATORY PATTERN IN THE HUMAN

Use the diagram in Figure 36-8, to trace blood from one part of the body to another. On the diagram, red indicates oxygenated blood, blue unoxygenated blood, and purple a capillary bed where oxygen and carbon dioxide are being exchanged at the tissue level. Arrows indicate the direction of flow. Figures 36-9, 36-10, 36-11, and 36-12 show in detail the vessels of the appendages, head, neck, and brain. Become familiar with the flow pattern in these areas.

## MAJOR HUMAN ARTERIES AND VEINS

Compare the human blood vessels shown in this exercise with those you have dissected in your specimen. All major branches of the aorta that you exposed in the cat should be compared with human vessels. Trace the branches and tributaries of all major vessels to the specific organs they supply and drain. This will prepare you for the tracing questions in the laboratory review section at the end of this exercise.

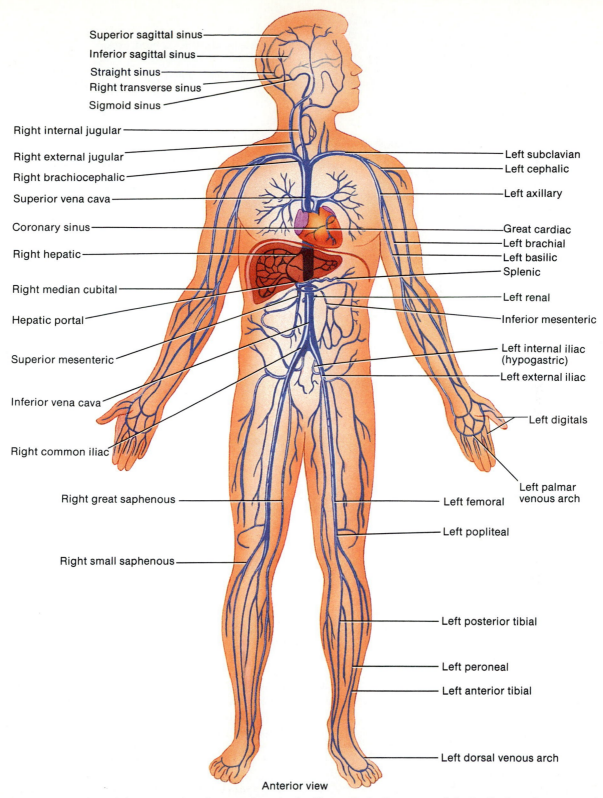

Superior sagittal sinus
Inferior sagittal sinus
Straight sinus
Right transverse sinus
Sigmoid sinus
Right internal jugular
Right external jugular
Right brachiocephalic
Superior vena cava
Coronary sinus
Right hepatic
Right median cubital
Hepatic portal
Superior mesenteric
Inferior vena cava
Right common iliac
Right great saphenous
Right small saphenous

Left subclavian
Left cephalic
Left axillary
Great cardiac
Left brachial
Left basilic
Splenic
Left renal
Inferior mesenteric
Left internal iliac (hypogastric)
Left external iliac
Left digitals
Left palmar venous arch
Left femoral
Left popliteal
Left posterior tibial
Left peroneal
Left anterior tibial
Left dorsal venous arch

Anterior view

**Figure 36-7** Principal human veins (in anterior view). (From G. J. Tortora and S. R. Grabowski, *Principles of Anatomy and Physiology*, 7th ed., New York: HarperCollins, 1993.)

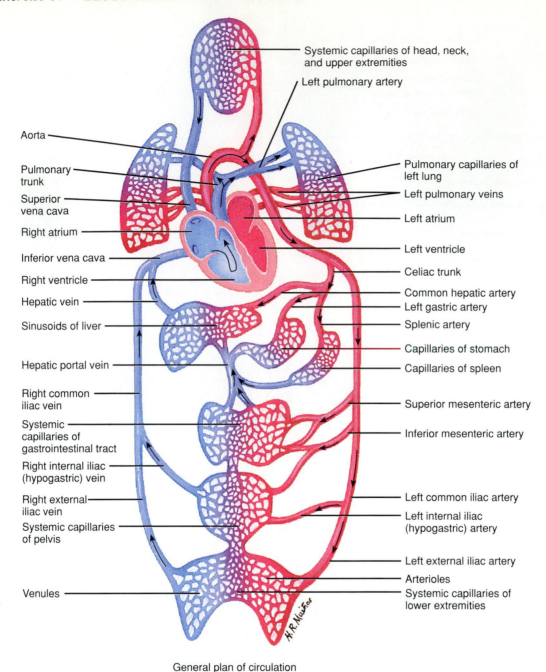

Aorta

Pulmonary trunk

Superior vena cava

Right atrium

Inferior vena cava

Right ventricle

Hepatic vein

Sinusoids of liver

Hepatic portal vein

Right common iliac vein

Systemic capillaries of gastrointestinal tract

Right internal iliac (hypogastric) vein

Right external iliac vein

Systemic capillaries of pelvis

Venules

Systemic capillaries of head, neck, and upper extremities

Left pulmonary artery

Pulmonary capillaries of left lung

Left pulmonary veins

Left atrium

Left ventricle

Celiac trunk

Common hepatic artery

Left gastric artery

Splenic artery

Capillaries of stomach

Capillaries of spleen

Superior mesenteric artery

Inferior mesenteric artery

Left common iliac artery

Left internal iliac (hypogastric) artery

Left external iliac artery

Arterioles

Systemic capillaries of lower extremities

General plan of circulation

**Figure 36-8** Human circulatory routes: Systemic circulation is indicated by heavy black arrows, pulmonary circulation by thin black arrows, and hepatic portal circulation by thin red arrows. (From G. J. Tortora and S. R. Grabowski, *Principles of Anatomy and Physiology,* 7th ed., New York: HarperCollins, 1993.)

RIGHT ARM

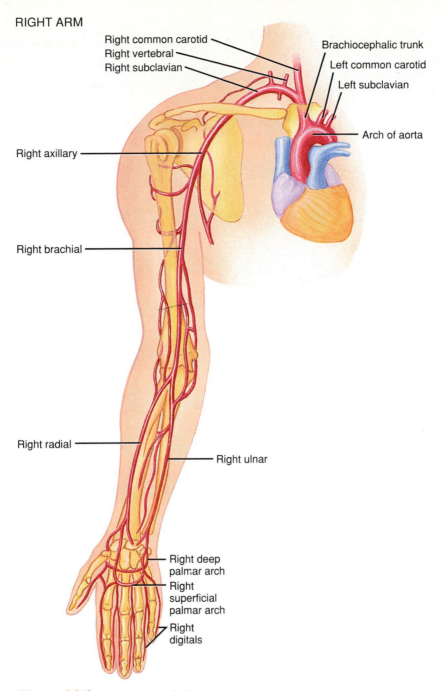

**Figure 36-9** (a) Arteries of the human right upper extremity (in anterior view). (From G. J. Tortora and S. R. Grabowski, *Principles of Anatomy and Physiology,* 7th ed., New York: HarperCollins, 1993.)

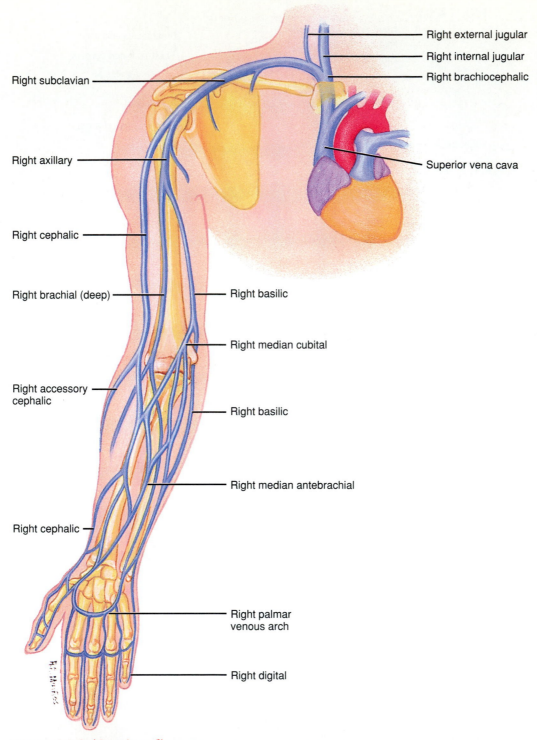

**Figure 36-9 *(Continued)*** (b) Veins of the human right upper extremity (in anterior view). (From G. J. Tortora and S. R. Grabowski, *Principles of Anatomy and Physiology,* 7th ed., New York: HarperCollins, 1993.)

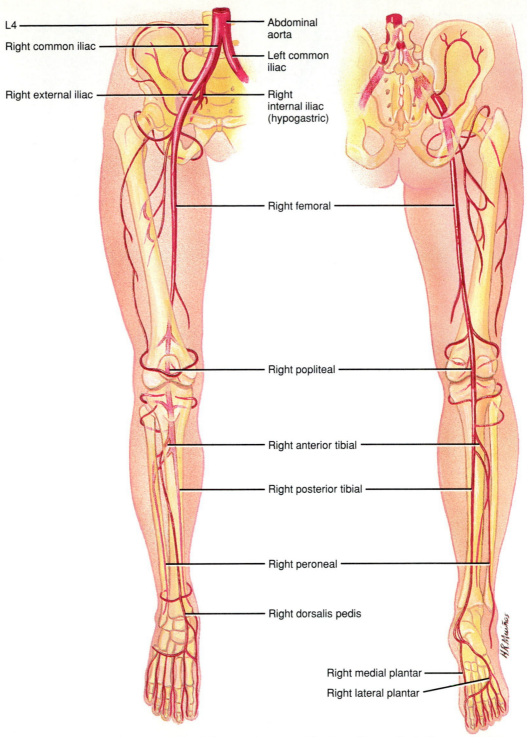

**Figure 36-10** (a) Arterial blood flow in human right leg. (From G. J. Tortora and S. R. Grabowski, *Principles of Anatomy and Physiology,* 7th ed., New York: HarperCollins, 1993.)

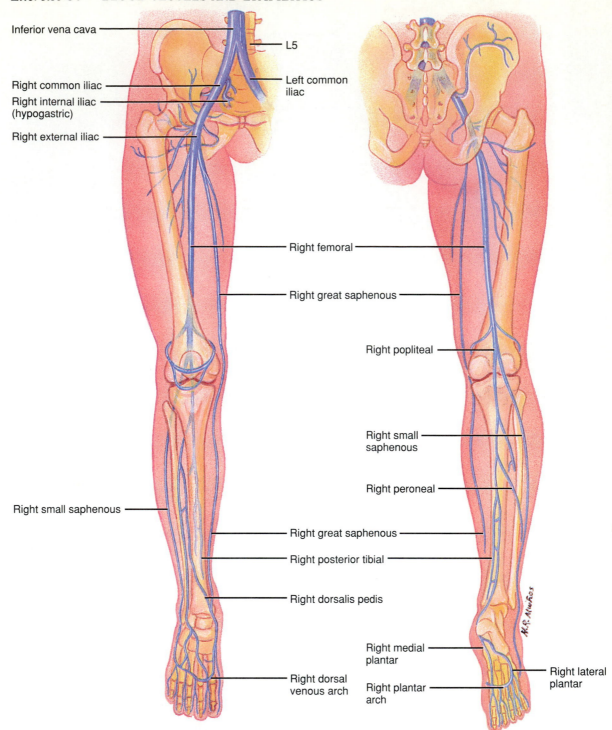

**Figure 36-10 *(Continued)*** (b) Venous blood flow in human right leg. (From G. J. Tortora and S. R. Grabowski, *Principles of Anatomy and Physiology,* 7th ed., New York: HarperCollins, 1993.)

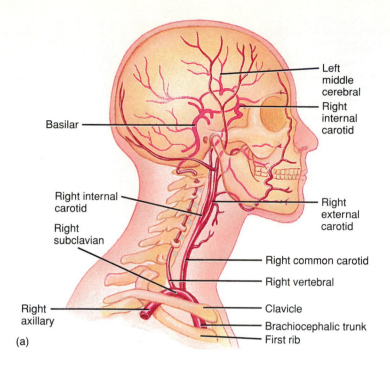

Left middle cerebral

Right internal carotid

Basilar

Right internal carotid

Right subclavian

Right external carotid

Right common carotid

Right vertebral

Right axillary

Clavicle

Brachiocephalic trunk

First rib

(a)

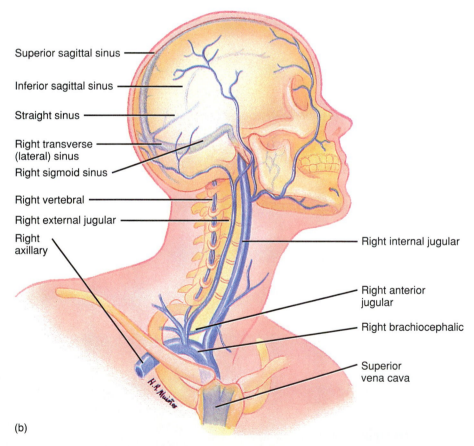

Superior sagittal sinus

Inferior sagittal sinus

Straight sinus

Right transverse (lateral) sinus

Right sigmoid sinus

Right vertebral

Right external jugular

Right axillary

Right internal jugular

Right anterior jugular

Right brachiocephalic

Superior vena cava

(b)

**Figure 36-11** (a) Arterial blood flow to the head. (b) Venous blood flow to the head. (From G. J. Tortora and S. R. Grabowski, *Principles of Anatomy and Physiology,* 7th ed., New York: HarperCollins, 1993.)

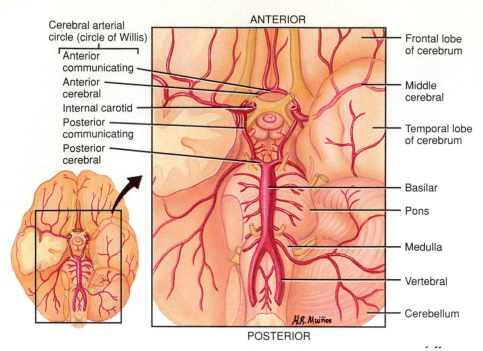

**Figure 36-12** Arteries of the base of the human brain: Note the arteries that the cerebral arterial circle (circle of Willis) comprises. (From G. J. Tortora and S. R. Grabowski, *Principles of Anatomy and Physiology,* 7th ed., New York: HarperCollins, 1993.)

## THE LYMPHATIC SYSTEM

The lymphatic system is an accessory circulatory system composed of fluid (lymph), vessels (lymphatic capillaries and thin-walled conducting vessels), and lymph nodes. There is no pumping organ (heart), and the fluid moves by means of skeletal muscle contraction in the appendages and breathing movements in the thorax. All vessels in this system are efferent and lymph eventually returns to the heart through veins (Figures 36-13 and 36-14). Numerous tiny valves along the length of the major vessels close behind the moving fluid and prevent backflow. This is also the case with many veins in the blood circulatory system. The walls of the lymphatic vessels are so thin that the presence of the valves causes beadlike appearance in the major vessels, which can be seen with the naked eye. The presence of these valves will help you to identify the major lymphatic vessels of the cat.

### Location of Lymph Vessels

Vessels in the lymphatic system (Figure 36-13) are ubiquitous, but most of them are too small to be seen with the naked eye. There are, however, two main ducts that can be found in your cat dissections:

1. *Left thoracic duct*—This is the main duct of the system (Figures 36-13 and 36-14). It can be seen dorsolateral to the thoracic aorta. It presents a beadlike appearance and is usually yellowish in color. It is only a few inches long and empties into the venous system at the left internal jugular vein, or near its junction with the left subclavian vein.

2. *Right lymphatic duct*—This duct (Figures 36-13b and 36-15) is similar in appearance to the thoracic duct, but it is shorter and more difficult to find. It empties into the right external jugular vein in the cat. Figure 36-13b shows the condition in the cat, and Figure 36-14 diagrams the human anatomy.

### Lymph Nodes

These are large- to medium-sized nodules (Figures 36-14 and 36-15) located in all parts of the body. They act as filters of the lymph that passes through them. Some prominent nodes in the cat are

1. *Cervical nodes*—These large masses have already been noted in the dissection of the neck and jaw muscles (Figure 18-1).

2. *Inguinal nodes*—These nodes vary in size and number. Most of them are small masses found in the groin region, near the inguinal canals. They are dissected in a subsequent exercise on the reproductive system (Figure 36-13).

3. *Pancreas of Aselli*—This is a single, very large lymph node, which represents the fusion of many smaller

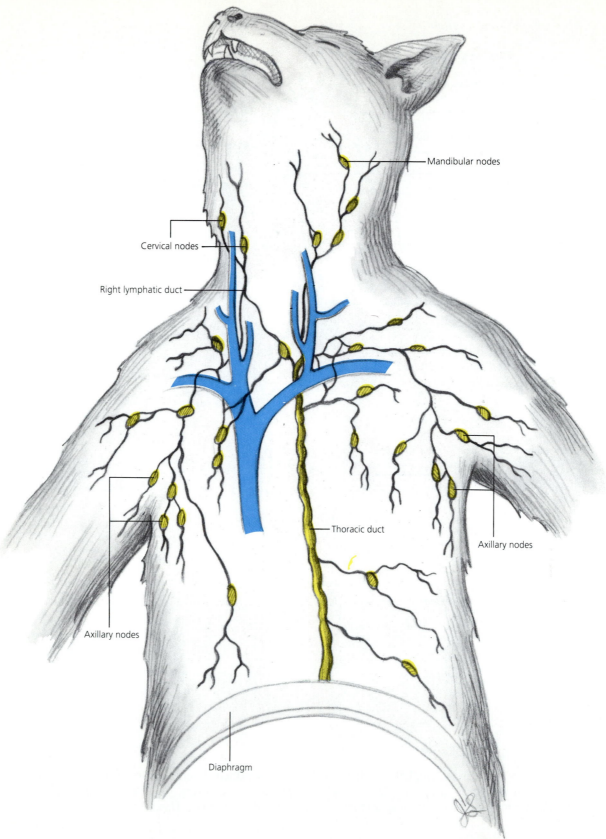

**Figure 36-13** Details of the left thoracic and right lymphatic ducts in the cat thorax. (From P. J. Donnelly, *Laboratory Manual for Anatomy,* 2nd ed., New York: HarperCollins, 1993.)

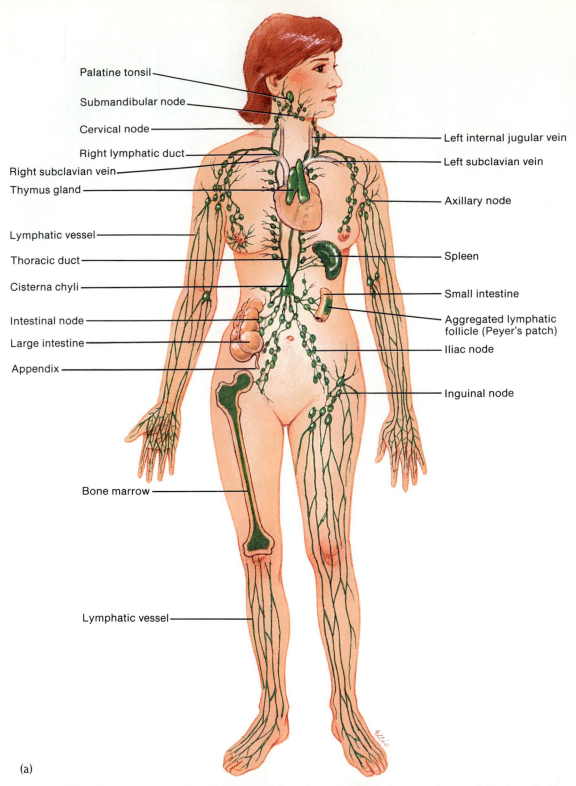

Palatine tonsil

Submandibular node

Cervical node

Right lymphatic duct

Right subclavian vein

Thymus gland

Lymphatic vessel

Thoracic duct

Cisterna chyli

Intestinal node

Large intestine

Appendix

Bone marrow

Lymphatic vessel

Left internal jugular vein

Left subclavian vein

Axillary node

Spleen

Small intestine

Aggregated lymphatic follicle (Peyer's patch)

Iliac node

Inguinal node

(a)

**Figure 36-14** Human lymphatic system: (a) location of principal components of the lymphatic system; (b) details of the thoracic and right lymphatic ducts. (From G. J. Tortora and S. R. Grabowski, *Principles of Anatomy and Physiology,* 7th ed., New York: HarperCollins, 1993.)

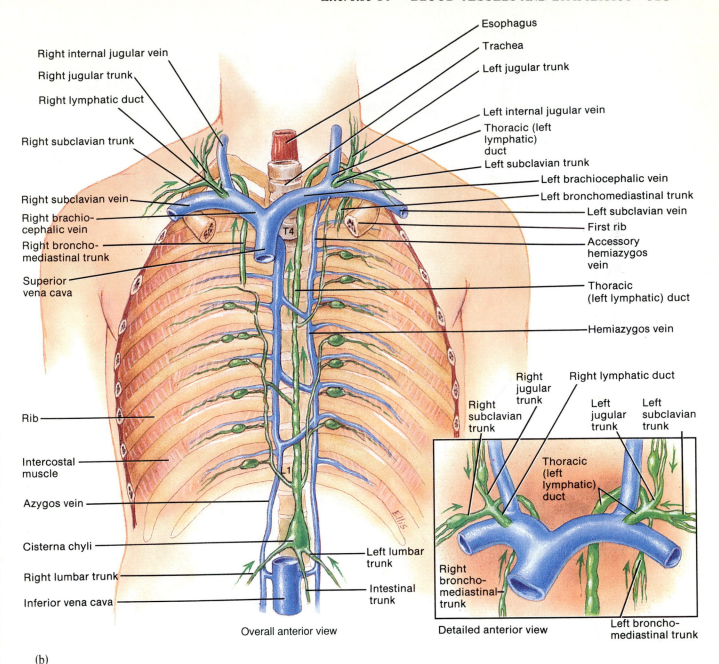

Right internal jugular vein

Right jugular trunk

Right lymphatic duct

Right subclavian trunk

Right subclavian vein

Right brachio-cephalic vein

Right broncho-mediastinal trunk

Superior vena cava

Rib

Intercostal muscle

Azygos vein

Cisterna chyli

Right lumbar trunk

Inferior vena cava

Esophagus

Trachea

Left jugular trunk

Left internal jugular vein

Thoracic (left lymphatic) duct

Left subclavian trunk

Left brachiocephalic vein

Left bronchomediastinal trunk

Left subclavian vein

First rib

Accessory hemiazygos vein

Thoracic (left lymphatic) duct

Hemiazygos vein

T4

L1

Left lumbar trunk

Intestinal trunk

Overall anterior view

Right jugular trunk

Right lymphatic duct

Right subclavian trunk

Left jugular trunk

Left subclavian trunk

Thoracic (left lymphatic) duct

Right broncho-mediastinal trunk

Left broncho-mediastinal trunk

Detailed anterior view

(b)

nodes, located in the mesentery of the small intestine. It is about 5 cm in diameter and is the largest lymph node in the cat's body. It was named for its discoverer, Gasparo Aselli, a seventeenth-century Italian anatomist. There are comparable lymph nodes in the human, but they do not fuse into a large mass.

4. *Axillary nodes*—These are numerous small nodules in the underarm region. These nodes are usually removed during surgery for breast cancer in the human, to prevent spread (metastasis) of the

cancer cells from the mammary glands via the lymphatics to other parts of the body.

## Lymph Flow

An example of lymph flow is shown diagramatically in Figure 36-15. Lymph is taken up into lymph capillaries from the surrounding tissue cells in all organs. It then passes through *lymph vessels* and *lymph nodes,* which filter the fluid until it reaches either the *left thoracic*

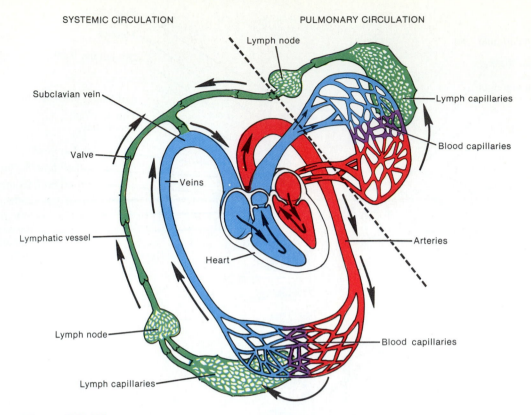

SYSTEMIC CIRCULATION          PULMONARY CIRCULATION

**Figure 36-15** Schematic representation of the relationship (in humans) of the lymphatic system to the cardiovascular system, showing the flow pattern. (From G. J. Tortora, *Introduction to the Human Body,* 2nd ed., New York: HarperCollins, 1991.)

*duct* or the *right lymphatic duct,* which flow into the venous system at or near the subclavian vein.

## Path of Lymph

(See Figure 36-15.)

Lymph capillaries

Lymph vessels

Lymph nodes

Left thoracic duct (see Figure 36-14), which collects lymph from:
1. Entire body posterior to the shoulder regions
2. Left forelimb
3. Left thorax, head, and neck

Right lymphatic duct, (Figure 36-14), which collects lymph from:
1. right thorax, head, and neck
2. right forelimb

## Related Lymphatic Organs

These are large masses of lymphoid tissue that often have multiple functions.

## Tonsils

(See Exercise 42, on the respiratory system.)

1. *Palatine tonsils*—Paired masses which hang from the soft palate in the oropharynx
2. *Pharyngeal tonsil*—one large node located medially within the nasopharynx; also called the *adenoid*
3. *Lingual tonsils*—paired masses located at the base of the tongue in the oropharynx

## Spleen

(Refer to Figures 34-2 and 36-14a, as well as Exercise 42, on the respiratory system.) In the cat, the spleen is a very large dark organ located below the diaphragm on the left side of the body. It is often so large that its length spans the length of the abdominal cavity. (Refer to Exercise 28, on the endocrine glands.) In the human, the spleen is also located on the left side of the body, directly beneath the diaphragm (Figure 36-16), but it is much smaller than the cat spleen. The spleen has multiple functions:

1. It destroys red blood cells that are old (average length of life is 120 days) and worn out. It passes

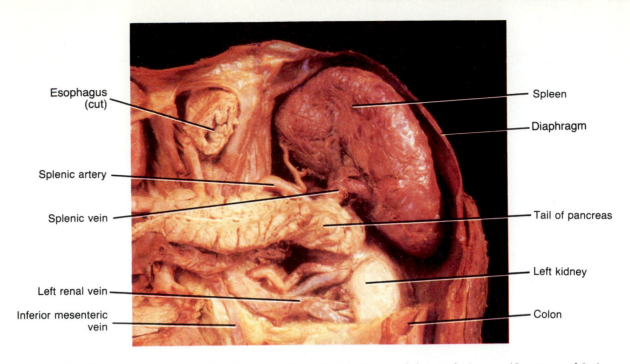

Esophagus (cut)

Splenic artery

Splenic vein

Left renal vein

Inferior mesenteric vein

Spleen

Diaphragm

Tail of pancreas

Left kidney

Colon

**Figure 36-16** Gross structure of the human spleen, in relation to abdominal viscera. (Courtesy of J. A. Gosling, P. F. Harris et al., *Atlas of Human Anatomy,* Gower Medical Publishing Ltd., 1985.)

the breakdown products on to the liver, through the portal system (Figures 36-6 and 36-8), for further processing. Consult your text for more detail on the physiology of the spleen.
2. It serves as a blood reservoir. It releases blood into the general circulation in times of stress.
3. It contains phagocytes, which cleanse the blood that passes through it.

## Thymus Gland

(Refer to Exercise 28, on the endocrine glands.) This organ is located anteroventral to the heart and to the major blood vessels in the thoracic cavity, and it plays a dual role, performing the following two functions:

1. It produces the hormone *thymosin,* which is responsible for the maturation of T lymphocytes.
2. It plays an important part in setting up the immune system before birth, by producing antibodies.

Examine the dissected specimens in your class, and note the differences in the size of this organ. A very young cat will have a large thymus gland, and a very old cat will appear to have none. In the human, the thymus is prominent in youngsters and begins to recede as the child matures. It becomes very small in the adult but begins to increase in size again in old age.

# BLOOD VESSELS AND LYMPHATICS

1. Trace the pathway of a drop of blood from the head to the leg. Some of the steps are provided. You fill in the rest. Include the pulmonary circuit, cavities, and valves.

   a. External jugular vein _____

   b. _____

   c. _____

   d. _____

   e. _____

   f. Right ventricle _____

   g. _____

   h. _____

   i. Lung capillaries _____

   j. _____

   k. _____

   l. _____

   m. Left ventricle _____

   n. _____

   o. _____

   p. Arch of aorta _____

   q. _____

   r. _____

   s. _____

   t. _____

   u. Femoral artery _____

2. Trace the pathway of a drop of blood from the myocardium of the heart to the arm.

   a. Coronary vein _____

   b. _____

   c. _____

   d. _____

   e. Right ventricle _____

   f. _____

   g. _____

   h. _____

   i. _____

   j. Left atrium _____

   k. _____

   l. _____

   m. _____

   n. _____

   o. _____

   p. Brachiocephalic artery _____

   q. _____

   r. _____

   s. _____

   t. Radial artery _____

3. Define these terms:

   a. Lymph _____

   _____

   b. Lymph vessel _____

   _____

   c. Lymph node _____

   _____

   d. Right lymphatic duct _____

   _____

   e. Left thoracic duct _____

   _____

4. Fill in the chart below: Choose an organ and name the artery which supplies it and the vein which drains it.

| Supplied by: | Organ | Drained by: |
|---|---|---|
| | | |
| | | |
| | | |
| | | |
| | | |
| | | |
| | | |
| | | |
| | | |
| | | |
| | | |

# Demonstration of the Normal Cardiac Cycle

The heart is a specialized muscle that has a fundamental physiological property of rhythmicity. This feature accounts for the uniformly paced alternating periods of contraction and relaxation. In the human, the heartbeat is regulated by the *sinoatrial node,* known as the *SA node* or *pacemaker,* located in the wall of the right atrium.

In this exercise, we are concerned with the electrical activity associated with the contraction of heart muscle. Records of such electrical activity, referred to as *bioelectrical potentials,* may be made by suitable equipment and are of importance in the diagnosis and treatment of various disorders and disease states.

The *electrocardiograph* is the instrument that monitors electrical activity of the heart. It amplifies the impulses so that they are strong enough to activate a recording device. Electrodes attached to the surface of the body receive impulses from the heart. The amplified impulse causes a recording needle to move up and down on the recording paper. The needle may be a *pen,* which marks the paper with ink, or a *stylus,* which inscribes on special paper that does not require ink. The paper moves under the needle at a rate determined by the scale of the graph, and therefore the record appears as a series of waves (Figure 37-1). The record produced by the electrocardiograph is an *electrocardiogram,* abbre-

viated *ECG* or *EKG.* An ECG often is obtained with a *polygraph,* a recording device that operates on the principle just described but that is capable of monitoring impulses from several different physiological activities, in addition to that of the heart. Depending on the instrument used, the record is superimposed on either a standardized graph or other appropriate material. The vertical lines indicate time in seconds, and the horizontal lines amplitude in millivolts (mV). The space between two vertical lines represents 0.04 second; the space between two horizontal lines represents 0.1 mV.

The procedure used in taking an ECG involves the application of a jelly containing an electrolyte, which forms a low-resistance surface between the skin and the electrode. This material aids the conduction of electrical signals. Paper pads presaturated with an electrolyte solution also can be used in place of the jelly. The electrodes are held securely in place by means of straps. The particular arrangement of two electrodes is a *lead.* The position of the electrodes on the body, the manner in which they are wired to the recording system, and any and all extra movements of the individual undergoing testing will influence the final record.

The component waves of a normal ECG are shown in Figure 37-1. The major components include the *P wave,* which indicates the spread of excitation from the pacemaker over the atrial musculature; the *QRS wave (complex),* which represents the depolarization of the ventricular heart muscle that immediately precedes ventricular systole; and the *T wave,* which represents the repolarization of the ventricular muscle, which occurs just before ventricular diastole. A *U wave* is sometimes seen following the T wave and is thought to represent late repolarization of the ventricles. This exercise considers the equipment and procedure used in obtaining an ECG.

The direction of depolarization depends on the heart's orientation and on the particular instant the heart cycle is being recorded or examined. From a med-

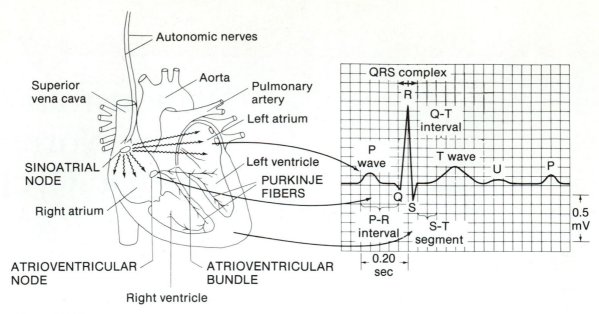

**Figure 37-1** The relationship of selected heart components to the conduction of electricity in the heart, as indicated by a normal electrocardiogram.

ical standpoint, it is sometimes desirable to record from more than two sites and even in two or three dimensions. With the particular technique of *vectorcardiography,* the continuous recording of the heart's electrical activity can be achieved. This type of continuous display of the depolarization–repolarization process as it moves over the heart may reveal important abnormalities. For example, an increase in the size (hypertroph) of one ventricle shifts the average or mean direction of depolarization during the cardiac cycle. This change occurs because it takes longer for the ventricle to depolarize and may be the result of a narrowing of the aortic semilunar valve.

In situations where there is a narrowing of the pulmonary semilunar valve or congenital defects such as the tetralogy of Fallot, a shift of the depolarization toward a hypertrophied right ventricle would occur.

Depolarization-pattern abnormalities also occur with a heart block, which is a delay of the electrical impulse or even a complete blockage of an impulse in its progress over the conduction system, or as a result of a heart attack.

This exercise also considers the determination of the mean electrical axis (direction) of depolarization of the cardiac cycle. This determination can be made by recording the QRS complex voltages from two different dimensions, using two different leads—namely, Leads I and III. Lead I provides a horizontal axis of observation (from left arm to right arm), while Lead III has an axis of about 120° (from left arm to left leg). The recordings from Leads I and III should show the normal mean

electrical axis of the ventricles to be about 59° (Figure 37-2a).

## Materials

The following materials should be provided for class use or demonstration purposes:

1. Electrocardiograph systems and appropriate cables
2. ECG electrodes (metal or disposable paper pads)
3. Tubes of electrode jelly
4. Disposable abrasive pads or brushes
5. Commercially packaged sterile alcohol prep-pads
6. Sterile gauze pads (4 × 4 inch)
7. Paper towels
8. Examples of normal and abnormal electrocardiograms

## Procedure 1: The Normal ECG

The instructor will demonstrate the principles, calibration, and correct use of the electrocardiograph system. This procedure should be performed on a group basis.

1. Before connecting the electrodes and taking an electrocardiogram, check the following items.
   a. Sufficient paper and recording ink (if needed) are available.
   b. The recording pen or device is properly centered.
   c. The recording speed has been set for 25 mm/second.
   d. The preamplifier has been set at 10 mm/mV.

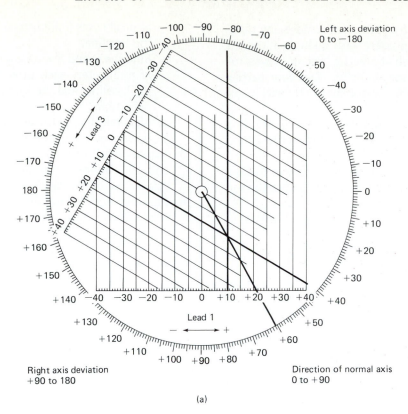

(a)

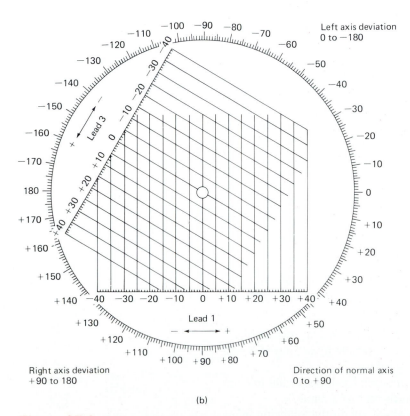

(b)

**Figure 37-2** Grids used for the determination of the mean electrical axis of the ventricles: (a) grid showing an example of how to determine the mean electrical axis of the ventricles (results showing a value of 59°); (b) a grid to be used by each student, for Procedure 2.

2. Ask your laboratory partner to remove any materials that could interfere with taking a recording (remove wristwatch or bracelet, roll up sleeves, roll down stockings, and so forth). Have your partner lie down on a table or cot in a comfortable and relaxed position.

3. Disinfect both wrists and the area on the inside of the legs just above both ankles with sterile alcohol prep-pads.

4. Apply a small amount of electrode jelly to the skin where the electrodes will be placed and to the surface of the electrodes.

5. The following three leads will be used in this exercise.
   a. *Lead I*—electrodes attached to the left and right wrists
   b. *Lead II*—electrodes attached to the right wrist and left leg
   c. *Lead III*—electrodes attached to the left wrist and left leg

6. Obtain a record with each lead. Usually, six ECG complexes (six complete cycles of P wave, QRS wave, and T wave) for each lead will be sufficient. Be sure to mark the record with your partner's name.

7. Have your partner exercise vigorously for 3–5 minutes. Jogging, running in place, or going up and down stairs are suitable forms of exercise. Again, obtain a record with each lead.

8. After the recording is complete, remove the leads, and clean the electrodes. The subject should remove the electrode paste with gauze or paper towels.

9. Obtain representative segments of the ECG. Attach and label them in the results and observations section.

10. Change places with your partner, and repeat Steps 1–9.

### Procedure 2: Mean Electrical Axis of the Ventricles

This procedure should be performed on a group basis, after the instructor has demonstrated the correct use of the electrocardiograph system.

1. Repeat Steps 1–4 of Procedure 1 in this exercise.
2. Leads I and III, with a sensitivity setting of 1, will be used in this procedure.
3. Obtain a record with each label. Be certain to mark the record with your partner's name.
4. With the Lead I record, obtain the following information items, and enter your findings and calculations in the appropriate spaces of the results and observations section.
   a. Find the QRS complex, and count the number of small boxes (millimeters) that it projects above the baseline.
   b. Find the Q and S waves, and count the number of millimeters each projects below the top of the baseline. Add the sum of the two downward deflections.
   c. Algebraically add the two values obtained for the upward and downward deflections.

5. Repeat Step 4 with Lead III, and enter your findings and calculations in the appropriate spaces of the results and observations section.

6. Find and mark the number that corresponds to the total value obtained with Lead I on the horizontal axis in Figure 37-2b. With the aid of a ruler, make and extend a vertical line from your mark to the top of the figure.

7. Find and mark the number that corresponds to the total value obtained with Lead III on the slanted axis located on the left-hand side of Figure 37-2b. With the aid of a ruler, make and extend a straight line down to the right-hand side of this figure.

8. With the aid of a ruler, draw an arrow line from the centrally located circle in Figure 37-2b, which intersects the lines made earlier on the figure. Extend the arrow to the edge of the outer circle, and record the value that represents the mean electrical axis of the ventricles. Enter your finding in the results and observations section.

9. An example of this procedure is shown in Figure 37-2a. The values for Leads I and III were 10 and 10, respectively.

## RESULTS AND OBSERVATIONS

### The Normal ECG

1. Attach labeled segment of electrocardiograms made during relaxation period.
   a. Lead I

   b. Lead II

c. Lead III

3. Are there any noticeable differences in the records taken with the three leads? If so, how would you explain the differences? _____

_____

_____

## The Mean Electrical Axis of the Ventricles

2. Attach labeled segment of electrocardiograms made immediately after vigorous exercise.
   a. Lead I

1. Enter your findings for Lead I in the spaces provided.
   a. Number of millimeters that the QRS complex projects *above* the baseline; upward deflections _____
   b. Sum in millimeters of the Q and S waves that project *below* the baseline; downward deflections _____
   c. The algebraic sum of the upward and downward deflections (if the resulting total is a negative number, enter it as such) _____

   b. Lead II

2. Enter your findings for Lead III in the spaces provided.
   a. Number of millimeters that the QRS complex projects *above* the baseline; upward deflections _____
   b. The sum in millimeters of the Q and S waves that project *below* the baseline; downward deflections _____
   c. The algebraic sum of the upward and downward deflections (if the result is a negative number, enter it as such)
   
   c. Lead III
   
   Total _____

3. Mean electrical axis of the ventricles was found to be _____

4. List four heart conditions that could produce depolarization-pattern abnormalities.

   a. _____

   b. _____

   c. _____

   d. _____

# LABORATORY REVIEW *37*

## *DEMONSTRATION OF THE NORMAL CARDIAC CYCLE*

### COMPLETION, DEFINITION, IDENTIFICATION, AND INTERPRETATION

1. What is an electrocardiogram? _____
   _____

2. What is a lead? _____
   _____

3. Identify the specific components of the electrocardiogram shown in Figure Q37-1.

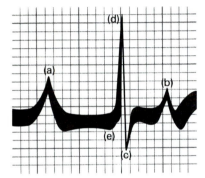

   a. _____

   b. _____

   c. _____

   d. _____

   e. _____

**Figure Q37-1** ECG components.

4. With the aid of your textbook or other reference, indicate what each of the following components of an electrocardiogram represents.

   a. P wave _____

   b. QRS complex _____

   c. T wave _____

   d. Q–T interval _____

5. What is a normal cardiac cycle? _____
   _____

6. With the aid of your textbook or other reference, determine the effect of the following chemical or physical factors on rate and strength of the heartbeat.

   a. Temperature _____

   b. Acetylcholine _____

   c. Epinephrine _____

   d. Digitalis _____

7. Define or explain the following:

   a. Fibrillation _____

   _____

   b. Heart block _____

   _____

8. Distinguish between diastole and systole. _____

   _____

9. a. What is a myocardial infarction? _____

   _____

   _____

   b. Can a myocardial infarction be detected in an ECG? If so, how is it indicated?

   _____

   _____

10. a. Would you consider the ECG in Figure Q37-2 to be normal? _____

   b. If not, what features are abnormal? _____

   _____

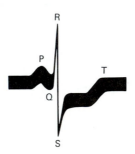

**Figure Q37-2**

11. Using the following recordings from Leads I and III of an electrocardiogram and the grid chart provided in Figure Q37-3, determine the electrical axis of the ventricles. Show your calculations.

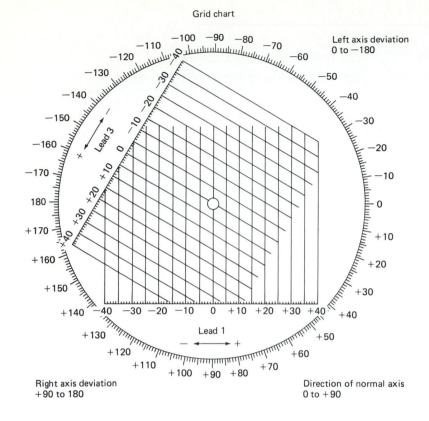

Grid chart

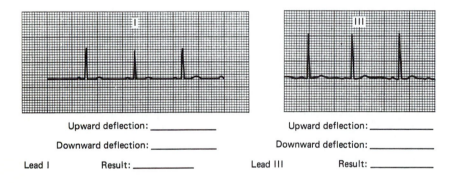

Upward deflection: _____

Downward deflection: _____

Lead I        Result: _____

Upward deflection: _____

Downward deflection: _____

Lead III        Result: _____

**Figure Q37-3**

12. Considering that the normal mean electrical axis of the ventricles is 59°, was the value obtained in Question 11 normal? _____

## ANSWERS

1. _____
2. _____
3. _____
4. _____
5. _____
6. _____
7. _____
8. _____
9. _____
10. _____

## MULTIPLE CHOICE

Indicate the correct answers in the spaces provided.

1. In the human, which of the following regulates the heartbeat?
   a. T wave
   b. right atrium
   c. left ventricle
   d. QRS complex
   e. sinoatrial node

2. Where is the pacemaker of the heart located?
   a. T wave
   b. right atrium
   c. left ventricle
   d. QRS complex
   e. sinoatrial node

3. What is the record produced by an electrocardiograph?
   a. pulse
   b. lead
   c. bioelectrical potential
   d. electrocardiogram
   e. QRS complex

4. Which of the following is the particular arrangement of two electrodes in an EKG procedure?
   a. pulse
   b. lead
   c. bioelectrical potential
   d. electrocardiogram
   e. QRS complex

5. Which of the following will influence the EKG?
   a. extra movements by the individual being tested
   b. the position of electrodes on the body
   c. the way in which leads are wired to the recording system
   d. all of these
   e. choices *a* and *c* only

6. Which of the following component waves of a normal EKG indicates the spread of excitation from the pacemaker over the atrial musculature?
   a. U wave
   b. QRS wave (complex)
   c. T wave
   d. P wave
   e. R wave

7. Which of the following component waves of a normal EKG represents the depolarization of the ventricular heart that immediately precedes ventricular systole?
   a. U wave
   b. QRS wave (complex)
   c. T wave
   d. P wave
   e. R wave

8. Which of the following in a normal EKG represents the repolarization of the ventricular muscle that occurs just before ventricular diastole?
   a. U wave
   b. QRS wave (complex)
   c. T wave
   d. P wave
   e. R wave

9. When is a U wave seen?
   a. at same time as a T wave
   b. following a T wave
   c. at same time as a P wave
   d. following a P wave
   e. at same time as the QRS complex
10. With which of the following conditions can depolarization-pattern abnormalities occur?
   a. heart blocks
   b. tetralogy of Fallot
   c. narrowing of the pulmonary semilunar valve
   d. heart attack
   e. all of these

# Blood Pressure and Pulse Determination

When the ventricles of the heart contract, blood is pumped under pressure into the pulmonary artery and aorta. This pressure exerts a force against the walls of the vessels through which blood flows. Within any blood vessel, artery, vein, or capillary, the pressure depends on the volume of blood present, the size of the vessel, the rate of blood flow, and any resistance to blood flow. Blood pressure is the basis for maintaining a functional and stable movement of blood from the heart, to the various regions of the body, and back into the heart. Because of its importance, a blood pressure determination is a major part of any physical examination. In a medical context, *blood pressure* refers to the pressure of the arterial system.

The routine method for measuring arterial blood pressure is indirect and usually employs a *sphygmomanometer*. This device basically consists of an inflatable cuff, an inflation pumping device, a pressure gauge, and an exhaust control valve.

A *stethoscope* is used to amplify the sounds within the artery used for the determination. Usually, the brachial artery is chosen, for convenience. In the procedure, the pressure cuff is first inflated around the upper arm, to stop the flow of blood temporarily (Figure

38-1). This collapses the artery. When a stethoscope is placed above the brachial artery, no sound can be heard (Figure 38-2a). The pressure in the cuff is gradually reduced until the blood flows through the vessel with each systole of the heart. When the cuff pressure is equal to the systolic pressure, the first *Korotkoff sound* is heard (Figure 38-2b). Korotkoff sounds are produced by the turbulent flow of blood through the partially constricted brachial artery. The reading on the pressure gauge when the first Korotkoff sound is detected corresponds to the individual's *systolic blood pressure*. The artery will now remain open during the systole of the heart. The sounds continue with each heart cycle (Figure 38-2c). Reducing the cuff pressure further enables the artery to remain open even during diastole. Blood flows freely now, and the thumping sound disappears. The reading at which the sound disappears corresponds to the *diastolic blood pressure* (Figure 38-2d).

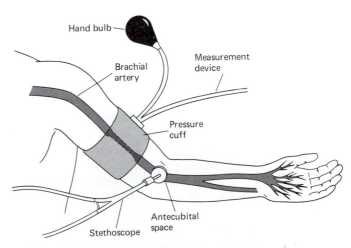

**Figure 38-1** Measuring blood pressure: position of a sphygmomanometer cuff and stethoscope during measurements using the brachial artery.

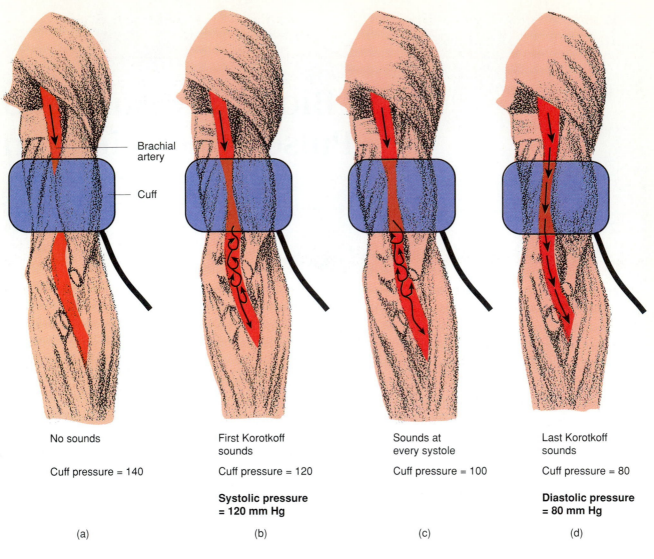

| | | | |
|---|---|---|---|
| No sounds | First Korotkoff sounds | Sounds at every systole | Last Korotkoff sounds |
| Cuff pressure = 140 | Cuff pressure = 120 | Cuff pressure = 100 | Cuff pressure = 80 |
| | **Systolic pressure = 120 mm Hg** | | **Diastolic pressure = 80 mm Hg** |
| (a) | (b) | (c) | (d) |

**Blood pressure = 120/80**

**Figure 38-2** Diagrammatic representation of the appearance of the brachial artery at different pressures in the cuff. Korotkoff sounds are produced by the turbulent flow of blood through the partially constricted brachial artery. These sounds will continue to be heard at every systole as long as the cuff pressure remains greater than the diastolic pressure.

Blood pressure usually is measured in millimeters (mm) of mercury (Hg). The arterial blood pressure is expressed as the systolic blood pressure value over the diastolic pressure value. Table 38-1 lists values for males and females. The difference between the systolic and diastolic pressures is known as the *pulse* pressure. Various factors influence arterial blood pressure, including age, emotional state, sex, weight, and strenuous exercise.

Today, additional methods have been developed for arterial pressure determinations. Electronic devices incorporating both sound and light indicators are available for rapid determinations (see Figure 38-3).

The rate and characteristics of the arterial pulse are extremely important because it is indicative of heart action. This pulsation of the blood through the arteries is in response to the contraction of the heart. The number of pulsations per minute is the *pulse rate* or *heart rate*. It can be felt in any artery that lies near the surface of the body or over a bone or other firm tissue. One of the sites commonly used to measure pulse rate is the undersurface of the wrist behind the thumb, where the radial artery pulsates against the radius bone. Individuals with a rapid heart rate are said to be exhibiting *tachycardia,* while those persons with a very slow rate exhibit *bradycardia.*

This exercise considers several approaches to determination of blood pressure and pulse rate.

## Table 38-1 Selected Blood Pressure Values, Normal Range

| Sex | Age | Systolic | Diastolic |
|-----|-----|----------|-----------|
| Females | 16–24 | 100–130 | 60–85 |
| | 25–29 | 102–130 | 60–86 |
| | 30–34 | 102–135 | 60–88 |
| | 35–39 | 105–140 | 65–90 |
| | 40–44 | 105–150 | 65–92 |
| | 45–49 | 105–151 | 65–96 |
| | 50–54 | 110–165 | 70–100 |
| | 55–59 | 110–170 | 70–100 |
| | 60–64 | 115–175 | 70–100 |
| Males | 16–18 | 105–135 | 60–86 |
| | 19 | 105–140 | 60–88 |
| | 20–24 | 105–140 | 62–88 |
| | 25–29 | 108–140 | 65–90 |
| | 30–39 | 110–145 | 68–92 |
| | 40–44 | 110–150 | 70–94 |
| | 45–49 | 110–155 | 70–96 |
| | 50–54 | 115–160 | 70–98 |
| | 55–59 | 115–165 | 70–98 |
| | 60–64 | 115–170 | 70–100 |

## BLOOD PRESSURE DETERMINATIONS

### Materials

The following materials should be provided for class use:

1. Stethoscope
2. Standard sphygmomanometer
3. Electronic sphygmomanometer
4. Astropulse 10 blood pressure monitor
5. Digimax 5000 electronic blood pressure monitor
6. Disinfectant
7. Sterile gauze pads or cotton

### Procedure 1: Examination of Blood Pressure Determination Devices

This exercise is to be performed by students individually. Knowing the parts of equipment and their normal positions is extremely important if equipment is to be used effectively.

1. Examine the various devices provided for the measurement of blood pressure and pulse determinations.
2. With the aid of Figure 38-3, identify the parts and note the positions of these devices.

### Procedure 2: Auscultation with Standard Sphygmomanometer

This procedure is to be performed by students in pairs.

1. Ask your laboratory partner to sit comfortably, and place the left forearm, palm up, on a table. The upper arm should be at heart level.
2. With the aid of Figure 38-3, identify the parts and note the positions of these devices.
3. Examine the *sphygmomanometer cuff* to see whether the area to be placed against the brachial artery is marked. If the cuff is not marked, hold the inflatable portion of the cuff on the inside of the upper arm, while you wrap the long end of the cuff around the arm. The cuff should fit snugly. Inflate the cuff until it does not fall. Ask the instructor for assistance if you have any problems with the placement of the cuff. Place the measurement gauge or mercury column in a position close enough to be read easily.
4. Insert the earpieces of the stethoscope into your ears, and place the stethoscope bell or diaphragm above the brachial artery at the antecubital space (see Figure 38-1). What do you hear? (Have a pencil and paper handy.)
5. Open the valve on the rubber bulb, and inflate the cuff to 150 mm Hg. (Do not go higher without the recommendation of your instructor.) Close the valve. (*Note:* You should work quickly from this point.)
6. Open the valve on the rubber bulb slightly and listen carefully for pressure sounds. As the cuff pressure is decreased, vibrations of the arterial wall are caused by the spurting of blood through the narrowed artery under the cuff into the fully opened artery below the cuff. These vibrations are *Korotkoff* sounds (Figure 38-2). The pressure at which these sounds are detected is the systolic pressure. Record this pressure in the results and observations section (or make note, for later reference).
7. Slowly open the valve farther while you listen for the total disappearance of all sounds. The point at which no sound can be detected is the *diastolic pressure*. Record this value in the results and observations section (or make yourself a note).
8. Determine the systolic and diastolic pressures with your partner standing and after running up and down several flights of stairs. Record all of your findings in the results and observations section.
9. Calculate the pulse pressure for each situation, using the following formula:

$$\text{Pulse pressure} = \text{systolic pressure} - \text{diastolic pressure}$$

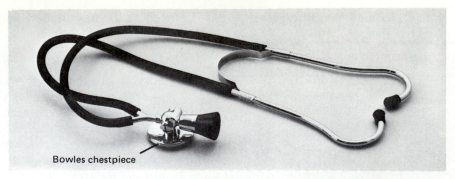

Bowles chestpiece

(a)

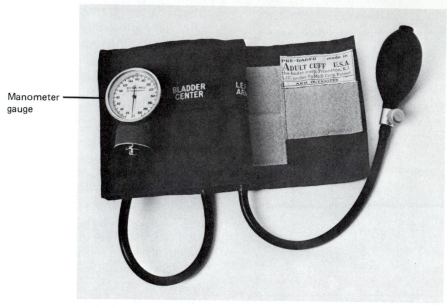

Manometer gauge

BLADDER CENTER

(b)

**Figure 38-3** Instruments used for blood-pressure measurement: (a) combination stethoscope with a *Bowles chestpiece;* (b) *Hare Deluxe Sphygmomanometer,* which offers several convenient features, including a cuff that indicates the right size for different individuals and a clearly marked bladder center, which shows where to position the cuff over the artery; (c) *electronic sphygmomanometer*—solid state electronic device that can be used to read systolic and diastolic blood pressure using both a flashing light and a sound, and which includes a highly sensitive and functional microphone in the cuff, replacing the use of the stethoscope (courtesy of Dyna-Med, Inc., Carlsbad, CA); (d) *Digimax 500 electronic blood pressure monitor,* which can provide a simultaneous digital readout of diastolic and systolic pressure and the pulse rate; (e) gauge of the *Astropulse 10 blood pressure monitor,* which uses electronic sound and light indicators, a standard needle indicator, and a microphone in the cuff, eliminating the need for a stethoscope (courtesy of Dyna-Med, Inc., Carlsbad, CA).

Enter your findings, and answer the questions in the results and observations section.

10. Repeat the entire procedure using the right arm. Record your findings in the results and observations section.

11. Change places with your laboratory partner, wipe the stethoscope with disinfectant, and repeat Steps 1–10.

**Procedure 3: Astropulse 10 Blood Pressure Monitor**

This procedure is to be performed by students in pairs.

1. Check to see whether the unit is ready to use. The battery should be in place and the microphone jack plugged into the receptacle at the bottom of the unit.

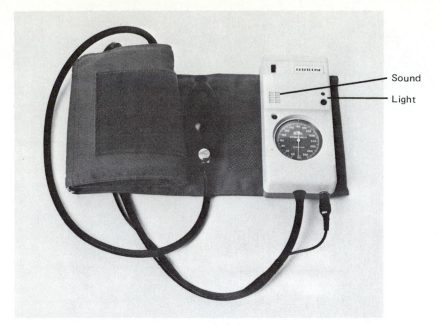

Sound

Light

(c)

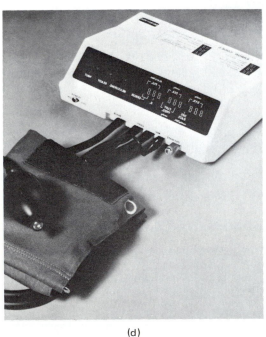

(d)

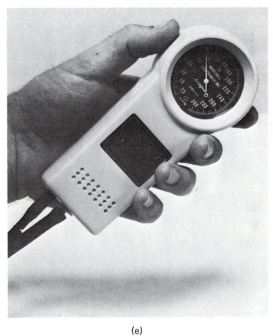

(e)

2. Ask your laboratory partner to sit comfortably with the left palm facing up and the upper arm in line with the heart.

3. Be certain that the end of the pressure cuff is inserted through the D ring, and slip the cuff over the left arm (Figure 38-4).

4. Turn the cuff so that its blue dot is on the inside of the arm just above the elbow. Pull the end of the cuff so that it fits firmly on the arm, and wrap any extra cuff under and around the arm. Press the Velcro tape together to secure the cuff.

5. Place or hold the pressure gauge so that it can be easily read. Now you are ready to determine the systolic and diastolic pressures.

6. With the power off, close the air-flow valve on the bulb by turning it in a clockwise direction. Inflate

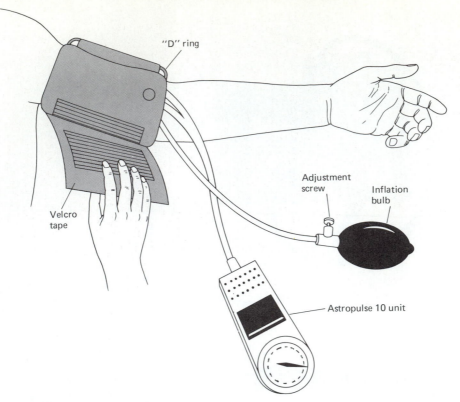

**Figure 38-4** The pressure cuff and its application: Various parts of the pressure cuff, the air flow system, and the positioning of the arm are shown; the entire procedure can be performed by one individual.

the cuff quickly by repeatedly squeezing the bulb until the needle reaches the point marked 150 mm on the gauge. (In some cases, it may be necessary to inflate the cuff beyond 150 mm. Consult with your instructor.)

7. Turn on the power switch. You will hear a beep and see a light flash on the gauge as the power enters the unit. (The beep and light will not occur again until the pressure is reduced.)

8. Open the valve slightly in a counterclockwise direction so that the pressure falls at the rate of 2–4 mm/second (one mark on the gauge equals 2 mm).

9. Continue to deflate the cuff at the rate of 2–4 mm/ second. When the pressure in the cuff is equal to the *systolic pressure* in the artery, the beep will sound, and the light will flash. Note the reading on the gauge.

10. Allow the pressure to continue to drop at the same rate by opening the valve. The beeping and flashing will continue. Watch the gauge. At the moment the beeping and flashing stop, take a reading. This is the *diastolic pressure value.*

11. Repeat Steps 1–10 until you can carry out the procedure easily.

12. Obtain readings for systolic and diastolic pressures in sitting and standing positions and after exercise. Record the values, and answer the questions in the results and observations section.

13. Change places with your partner, and repeat Steps 1–12.

14. Experiment and carry out the procedure on yourself.

## PULSE-RATE DETERMINATIONS

### Materials

The following materials should be provided for class use:

1. Electronic pulse monitor
2. Paper towels or other material to wipe fingers
3. Stopwatch

### Procedure 1: Examination of Pulse-Rate Determining Device

This procedure is to be performed by students individually.

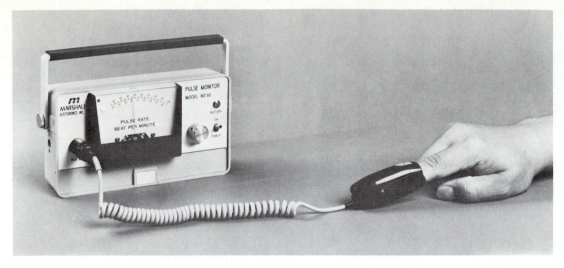

**Figure 38-5** *Electronic pulse monitor:* The pulse rate of an individual can be determined simply by inserting the index finger into the sensitive photoelectric sensor(s); the pulse rate registers on the scale dial. (Courtesy of Dyna-Med, Inc., Carlsbad, CA.)

1. Examine the electronic pulse monitor provided.
2. With the aid of Figure 38-5, familiarize yourself with the parts of this device.

## Procedure 2: Radial Pulse Determinations

This procedure is to be performed by students in pairs.

1. Ask your laboratory partner to sit comfortably, and hold the forearm of one hand with palm up toward you.
2. Take the wrist of this hand, and lightly place your index and middle fingers on the surface facing you in the area behind your thumb (Figure 38-6). Your thumb should be holding the underside of the wrist.
3. Feel for the beating sensation in this area. Do not use your thumb because it has a pulse of its own.
4. Count the number of beats that occur per 15 seconds. Multiply the beats by 4, to obtain the pulse rate for a minute.
5. Repeat this procedure three times, and obtain an average value. Record your findings in the results and observations section.
6. Take the pulse rate of your partner in the standing position and after exercise. Running up and down a flight of stairs three times or running rapidly in place for 2 minutes would be sufficient exercise.
7. Enter your findings, and answer the questions in the results and observations section.
8. Change places with your laboratory partner, and repeat Steps 1–7.

## Procedure 3: Electronic Pulse Monitoring

This procedure is to be performed by students in pairs.

1. Turn on the electronic pulse device.
2. Ask your laboratory partner to sit comfortably next to the measuring device.
3. Insert your partner's index finger into the photoelectric sensor, and obtain a reading.
4. Repeat this step. Was the reading the same? Enter your findings in the results and observations section.
5. Obtain pulse rates with your partner standing and after exercise. Enter your findings in the results and observations section.
6. Change places with your partner, and repeat Steps 1–5.
7. Turn off the electronic device after you have completed this procedure.

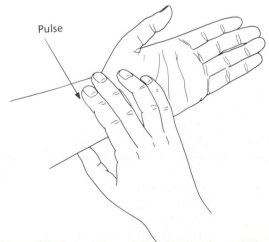

Pulse

**Figure 38-6** Position of the fingers for radial pulse determination.

## Procedure 4: Carotid Pulse Determinations

This procedure is to be performed by students in pairs.

1. Ask your partner to run in place for 2 minutes and then sit down.
2. Place your fingers (no thumb) on either side of the partner's larynx. Gently press inward and toward the back of the neck. Continue to apply gentle pressure until a pulse is felt.
3. Once a pulse region is located, practice the procedure until only two fingers are necessary to make a pulse determination. Record your findings, and answer the questions in the results and observations section.
4. Repeat Step 1, and determine the radial pulse rate (refer to Procedure 2). Enter your findings in the results and observations section.
5. Change places with your partner, and repeat Steps 1–4.

## RESULTS AND OBSERVATIONS

### Blood Pressure Determinations

1. Enter the blood pressure readings determined by auscultation in Table 38-2. Record your own and your partner's systolic (S) and diastolic (D) pressure. Space is provided for measurements of other students.
2. Enter pulse-pressure results in Table 38-3.
3. Compare your pressures with those listed for your age group in Table 38-1. Are your values similar to those listed? _____
4. Are the values for your partner similar to those listed? _____
5. Did this procedure cause you any problems? Explain. _____
   _____
6. Enter the blood pressure readings determined by the Astropulse 10 blood pressure monitor in Table 38-4.
7. Was the electronic device easier to use than the standard sphygmomanometer? Explain. _____
   _____
8. Were the determinations comparable to those obtained by other methods used? _____

**Table 38-2**

| Student | Seated | | | | Standing | | | | After Exercise | | | |
|---|---|---|---|---|---|---|---|---|---|---|---|---|
| | Left Arm | | Right Arm | | Left Arm | | Right Arm | | Left Arm | | Right Arm | |
| | S | D | S | D | S | D | S | D | S | D | S | D |
| | | | | | | | | | | | | |
| | | | | | | | | | | | | |
| | | | | | | | | | | | | |
| | | | | | | | | | | | | |

**Table 38-3**

| Student | Seated | | Standing | | After Exercise | |
|---|---|---|---|---|---|---|
| | Left Arm | Right Arm | Left Arm | Right Arm | Left Arm | Right Arm |
| | | | | | | |
| | | | | | | |
| | | | | | | |
| | | | | | | |

9. Was your blood-pressure value within the normal range? _____

10. Was your laboratory partner's value within the normal range? _____

## Pulse-Rate Determinations

1. Enter your findings obtained with the radial pulse determinations (R) and with the electronic monitor (E) in Table 38-5.

2. If an electronic monitoring device was provided, were the determinations comparable to those obtained by other methods used? _____

3. Enter your findings comparing radial and carotid pulse determinations in Table 38-6.

### Table 38-4

| Student | Seated | | Standing | | After Exercise | |
|---|---|---|---|---|---|---|
| | Systolic | Diastolic | Systolic | Diastolic | Systolic | Diastolic |
| | | | | | | |
| | | | | | | |
| | | | | | | |
| | | | | | | |

### Table 38-5

| Student | Seated | | Standing | | After Exercise | |
|---|---|---|---|---|---|---|
| | R | E | R | E | R | E |
| | | | | | | |
| | | | | | | |
| | | | | | | |
| | | | | | | |

### Table 38-6

| Student | Radial Pulse Determination | Carotid Pulse Determination |
|---|---|---|
| | | |
| | | |
| | | |
| | | |

# LABORATORY REVIEW *38*

## *BLOOD PRESSURE AND PULSE DETERMINATION*

### COMPLETION, IDENTIFICATION, DEFINITION, AND INTERPRETATION

1. List three factors that affect heart rate.

   a. _____

   b. _____

   c. _____

2. a. What is pulse rate? _____

   _____

   b. What is the average value for an adult? _____

   _____

3. What sound is heard with a stethoscope when the brachial artery is collapsed?

   _____

4. List the major conditions that should be met while taking an individual's blood pressure. _____

   _____

   _____

5. Identify the device shown in Figure Q38-1 and identify the specific labeled parts.

   a. Device _____

   b. _____

   c. _____

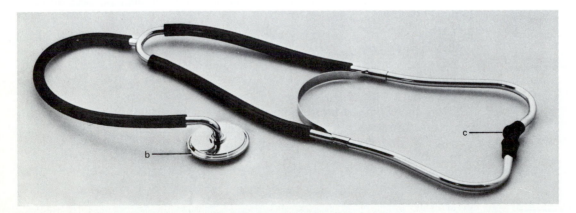

**Figure Q38-1** (Courtesy of Dyna-Med, Inc., Carlsbad, CA.)

**ANSWERS**

6. Which artery is usually used to take an individual's pulse rate? _____

7. Complete the following formula.

   Pulse rate (beats/minute) × stroke volume (ml) = _____ (liters/minute)

8. What are Korotkoff sounds? _____

_____

9. Define the following:

   a. Bradycardia _____

   b. Tachycardia _____

10. With the aid of your textbook and other references, indicate the effects of the following conditions on arterial blood pressure:

   a. Vasoconstriction _____

   b. Increased blood volume _____

_____

   c. Arteriosclerosis _____

_____

   d. Increased aldosterone concentration _____

_____

11. What is hypertension? _____

_____

## MULTIPLE CHOICE

Insert the correct answers in the spaces provided.

1. In a medical context, to which system does *blood pressure* refer?
   a. venous
   b. arterial
   c. capillary
   d. diastolic
   e. systolic

2. Through which of the following does the turbulent flow of blood produce Korotkoff sounds?
   a. aorta
   b. subclavian artery
   c. pulmonary arteries
   d. brachial artery
   e. pulmonary veins

3. Which of the two blood-pressure measurements is the reading on a blood pressure gauge measuring when the first Korotkoff sound is detected?
   a. systolic
   b. diastolic
   c. brachial
   d. venous
   e. arterial

1. _____
2. _____
3. _____
4. _____
5. _____
6. _____
7. _____
8. _____

4. Which of the two blood-pressure measurements is the reading on a blood pressure gauge measuring when the Korotkoff sound disappears?
   a. systolic
   b. diastolic
   c. brachial
   d. venous
   e. arterial

5. Which is correct in an individual's blood pressure result?
   a. the systolic pressure is listed above or before the diastolic
   b. the diastolic pressure is listed above or before a systolic

6. Is the systolic pressure greater or less than the diastolic pressure when the Korotkoff sounds continue to be heard at every systole?
   a. less
   b. greater

7. What is the difference between the systolic and the diastolic pressure called?
   a. Korotkoff pressure
   b. pulse pressure
   c. heart pressure
   d. arterial pressure
   e. choices *a* and *b* only

8. What is the term used for individuals exhibiting a rapid heart rate?
   a. tachyardia
   b. bradycardia
   c. neutrocardia
   d. diastole
   e. systole

# Blood and the Immune System

*T*he circulatory system, which consists of the blood, heart, and blood vessels, provides a link between the body's internal components and its external environment. The exercises in this section deal specifically with the blood and its components and demonstrate their involvement in providing vital support for cellular activities and the maintenance and protection of a favorable cellular environment. In addition, attention is given to conditions that interfere with the production of red blood cells or that lead to an increased level of red-blood-cell destruction.

The two major components of blood are *formed elements*—including erythrocytes (red blood cells), leukocytes (white blood cells), and platelets—and *plasma*, a pale yellow fluid. Plasma contains about 90 percent water, with the remaining 10 percent consisting of carbohydrates, hormones, lipids, proteins (albumin, enzymes, fibrinogen, immunoglobulins or antibodies, and other globulins), salts, and vitamins. Several laboratory procedures are used to determine the concentrations of blood components (Figure XI-1). Because the results of such tests are

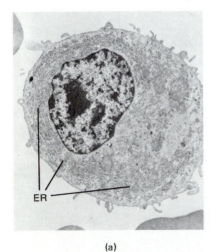

(a)

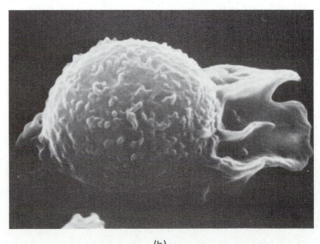

(b)

**Figure XI-1** Normal and abnormal blood cells: (a) This electron micrograph shows an activated plasma cell; this cell type is the immunoglobulin producer of the body; note the presence of the large quantity of endoplasmic reticulum (ER) within the cell (From D. Douer, A. Weinberger, M. Djaldetti, J. Asherov, I. Pick, and J. Pinkhas. *Acta Hematologica*, 62: 81–85, 1979). (b) This scanning micrograph shows a leukemia cell originating from lymphoid tissue.

of diagnostic importance, instrumentation improvements have been made so that procedures can be performed more rapidly and accurately.

Several exercises consider procedures that are used for demonstrating the presence and functions of the components of blood. In addition, attention is given to conditions that interfere with the production of red blood cells or that lead to an increased level of red-blood-cell destruction.

## GENERAL REFERENCES

Hillman, R. S., and C. A. Finch. *Red Cell Manual,* 6th ed. Philadelphia: F. A. Davis Company, 1992.

Harmening, D. M. (Ed.). *Clinical Hematology and Fundamentals of Hemostasis,* 2nd ed. Philadelphia: F. A. Davis Company, 1991.

Koepke, J. A. (Ed.). *Practical Laboratory Hematology.* New York: Churchill Livingston, 1991.

Tortora, G. J., and S. R. Grabowski. *Principles of Anatomy and Physiology,* 7th ed. New York: HarperCollins, 1993.

## SPECIFIC REFERENCES

Dvorak, A. M., and H. F. Dvorak. "The Basophil." *Archives of Pathology and Laboratory Medicine,* 103: 551–557, 1979.

Evans, V. J. "Platelet Morphology and the Blood Smear." *Journal of Medical Technology,* 1: 689–695, 1984.

Golde, D. W., and J. C. Grasson. "Hormones That Stimulate the Growth of Blood Cells." *Scientific American,* 259: 62–70, 1988.

Metcalf, D. "Control of Granulocytes and Macrophages: Molecular, Cellular, and Clinical Aspects." *Science,* 254: 529–533, 1991.

Morse, E. E., and B. Jacobs. "Blood Donation and Its Aftereffects." *American Scientist,* 73: 68–69, 1985.

Ranney, H. M. "The Spectrum of Sickle Cell Disease." *Hospital Practice,* 27: 133–163, 1992.

Roberts, H. R., and J. N. Lozier. "New Perspectives on the Coagulation Cascade." *Hospital Practice,* 27: 97–112, 1992.

Von Boehmer, H., and P. Kisielow. "How the Immune System Learns About Self." *Scientific American,* 265: 74–81, 1991.

# Blood Smear Preparation and Differential Staining

1. Prepare and stain blood smears
2. Distinguish among blood cells
3. Perform a differential blood count
4. Recognize and identify certain abnormalities and disease agents in blood smears

Blood cells must go through a development cycle in the same way that a human matures during development. In healthy individuals, only mature cells are seen, upon examination of blood specimens. Immature or abnormal cells can be observed in various disease states. Thus, the examination of a patient's blood may be extremely important in either the detection of abnormalities or the diagnosis of disease.

An essential part of a hematological examination is a properly prepared *blood smear* (or *spread*). Such a smear is made by obtaining a drop of blood from the patient, placing it on a slide, and then spreading it, with the aid of a second slide, into a thin layer. After the preparation dries, any one of several staining procedures, such as Wright's or Giemsa, can be applied.

The examination of blood for abnormalities and the determination of the percentage of various types of cells may be diagnostic in themselves (Figure 39-1), or they may serve as a basis for additional tests. For example, the presence of sickled cells in the blood is diagnostic for the condition of sickle-cell anemia (Figure 39-1), whereas an elevated white-cell count may be a sign that a disease exists, but not an identification of the disease.

This exercise concentrates on formed elements of

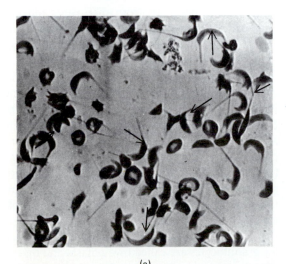

(a)  (b)

**Figure 39-1** The appearance of sickle cells: (a) a light microscopic view of a red-blood-cell preparation, showing a large number of sickled cells (arrows) (from N. M. Rumen, *Blood*, 45: 45–48, 1975); (b) a scanning micrograph of sickled cells.

blood. The formed elements of blood include red blood cells *(erythrocytes)*, white blood cells *(leukocytes)*, and platelets *(thrombocytes)*. *Erythrocytes* carry oxygen to tissues and remove carbon dioxide from them. *Leukocytes* function in inflammation and immune responses of the body. There are five types of leukocytes, divided into two general categories: granulocytes and agranulocytes. The *granulocytes,* cells containing cytoplasmic granules, consist of neutrophils, eosinophils, and basophils (so named for the staining characteristics of their granules; see Color Plate 36). *Neutrophils* serve as essential agents in an individual's defense against various bacteria and some fungi. *Eosinophils* have a role in defending against disease agents *(pathogens)* such as protozoa and worms. The exact function of *basophils* is not well understood. These cells are believed to be associated with allergic reactions and inflammation. The *agranulocytes,* cells without granules, consist of lymphocytes and monocytes (Color Plate 37). *Monocytes* are critical in the body's defense against intracellular disease-causing agents such as viruses and some bacteria. *Lymphocytes* also help in recognizing and destroying several types of pathogens. The platelets, together with coagulation factors in plasma, control the clotting of blood. The diameters of the formed elements are as follows: erythrocytes, approximately 7 μm; lymphocytes, 9–18 μm; monocytes, 24 μm; granulocytes, 10–14 μm; and platelets, 2–4 μm.

Determination of the proportion of each type of leukocyte in the blood is made by means of the *differential count* procedure. Out of a sample of 100 white blood cells observed on a stained blood smear, the normal number for each type of cell is

| | |
|---|---|
| Basophils | 0–1 |
| Eosinophils | 1–4 |
| Neutrophils | 60–70 |
| Large lymphocytes | 0–3 |
| Small lymphocytes | 25–30 |
| Monocytes | 4–8 |
| | 90–116 |

Detection of abnormalities associated with erythrocytes, such as anemias and malignancies, can be made from examination of blood smears. *Anemia* is a lower-than-normal number of erythrocytes or an abnormally low hemoglobin content in the cells. In general, anemia is caused by blood loss, by an interference with erythrocyte production in the bone marrow, by accelerated erythrocyte destruction, or by combinations of these factors. The appearance of the erythrocytes is one method of classifying anemias. The cells may be described in terms of size and hemoglobin content. Combinations of the following properties are encountered in examination of blood smears.

## ERYTHROCYTE SIZE

Macrocytic—cells are larger than normal

Normocytic—cells are normal size

Microcytic—cells are smaller than normal

## HEMOGLOBIN CONTENT

Hyperchromic—cells contain more than normal amount

Normochromic—cells contain normal amount

Hypochromic—cells contain less than normal amount

Anemias resulting from the decreased production of erythrocytes can be caused by various deficiencies, including iron or vitamin $B_{12}$; by destruction of bone marrow by chemical and physical agents; and by malignancies of different types. Anemias caused by the increased destruction of red blood cells occur in cases of severe infections, sickle-cell anemia, incorrect transfusions, and hemolytic disease of the newborn.

Various conditions also are associated with leukocytes. *Leukopenia,* or low white-blood-cell count, often is induced by viral infections and by treatments of malignancies by radiation or chemicals. An above-normal white-blood-cell count may be provoked by several conditions (see Table 39-1).

The leukemias are a group of diseases primarily affecting the lymphatic, mononuclear, and phagocytic systems and the bone marrow. Overproduction of leukocytes and the presence of great numbers of immature and abnormal forms of white blood cells in the bone marrow (Color Plate 38) and blood-forming organs are characteristic features of these diseases. Many of the immature and abnormal leukemic cells gain access to the circulating bloodstream. These rapidly reproducing leukemic cells fill the bone marrow spaces, prevent immature red blood cells from maturing, and interfere with platelet formation. Leukemic cells also spill into regions of the body and accumulate in organs with no blood-forming capability.

Three types of leukemia are recognized: granulocytic (myelogenous), lymphatic, and monocytic (Color Plates 38–41). This classification is based on the particular leukocyte involved. Leukemias also are grouped according to other factors, such as the duration of the disease—acute (rapid onset) or chronic (of long duration)—and the prominence of immature and abnormal cell types in the bone marrow and blood. In granulocytic (myelogenous) leukemia, an excessive number of granulocytes and the bone marrow cells that give rise to them (precursors) appear in the blood. The precursors for granulocytes are called myeloblasts and myelocytes. With lymphatic leukemia, an abnormal number of lymphocytes and their precursors (lymphoblasts and lymphoid cells) appear in the blood and bone marrow.

**Table 39-1   Conditions Associated with White-Blood-Cell Number**

| Cell Type | Condition Causing Increase in Cell Number |
|---|---|
| Basophil | Granulocytic leukemia |
| Eosinophil | Allergies, worm infections, Hodgkin's disease, some leukemias |
| Lymphocyte | Hepatitis, infectious mononucleosis, measles, mumps, tuberculosis, lymphocytic leukemia |
| Monocyte | Hodgkin's disease, monocytic leukemia, tuberculosis |
| Neutrophil | Granulocytic leukemia, pus-producing bacterial infections, poisons |

In monocytic leukemia, an abnormal number of immature monocytic leukocytes are found in the blood (Figure 39-2a). One other type of leukemia, known as plasma-cell (multiple) myeloma, is included here, even though the cells involved normally do not circulate in the bloodstream. Plasma cells (described in the section introduction) are immunoglobulin-producing cells. When abnormal plasma cells spill over from locations such as the marrow of the ribs and skull into the peripheral blood, a condition of plasma-cell leukemia exists (Figure 39-2b). In this exercise, slides from several types of leukemia states are placed on demonstration, to emphasize the appearance of abnormal and immature leukocytes.

This exercise considers preparation of a blood smear, a differential-count procedure, and examination of normal and abnormal blood specimens. Exercise 40 contains procedures using blood-cell-counting equipment.

## Materials

1. The following materials should be provided per each four students:
   a. Commercially prepared sterile alcohol prep-pads
   b. Sterile, disposable blood lancets
   c. Sterile gauze pads (2 × 2 inch)
   d. Wright's stain and buffer solution for Wright's stain (in dropper bottles or other suitable containers)
   e. Staining rack

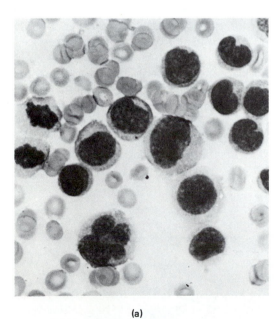

(a)

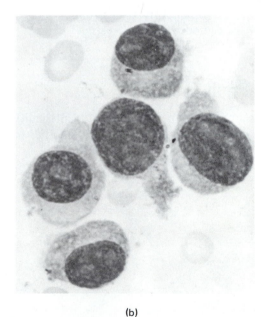

(b)

**Figure 39-2** Examples of the leukemias: (a) leukemic cells with distinct monocytic morphology (from I. Mirchandani, P. Tabaczka, and M. Palutka. *American Journal of Hematology,* 12: 139–147, 1982); (b) the appearance of plasma-cell leukemia, in which the majority of the cells shown are plasma cells (from D. Douer, A. Weinberger, M. Djaldetti, J. Asherov, I. Pick, and J. Pinkhaus. *Acta Hematologica,* 62: 81–85, 1979).

f. Containers for the disposal of lancets, pads, and used slides
g. Disposable surgical gloves
h. Protective eyewear (optional)
2. The following materials should be provided for class use:
   a. Clean, lint-free glass slides
   b. Immersion oil
   c. Xylene and lens paper
   d. Demonstration slides of blood smears representing normal blood and noninfectious anemic states
      i. Normal blood smears
      ii. Sickle-cell anemia
      iii. Macrocytic, hyperchromic anemia
      iv. Microcytic, hypochromic anemia
      v. Hemolytic anemia
   e. Demonstration slides of blood smears from selected microbial infections
      i. Representative blood stages of malaria parasites, *Plasmodium falciparum* or *P. vivax* (see Color Plate 42)
      ii. African sleeping sickness (*Trypanosoma brucei* variety *gambiense*) infection (see Color Plate 43)
      iii. Infectious mononucleosis
   f. Demonstration slides of blood smears from selected leukemic conditions
      i. Acute myelomonocytic leukemia
      ii. Chronic granulocytic leukemia
      iii. Acute lymphoblastic leukemia
   g. Twelve demonstration microscopes

## Procedure 1: Blood Smear Preparation

This procedure is to be performed by students individually, but it will require the assistance of another student.

**CAUTION**
Working with blood or other body fluids will not pose any danger if the following precautions are observed:

1. Place any and all blood, or materials containing blood, into the disinfectant containers provided.
2. Dispose of any blood containing materials as indicated by the instructor.
3. In cases of blood spillage, wipe the area with a disinfectant-soaked paper towel, and dispose of all materials as indicated by the instructor.
4. Always wash your hands after handling blood and associated materials.
5. The use of surgical gloves and protective eye wear.

1. Wipe the finger from which a sample is to be taken with an alcohol-moistened (but not saturated) gauze pad. Do not use a finger that will be involved later in holding a pencil or other such activity. (See Exercise 5 for this procedure.)
2. Unwrap the blood lancet.
3. Hold the finger in a downward position, and squeeze it gently.
4. Have your laboratory partner puncture the ball of the finger with the lancet.
5. Discard the first two drops of blood by wiping them off with a fresh pad.
6. Keeping the finger in a downward position, touch one end of a clean glass slide to a newly formed blood drop. Put the slide on a flat surface, wet side up.
7. Place a fresh, sterile pad on the punctured area, and hold it tightly in place until the bleeding stops.
8. Pull one edge of a second slide into the blood drop (Figure 39-3a). Allow the blood to run across its width (Figure 39-3b).
9. With the second slide at about a 30° angle, move

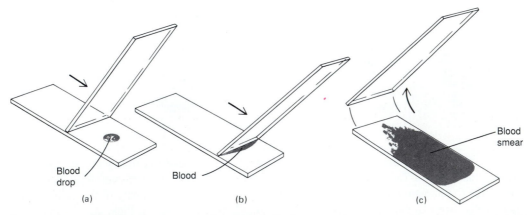

**Figure 39-3** Blood smear preparation: (a) Pull the second slide into the drop of blood; (b) allow blood to spread across its width; (c) push the slide toward the end of the first slide.

it toward the opposite end of the first slide (Figure 39-3c). A good smear should have a feathered edge.

10. Scratch your initials in the center of the preparation, and allow it to air dry.

## Procedure 2: Staining and Differential Count

1. Place your blood smear on the staining rack, or as otherwise directed by your instructor.
2. Apply enough Wright's stain to cover the entire preparation.
3. Allow the stain to remain in contact with the smear for approximately 1 minute.
4. Add the buffer solution (in drops) until the buffer and stain are in equal proportion. Blow gently on the slide to mix the two solutions.
5. Allow the resulting mixture to stay in contact with the smear for 4 minutes.
6. Pour off the solution, and flood the slide with buffer solution, in order to wash the preparation.
7. Air dry or blot dry the smear. No cover slip is needed.
8. Examine the demonstration slides of normal blood smears.
9. Perform a differential count—that is, count the first 100 white blood cells observed, and note the type of each cell. Enter your findings in the results and observations section.
10. Examine the red blood cells in your smear and in your laboratory partner's. Answer the relevant questions in the results and observations section. (Refer to Color Plates 3, 36, and 37.)

## Procedure 3: Demonstration of Normal Blood and Noninfectious Anemic States

1. Your instructor will emphasize the distinguishing features of a normal blood smear, as compared to those of anemic states.
2. Examine all the demonstration slides provided for this portion of the exercise. Refer also to Color Plates 3, 36, and 37.

3. Make sketches of the distinguishing features noted in the results and observations section, and answer all questions pertaining to this exercise in the laboratory review section.

## Procedure 4: Demonstration Slides of Blood Smears from Selected Microbial Infections

1. Your instructor will emphasize the major distinguishing features of the preparations.
2. Examine all the demonstration slides. Refer to Color Plates 42–44.
3. Note all distinguishing features. Make sketches of these in the results and observations section.
4. Note whether the infectious disease agents are found inside or outside the erythrocytes.

## Procedure 5: Demonstration Slides from Selected Leukemic Conditions

1. Your instructor will emphasize the major features of the preparations. Also refer to Color Plates 38–41.
2. Pay particular attention to the presence of immature blood cells, as well as the smaller-than-normal size of others.
3. Make sketches of typical leukemic cells in the results and observations section, and answer questions associated with this portion of the exercise.
4. Answer any items pertaining to this and previous portions of this exercise in the laboratory review section.

## RESULTS AND OBSERVATIONS

1. Enter your findings for the differential count in Table 39-2.

### Table 39-2

| Type of White Cell | Number per 100 | Average Number per 100* |
|---|---|---|
| Lymphocyte | | |
| Monocyte | | |
| Eosinophil | | |
| Basophil | | |
| Neutrophil | | |

*Enter normal values here, as indicated in your textbook or other reference.

2. Did you observe any unusual cells in the demonstration slide? If so, describe them. _____

   _____

   _____

3. Were your red blood cells and those of your partner similar to those in the demonstration slides of a normal blood smear? Describe any differences.

   _____

   _____

4. Sketch distinguishing features of the blood preparations representing normal blood and noninfectious anemic states.

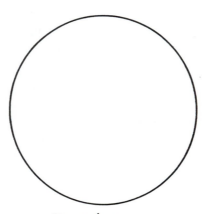

**Normal smear**

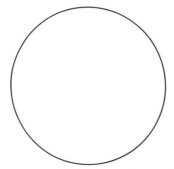

Macrocytic, hyperchromic anemia

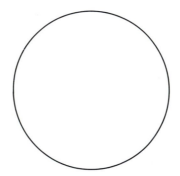

Microcytic, hypochromic anemia

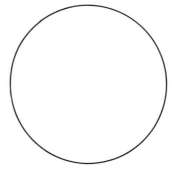

Sickle cell anemia

5. Sketch distinguishing features of blood smear preparations from selected microbial infections.

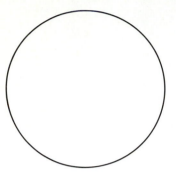
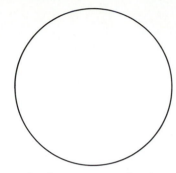
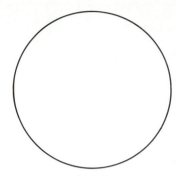

African sleeping sickness    Infectious mononucleosis    Malaria infection with
infection                                                *Plasmodium falciparum*

6. In which of the demonstration-slide preparations
   did you find disease agents actually inside erythro-

   cytes? _____

7. Sketch distinguishing features of leukemic blood
   cells.

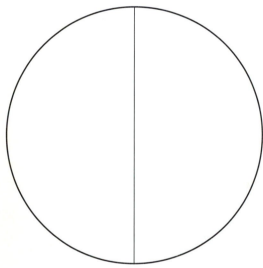
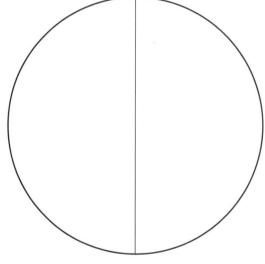

Acute myelomonocytic   Chronic granulocytic        Acute lymphoblastic   Chronic lymphoblastic
leukemia               leukemia                    leukemia              leukemia

# LABORATORY REVIEW *39*

## *BLOOD SMEAR PREPARATION AND DIFFERENTIAL STAINING*

### COMPLETION, IDENTIFICATION, AND COMPARISON

1. Identify the blood cells shown in the blood smear in Figure Q39-1.

   a. _____

   b. _____

   c. _____

   d. _____

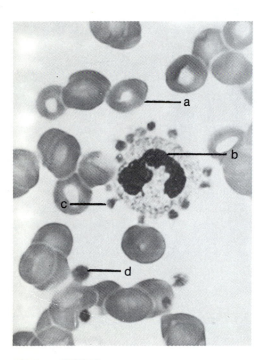

**Figure Q39-1** A blood smear.

2. Give the function or explain the activity of the following:

   a. Differential stain _____

   _____

   b. Lymphocyte _____

   c. Monocyte _____

   d. Eosinophil _____

   e. Neutrophil _____

   f. Basophil _____

   g. Platelet _____

**ANSWERS**

3. Would you consider the micrograph in Figure Q39-2 typical of a normal blood smear? If not, explain your answer. _____

_____

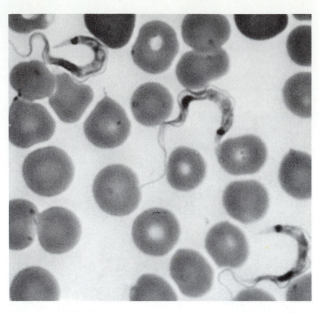

**Figure Q39-2**

4. List three infectious diseases of the circulatory system.

   a. _____ ;

   b. _____ ;

   c. _____

5. Name two hereditary diseases of the blood.

   a. _____ ;

   b. _____ ;

6. Which of the micrographs in Figure Q39-3 contain macrocytic, normocytic, or microcytic erythrocytes? Use a micrograph of a normal smear as a guide.

   a. _____ ;

   b. _____ ;

   c. _____

## MULTIPLE CHOICE

Insert the correct answers in the spaces provided.

1. Which of the following cell types is known to carry oxygen to tissues and to remove carbon dioxide from them?
   a. basophils
   b. monocytes
   c. neutrophils
   d. platelets
   e. erythrocytes

1. _____

2. _____

3. _____

4. _____

5. _____

6. _____

7. _____

8. _____

9. _____

10. _____

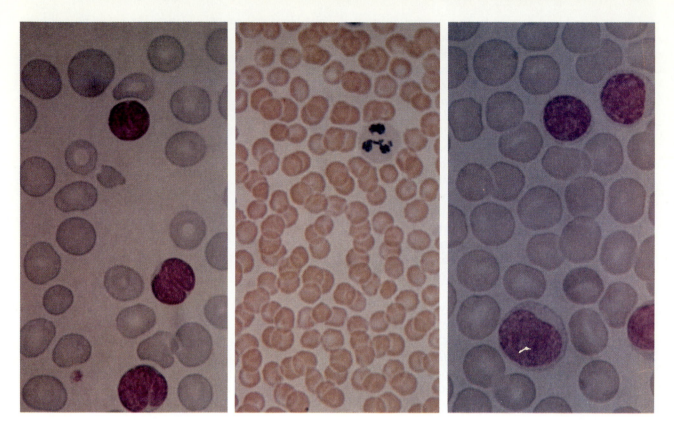

**Figure Q39-3**

2. Which of the following cell types is not an example of a leukocyte?
   a. basophil
   b. monocyte
   c. neutrophil
   d. lymphocyte
   e. platelet
3. Which of the following would be an example of a granulocyte?
   a. lymphocyte
   b. erythrocyte
   c. platelet
   d. basophil
   e. megakaryocyte
4. Which of the following would be an example of an agranulocyte?
   a. lymphocyte
   b. erythrocyte
   c. platelet
   d. basophil
   e. megakaryocyte
5. In a normal differential white-blood-cell count, which of the following cell types
   would be the most numerous?
   a. basophils
   b. eosinophils
   c. neutrophils
   d. lymphocytes
   e. monocytes

6. In a normal differential white-blood-cell count, which of the following cell types would be the least numerous?
   a. basophils
   b. eosinophils
   c. neutrophils
   d. lymphocytes
   e. monocytes

7. Which of the following terms describes erythrocytes that are larger than normal?
   a. macrocytic
   b. normocytic
   c. microcytic
   d. hypochromic
   e. hyperchromic

8. Which of the following terms describes erythrocytes that contain more than the normal hemoglobin content?
   a. macrocytic
   b. normocytic
   c. microcytic
   d. hypochromic
   e. hyperchromic

9. Which of the following terms describes red blood cells that contain less than the normal level of hemoglobin?
   a. macrocytic
   b. normocytic
   c. microcytic
   d. hypochromic
   e. hyperchromic

10. What are antibody-producing cells known as?
    a. erythrocytes
    b. lymphocytes
    c. plasma cells
    d. platelets
    e. monocytes

# Blood Cell Counts, the Hematocrit (Packed Cell Volume), Coagulation Time, and Hemoglobin Determinations

**After completing this exercise,
you should be able to**

1. Describe and use blood-cell-counting equipment
2. Perform standard total red-blood-cell and white-blood-cell counts
3. Carry out the calculations necessary to determine the numbers of red and of white blood cells in blood samples
4. Perform and interpret a standard hematocrit procedure
5. Describe the general stages in coagulation
6. Distinguish between serum and plasma
7. Perform and interpret standard clotting tests and hemoglobin determinations
8. Detect selected clotting and hemoglobin abnormalities

The total blood volume in an adult human is about 6.1 liters, or approximately 7.5 percent of body weight. About 45 percent of blood consists of formed elements: erythrocytes, leukocytes, and platelets. The remaining 55 percent, the fluid portion, is plasma.

The number of the formed elements bears a direct relationship to an individual's state of health. This exercise emphasizes both red and white blood cells. The total number of red blood cells in an adult human is approximately 25 trillion. These cells survive for 100–120 days within the circulatory system. In laboratory tests, the normal average red-blood-cell count is usually in the range of 4.5–5.5 million cells per cubic millimeter ($mm^3$) of blood. Males tend to have a red-blood-cell count about 20 percent higher than females at the same age and level of physical activity. The red-cell count increases proportionately as the level of physical activity increases over a period of time. Thus, a well-trained athlete can be expected to have a higher count than a bedridden hospital patient.

In the healthy adult, the normal number of circulating white blood cells is about 7000/$mm^3$ of blood. This value is somewhat higher in children. Newborns have counts of about 16,000/$mm^3$. If the white-blood-cell count is quite high (i.e., over 10,000), the condition of *leukocytosis* exists. If, on the other hand, the final count is 5000 or below, a condition of *leukopenia* exists. Leukocytosis occurs in situations of acute infections. It is also evident in leukemia, but in this case, white blood cells are abnormal. Leukopenia generally is associated with viral infections such as chickenpox, measles, mumps, and poliomyelitis. The survival or life span of white blood cells in the blood is difficult to determine exactly.

One laboratory method used to determine the number of blood cells incorporates the use of an instrument termed a *hemocytometer,* a counting chamber, and diluting pipettes (Figure 40-1). The counting chamber is designed to hold exactly 1 $mm^3$ of a diluted blood sample. On the bottom of the chamber is a glass surface

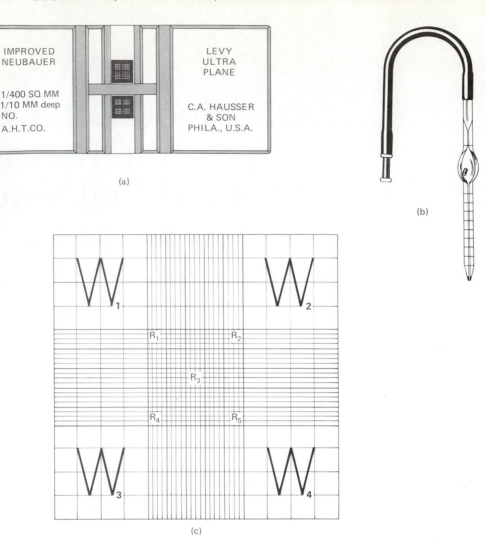

**Figure 40-1** Equipment for blood cell counting: (a) hemocytometer counting chamber; (b) red-blood-cell pipette and tubing—note the presence of a mixing bead in the bulb portion; (c) enlargement of grid on the glass surface of the counting chamber. The areas used for red-cell counts are labeled $R_1$ through $R_5$ and those for white-cell counts are labeled $W_1$ through $W_4$.

etched in a precise grid. The blood cell determination involves counting cells in several squares of the grid and obtaining an average number. That number is multiplied by a factor that takes into account the amount of dilution.

More advanced approaches to making blood cell counts incorporate an electronic counting device. As a suspension containing either red or white blood cells passes through a narrow opening at high speed, each cell is counted automatically. Excellent reproducibility and accuracy generally are obtained with electronic counting devices.

The laboratory offers a number of procedures that serve as valuable aids in the diagnosis and management of various disease states. The *hematocrit*, or *packed cell volume*, is one such procedure. It is an important measurement of red-blood-cell concentration and is used as a test for the detection of an anemic condition. The hematocrit is the ratio of red blood cells to whole blood by volume. When subjected to centrifugal force, red blood cells separate from plasma and the other constituents in blood; they become "packed" together. Once separated, the volumes can be measured to obtain the ratio (Figure 40-2).

In a normal condition, such a preparation consists of approximately 45 percent packed red blood cells, 1 percent white blood cells and platelets, and 54 percent plasma. The normal average hematocrit values for adult males is 47 percent (range 40–54) and for adult females (range 37–47). Hematocrits of 15 percent or lower are

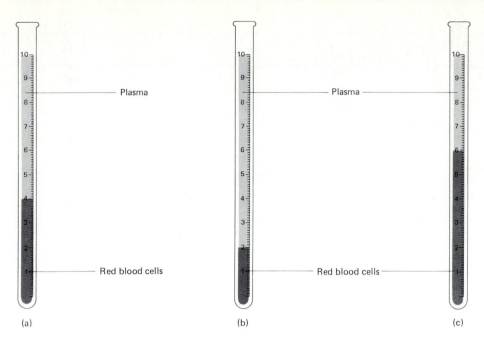

**Figure 40-2** Hematocrit determination: A hematocrit is the ratio of red blood cells to whole blood. The appearance of three specimens after centrifuging in Wintrobe tubes is shown: (a) normal blood; (b) low hematocrit or serious anemia; (c) elevated hematocrit, as in a condition of dehydration resulting from a severe burn.

characteristic of anemia and significant red-blood-cell loss. Conditions such as polycythemia and malignancies involving the red bone marrow, in which there is an overproduction of red blood cells, produce hematocrits of 65 percent and higher. The hematocrit is also performed in this exercise.

Blood normally is maintained in liquid form as long as it remains in blood vessels. However, when blood comes out of a vessel, it is changed within a short time to a soft, jellylike mass, a *blood clot.* The liquid released by the clot is termed *serum.* The physiological process whereby blood is so changed is termed *coagulation.* Clot formation is a vital mechanism that prevents excessive blood loss from the body.

Examination of a magnified blood clot shows that it is composed of a network of delicate fibers, in which both red and white blood cells are enmeshed. The fibers consist of *fibrin,* a substance derived from one of the plasma proteins, *fibrinogen.* The conversion of fibrinogen to the solid fibrin is a complicated chemical process involving enzyme systems.

In recent years, the enzymatic reactions associated with clotting have been formalized into a "cascade" scheme involving at least 13 factors (Table 40-1). It is widely held that the different coagulation factors exist in plasma as inactive proenzymes that are converted to active enzymes during the clotting process. The reactants appear in sequence, with the product of the first reaction operating as an enzyme to catalyze the second reaction. The second reaction yields a product that is an enzyme to catalyze the third reaction, and so on. Once activated, the sequence must continue to completion.

The complex process of coagulation is often greatly simplified into five stages, with the first three related to the formation of the visible blood clot (Figure 40-3), the fourth to clot retraction, and the fifth to dissolution of the clot.

Stage 1. Formation or release of tissue factor (TF), the initiator of clotting

Stage 2. Prothrombin converted to thrombin (by thromboplastin and other coagulation factors)

Stage 3. Fibrinogen converted to fibrin (by thrombin and other coagulation factors)

Stage 4. Clot retraction or *syneresis* (consolidation and contraction of fibrin threads)

Stage 5. Fibrinolysis (clot dissolution)

All five stages may be occurring continuously in minute amounts in the normal coagulation process. Table 40-1 lists the plasma clotting factors.

Depending on the manner in which thromboplastin is introduced, the clotting process may proceed along one of two routes, the *extrinsic* or the *intrinsic* pathway (as shown in Figure 40-3). The extrinsic pathway begins

**Table 40-1**

| Number* | Name |
|---|---|
| Plasma clotting factors | |
| I | Fibrinogen |
| II | Prothrombin |
| III | Tissue thromboplastin (extrinsic or intrinsic), the tissue factor (TF) |
| IV | Calcium ions ($Ca^{2+}$) |
| V | Proaccelerin, or labile factor |
| VII | Proconvertin, serum prothrombin conversion accelerator (SPCA), or stable factor |
| VIII | Antihemophilic factor (AHF) |
| IX | Plasma thromboplastin component (PTC), or Christmas factor |
| X | Stuart–Prower factor, or Stuart factor |
| XI | Plasma thromboplastin antecedent (PTA) |
| XII | Hageman factor |
| XIII | Fibrin stabilizing factor (FSF) |
| Platelet factors (Pf) | |
| $Pf_1$ | Platelet accelerator |
| $Pf_2$ | Thrombin accelerator |
| $Pf_3$ | Platelet thromboplastin factor |
| $Pf_4$ | None assigned formally |

*The substance that had been assigned number VI is no longer used in coagulation theory.

when a blood vessel is ruptured, and damaged tissues surrounding it or the ruptured area of the blood vessel itself release a lipoprotein termed *tissue thromboplastin*. This substance reacts with coagulation factors in plasma to form *extrinsic thromboplastin (extrinsic prothrombin activator)*.

The intrinsic pathway involves the clumping of platelets, their disintegration, and the release of platelet coagulation factors into the plasma. These factors in turn react with certain other coagulation factors to form *intrinsic thromboplastin* or tissue factor *(intrinsic prothrombin activator)*.

Coagulation or bleeding disorders may be the result of a variety of factors, such as platelet abnormalities, blood vessel (vascular) abnormalities, or coagulation-factor deficiencies. The laboratory tests available to detect or screen for these disorders are described in Table 40-2.

Three tests associated with clotting and related disorders, clotting time, prothrombin time (PT), and activated partial thromboplastin time (APTT), are considered in this exercise. The PT and APTT tests complement one another and can be used together with other laboratory tests to determine the basis of bleeding

**Table 40-2   Coagulation-Disorder Screening Tests**

| Test | Brief Description |
|---|---|
| Clotting or bleeding time | A check for platelet aggregation and primary clot formation. This test is a helpful indication of platelet abnormality, whether in number or in function. Individuals with vascular problems will have an abnormal bleeding time. |
| Platelet count | Determination of the quantity of platelets. |
| Prothrombin time (PT) | A test to detect deficiencies in the extrinsic clotting system (Figure 40-3) and defects with Factors I, II, VII, or X in the presence of tissue thromboplastin. Individuals on anticoagulant therapy will have a prolonged PT. |
| Activated partial thromboplastin time (APTT) | A test to detect deficiencies in the intrinsic clotting system (Figure 40-3) and defects with Factors I, V, VIII, IX, X, XI, and XII. Individuals on anticoagulant therapy will have a prolonged APTT test result. |

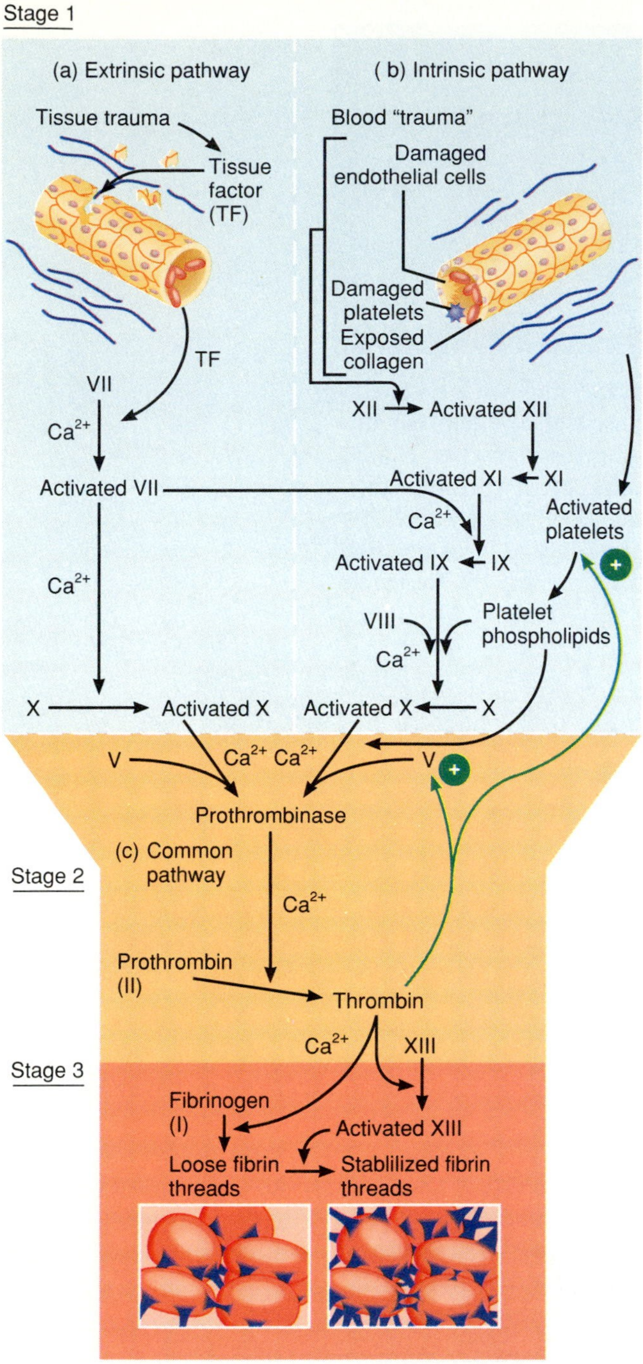

Stage 1

**Figure 40-3** Blood-clotting mechanism: Both extrinsic (a) and intrinsic (b) pathways are shown. Roman numerals are plasma coagulation factors. (From G. J. Tortora and S. R. Grabowski, *Principles of Anatomy and Physiology,* 7th ed., New York: HarperCollins, 1993.)

**Table 40-3  Clotting Factor Defects: Results with PT and APTT**

| Clotting Factor Defect | PT Result | APTT Result |
|---|---|---|
| V | Abnormal | Abnormal |
| VII | Abnormal | Normal |
| VIII | Normal | Abnormal |
| IX | Normal | Abnormal |
| X | Abnormal | Abnormal |
| XI | Normal | Abnormal |
| XII | Normal | Abnormal |

The major fibrinolytic system, known as the *plasminogen–plasmin system,* consists of five components—plasminogen, plasmin, plasminogen activators, plasminogen activation inhibitors, and plasmin inhibitors. Normal serum contains less than 10 micrograms per milliliter (μg/ml) of FDP. Tests used to measure FDP levels are of value in cases involving recovery from abdominal operations, kidney transplant, or childbirth. Such tests also serve as an aid in the diagnosis of pulmonary embolism and renal disease. The Thrombo-Wellco test is a rapid and reliable procedure for the semiquantitative determination of FDP (Figure 40-4).

Hemoglobin is responsible for gaseous transport in the body. The hemoglobin content of the blood usually is expressed in grams per 100 milliliters (g/100 ml) or sometimes as a percentage of normal. The normal hemoglobin content of the blood is about 14.5–15 g/100 ml. Tests for the determination of hemoglobin content are included in this exercise.

Normal hemoglobin (HbA) is a conjugated protein consisting of *heme,* an iron-containing nonprotein, and a globin (protein) portion composed of two pairs of polypeptide chains. The amino acids and the sequence of their arrangement in each polypeptide are under genetic control. If a gene causes a substitution for an amino acid or a change in the amino-acid sequence in one of the polypeptides, an abnormal molecule of hemoglobin forms. Such genetic changes (mutations) may result in an anemia.

One of the better-known gene-controlled defects is sickle-cell anemia. In sickle-cell anemia, hemoglobin S (HbS) is produced; under conditions of low oxygen tension, HbS tends to form crystals in red blood cells that result in the characteristic sickle-shaped distortion. In *homozygous individuals* (individuals in whom both genes for hemoglobin code for HbS), severe hemolytic sickle c-cell anemia occurs. In *heterozygous individuals* (individuals who have one gene for HbS and one for HbA), the hemolytic anemia is comparatively mild.

disorders associated with prolonged clotting time. Table 40-3 shows test results that can occur in case of clotting factor defects.

This exercise also considers *fibrinolysis,* or clot dissolution. *Fibrinolysis* is the process responsible for the enzymatic degradation of fibrin thrombi, which results in the formation of *fibrin degradation products* (FDP).

**Figure 40-4** Result patterns with the Thrombo-Wellco test/ Latex test. A negative result appears on the left. On the right, a positive clumping or agglutination reaction can be seen. In a positive test, latex particles coated with antibodies against *fibrin degradation products* (FDP) react with FDPs in serum from a blood sample. (Courtesy of Wellcome Reagents Division, Burroughs Wellcome Company, Research Triangle Park, NC.)

Screening tests for HbS are available. This exercise demonstrates this type of procedure.

## BLOOD CELL COUNTS

### Materials

1. The following materials should be provided per each two students:
   a. One *hemocytometer* and cover glass (a Neubauer hemocytometer is recommended)
   b. Two Unopette Reservoirs (Becton-Dickinson) for red-blood-cell determinations
   c. Two Unopette Reservoirs (Becton-Dickinson) for white-blood-cell determinations
   d. Four capillary pipette assembly systems (Becton-Dickinson)
   e. One microscope
   f. One hand counter
2. The following materials should be provided for class use:
   a. Individual, sterile alcohol prep-pads
   b. Sterile blood lancets
   c. Kim wipes or other laboratory tissue material
   d. Disposal container for blood contaminated items
   e. Surgical gloves
   f. Protective eyewear (optional)

### CAUTION

Working with blood or other body fluids will not pose any danger if the following precautions are observed:

1. Place any and all blood or materials containing blood into the disinfectant containers provided.
2. Dispose of any blood-containing materials as indicated by the instructor.
3. In cases of blood spillage, wipe the area with a disinfectant-soaked paper towel, and dispose of all materials as indicated by the instructor.
4. Always wash your hands after handling blood and associated materials.
5. The use of surgical gloves and protective eyewear.

### Procedure 1: Red-Blood-Cell Count

This procedure is to be performed by students in pairs, as in other procedures where partners change places, so that each does the experiment (still working in pairs, with materials distributed to pairs). Blood samples for both red-blood-cell and white-blood-cell counts can be obtained at the same time. (*Note:* Different-colored reservoirs are needed.)

1. Examine the Unopette Reservoir for red-blood-cell determinations and the capillary pipette assembly (Figure 40-5). Identify the reservoir chamber, di-

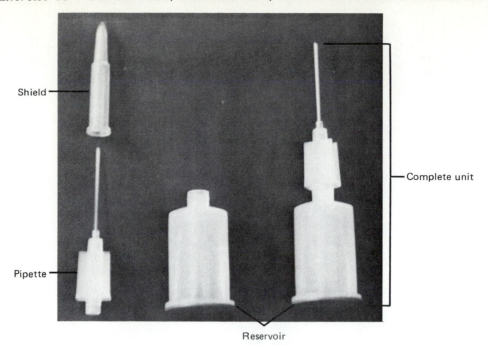

**Figure 40-5** Components of a Unopette Reservoir System: The reservoirs used for red-blood-cell counts have a red undersurface; those used for white-blood-cell counts have a white undersurface.

luent fluid, protective shield on the capillary pipette, and the pipette itself. Note the color of the reservoir's bottom surface. It should be red.

2. Hold the reservoir on a flat surface in one hand, and hold the pipette assembly in the other hand. Push the tip of the pipette shield firmly through the diaphragm in the reservoir's neck (Figure 40-6a). Pull out the assembly unit and remove the protective shield from the pipette assembly with a twist.

3. Wipe the finger of your partner from which the blood sample is to be taken with an alcohol-moistened gauze square.

4. Hold the finger in a downward position, and prick it with a sterile lancet. Wipe away the first drop and gently squeeze the finger to provide a free flow of blood.

5. Holding the pipette almost horizontally, touch the tip of it to the forming blood drop (Figure 40-6b). The pipette will fill by capillary action. When the blood reaches the end of the capillary bore in the neck of the pipette, the filling action will stop. Carefully wipe any excess blood from the pipette's surface with a Kim wipe or similar wiping tissue.

6. Squeeze the reservoir slightly, to force out a small amount of air. While still maintaining pressure on the reservoir, cover the opening of the overflow chamber of the pipette with your index finger, and push the pipette securely into the reservoir neck (Figure 40-6c).

7. Release the pressure on the reservoir, and remove your finger from the pipette opening. Negative pressure will draw the blood into the diluent fluids.

8. Mix the contents of the reservoir chamber by squeezing the reservoir gently two or three times. (Squeeze gently so that the diluent is not forced out of the chamber.) Also invert the reservoir gently a few times (Figure 40-6d).

9. Remove the pipette from the chamber, reverse its position, and replace it on the reservoir (Figure 40-6e). This converts the system into a dropper assembly (Figure 40-6f).

10. Squeeze a few drops out of the system into a container, or wipe with tissue to clean the capillary bore. Now you are ready to fill the blood-counting chamber.

11. Wipe the glass cover of the hemocytometer, and position it as shown in Figure 40-7a.

12. Hold the end of the pipette, and squeeze the sides of the reservoir so that you can deposit a small drop of the diluted specimen onto the polished surface of the counting chamber next to the edge of the cover glass. *Be careful not to allow the chamber to overfill.* If the chamber overfills, charge (fill) the other side of the chamber.

13. Carefully place the charged hemocytometer on the microscope stage, and focus with the low-power objective, to bring the small (R) grid areas into clear view (Figure 40-1).

14. Then move the high-power objective into place,

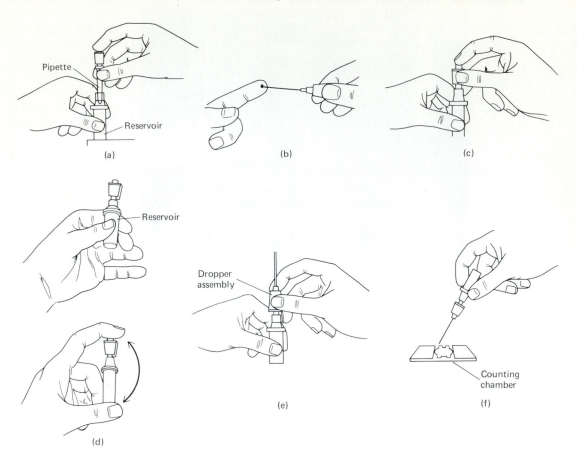

**Figure 40-6** Preparation of the Unopette Reservoir System: (a) puncturing the diaphragm; (b) obtaining a blood sample; (c) introducing the sample; (d) mixing the contents of the system by squeezing and inverting; (e) converting to a dropper assembly; (f) filling (charging) the blood-counting chamber. (Modified from *RBC Determination for Manual Methods and WBC Determination for Manual Methods,* Becton-Dickinson, Division of Becton, Dickinson and Company.)

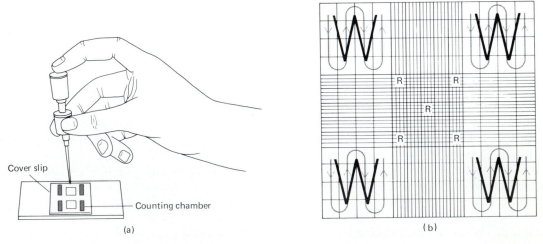

**Figure 40-7** The use of the hemocytometer: (a) Positioning the Unopette and filling the chamber—it is important not to overfill the chamber; a properly filled chamber has a blood specimen only within the space between the cover glass and the counting area. (b) This shows the direction to follow for white-cell counts.

and, with a hand counter or by other means of counting, determine the number of cells in each of the five specified areas. On the edges of the squares, count only the cells that touch the lines on the *left* and *top* sides.

15. Multiply the total number of cells counted in the five squares by 10,000, to obtain the red-blood-cell count for 1 mm.$^3$ Record your findings in the results and observations section.
16. Rinse the hemocytometer, wipe it dry, and repeat Steps 11–15. Was your second count similar to the first?
17. Repeat Step 16.
18. Have your partner repeat the entire procedure.
19. Dispose of all materials as directed by your instructor.
20. Answer the questions in the results and observations section.

## Procedure 2: White-Blood-Cell Count

This procedure is performed by students in pairs.

1. Follow steps 1–12 of Procedure 1, but use the Unopette chamber for white-blood-cell determinations. It has a white bottom surface.
2. Place the hemocytometer on the microscope stage. With the low-power objective, focus on the chamber area to bring the four large (W) corner regions into view (see Figure 40-1).
3. With a hand counter or by other means, determine the number of cells in each of the four specified areas. On the edges of the squares, count only the cells that touch the lines on the *left* and *top* sides.
4. Multiply the total number of cells by 50 to obtain the white-blood-cell count for 1 mm$^3$. Record your findings in the results and observations section.
5. Rinse the counting chamber, wipe it dry, and repeat Steps 1–4. Was your second count similar to the first?
6. Repeat Step 5.
7. Have your partner repeat the entire procedure.
8. Dispose of all materials as directed by your instructor.
9. Answer the questions in the results and observations section.

## THE HEMATOCRIT (PACKED CELL VOLUME)

### Materials

1. The following materials should be provided per four students:

   a. Commercially prepared individual sterile alcohol prep-pads
   b. Sterile, disposable blood lancets
   c. Sterile gauze pads (2 × 2 inch)
   d. Heparinized capillary tubes
   e. Clay-Adams Seal-Ease

2. The following materials should be provided for class use:

   a. Microhematocrit centrifuge
   b. Microhematocrit tube reader (Clay-Adams) or a substitute
   c. Containers of disinfectant
   d. Disposable surgical gloves

### Procedure

This procedure is to be performed by students in groups of four.

1. Laboratory partners should obtain blood specimens from one another. Wipe the finger from which the sample is to be taken with an alcohol-moistened sterile pad. Hold the finger in a downward position, and gently squeeze it. Puncture the ball of the finger with the lancet. Discard the first two drops of blood. (Refer to Exercise 5 for a description of the procedure.)
2. Place the marked (red) end of a heparinized capillary tube in the forming blood drop, and allow the blood to flow downward to fill about two-thirds of the tube (Figure 40-8a).
3. Put a finger over the unmarked end of the tube, and gently stick the red end into the Seal-Ease (Figure 40-8b). This seals the blood-containing end of the tube.
4. Place the prepared blood specimen into a slot of the microhematocrit centrifuge, with the sealed end of the capillary tube next to the rubber ring guard at the outer rim of the instrument (Figure 40-8c). Do not centrifuge the specimen until there are additional specimen tubes. There must be an even number of tubes for proper balance. Use the specimens of your laboratory partners.
5. Check to be sure that the centrifuge lid is properly fastened. A wrench is usually provided for this purpose.
6. Set the centrifuge time for 4 minutes, and turn on the instrument.
7. After centrifugation, place your capillary tube on or next to the tube reader, and determine the hematocrit reading. There are usually directions on the instrument to help you make your determination.
8. Record your findings, and answer the questions in the results and observations section.

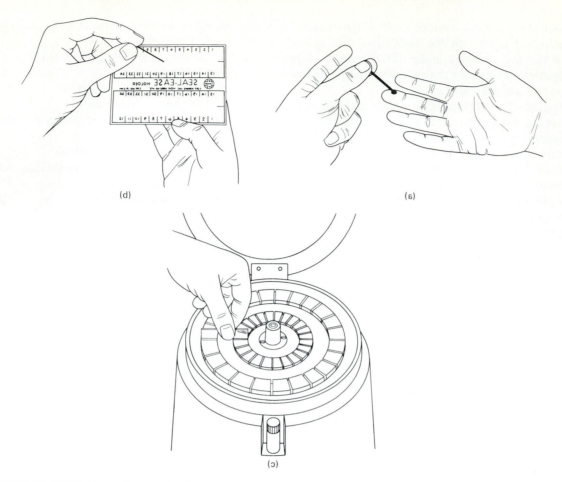

**Figure 40-8** General steps in the hematocrit procedure: (a) specimen collection; (b) sealing the capillary tube; (c) inserting the specimen into the hematocrit for centrifugation.

## RESULTS AND OBSERVATIONS

1. Enter the results of both red-blood-cell (RBC) and white-blood-cell (WBC) counts in Table 40-4.
2. Indicate the calculation formula for the following:
   a. RBC determinations

   _____

   b. WBC determinations

   _____

3. Enter the hematocrit determinations for members of your laboratory group in Table 40-5.

## COAGULATION (CLOTTING) TIME

### Materials

The following materials should be provided for class use:

1. Sterile alcohol prep-pads
2. Sterile, disposable blood lancets
3. Sterile gauze pads
4. Nonheparinized capillary tubes (0.5-mm diameter)
5. Stopwatch or other timing device
6. Glass files
7. Container with disinfectant for disposal of glass and other blood-containing items
8. Disposable surgical gloves

### Procedure

This procedure is to be performed by students in pairs.

1. Disinfect the surface of a finger with an alcohol prep-pad, and puncture it with the lancet, to produce a free flow of blood (as described in Exercise 5). Wipe away the first blood drop.
2. Hold the finger in a downward position, and place a nonheparinized capillary tube into and below the newly forming blood drop (Figure 40-9a). Allow the

**Table 40-4**

| Student | Sex | Number of RBCs Counted | RBC Count per mm³ | Number of WBCs Counted | WBC Count per mm³ |
|---------|-----|------------------------|-------------------|------------------------|-------------------|
|  |  |  |  |  |  |
|  |  |  |  |  |  |
|  |  |  |  |  |  |
|  |  |  |  |  |  |
|  |  |  |  |  |  |
|  |  |  |  |  |  |

tube to fill as completely as possible by capillary action. (Prevent air bubbles from forming by keeping the open end of the tube completely submerged in the blood drop.)

3. Have your partner start timing as soon as the tube is filled to capacity.
4. After 1 minute, use the file to make a small surface scratch near one end of the tube. Break the tube on the score line, and separate the small piece from the rest of the tube. Look for the presence of coagulation (Figure 40-9b). Continue to break off pieces of the tube at 1-minute intervals until coagulation is observed. Record your findings in the results and observations section.

**Table 40-5**

| Student | Sex | Hematocrit | Normal or Abnormal |
|---------|-----|------------|--------------------|
|  |  |  |  |
|  |  |  |  |
|  |  |  |  |
|  |  |  |  |

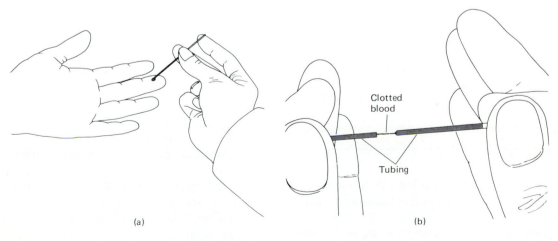

**Figure 40-9** Steps in determining coagulation time: (a) obtaining a specimen; (b) breaking portions of capillary tube to find coagulation time.

5. Change places with your laboratory partner, and repeat Steps 1–4. Answer the questions in the results and observations section.
6. Dispose of the capillary tube as directed by your instructor.

## PROTHROMBIN TIME (PT)

### Materials

The following materials should be provided for class use:

1. Simplastin or other commercial source of tissue thromboplastin
2. Plasma samples kept at 37°C
3. Plastic test tubes
4. 1-ml pipettes, graduated in 0.1 ml, with rubber bulbs
5. 37°C water bath with a rack for tubes
6. Stopwatch or other timing device
7. Container with disinfectant for disposal of pipettes and other blood-containing items
8. Disposable surgical gloves

### Procedure

This procedure is to be performed by students in pairs or in groups of four.

---

**CAUTION**

Working with blood or other body fluids will not pose any danger if the following precautions are observed:

1. Place any and all blood or materials containing blood into the disinfectant containers provided.
2. Dispose of any blood-containing materials as indicated by the instructor.
3. In cases of blood spillage, wipe the area with a disinfectant-soaked paper towel, and dispose of all materials as indicated by the instructor.
4. Always wash your hands after handling blood and associated materials.
5. The use of surgical gloves and protective eyewear.

---

1. Your instructor will review the use of a pipette.
2. Pipette 0.2 ml of the thromboplastin into a small test tube. Place the tube in the 37°C water bath.
3. Remove the tube containing the thromboplastin from the water bath, and pipette 0.1 ml of warmed plasma sample into it. Shake gently to mix.

4. Have your partner start timing as soon as the two materials are mixed.
5. Place the tube in the water bath again.
6. At least every 4 seconds, remove the tube, tilt it, and look for the formation of a fibrin clot.
7. Stop the procedure as soon as a clot appears. Record the time in the results and observations section.
8. Repeat Steps 2–7, and try to obtain a more accurate time for formation of the fibrin clot.
9. Answer the questions in the results and observations section.

## ACTIVATED PARTIAL THROMBOPLASTIN TIME (APTT)

### Materials

The following materials should be provided for class use:

1. Activated cephaloplastin
2. Calcium chloride ($CaCl_2$) 0.02 molar solution
3. Two plasma samples
4. Plastic test tubes
5. Test-tube racks
6. 1-ml pipettes, graduated in 0.1 ml, with rubber bulbs
7. 37°C water bath with a rack for tubes
8. Stopwatch or other timing device
9. Marking pens
10. Container for disposal of pipettes and other blood-containing items
11. Disposable surgical gloves

### Procedure

This procedure is to be performed by students in pairs or in groups of four.

1. Your instructor will review the use of pipettes.
2. Obtain three test tubes. Label two of the tubes "1" and "2," to correspond to the labels on the plasma samples; label the third tube "control."
3. With a 1-ml pipette, introduce 0.1 ml of activated cephaloplastin into each tube. Dispose of the pipette as indicated by the instructor.
4. With separate pipettes, introduce 0.1 ml of the plasma samples into the correspondingly labeled tubes. Dispose of the pipettes as indicated.
5. Incubate all tubes at 37°C for 3 minutes.
6. With another 1-ml pipette, introduce 0.1 ml of warmed $CaCl_2$ into each tube, and immediately start the timer.
7. Observe the tubes for the appearance of fibrin-thread formation.

8. Place the tubes in the 37°C water bath, and gently shake them continuously. Examine the tubes at 20-second intervals. Holding and tilting the tubes against a bright light helps to detect the fibrin threads.
9. Note the time required for the reaction to occur. Normal reaction time is approximately 40 seconds or less.
10. Record your findings in the results and observations section, and answer the questions pertaining to this exercise in the laboratory review section.

## RESULTS AND OBSERVATIONS

### Coagulation (Clotting) Time

1. Enter your findings and those of your laboratory partner in Table 40-6. Indicate the presence of coagulation with a "+" and its absence by a "−."
2. What is a normal clotting time for blood? _____

_____

3. Were the coagulation times of most students within normal limits? (Check the clotting times of at least 12 students before answering this question.) What

   variations did you observe? _____

   _____

4. List three factors that increase clotting time.
   a. _____
   b. _____
   c. _____

### PT

1. How long did it take for the plasma to clot? _____

_____

2. How would this test be performed with a blood sample? _____

_____

_____

3. List two factors that increase prothrombin time.
   a. _____
   b. _____

### APTT

1. Enter your findings in Table 40-7.
2. Were the test results easy to read and interpret?

_____

3. Were any abnormal reactions observed? If so, which

   ones? _____

   _____

4. If an individual had a defect in Factor X, would you

   expect to see an abnormal APTT reaction? _____

**Table 40-7**

| Reaction Time | Sample 1 | Sample 2 | Control |
|---|---|---|---|
|  |  |  |  |

## DETECTION OF FIBRIN DEGRADATION PRODUCTS (FDP)

### Materials

The following materials should be provided for the instructor's use:

1. Latex test suspension (contains polystyrene latex particles, coated with sheep antibodies against FDP)
2. Positive control serum
3. Negative control serum
4. Two unknown serum preparations, diluted 1:5 in buffer
5. Glycine saline buffer
6. Glass slides
7. Wooden applicator sticks
8. Clean Pasteur pipettes with rubber bulbs

**Table 40-6**

| Student | Coagulation Time (in minutes) | | | | | |
|---|---|---|---|---|---|---|
|  | 1 | 2 | 3 | 4 | 5 | 6 |
|  |  |  |  |  |  |  |
|  |  |  |  |  |  |  |

9. Wax pencil or other marking device
10. Container with disinfectant for blood-containing items

## Procedure

This procedure is to be demonstrated by the instructor.

1. Use the marking device to make and number four separate circles on a glass slide.
2. With separate Pasteur pipettes, place 1 drop of each specimen in the circles, as follows:
   a. Circle 1, unknown specimen 1
   b. Circle 2, unknown specimen 2
   c. Circle 3, known positive serum
   d. Circle 4, known negative serum
3. Gently shake the container of latex test suspension. Add 1 drop of the suspension to each circle.
4. Stir each combination with separate applicator sticks.
5. Rock the slide back and forth for 2 minutes, and observe for clumping (agglutination). Agglutination indicates the presence of more than 10 μg/ml of FDP.
6. Record your findings, and answer the questions in the results and observations section.
7. Dispose of slides and other items, as directed by your instructor.

## RESULTS AND OBSERVATIONS

1. Enter your findings in Table 40-8. Indicate the presence of clumping by a " + " and its absence by a " − ."
2. Which of the unknown samples produced a positive result? _____
3. What does FDP represent? _____
   _____

### Table 40-8

| Specimen | Presence of Agglutination |
|----------|---------------------------|
| Unknown 1 | |
| Unknown 2 | |
| Known positive serum | |
| Known negative serum | |

4. List three situations in which FDP determinations are useful.
   a. _____
   b. _____
   c. _____

## HEMOGLOBIN DETERMINATIONS

### Materials

The following materials should be provided for class use:

1. Commercial sterile alcohol prep-pads
2. Sterile, disposable blood lancets
3. Sterile gauze pads
4. Tallquist test paper
5. Tallquist reference scale
6. Hemo Cue hemoglobinometer
7. Hemo Cue microcuvettes
8. Kim wipes or other tissue material or lens paper
9. Isopropyl alcohol
10. Rubber globes (disposable surgical gloves)
11. Container of disinfectant

### Procedure 1: Tallquist Method

This procedure is to be performed by students in pairs.

1. Remove a piece of test paper from a Tallquist booklet and place it on a flat surface before you.
2. Obtain a blood sample (as described in Procedure 1, Exercise 5). Wipe away the first drop of blood.
3. Touch the second forming blood drop to the center of the Tallquist paper.
4. Wipe the finger, and apply a sterile gauze pad to stop any additional bleeding.
5. After 15 seconds, compare your blood sample with the Tallquist reference scale. Record your hemoglobin estimate in grams of hemoglobin per 100 ml of blood in the results and observations section. (Refer to Color Plate 45.)
6. Repeat Steps 1–5 with your laboratory partner.
7. Answer the questions in the results and observations section.

### Procedure 2: Hemoglobinometer Method

This procedure is to be performed by students in pairs.

1. Examine the hemoglobinometer and its parts (Figure 40-10a). Note the display window for hemoglobin measurements.

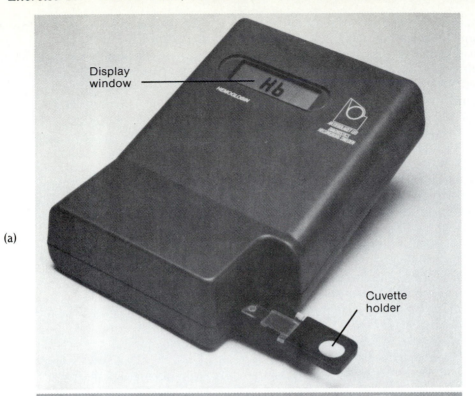

(a)

Display window

Cuvette holder

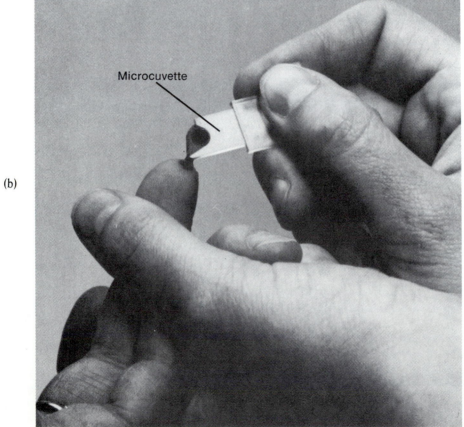

(b)

Microcuvette

**Figure 40-10** The *Hemo Cue*™ hemoglobinometer and specimen preparation: (a) the Hemo Cue and its parts (courtesy of Medical Equipment Designs, Inc., Laguna Hills, California); (b) filling the Microcuvette—note the positions of the fingers (courtesy of Medical Equipment Designs, Inc., Laguna Hills, California).

2. Have your partner put on a pair of rubber gloves before performing any blood work.
3. Obtain a blood sample (as described in Procedure 1, Exercise 5). Wipe away the first two drops.
4. Position the finger so that one large drop of blood forms. It should be big enough to fill the microcuvette completely.
5. Introduce the microcuvette's tip into the center of the blood drop, as shown in Figure 40-10b.
6. Fill the cuvette in one continuous process. Wipe off any excess blood on the outside surface of the cuvette. (*Note:* If a second sample is to be taken from the same finger, it should be done immediately after the first has been taken. Wipe away the remains of the first drop of blood, and take a second sample from a second drop.
7. Place the microcuvette in the cuvette holder, and push it into the Hemo Cue, to the stop point.
8. Wait 45 seconds, and read the hemoglobin value shown on the display window.
9. Enter your findings in the results and observations section. Note that the value is given in grams/liter.
10. Change places with your partner, and repeat Steps 2 through 9.

## RESULTS AND OBSERVATIONS

1. Enter your findings and those of your laboratory partner in Table 40-9.
2. Did the hemoglobin values obtained with both methods agree? _____
3. If not, what factors could account for the differences? _____

## SICKLE HEMOGLOBIN (HbS) ASSAY DEMONSTRATION

### Materials

The following items should be provided for the instructor's use:

1. One HemoCard for an Hbs assay (see Figure 40-11)
2. Three whole blood samples, to include
   a. One containing hemoglobin AS (heterozygous for sickle-cell trait)
   b. One containing hemoglobin SS (homozygous for sickle-cell trait)
   c. One normal specimen
   d. Two unknowns
3. Controls
   a. Hb AA Control (HbS negative)
   b. Hb AS Control (HbS positive)
4. Containers of the following preparations
   a. Sample conditioner
   b. HbS reagent
   c. Wash solution
5. Seven test tubes
6. Seven sample cups
7. One test tube rack or HemoCard HbS Assay Acrylic Workstation
8. One 1-ml pipette
9. Eleven pipettes (5 μl capacity) with delivery device
10. Eight disposable transfer pipettes (1 ml capacity)
11. One pair of disposable surgical gloves
12. Container for pipette disposal
13. Container of disinfectant
14. Marking pen

### Procedure

This procedure is to be demonstrated by the instructor (Figure 40-12).

1. Label one sample cup and one test tube with the following *codes,* to represent each of the different preparations to be used in the assay.
   a. *AS* (for heterozygous sickle-cell-trait blood sample)
   b. *SS* (for homozygous sickle-cell-trait blood sample)
   c. *N* (for normal sample)
   d. *1* (for unknown blood, Sample 1)
   e. *2* (for unknown blood, Sample 2)
   f. *CAA* (for HbAA control)
   g. *CAS* (for HbAS control)

**Table 40-9**

| Student | Sex | Hemoglobin Value, Tallquist (in g/100 ml) | Hemoglobin Value, Hemoglobinometer (g/l) |
|---|---|---|---|
|  |  |  |  |
|  |  |  |  |

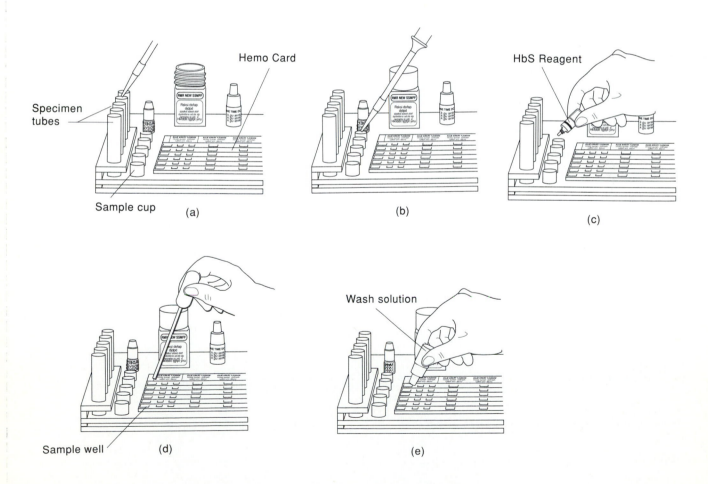

**Figure 40-11** The HemoCard. (After IsoLab Inc. Akron, Ohio, 44321.)

**Figure 40-12** Steps in the sickle hemoglobin S(HbS) detection procedure.

2. With a 5-ml pipette, add 2 ml of the sample conditioner to each test tube (Figure 40-12a).
3. With separate μl pipettes, add 2.0 μl of each blood sample and each control to their respectively labeled test tubes (Figure 40-12a).
4. With separate μl pipettes, transfer 5 μl of each sample and control to their corresponding sample cups (Figure 40-12b).
5. Next, squeeze 1 drop of the HbS reagent into each sample cup (Figure 40-12c).
6. Allow the samples to stand for 2 minutes.
7. During this time, fill out the HemoCard by entering the date and the code for each sample onto their respective places.
8. With separate disposable plastic pipettes, transfer the contents of the individual sample cups to the appropriate sample well on the HemoCard (Figure 40-12d).
9. Place the pipettes into the container of disinfectant.
10. Add 1 drop of wash solution to each sample well, and observe the results (Figure 40-12e). A pink to red color is a positive result for HbS.
11. Record your findings, and answer the questions in the results and observations section.

## RESULTS AND OBSERVATIONS

1. Enter your findings in Table 40-10. Indicate samples positive for HbS by a " + ," and with " − " for a negative one.
2. Which, if any, of the unknown blood samples showed

   a positive result for HbS? _____

**Table 40-10**

| Sample | Color Reaction | HbS (+/−) |
|--------|----------------|-----------|
| AS | | |
| SS | | |
| 1 | | |
| 2 | | |
| N | | |
| CAA | | |
| CAS | | |

# LABORATORY REVIEW 40

NAME _____

LAB SECTION _____ DATE _____

## BLOOD CELL COUNTS, THE HEMATOCRIT (PACKED CELL VOLUME), COAGULATION TIME, AND HEMOGLOBIN DETERMINATIONS

### COMPLETION, DEFINITION, AND INTERPRETATION

1. a. Are human red blood cells nucleated? _____

   b. Are all mammalian red blood cells nucleated? _____

2. Describe human erythrocytes by filling in the following information.

   a. Function _____

   b. Shape _____

   c. Diameter _____

   d. Average life span _____

   e. Normal number per mm$^3$ in the adult female _____

   f. Normal number per mm$^3$ in the adult male _____

3. a. What is hemoglobin? _____

   _____

   b. What heavy metal has as its chief function the synthesis of hemoglobin? _____

   _____

4. What is anemia? _____

   _____

5. List three causes of anemias.

   a. _____

   b. _____

   c. _____

6. Where are red blood cells formed in the human adult? _____

   _____

7. What is the study of blood called? _____

8. What is a hemocytometer? _____

   _____

9. In Table Q40-1, determine the final RBC counts, and indicate whether the results are normal or abnormal.

**Table Q40-1**

| Microscopic Count | Sex | RBC Count (per mm³) | Normal or Abnormal |
|---|---|---|---|
| 1000 | Male | a. | b. |
| 300 | Female | c. | d. |
| 150 | Female | e. | f. |
| 195 | Male | g. | h. |

10. In Table Q40-2, determine the final WBC counts, and indicate whether the results are normal or abnormal.

**Table Q40-2**

| Microscopic Count | Sex | WBC Count | Normal or Abnormal |
|---|---|---|---|
| 900 | Male | a. | b. |
| 100 | Female | c. | d. |
| 500 | Female | e. | f. |
| 62 | Male | g. | h. |

11. Are the dilution pipettes for RBC and WBC counts different? Explain. _____
_____

12. How many and which squares are counted in an RBC count? _____
_____

13. How many and which squares are counted in a WBC count? _____
_____

14. What is a hematocrit? _____
_____

15. What are the percentages of the components of a blood sample in a normal hematocrit? _____
_____

16. The results of 12 hematocrit determinations are listed in Table Q40-3. Indicate whether they are normal or abnormal.

**Table Q40-3**

| Adult Male Patient | Hematocrit Value | Normal or Abnormal | Adult Female Patient | Hematocrit Value | Normal or Abnormal |
|---|---|---|---|---|---|
| 1 | 36 | a. | 1 | 32 | g. |
| 2 | 45 | b. | 2 | 40 | h. |
| 3 | 54 | c. | 3 | 61 | i. |
| 4 | 63 | d. | 4 | 47 | j. |
| 5 | 42 | e. | 5 | 22 | k. |
| 6 | 40 | f. | 6 | 37 | l. |

17. Distinguish between plasma and serum. _____

_____

_____

18. What is the normal hematocrit value for the following?

   a. Adult female _____

   b. Adult male _____

19. Describe four laboratory tests associated with the clotting activity of blood.

   a. _____

   b. _____

   c. _____

   d. _____

20. Distinguish between the intrinsic and the extrinsic pathways of clotting. _____

_____

_____

_____

21. How many factors are known to be involved with the blood-coagulation process?

_____

22. List the stages of blood coagulation, and indicate which factors are associated with each.

   a. _____

   b. _____

   c. _____

23. Which of the blood-clotting factors require vitamin K for their formation? _____

_____

24. What are platelets, and where are they formed? _____

_____

25. Indicate the PT and APTT reactions that can be expected with defects of the clotting factors listed in Table Q40-4.

**Table Q40-4**

| Defective Clotting Factor | PT Reaction | APTT Reaction |
|---|---|---|
| VIII | | |
| V | | |
| X | | |
| XI | | |
| XII | | |
| IX | | |
| VII | | |

26. What is the physiological function of hemoglobin? _____

27. How is the hemoglobin content of blood usually expressed? _____

28. Give the normal hemoglobin content ranges for the following:

    a. Adult female _____

    b. Adult male _____

    c. 4-year-old child _____

    d. 1-year-old child _____

29. What is sickle-cell anemia? _____

**MULTIPLE CHOICE**

Insert the correct answer in the spaces provided:
1. What is the total blood volume of an adult human (in liters)?
    a. 5
    b. 6
    c. 8
    d. 2
    e. 2
2. What is the percentage of normal blood that consists of formed elements?
    a. 55
    b. 75
    c. 45
    d. 61
    e. 16

1. _____
2. _____
3. _____
4. _____
5. _____
6. _____
7. _____
8. _____
9. _____
10. _____
11. _____
12. _____

3. What is the percentage of plasma found in blood?
   a. 55
   b. 75
   c. 45
   d. 61
   e. 16

4. What is the usual range for the normal average RBC count, per cubic millimeter?
   a. 4.5–5.5 million
   b. 25–30 trillion
   c. 7000–8000
   d. 10,000–11,000
   e. 5000–6000

5. In a healthy adult human, what is the normal number of circulating WBCs per mm$^3$?
   a. 4.5–5.5 million
   b. 25 trillion
   c. 7000
   d. 10,000
   e. 5000

6. Which of the following procedures is used to determine the ratio of RBCs to the whole body by volume?
   a. hemocytometer
   b. coagulation time
   c. activated partial thromboplastin time
   d. hematocrit
   e. fibrin degradation products determination

7. What would a packed-cell-volume value of 49 for an adult male indicate?
   a. a significant blood loss
   b. an anemia
   c. polycythemia
   d. a malignancy
   e. a normal result

8. In the coagulation process, which of the following substances is converted to fibrin?
   a. thromboplastin
   b. fibrinogen
   c. thrombin
   d. prothrombin
   e. choices *a* and *b* only

9. In the coagulation process, which of the following substances is converted to thrombin?
   a. thromboplastin
   b. fibrinogen
   c. prothrombin
   d. fibrin
   e. proconvertin

10. Which of the following tests is used to detect deficiencies in the extrinsic clotting system?
    a. bleeding time
    b. platelet count
    c. prothrombin time
    d. activated partial thromboplastin time
    e. clot dissolution

11. Which of the following tests is used to detect deficiencies in the intrinsic clotting system?
    a. bleeding time
    b. platelet count

    c. prothrombin time

    d. activated partial thromboplastin time

    e. clot dissolution

12. Which of the following tests is used to check primary clot formation?

    a. bleeding time

    b. platelet count

    c. prothrombin time

    d. activated partial thromboplastin time

    e. clot dissolution

# Lymphatics and the Immune Response

**After completing this exercise,
you should be able to**

1. Perform the slide blood-typing procedure
2. Understand the basis of blood typing and cross-matching
3. Identify major blood group antigens in unknown blood samples
4. Define *electrophoresis*
5. Outline a general procedure for electrophoresis
6. Use electrophoresis to separate and to identify some specific serum proteins
7. Use electrophoresis to identify abnormal hemoglobin molecules

The human body defends itself against foreign and invading agents by means of its immune system, which consists of two closely allied forms of immune responses. One of these responses, known as *cellular (cell-mediated) immunity,* involves the formation of special lymphocytes, termed T cells, that are capable of attacking foreign factors and cells and destroying them. Cell-mediated immunity is particularly effective against various types of microorganisms, worms, virus-infected cells, cancer cells, and foreign (to the body) tissue transplants. The second type of immune response, known as *humoral (antibody-mediated) immunity* involves antibodies that are present in blood plasma and lymph (the body's fluids or *humors*). In humoral immunity, another type of specialized lymphocytes, termed B cells, develop into plasma cells that produce antibodies known as *immunoglobulins.* This form of immunity is particularly effective against bacterial and viral disease agents.

Both types of immunity are essential for an effective immune system, and both originate in lymphoid tissue located in the lymph nodes, spleen, gastrointestinal tract, and bone marrow (Color Plate 46). Lymphoid tissue is strategically located to intercept a foreign and/or invading agent before it can spread too far in the body and cause extensive injury. This exercise uses blood typing to show antibodies in action against foreign factors known as antigens. An *antigen* is any chemical substance recognized by the immune system as being foreign and which causes the body to produce specific immunoglobulins and/or T cells that react with the antigen.

The second portion of this exercise uses electrophoresis to show the presence of antibodies in a blood specimen. Antibodies belong to a group of plasma *globulins,* specifically termed gamma globulin.

It is for this reason that antibodies are also referred to as immunoglobulins *(Igs).* There are five different classes of immunoglobulins, designated as *IgG, IgA, IgM, IgD,* and *IgE.* Each has a distinct chemical structure and a specific biological role.

In 1900, Karl Landsteiner observed that the mixing of red blood cells and sera from two different human donors, followed by incubation at body temperature, resulted in the clumping or agglutination of the red cells (Figure 41-1). This phenomenon, known as *isohemagglutination,* enabled Landsteiner and others to demonstrate the presence of every individual's red blood cells' specific, genetically determined antigens, termed *agglutinogens.* At present, more than 300 blood factors are known to exist in the human species. Examples include A, B, O, AB, M, N, S, $Rh_o$, $Hr_o$, and rh'. The major classification of blood types is the ABO blood grouping system (Figure 41-2). Blood types are based on the presence of these factors. For example, an individual whose red cells contain only agglutinogen B (of the ABO group of factors) is said to have blood type B.

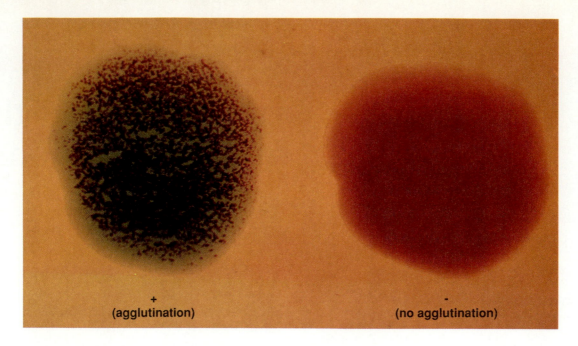

**Figure 41-1** Blood typing results, showing agglutination (positive) or lack of agglutination (negative).

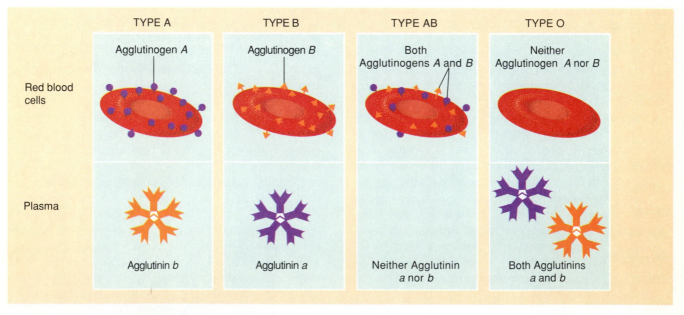

**Figure 41-2** Agglutinogens (antigens) and agglutinins (antibodies) involved in the ABO blood-grouping system. (From G. J. Tortora and S. R. Grabowski, *Principles of Anatomy and Physiology,* 7th ed., New York: HarperCollins, 1993.)

In addition to having specific blood-group antigens on red cells, the blood of some humans contains antibodies that react with the erythrocytes from individuals of the other blood types, causing them either to agglutinate or to lyse. These antibodies, termed *agglutinins,* are found in the blood plasma and/or serum. Usually, the antibodies of an individual are not directed against the factors normally present in that individual's blood. Most individuals have antibodies against factors that are absent from their blood cells. For instance, a person of blood type O has antibodies against agglutinogen A (termed *anti-A, antibody a*) and against agglutinogen B (*anti-B, antibody b*). Figure 41-2 shows the characteristics of the major blood types. Antibody concentrations can be increased as a result of transfusions or in the case of women who have borne children possessing different blood characteristics.

Because AB individuals do not possess antibodies against any of the other major blood types, they are referred to as *universal recipients.* Because type O blood cells have no agglutinogens to serve as antigens in another person's blood, members of this group are referred to as *universal donors.*

In 1937, Landsteiner and A. S. Wiener uncovered antisera that clumped (agglutinated) rhesus monkey blood, as well as the cells of 85 percent of Caucasians. This phenomenon was independent of the reactions obtained with the four major blood types. The factor was designated *Rh* for "rhesus." Several other factors subsequently were found to be involved in a complex Rh–Hr system. These factors are now recognized as the principle cause of *erythroblastosis fetalis* (hemolytic disease of the newborn) and as the cause of some other transfusion reactions.

The determination of an individual's blood group can be made by testing his or her red cells with standardized anti-A or anti-B sera. The mixtures are observed for the presence or absence of agglutination. Table 41-1 shows the patterns of reactions that are used for blood-type determinations.

Determining the Rh factor of an individual's red cells is obtained by agglutination tests using the appropriate antisera. The classic and clinically most important of the factors is $Rh_o$ (also called "D," using a different method of categorizing blood groups).

Both false-positive and false-negative reactions may be encountered in testing. To prevent this, suitable control measures are used; These include (1) pretesting of sera for activity, (2) dilution of cells or sera in physiological saline if a reaction has not occurred after a sufficient length of time, and (3) incorporation of tests with known cells and sera.

The blood-group antibodies are of significant importance in cases of transfusions. The procedure known as *compatibility testing* or *cross-matching* is carried out to prevent a transfusion reaction as a consequence of blood incompatibilities between the recipient and the donor. Specifically, the cross-matching technique is designed to detect any incompatibility between the recipient's serum and the donor's cells (known as the *major cross-match*) or between the recipient's cells and the donor's serum (the *minor cross-match*). After the blood specimens have been typed as to the ABO, Rh (or D), and any other factors that appear to be indicated, different mixtures of donor's and recipient's cells and sera are made. Agglutination or lysis of the red cells from either the donor or the recipient is considered to be an incompatible situation.

In certain clinical conditions, which include hemolytic anemia, incompatible transfusions, and hemolytic disease of the newborn, unusual antibodies are present on the surface of blood cells. These globulins are not detected in major or minor cross-matches. The detection of such antibodies is accomplished by the Coombs' or antiglobulin test. The test employs the use of antihuman globulins (Coombs' reagent). This diagnostic material is prepared by successive injections of human serum into a suitable laboratory animal, such as a rabbit or goat. The serum of the immunized animal will form antibodies against the globulin of human serum. This material is collected and used in the detection of human antibodies on the surface of RBCs. The procedure known as the direct Coombs' test is performed by mixing RBCs with the Coombs' reagent. Agglutination of the cells is a positive test demonstrating the presence of human antibodies. The direct Coombs' test is demonstrated in this exercise.

Medicolegal aspects of blood grouping are important. Standard blood-grouping procedures provide information, such as the compatibility of blood donors and recipients, identification of blood stains, detection of rare blood types, potential dangers in pregnancy, and facts that help decide in cases of disputed parentage. In this exercise, the procedures for blood typing and cross-matching are considered.

**Table 41-1  Reaction Patterns for Blood-Type Determinations**

| Blood Type | Agglutination When Mixed with Antiserum | |
| --- | --- | --- |
| | anti-A | anti-B |
| A | + | − |
| B | − | + |
| AB | + | + |
| O | − | − |

Electrophoresis is an analytical technique in which charged molecules in solution, mainly proteins and nucleic acids, migrate in response to an electrical field (Figure 41-3). The rate of movement of the molecules depends on several factors, including the net electrical charge, the size and shape of the molecules, the strength of the electrical field, and the ionic, strength, thickness, and temperature of the *medium* (material substance) through which the molecules are moving. Electrophoresis is a simple, rapid, and highly sensitive technique used to identify and study the properties of a single charged molecule and to separate a mixture of charged molecules. Refer to Color Plate 47. Proteins or unusual molecules found in serum are the most frequently studied substances.

Electrophoresis may be carried out in a supporting medium or matrix, such as paper, cellulose acetate, or gels formed in tubes, slabs, or flat beds. In most electrophoretic units, the gel is mounted between two buffer chambers containing separate electrodes (see top left of Figure 41-4). With this arrangement, the only electrical connection between the chambers is through the gel.

After the electrophoresis of a protein or nucleic acid sample is completed, the gel or other supporting matrix can be analyzed by one of several methods. The most common analytical procedure is staining. Protein gels are frequently stained by Coomassie blue or Ponceau S. Nucleic acids are usually stained with ethidium bromide, a fluorescent dye that produces an orange glow when exposed to ultraviolet light. This exercise considers the use of electrophoresis with serum proteins.

---

### CAUTION

Working with blood or other body fluids will not pose any danger if the following precautions are observed:

1. Place any and all blood or materials containing blood into the disinfectant containers provided.
2. Dispose of any blood-containing materials as indicated by the instructor.
3. In cases of blood spillage, wipe the area with a disinfectant-soaked paper towel, and dispose of all materials as indicated by the instructor.
4. Always wash your hands after handling blood and associated materials.
5. The use of disposable surgical gloves

---

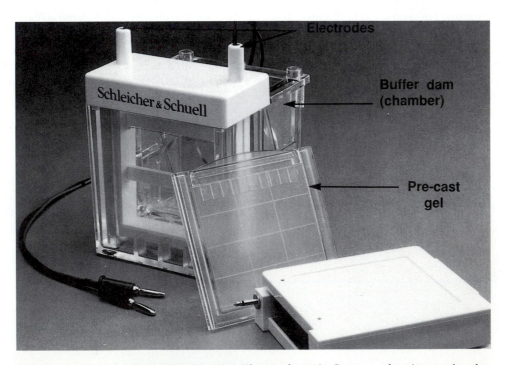

**Figure 41-3** Complete S&S EP Mini-Electrophoresis System, showing tank, electrodes, and precast gel. (Courtesy of Schleicher & Schuell Inc., Keene, New Hampshire.)

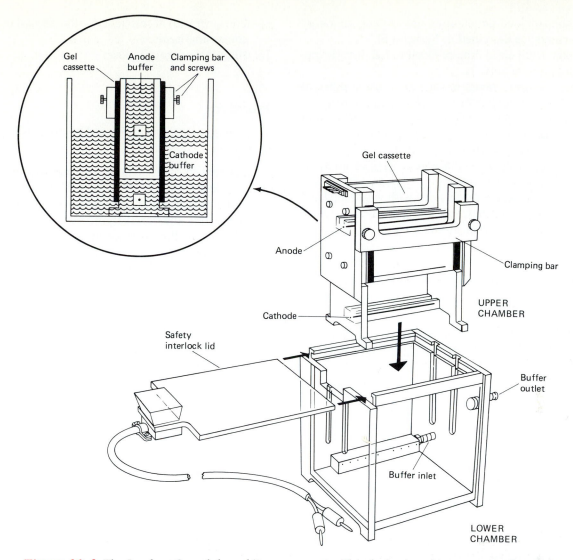

**Figure 41-4** The Isophore® module and its components. This device is quite compact and straightforward in operation. Note the two chambers and their respective parts. The insert at the upper left-hand corner shows a side view of two gel cassettes held in place by clamping bars and screws. (After Isolab, Inc.; RESOLVE®, Isophore®, Isofocus®, and Isolytes® are registered trademarks of Isolab, Inc.)

## BLOOD TYPING

### Materials

1. The following materials should be provided per each four students:
   a. Commercially prepared sterile alcohol prep-pads
   b. Sterile gauze pads (2 × 2 inch)
   c. Sterile blood lancets
   d. Eight glass slides
   e. Wooden applicator sticks
   f. Container for the disposal of the lancets and applicator sticks

2. The following materials should be provided for class use:
   a. Blood typing antisera—anti-A, anti-B, and anti-D ($Rh_o$)
   b. Unknown blood specimens
   c. Disposable surgical gloves

### Procedure 1: ABO and $Rh_o$ Determinations

This procedure is to be performed by students in groups of four.

1. Place 1 drop of each antiserum on separate areas of the blood-typing slide.

2. Obtain a blood sample from one of your laboratory partners (as described in Exercise 5).
3. Allow 1 full-sized drop of blood to fall directly into each drop of antisera.
4. Mix the components in each area, using different ends of the applicator sticks for each mixture.
5. If a reaction does not develop after 2 minutes, warm the slide over a desk lamp or similar device, and gently shake the slide from side to side.
6. Refer to Table 41-1 and Figure 41-1 to interpret the reactions obtained.
7. Record the reactions obtained with your blood and your laboratory partner's in Table 41-2 of the results and observations section.

### Procedure 2: Unknown Blood Sample

1. Obtain an unknown blood sample from your instructor.
2. Determine the major blood type and Rh factor or factors, using Procedure 1 of this exercise and Figure 41-1 as a guide.
3. Record your findings in the results and observations section. Check your results with the instructor.

## CROSS-MATCHING

### Materials

The following materials should be provided for class use: (*Note:* All plasma or serum preparations should be heated to 56°C for 30 minutes prior to use.)

1. Donor blood cell preparation, type A, in physiological saline
2. Donor blood cell preparation, type O, in physiological saline
3. Recipient plasma or serum preparation, type A
4. Recipient plasma or serum preparation, type O
5. Two recipient plasma or serum preparations, blood type unknown
6. Recipient blood-cell preparation, type A, in physiological saline
7. Recipient blood-cell preparation, type O, in physiological saline
8. Donor plasma or serum preparation, type A
9. Donor plasma or serum preparation, type O
10. Two donor plasma or serum preparations, blood type unknown
11. Glass slides
12. Pasteur pipettes with rubber bulbs
13. Wooden applicator sticks

14. Container with disinfectant, for the disposal of blood-containing items.
15. Disposable surgical gloves

### Procedure

This procedure is to be performed by students in groups of four.

1. Perform the four major cross-matches, using the combinations of donor-cell preparations and recipient serum preparations indicated in Table 41-3 in the results and observations section.
2. Observe for the presence of agglutination. Record your findings, and answer the questions in the results and observations section.
3. Perform the four minor cross-matches using the combinations of recipient cell preparations and donor serum preparations indicated in Table 41-4 in the results and observations section.
4. Observe for the presence of agglutination. Record your findings, and answer the questions in the results and observations section.

## DEMONSTRATION OF THE DIRECT COOMBS' TEST

### Materials

The following materials should be provided for class demonstration:

1. One saline-washed suspension of sensitized RBCs (coated with human gamma globulins)
2. One saline-washed suspension of nonsensitized RBCs
3. One saline-washed suspension of unknown RBCs
4. Coombs' reagent
5. Glass slides
6. Pasteur pipettes with bulbs
7. Wooden applicator sticks
8. Marking pen
9. Container for the disposal of applicator sticks and pipettes
10. Disposable surgical gloves

### Procedure

The instructor will demonstrate this procedure.

1. Label three separate areas of a glass slide, "+," "−," and "X" (for an unknown). Make the markings near the edge of the slide.

2. Place 1 drop of each blood-cell suspension near its labeled area.
3. Add 1 drop of the Coombs's reagent to each blood drop.
4. Mix the components in each area, using a different applicator stick for each combination.
5. Examine the slide for agglutination during a 3-minute period.
6. Answer the questions in the results and observations section.

## RESULTS AND OBSERVATIONS

### Blood Typing

1. Enter your findings regarding the ABO and $Rh_o$ determinations in Table 41-2. Indicate agglutination by a plus sign and no agglutination by a minus sign.
2. Enter your findings on the unknown blood sample.

   a. Sample number _____

   b. ABO type _____

   c. $Rh_o$ type _____

### Cross-Matching

1. Enter your findings for the major cross-matches in Table 41-3. Indicate an incompatibility by a plus sign and compatibility by a minus sign.
2. Which of the unknown combinations for the major cross-matches showed incompatibility? _____
3. Enter your findings for the minor cross-matches in Table 41-4.
4. Which of the combinations for the minor cross-matches showed incompatibility? _____
5. Give at least two possible causes for the incompatibilities observed. _____
_____

### Direct Coombs' Test

1. What reaction indicates a positive reaction?
_____
2. Was the unknown specimen positive or negative?
_____
3. What is the purpose of the Coombs' test?
_____
_____

**Table 41-2**

| Student | Reactions with Antisera | | | Blood Type |
|---|---|---|---|---|
| | anti-A | anti-B | anti-D ($Rh_o$) | |
| | | | | |
| | | | | |
| | | | | |
| | | | | |

**Table 41-3**

| Donor Cells Blood Type | Recipient Blood Type | | | |
|---|---|---|---|---|
| | A | O | Unknown 1 | Unknown 2 |
| A | | | | |
| O | | | | |

**Table 41-4**

| Recipient Cells Blood Type | Donor Blood Type | | | |
|---|---|---|---|---|
| | A | O | Unknown 1 | Unknown 2 |
| A | | | | |
| O | | | | |

## PLASMA PROTEIN DETERMINATIONS

### Materials

The following materials should be provided for general class use:

1. Isophore® Universal Gel Electrophoresis system or other electrophoresis unit
2. Silicon grease or similar lubricant
3. Electrophoresis basic buffer (Isophore®)
4. Electrophoresis acidic buffer (Isophore®)
5. Stacking buffer (Isophore®)
6. Application buffer (Isophore®)
7. Fixative Ponceau S dye solution
8. Acetic acid rinse solution
9. Cleaning solution
10. Phenol red pH indicator (0.1 percent)
11. Power source constant current
12. Separate trays for butter, dye, rinse, and cleaning solutions
13. Glass vials or tubes (small capacity)
14. Test tube racks
15. Disposal surgical gloves
16. 1-ml pipettes, graduated in 0.1 ml, with rubber bulbs
17. Pasteur pipettes with rubber bulbs
18. Single-edge razor blades or scalpels
19. Transparent tape (1-inch wide)
20. Spatulas
21. Flat, blunt-end forceps
22. Filter paper
23. Marking pencil

The following materials should be provided for students in groups of four, or as indicated by the instructor:

1. Normal plasma sample (0.5 ml)
2. Three unknown plasma samples (0.5 ml each)
3. Nine 5-lambda pipettes or microsyringes or micropipettes
4. One gel cassette (Isofocus® survey gel, pH 3–10)
5. One sample spacer comb (Isophore®)
6. One set of drying weights (to be used during drying of gel after electrophoresis run)

### Procedure 1: Electrophoresis Unit and Accessories

This procedure is to be performed by students in groups of four.

1. Examine the Isophore® unit (or other electrophoresis system provided), power unit, and accessories.
2. Refer to Figure 41-4, and identify the following components of the Isophore® unit:
   a. Anode in the upper chamber
   b. Upper chamber support containing the cathode
   c. Safety-interlocking lid (prevents contact with the buffers and platinum electrodes)
   d. Gel cassettes that form the side walls of the upper chamber
3. Examine the power unit and accessories provided.

### Procedure 2: Gel Preparation

This procedure is to be performed by students in groups of four.

1. Remove one Isofocus® gel from its package. Shake any excess liquid from the surface. (Do not separate the glass or plastic plate from the gel before performing the electrophoresis run. Note that the gel is held between a glass plate and a notched plastic plate. A notched line, which extends along the lower edge of the plastic plate, prevents the gel from sliding out of its position.
2. Examine the gel, to be sure that the plastic spacers on both sides of the gel are in direct contact with it. If they are not, make a small slit in the tape, and gently push the spacer with a spatula until it touches the gel (Figure 41-5a).
3. Use a Pasteur pipette to fill the space above the gel with a stacking buffer (see Figure 41-5b).
4. Place the sample spacer comb on top of the gel so that the tips of the comb enter the gel surface 1 to 2 mm (Figure 41-5c). When the sampler spacer is pressed into the gel, 9 separated sample wells are formed.

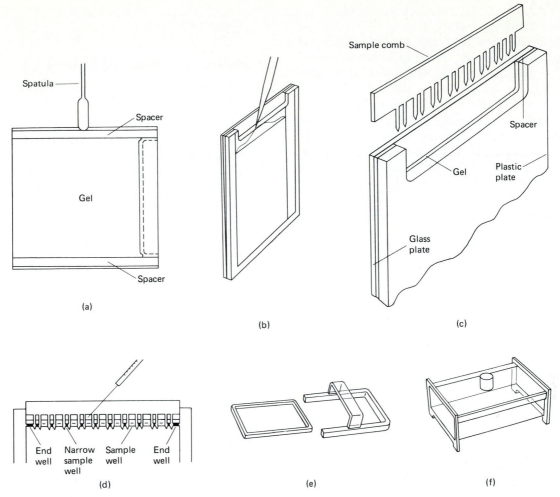

**Figure 41-5** Steps in sample application and an electrophoresis run: (a) correcting the spacers; (b) applying the stacking buffer; (c) the Isofocus® gel and sample-containing comb; (d) applying a sample with a microsyringe—each end well and each narrow sample well receives application buffer only; (e) transfering gel to transporter and tray; (f) fixation and staining. (From G. A. Wistreich, *Laboratory Manual for Human Physiology,* New York: Harper & Row, 1986.)

5. Place the upper chamber of the electrophoresis system on a convenient working surface. Loosen the screws on each side of the machine so that there is space between the gaskets. (Refer to Figure 41-4.)

6. Coat the gaskets with silicon grease.

7. Place one gel cassette between each pair of gaskets. If only one gel is to be run, remove one clamping bar, and replace it with a blanking plug. The notched plastic plate should be placed so that it faces toward the center of the upper chamber.

8. Tighten the screws in the clamping bar until each cassette is held snugly in place. Use a light touch when clamping—too much pressure will crack the glass plate.

9. Fill the lower chamber with the cathode (basic) buffer. The buffer level should be about 8.0 cm from the top of the chamber. About 1500 ml of buffer will be needed.

10. Carefully slide the upper chamber into the lower chamber.

11. Check the top and bottom of the gel cassette for air bubbles. Remove any bubbles by stirring the buffer or flushing with buffer (a syringe works well for this). The buffer must be in direct contact with the entire bottom edge of the gel. *It is important that all air bubbles be eliminated.*

12. Pour the anode (acidic) buffer into the upper chamber.

## Procedure 3: Sample Preparation

This procedure is to be performed by students in groups of four.

1. Obtain 4 small vials, and label them "N" (for normal plasma sample) and "1," "2," and "3" (for each of the unknown samples).

2. With a 1-ml pipette, introduce 0.2 ml of the application buffer to each vial.
3. With individual microsyringes or micropipettes, add 50 μl of each protein sample to its labeled tube.
4. With a new microsyringe, add 4 to 5 μl of application buffer to each narrow well and to both end wells of the sample comb (Figure 41-5d).
5. With individual micropipettes, introduce the protein sample into the sample wells, as follows: To introduce a sample, insert the micropipette or syringe between the teeth of the sample comb so that the tip of the micropipette or syringe is immediately above the gel. Gently express the sample onto the gel. In this electrophoretic gel, Samples 1, 2 and 3 will be tested in duplicate.
   a. Add 5 μl of Sample N to the first sample well on the left. (See Figure 41-5d. The end well and first narrow well have been filled with application buffer.)
   b. Add 5 μl of Sample 1 to the next sample well (second from left).
   c. Add 5 μl of Sample 2 to the third sample well from left.
   d. Add 5 μl of Sample 3 to the fourth sample well from left.
   e. Repeat 5-μl additions of Samples 1, 2, and 3 to the next three sample wells.
6. With a new syringe, apply 1 drop of the 0.1 percent phenol-red dye solution to the sample well containing sample N (the normal plasma sample). This chemical is a tracking dye, to show the movement in the system.
7. The samples are now ready for electrophoresis.

## Procedure 4: Electrophoretic Run

This procedure is to be performed by students in groups of four.

1. Slide the safety interlock lid over the Isophore® unit, enclosing both the chambers (see Figure 41-4). Connect the power cables (generally, black = negative, cathode, and lower chamber; red = positive, anode, and upper chamber).
2. Turn on the power, and adjust the voltage to 200 volts on the power setting. The milliamperes (mA) should read 3.5 to 5.8 mA.
3. With no more than eight samples, constant voltage is applied for 60 minutes.
4. Continue the electrophoretic run until separation is complete. (Run times may generally be shortened by increasing the voltage.)
5. Upon completion of the run, turn off the power supply, remove the safety lid, and pull out the upper chamber.

6. Pour out the anode buffer, and remove the gel for fixation and staining.

## Procedure 5: Fixation and Staining

This procedure is to be performed by students in groups of four.

1. After electrophoresis, open the cassette by slitting the tape on one side with a single-edge razor blade.
2. With the aid of a spatula, pry apart the two plates. Remove the plastic plate, leaving the gel on the glass plate.
3. Remove the sample spacer comb and the side spacers. To remove the comb, loosen the top screws on each clamp. Then pull up gently on the comb while rocking it back and forth. Air bubbles should be visible underneath the comb teeth.
4. Float the gel off the glass plate onto the transporter (Figure 41-5e). Place the holding screen over the gel to prevent it from floating away during the staining and destaining steps.
5. Place the gel in the tray containing the fixative Ponceau S dye solution for 7 to 10 minutes (Figure 41-5f).
6. Lift the transporter from the stain tray. Allow excess stain to drain back into the tray.
7. Transfer the transporter to the acetic acid rinse tray. Change the acetic acid 2 to 4 times, or until the dye no longer comes off the gel.
8. Transfer the gel to the clearing solution, and gently agitate the gel for about 1 minute.
9. Remove the gel and place it on a glass plate. Squeegee the gel gently to remove excess solution, and allow it to air dry. Place a drying weight on each edge of the gel, to prevent curling.
10. Place a clean glass plate over the dried gel, making a sandwich.
11. Place tape over the edges of the plates.
12. Label the tape to identify the different specimens used.

## Procedure 6: Examination

This procedure is to be performed by students in groups of four.

1. Examine the stained gel, and identify in each sample the plasma proteins, albumin, and the alpha, beta, and gamma globulins. Refer to Color Plate 47.
2. Sketch and label a representative electrophoretic pattern for one plasma protein sample in the space provided in the results and observations section.
3. Answer the questions in the results and observations section.

## RESULTS AND OBSERVATIONS

1. Sketch an electrophoretic pattern for one plasma protein sample in the space provided below. Label the respective fractions. Refer to Color Plate 47.

2. Were all plasma fractions similar in appearance in duplicate samples? If not, which ones were not?

   _____

3. Which of the plasma protein fractions was the most

   abundant? _____

# LABORATORY REVIEW *41*

NAME _____

LAB SECTION _____ DATE _____

## LYMPHATICS AND THE IMMUNE RESPONSE

### COMPLETION, DEFINITION, AND INTERPRETATION

1. On each of the three slides shown in Figure Q41-1, the left side contains anti-A serum, the middle anti-B serum, and the right side anti-$Rh_o$(D) serum. One drop of an unknown blood sample was added to the antisera on each side. On the basis of the reactions shown, determine (i) the major blood type of each unknown blood sample and (ii) whether it contained the $Rh_o$ factor.

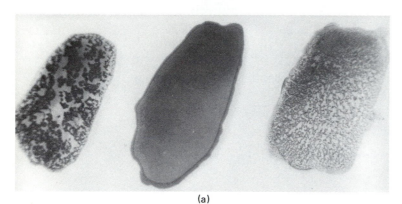

(a)

a. Major blood factor _____

Rh$_o$ factor _____

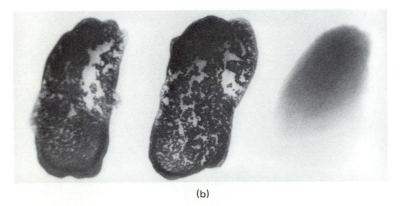

(b)

b. Major blood factor _____

Rh$_o$ factor _____

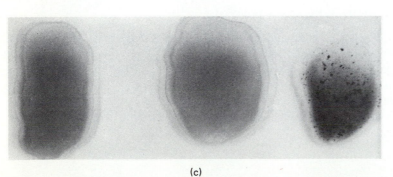

(c)

c. Major blood factor _____

Rh$_o$ factor _____

**Figure Q41-1**

595

**ANSWERS**

2. Define or explain the following:

   a. Cross-matching _____

   _____

   b. Coombs' test _____

   c. Hemolytic disease of newborn (Rh baby) _____

   _____

   _____

   d. ABO incompatibility _____

   _____

   e. Rh factor _____

   f. Agglutination _____

   _____

   g. Hemolysis _____

3. List five major blood types other than the ABO and Rh groups.

   a. _____

   b. _____

   c. _____

   d. _____

   e. _____

4. What is the value of electrophoresis? _____

   _____

5. Define or explain the following.

   a. Anode _____

   b. Cathode _____

   c. Isoelectric point _____

   d. Amino group _____

   e. Carboxyl group _____

## MULTIPLE CHOICE

Enter the correct answers in the spaces provided.

1. Which of the following are antibodies known to be present in the blood of individuals who are either blood type A, B, or O?
   a. agglutinogens
   b. agglutinins
   c. agglutinates

2. Against which of the following would an individual who is blood type A be expected to have antibodies?
   a. A
   b. B

1. _____

2. _____

   c. O
   d. $Rh_O$
   e. choices *a* and *b*

3. Against which of the following would an individual who is blood type B be expected to have antibodies?
   a. A
   b. B
   c. O
   d. $Rh_O$
   e. choices *a* and *b*

4. Against which of the following would an individual who is blood type O be expected to have antibodies?
   a. A
   b. B
   c. $Rh_O$
   d. choices *a* and *b*
   e. does not have antibodies against A, B, or O

5. Against which of the following would an individual who is blood type AB be expected to have antibodies?
   a. A
   b. B
   c. $Rh_O$
   d. choices *a* and *b*
   e. does not have antibodies against A, B, or O

6. Which of the following cells are known for their involvement with cell-mediated immune responses?
   a. B lymphocytes
   b. plasma cells
   c. erythrocytes
   d. lymphocytes
   e. T lymphocytes

7. How many different classes of immunoglobulins are recognized?
   a. 2
   b. 3
   c. 4
   d. 5
   e. 10

8. What is a substance that is recognized by the body as being foreign and that stimulates an immune response?
   a. immunoglobulin
   b. antibody
   c. antigen
   d. plasma cell
   e. foreign body

9. What is the term for cells from which plasma cells originate?
   a. T lymphocytes
   b. B lymphocytes
   c. antigens
   d. erythrocytes
   e. determinants

10. In which of the following do T lymphocytes originate?
   a. tonsils
   b. spleen
   c. liver
   d. thyroid gland
   e. thymus gland

**ANSWERS**

3. _____
4. _____
5. _____
6. _____
7. _____
8. _____
9. _____
10. _____

# The Respiratory System

When the term *respiration* is used alone, it generally refers to the overall metabolic activity that occurs within cells. When *respiration* is used in association with the *respiratory system,* it refers to the process of gaseous exchange between the atmosphere, the blood, and the cells.

During the exchange of gases, several processes are operating simultaneously. The movement of air between the atmosphere and the lungs is termed *ventilation.* It is achieved by *breathing,* a method of creating differences in pressure in the chest cavity. These pressure changes are a direct consequence of a coordinated contraction of muscles, especially the intercostals and those of the diaphragm and abdomen. *External respiration* is the gaseous exchange of carbon dioxide, oxygen, and some nitrogen that occurs between the lung cells and the lung capillary blood. *Internal respiration* is a similar exchange between the blood and cells throughout the body.

In addition to its major role in the process of gaseous exchange, the respiratory system plays a minor role in temperature regulation and water balance, based on the amount of water vapor lost during exhalation. It also participates—along with the kidney, liver, and gastrointestinal tract—in the overall regulation of blood pH.

The average normal individual has a lung capacity of about 6000 ml. This total is made up of several respiratory air components, including the expiratory reserve volume, inspiratory reserve volume, residual volume, and tidal volume (Figure XII-1). A variety of pulmonary-function tests are available to determine whether the lung tissues are functioning properly. In diseases such as carcinoma, emphysema, fibrosis, pneumonia, and pulmonary congestion, there is a substantial reduction in normally functioning lung tissue. Several pulmonary-function procedures are quite complex or require elaborate equipment. In Exercise 43, several of the more common tests are described and performed.

### GENERAL REFERENCES

Lehrer, S. *Understanding Lung Sounds* (with audiocassette). Philadelphia: Saunders, 1984.

Levitzky, M. G. *Pulmonary Physiology.* New York: McGraw-Hill, 1982.

Tortora, G. J., and S. R. Grabowski. *Principles of Anatomy and Physiology,* 7th ed. New York: HarperCollins, 1993.

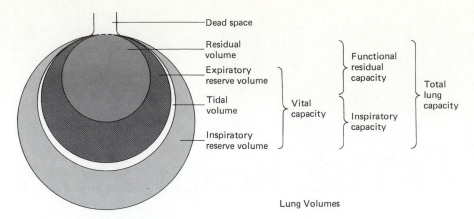

Lung Volumes

**Figure XII-1** Diagrammatic representation of respiratory air components.

### SPECIFIC REFERENCES

Laszlo, G., and Sudlow, M. F. (Eds.). *Measurement in Clinical Respiratory Physiology.* New York: Academic Press, 1983.

Naeye, R. L. "Sudden Infant Death." *Scientific American,* 242: 56–62, 1980.

Nero, A. V., Jr., "Controlling Indoor Air Pollution." *Scientific American,* 258: 42–48, 1988.

Schlesinger, R. B. "Defense Mechanisms of the Respiratory System." *BioScience,* 32: 45–50, 1982.

Weibel, E. R. "How Does Lung Structure Affect Gas Exchange?" *Chest,* 83: 657–665, 1983.

Whipp, B. J., and K. Wasserman (Eds.). *Lung Biology in Health and Disease: Vol. 52. Exercise—Pulmonary Physiology and Pathophysiology.* New York: Marcel Dekker, 1991.

# Anatomy of the Respiratory System

## After completing this exercise, you should be able to

1. Identify the organs of the respiratory tract of the cat
2. Name the organs of the human respiratory tract from charts, models, and other demonstrations
3. Describe the functions of the organs in internal and external respiration

## Materials

The following materials should be provided for class use:

1. Cat dissection specimens
2. Dissecting tools
3. Model of the human respiratory tract
4. Charts of the human respiratory system
5. Demonstration of a mammalian head (cut in sagittal section)

## ORGANS OF THE CAT RESPIRATORY SYSTEM

Locate the paired *external nares* of the cat. Air is taken into the nares, passes through the *nasal chamber* and the *internal nares,* and then enters the *pharynx*. The pharynx is divided into three regions: (1) *nasopharynx,* (2) *oropharynx,* and (3) *laryngopharynx*. From the laryngopharynx, the air passes through an opening called the *glottis* and enters the *larynx*. Beyond the larynx is the *trachea,* which divides into right and left *bronchi*. The bronchi penetrate the *lungs* and subdivide many

times into *bronchioles,* finally ending in blind sacs called *alveoli,* which are in contact with the blood. The gases $O_2$ and $CO_2$ are exchanged between an alveolus and the blood (Figures 42-1 and 42-6).

The mouth of the cat should be open. Observe as much as possible within the mouth cavity, such as teeth and tongue. Turn to the thoracic cavity and neck. Locate the *larynx,* and observe once again the muscles that move its cartilages. Make a midsagittal cut into the ventral surface of the larynx and part of the trachea. Spread the two sides of the larynx, and observe the narrow *vocal cords* within it. If there is dried blood in the larynx, remove it with a wet paper towel or cloth. Posterior to the larynx is the *trachea,* which runs the length of the neck and into the thoracic cavity between the lungs. This tube never flattens due to the presence of a series of cartilaginous rings that keep it open. At the lung, the trachea bifurcates to form paired *bronchi,* which then penetrate the lung tissue and subdivide many times.

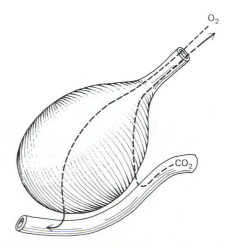

**Figure 42-1** Exchange of gases between alveolus and capillary.

Locate the *right* and *left lungs*. The right lung contains three lobes: anterior, medial, and posterior. The left lung has four lobes: anterior, medial, and posterior, and the *mediastinal lobe*. The *mediastinal cavity* between the lungs contains the *heart, esophagus, thymus gland, trachea,* and major blood vessels. Each lung is enclosed by membranes. The *inner pleural membrane* is the visceral peritoneum and adheres closely to the lung. The *outer pleural membrane* is the parietal peritoneum. Between the two is a potential space called the *pleural cavity*. The pleural membranes move with the lung as it expands and contracts during inhalation and exhalation.

Study Figure 42-2 for review.

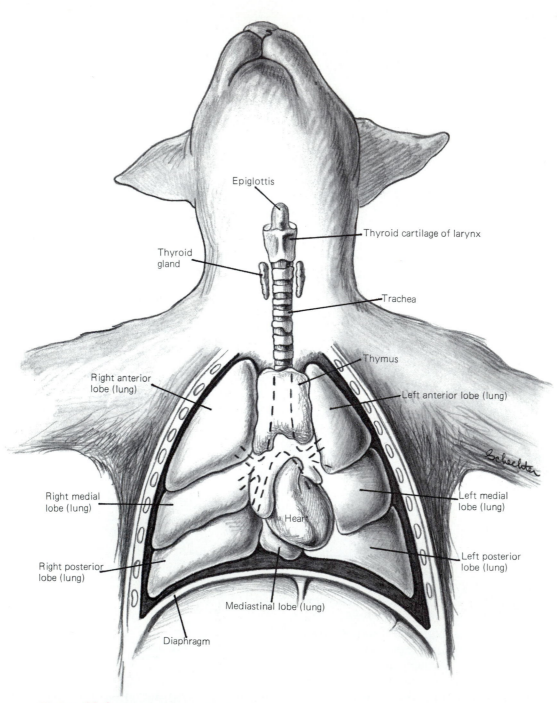

Epiglottis

Thyroid cartilage of larynx

Thyroid gland

Trachea

Thymus

Right anterior lobe (lung)

Left anterior lobe (lung)

Right medial lobe (lung)

Left medial lobe (lung)

Heart

Right posterior lobe (lung)

Left posterior lobe (lung)

Mediastinal lobe (lung)

Diaphragm

**Figure 42-2** Cat respiratory tract.

## THE HUMAN RESPIRATORY SYSTEM

### Lung

The human lung has five lobes:
Right superior lobe
Right medial lobe
Right inferior lobe
Left superior lobe
Left inferior lobe

The relationship of the human pleura to the lung in the human thoracic cavity is the same as it is in the cat. There are two pleural membranes (inner and outer), and there is a potential space between them, the *pleural cavity*.

### Nasal Cavity

The *nostrils*, or *external nares*, open into a large bony cavity of the skull, below the cranium and above the mouth (Figure 42-4). The bones involved are the *ethmoid, maxillae, palatines, vomer,* and the *inferior nasal conchae*. Review these bones in Exercise 11—The Skull.

The ethmoid bone (which contains the superior and middle nasal conchae) forms the lateral walls and roof of the nasal cavity along with the inferior nasal conchae, which are separate bones. The floor of the nasal cavity is formed by the maxillae and the palatine bones. These latter four bones also form the hard palate. The cartilaginous part of the *nasal septum* that divides the nasal cavity into right and left sides is contiguous with the *perpendicular plate* of the *ethmoid* and the *vomer* bone. In this cavity, air is warmed, filtered, and moistened. Sensory olfactory nerve endings are found in parts of the epithelial lining, which is of the *pseudostratified ciliated* type. (Exercise 6—Histology)

### Pharynx

Air from the nasal cavity passes through the *internal nares* to enter a large cavity at the back of the mouth (the throat). It is a common passageway for both food and air and is highly muscular. The cavity is divided into three regions:

1. Nasopharynx
2. Oropharynx
3. Laryngopharynx

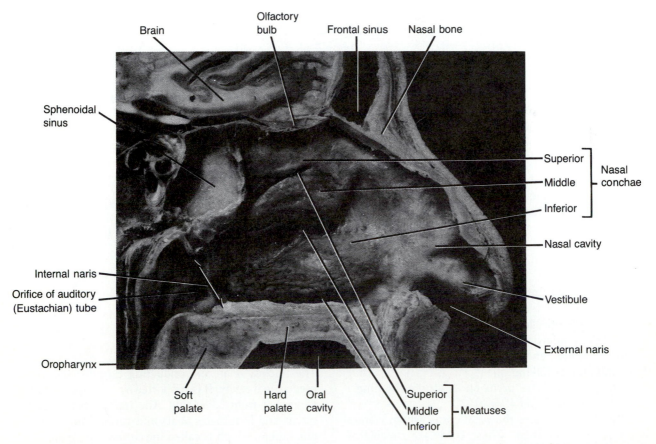

**Figure 42-3** Human respiratory tract—midsagittal section. (From J. A. Gosling, P. F. Harris et al., *Atlas of Human Anatomy,* Gower Medical Publishing Ltd., 1985.)

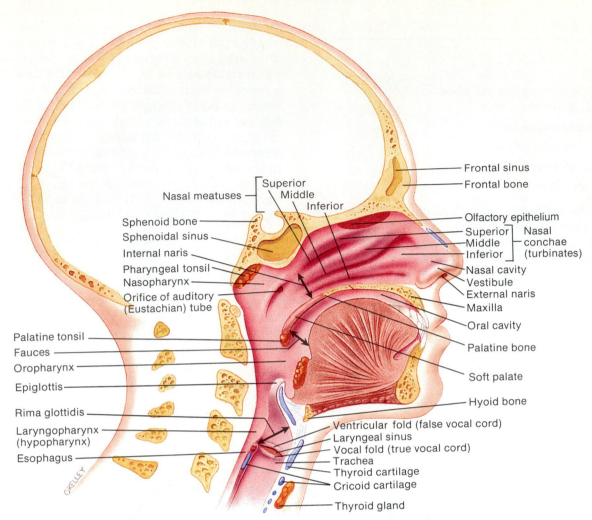

**Figure 42-4** Human nasal cavity: The left side of the head and neck are seen in sagittal section with the nasal septum. (From G. J. Tortora and S. R. Grabowski, *Principles of Anatomy and Physiology,* 7th ed., New York: HarperCollins, 1993.)

Lymphoid tissue, called *tonsils,* is found in the pharynx. Two types—the *palatine tonsils* and the *lingual tonsils*—are located in the oropharynx, and the *pharyngeal tonsils* (adenoids) are located in the nasopharynx.

## Larynx

The opening from the pharynx into the larynx between the vocal cords is termed the *glottis.* Guarding this opening is a large laryngeal cartilage, the *epiglottis.* When one is swallowing, the epiglottis covers the opening to the glottis, and the food enters the esophagus. When inhaling or talking, the glottis is open, to permit air to flow through the larynx on its way to the lungs. Study Figures 42-4 and 42-5.

The larynx contains the vocal cords, which vibrate as air passes over them. These cords are capable of producing sounds. The framework of the larynx consists of nine cartilages that are moved by muscles, many of which you have already observed in the cat:

Thyroid cartilage (forms the Adam's apple in males)

Epiglottis

Cricoid cartilage

Paired arytenoid cartilages (2)

Paired corniculate cartilages (2)

Paired cuneiform cartilages (2)

## Trachea, Bronchi, and Bronchioles

The trachea is an air-conducting tube located in the mediastinal cavity. It is supported by cartilagenous rings to keep it from collapsing. The trachea begins at the larynx and terminates at its bifurcation into the bronchial tubes, which is called the *carina.* There are many mucus-producing cells in its pseudo-stratified ciliated

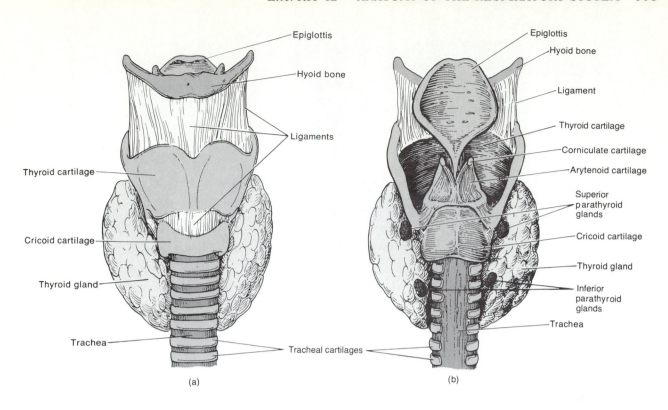

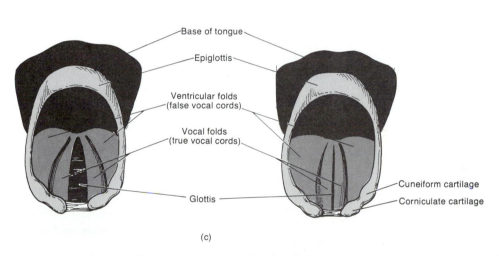

**Figure 42-5** Larynx: (a) anterior view; (b) posterior view; (c) viewed from above—in the figure on the left, the true vocal cords are relaxed; in the figure on the right, the vocal cords are pulled taut. (From G. J. Tortora and N. P. Anagnostakos, *Principles of Anatomy and Physiology*, 3rd, 4th, 5th, 6th eds., New York: HarperCollins, 1981, 1984, 1987, 1990.)

epithelial cells. These cells, which produce mucus, are called *goblet cells.* The bronchi and bronchioles branch about 24 times within the lung tissue and end in tiny air sacs, *alveoli,* through which gases are exchanged.

This exchange of gases between alveoli and blood is called *external respiration.* Study Figure 42-6, and be able to define all parts indicated by your instructor.

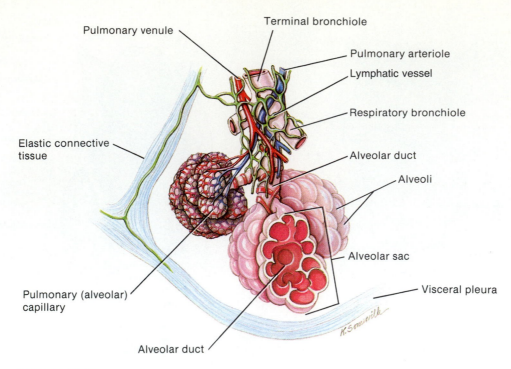

**Figure 42-6** Lung alveoli and blood supply. (From G. J. Tortora, *Introduction to the Human Body,* 2nd ed., New York: HarperCollins, 1991.)

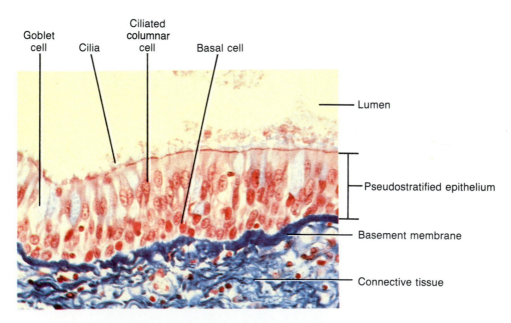

**Figure 42-7** Photomicrograph of enlarged aspect of the tracheal pseudostratified columnar epithelium at a magnification of 600×. (Courtesy of Andrew J. Kuntzman.)

# LABORATORY REVIEW *42*

## ANATOMY OF THE RESPIRATORY SYSTEM

### MATCHING AND FIGURE IDENTIFICATION

1. Match the phrases on the left with one of the correct terms on the right.

    a. _____ Cavity between membranes surrounding the lung

    b. _____ Name for two branches of the trachea

    c. _____ Respiratory passage that contains the vocal cords

    d. _____ Terminal parts of respiratory-duct system, in contact with blood

    e. _____ Opening into the larynx from the pharynx

    f. _____ Number of lobes in the human lung

    g. _____ In which part of the respiratory tract are the tonsils located?

    h. _____ A bone that makes up the hard palate

    i. _____ Another bone that makes up the hard palate

    j. _____ A prominent bone within the nasal chamber

2. Trace a molecule of $O_2$ from the air until it reaches the blood. Begin with the external nares and end with the pulmonary capillaries.

## ANSWERS

1. Bronchioles
2. Pharynx
3. Maxillary
4. 5
5. Mediastinum
6. Palatine
7. Ethmoid
8. Larynx
9. Epiglottis
10. Pleural
11. Glottis
12. 4
13. Bronchi
14. Alveoli
15. Pericardial
16. 6
17. Mandible

3. Identify the numbered items indicated in Figure Q42-1.

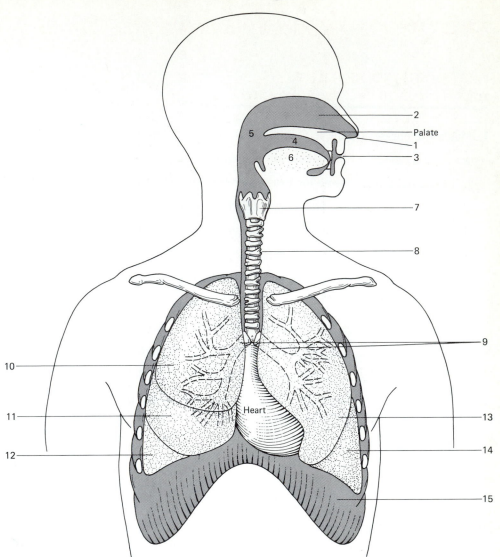

**Figure Q42-1** Human respiratory tract.

| | | |
|---|---|---|
| 1. _____ | | 9. _____ |
| 2. _____ | | 10. _____ |
| 3. _____ | | 11. _____ |
| 4. _____ | | 12. _____ |
| 5. _____ | | 13. _____ |
| 6. _____ | | 14. _____ |
| 7. _____ | | 15. _____ |
| 8. _____ | | |

# EXERCISE 43

# Pulmonary Function Tests

## After completing this exercise, you should be able to

1. Define *atmospheric pressure, intrapulmonary pressure,* and *intrapleural pressure*
2. Perform standard measurements and calculations used in evaluating the functioning of the respiratory system
3. Define *tidal volume, minute volume, vital capacity, residual air, total lung capacity, inspiratory reserve, expiratory reserve,* and *dead space*
4. Interpret respiratory volume measurements as being indicative of normal or abnormal functioning

---

The respiratory system consists of a number of anatomic passageways through which the air we breathe travels to the delicately structured lungs. During *inspiration* (the movement of air from the atmosphere into the lungs), air is moved through the nose, pharynx, larynx, trachea, and successively smaller divisions of the bronchial tree until it reaches the numerous tiny air sacs, the *pulmonary alveoli.* Conditions at the alveoli are ideal for the physiological exchanges of gases. During *expiration* (the movement of air from the lungs into the atmosphere), the path is reversed. Air flows through these passageways during respiration because of a pressure gradient involving three important pressures:

1. *Atmospheric pressure*—the pressure exerted by the air surrounding the body, which at sea level is equal to 760 mm of mercury (mm Hg)
2. *Intrapulmonary pressure*—the pressure within the respiratory space of the lungs (bronchi and alveoli)

3. *Intrapleural pressure*—the pressure within the pleural space between the lungs and the walls of the thoracic cage

Various factors are known to regulate respiration. For example, the temperature of blood flowing through the respiratory center may influence breathing. Above-normal body temperature, as during fever or severe muscular exercise, results in an increase in the respiratory rate (number of breaths per minute). Subnormal temperature, as during hypothermia, results in a decreased respiratory rate. Other respiration-regulating factors include pain and emotional state.

Before birth, the lungs contain no air, but after birth, once they have been filled with air, they are never completely empty. No expiration ever completely empties the alveoli; neither are they completely filled. The ability of an individual's respiratory system to exchange air with the outside environment is important in evaluating normal function or the presence of respiratory disease. A number of different measurements can be made in studying the various properties of the respiratory system (Figure 43-1). These are described in Table 43-1 (a graphic display of the relationships among these is shown in Figure XII-1). The instrument used to measure lung capacities, volumes of air exchanged, and rates of respiration is the *spirometer.* The spirometer can play an important role in the detection of restrictive or obstructive lung defects. This exercise describes procedures for making some of the respiratory measurements.

## Materials

The following materials should be provided for class use:

1. Spirometers, such as the Collins Survey Spirometer, the Breon Spirometer, or other appropriate device (with their operator's manuals, if possible)

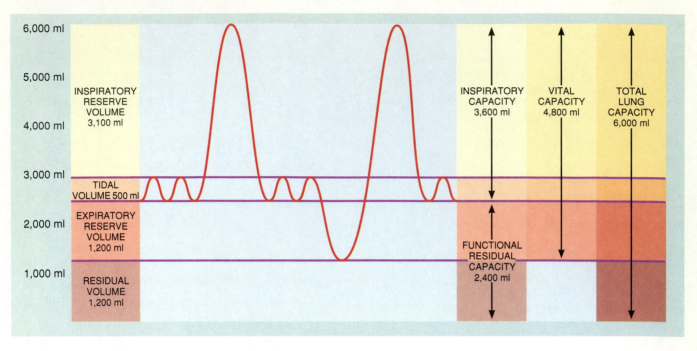

**Figure 43-1** Spirogram of pulmonary volumes and capacities. (From G. J. Tortora and S. R. Grabowski, *Principles of Anatomy and Physiology,* 7th ed., New York: HarperCollins, 1993.)

## Table 43-1   Respiratory Volume Measurements

| Measurement | Brief Description | Normal Values for Adults |
|---|---|---|
| Tidal volume (TV) | Quantity of air moved into and out of the lungs during a normal breath | 500 ml |
| Respiratory rate (RR) | Number of breaths per minute | 12–15 |
| Minute volume (MV) | Quantity of air moved into and out of the lungs in 1 minute (MV = TV × RR) | 6000 ml/min |
| Inspiratory reserve volume (IR) | Quantity of air above tidal volume that can be forcefully inspired after a normal inhalation | 3000 ml |
| Expiratory reserve volume (ERV) | Quantity of air above tidal volume that can be forcefully expired after a normal inhalation | 1200 ml |
| Vital capacity (VC) | Maximum quantity of air that can be moved into and out of the lungs (VC = IR + TV + ER) | 4500–5500 ml |
| Residual air (RA) | Quantity of air left in the lungs after maximum expiration | 1200 ml |
| Total lung capacity (TLC) | Quantity of air the lungs can hold (TLC = VC + RA) | 5900 ml |
| Dead space | Air that is not available for gas exchange | 100 ml (female) 150 ml (male) |

2. Breathing tube attachment
3. Disposable mouthpieces (adult)
4. Nose clips
5. Disinfectant and sterile gauze pads
6. Recording charts or paper
7. Recording pens and ink

## Procedure 1: Identification of Spirometer Parts and Accessories

This procedure is to be performed by students individually.

1. Examine the spirometer provided.
2. With the aid of Figure 43-2 and the operator's manual, identify the typical parts and accessories of the instrument.
3. Your instructor will demonstrate the correct usage of the spirometer.

## Procedure 2: Respiratory Rate

This procedure is to be performed by students in pairs.

1. Have your partner sit quietly for 5 minutes.

2. After the rest period, count the number of complete respiratory cycles (inspiration and expiration) in 60 seconds. Make two additional counts while your partner is still seated.
3. Make three counts with your partner standing and three counts while your partner is in the supine position.
4. Determine the average respiratory rate in each position by dividing the number of times counts were taken into the number of complete cycles. Enter your findings in the results and observations section.
5. Change places with your partner, and repeat Steps 1–4.

## Procedure 3: Tidal Volume (TV)

This procedure is to be performed by students in pairs.

1. Have your partner sit quietly for 5 minutes.
2. Disinfect a nose clip. Place it on your partner's nose so that the nasal passages are closed.
3. Place the disposable mouthpiece into your partner's mouth.

(a)

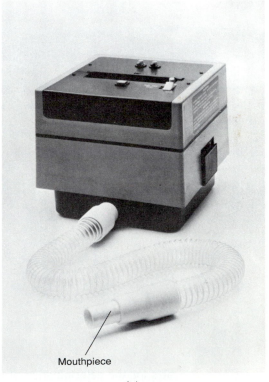

Mouthpiece

(b)

**Figure 43-2** Two instruments used for the measurement of lung volumes and capacities (spirometry)—note the various components of each: (a) Collins Respirometer (courtesy of Warren E. Collins, Inc., Braintree, MA); (b) Breon Spirometer (courtesy of Breon Laboratories, Inc., New York, NY).

4. Start the recording system, and have your partner breathe normally into the spirometer system for five complete respiratory cycles.
5. Record the time (in seconds) for each inspiration and the volume of air inspired. The average value for the inspiratory volume is the *tidal volume* at rest.
6. Repeat this procedure with your partner in the standing position and in the supine position.
7. Change places with your partner, and repeat Steps 1–6.
8. Record the measurements in the results and observations section.

**Procedure 4: Expiratory Reserve Volume (ERV)**

This procedure is to be performed by students in pairs.

1. Have your partner sit quietly and breathe normally.
2. Attach a nose clip, as described in Procedure 3.
3. Turn on the recording system.
4. Instruct your partner to breathe normally for four respiratory cycles, and then on the fifth normal expiration to exhale as much as possible into the spirometer mouthpiece. (Help your partner by counting the inspirations and expirations of the complete cycles.)
5. Record the volume.

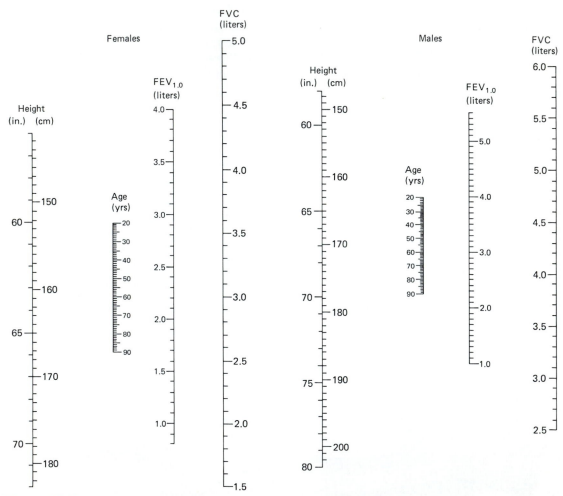

**Figure 43-3** Scales for the determination of normal spirometric values for females and males. These values are expressed in terms of a normal range for age, height, and sex. To use these scales, place the edge of a ruler at the correct point for the subject on the height scale and at the correct point on the age scale. The points at which the ruler intersects the $FEV_{1.0}$ (forced expiratory volume in 1 second) and FVC (forced vital capacity) scales give the normal values for those measurements. (Modified from T. L. Petty, *Office Spirometry*. New York: Breon Laboratories, 1980.)

6. Repeat Steps 4 and 5 two more times, and calculate an average value.
7. Repeat the procedure with your partner in the standing position and in the prone position.
8. Change places with your partner, and repeat Steps 1–7.
9. Record the measurements in the results and observation section.

## Procedure 5: Forced Expiratory Vital Capacity (FVC)

This procedure is to be performed by students in pairs.

1. Have your partner sit quietly and breathe normally.
2. Attach a nose clip, and turn on the recording system.
3. Instruct your partner to breathe normally for four complete respiratory cycles and then on the fifth cycle to inhale as deeply as possible and to exhale as forcefully and steadily as possible into the spirometer mouthpiece.
4. Record the volume, and time the expiratory portion of the last cycle. Determine the average FVC value.
5. Repeat Steps 3 and 4 two more times.
6. Change places with your partner, and repeat Steps 1–5.
7. Find the normal values for FVC on Figure 43-3.
8. Record the measurements in the results and observations section.

## RESULTS AND OBSERVATIONS

1. Enter your RR determinations in Table 43-2 (A = average).
2. Enter your TV determinations in Table 43-3 (T = time, V = volume, and A = average).
3. Enter your ERV determinations in Table 43-4 (A = average).
4. Enter your FVC determinations in Table 43-5 (V = volume and T = time for forced expiration).

**Table 43-2   RR**

| Student | Sitting | | | | Standing | | | | Prone | | | |
|---|---|---|---|---|---|---|---|---|---|---|---|---|
| | 1 | 2 | 3 | A | 1 | 2 | 3 | A | 1 | 2 | 3 | A |
| | | | | | | | | | | | | |
| | | | | | | | | | | | | |

**Table 43-3   TV**

| Student | Sitting | | | | | | | | Standing | | | | | | | | Prone | | | | | | | |
|---|---|---|---|---|---|---|---|---|---|---|---|---|---|---|---|---|---|---|---|---|---|---|---|---|
| | 1 | | 2 | | 3 | | A | | 1 | | 2 | | 3 | | A | | 1 | | 2 | | 3 | | A | |
| | T | V | T | V | T | V | T | V | T | V | T | V | T | V | T | V | T | V | T | V | T | V | T | V |
| | | | | | | | | | | | | | | | | | | | | | | | | |
| | | | | | | | | | | | | | | | | | | | | | | | | |

**Table 43-4   ERV**

| Student | Sitting | | | | Standing | | | | Prone | | | |
|---|---|---|---|---|---|---|---|---|---|---|---|---|
| | 1 | 2 | 3 | A | 1 | 2 | 3 | A | 1 | 2 | 3 | A |
| | | | | | | | | | | | | |
| | | | | | | | | | | | | |

**Table 43-5   FVC**

| Student | 1 | | 2 | | 3 | | Average | |
|---|---|---|---|---|---|---|---|---|
| | V | T | V | T | V | T | V | T |
| | | | | | | | | |
| | | | | | | | | |

# LABORATORY REVIEW *43*

## PULMONARY FUNCTION TESTS

### COMPLETION, DEFINITION, AND INTERPRETATION

1. Distinguish between inspiration and expiration. _____

   _____

   _____

2. Complete Table Q43-1 by briefly defining or explaining the terms listed and giving the normal values for each.

**Table Q43-1**

| Measurement | Definition | Normal Value |
|---|---|---|
| Tidal volume | a. | b. |
| Minute volume | c. | d. |
| Inspiratory reserve volume | e. | f. |
| Expiratory reserve volume | g. | h. |
| Vital capacity | i. | j. |
| Residual air | k. | l. |
| Total lung capacity | m. | n. |
| Dead space | o. | p. |
| Respiratory rate | q. | r. |

3. Define or explain the following:

   a. Atmospheric pressure _____

      _____

   b. Intrapulmonary pressure _____

      _____

   c. Intrapleural pressure _____

      _____

4. Indicate whether the measurement values in Table Q43-2 are normal or abnormal for adults.

**Table Q43-2**

| Measurement | Value(s) | Normal/Abnormal |
|---|---|---|
| Tidal volume | 200 ml | a. |
| | 500 ml | b. |
| Inspiratory reserve | 1000 ml | c. |
| | 2000 ml | d. |
| | 3000 ml | e. |
| Vital capacity | 2000 ml | f. |
| | 4500 ml | g. |
| | 5000 ml | h. |
| | 5500 ml | i. |
| Total lung capacity | 3500 ml | j. |
| | 6000 ml | k. |

5. List three major physiological functions of the respiratory system.

   a. _____

   b. _____

   c. _____

6. How would the respiratory rate be affected by the following situations?

   a. High fever _____

   b. Running _____

   c. Hypothermia _____

7. What is *hypoxia?* _____

   _____

8. Define or explain *hyperventilation.* _____

   _____

9. List four factors that regulate respiration.

   a. _____

   b. _____

   c. _____

   d. _____

10. With the aid of your textbook or other reference, complete the descriptions of respiratory disorders in Table Q43-3.

**Table Q43-3**

| Disorder | Description |
|---|---|
| a. | Collapse of lung tissue caused by extrinsic pressure or intrinsic pressure in pulmonary airways |
| Bronchitis | b. |
| Tuberculosis | c. |
| d. | Allergic reaction to proteins in foreign substances localized in respiratory membranes |
| Emphysema | e. |
| Pneumonia | f. |
| Pneumocystis infection | g. |

## MULTIPLE CHOICE

Enter the correct answers in the spaces provided.

1. Which of the following describes the movement of air between the atmosphere and the lungs?
   a. external respiration
   b. internal respiration
   c. breathing
   d. ventilation
   e. choices *a* and *c* only
2. What is the term for the gaseous exchange of carbon dioxide, oxygen, and some nitrogen that occurs between lung cells and lung capillary blood?
   a. external respiration
   b. internal respiration
   c. breathing
   d. ventilation
   e. choices *a* and *c* only
3. What is the term for the gaseous exchange between the blood and cells throughout the body?
   a. external respiration
   b. internal respiration
   c. breathing
   d. ventilation
   e. choices *a* and *c* only

1. _____
2. _____
3. _____
4. _____
5. _____
6. _____
7. _____
8. _____
9. _____
10. _____

4. Which of the following represents the pressure within the respiratory space of the lungs (bronchi and alveoli)?
   a. atmospheric pressure
   b. intrapulmonary pressure
   c. intrapleural pressure
   d. expiration
   e. inspiration

5. Which of the following represents the pressure exerted by the air surrounding the body, which at sea level is equal to 760 mm Hg?
   a. atmospheric pressure
   b. intrapulmonary pressure
   c. intrapleural pressure
   d. expiration
   e. inspiration

6. What is the term for the quantity of air moved into and out of the lungs during a normal breath?
   a. residual air
   b. total lung capacity
   c. vital capacity
   d. tidal volume
   e. minute volume

7. What would a tidal volume of 300 ml be considered?
   a. normal
   b. abnormal

8. What is the quantity of air moved into and out of the lungs in 1 minute?
   a. residual air
   b. total lung capacity
   c. vital capacity
   d. tidal volume
   e. minute volume

9. What is the term describing the quantity of air that the lungs can hold?
   a. residual air
   b. total lung capacity
   c. vital capacity
   d. tidal volume
   e. minute volume

10. What is the term for the quantity of air left in the lungs after maximum expiration?
    a. total lung capacity
    b. residual air
    c. vital capacity
    d. minute volume
    e. tidal volume

# The Digestive System

$A$s a group, nutrients such as carbohydrates, lipids, proteins, vitamins, and minerals are obtained from foods. Three processes take place in the body to convert the food consumed into usable forms. These processes are digestion, absorption, and metabolism. Exercises in this section consider selected components of the digestive system and emphasize the major types of normal activities, as well as some abnormal activities, associated with each one. Attention also is given to the role of microorganisms in tooth decay (caries).

Also examined are enzymes and the effects of factors such as pH and temperature on enzymatic activity. Tests used in the identification of proteins, carbohydrates, lipids, cholesterol, and other chemicals of importance in foods are demonstrated as well.

## GENERAL REFERENCES

Davenport, H. W. *Physiology of the Digestive Tract,* 5th ed. Chicago: Year Book Medical Publishers, 1982.

Netler, F. H. *The CIBA Collection of Medical Illustrations, Vol. 3, Part I, Upper Digestive System.* Summit, NJ: CIBA Pharmaceutical Products, 1959.

Sleisenger, M. H., and J. S. Fordtran (Eds.). *Gastrointestinal Disease: Pathophysiology, Diagnosis, Management,* 3rd ed., Vols. 1 and 2. Philadelphia: Saunders, 1984.

Tortora, G. J., and S. R. Grabowski. *Principles of Anatomy and Physiology,* 7th ed. New York: HarperCollins, 1993.

## SPECIFIC REFERENCES

Davenport, H. W. "Why the Stomach Does Not Digest Itself." *Scientific American,* 226: 86–93, 1972.

Martin, R. J., B. D. White, and M. G. Hulsey. "The Regulation of Body Weight." *American Scientist,* 79: 528–541, 1991.

Merlo, A., and S. Cohen. "Swallowing Disorders." *Annual Reviews Medicine,* 39: 17–28, 1988.

Moog, F. "The Lining of the Small Intestine." *Scientific American,* 245: 154–176, 1981.

Peters, M. N., and C. T. Richardson. "Stressful Life Events, Acid Hypersecretion and Ulcer Disease." *Gastroenterology,* 84: 114–119, 1983.

Sacks, F. M., and W. W. Willett. "More on Chewing the Fat—The Good Fat and the Good Cholesterol." *The New England Journal of Medicine,* 325: 1740–1742, 1991.

Track, N. S. "The Gastrointestinal Endocrine System." *Canadian Medical Association Journal,* 122: 287–292, 1980.

Wilkins, R. A., and H. B. Nunnerley. *Imaging of the Liver, Pancreas, and Spleen.* Boston: Blackwell Scientific, 1990.

# Organs of the Digestive System

## After completing this exercise, you should be able to

1. List the components of the mouth
2. Identify and give the functions of the four groups of teeth
3. List the organs in the digestive tract
4. Point out the associated digestive organs and the ducts connecting them to the digestive tract

## Materials

The following materials should be provided for class use:

1. Cat dissection specimens
2. Dissecting tools
3. Model of the human torso
4. Chart of the human digestive system

## INTRODUCTION: DIGESTION

Most foods we consume cannot be utilized directly and must be digested or chemically broken down into smaller components that can be absorbed into the blood and cells. This process is the major function of the *digestive system*. The chemical processes involved in the splitting of complex chemicals into simpler ones are *hydrolytic activity* (splitting in the presence of water molecules) and *enzymatic activity*. Hydrolysis is the basic mechanism for carbohydrate, protein, and fat digestion (the three major components of human diets). Specific enzymes are involved for each food substance. Digestion includes all the processes by which food is converted into end products that may be absorbed from the lumen of the intestine into the blood. Digestion begins in the mouth and continues in the stomach and small intestine. Elimination is the main function of the large intestine.

Many cell types contribute to digestion. Some produce mucus, which lubricates the tract and provides the medium for the action of hydrolytic enzymes, while others secrete various chemicals, including enzymes.

## THE MAMMALIAN DIGESTIVE TRACT

Digestion and the subsequent absorption and elimination of nonabsorbed foodstuffs involve a series of organs that form a long digestive (alimentary or gastrointestinal) tract. This is the series of organs through which the food passes. Each organ in it is specifically constructed and adapted for different stages in the digestive process.

The digestive system includes the digestive tract and associated organs, such as the salivary glands, liver, pancreas, and gallbladder. The salivary glands, liver, and pancreas are glandular and produce secretions. The gallbladder is a storage sac for bile.

## THE CAT MOUTH

### Teeth

Examine the cat's mouth. Locate the parts of the mouth in Figure 44-1. The mouth was opened for you before the cat was killed, to facilitate study of this area. If it is to be opened farther in order to see posterior parts, bone forceps will be needed, to cut back at the angles of the jaw. Follow directions given by your instructor.

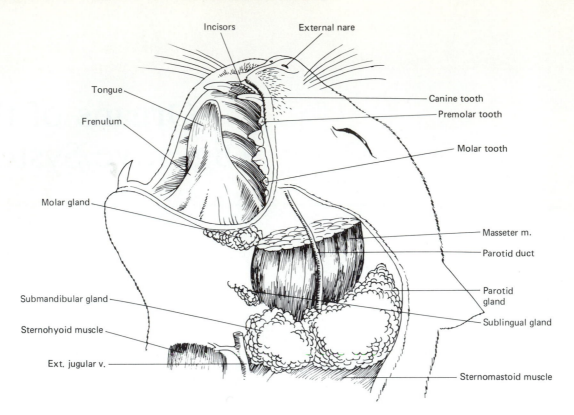

**Figure 44-1** Oral cavity of the cat.

Note the upper and lower lips. The mouth cavity (oral cavity) is divided into the *vestibule* and the mouth cavity proper by the gums and teeth. The vestibule is surrounded by lips and cheeks. Within the mouth cavity proper are the tongue and the teeth. Posterior to the mouth cavity is the pharynx. Observe that the tongue is tied to the floor of the mouth by a fold of tissue, the *frenulum linguae*. The tongue is a muscular organ (striated skeletal muscle), covered with a mucous membrane. On its rough outer surface are numerous papillae, many of which are horny and toothlike. In which direction are the points of these papillae directed? Cats groom and clean themselves with their tongues. The *hard palate* forms seven or eight transverse ridges. Locate them.

The cat has two sets of teeth in its lifetime. At birth, the kitten is toothless. The first or *deciduous* set of teeth are 26 in number. They are replaced by 30 *permanent* teeth, 14 in the lower jaw and 16 in the upper jaw. The teeth are of four types (see Figure 44-1, for an illustration of each of these in the cat):

Incisors—12 (3 in each quadrant)

Canines—4 (1 in each quadrant)

Premolars—10 (3 in each upper quadrant; 2 in each lower quadrant)

Molars—4 (1 in each quadrant)

## Salivary Glands

There are five pairs of salivary glands in the cat. Three of the more prominent and accessible pairs were discussed in Exercise 18. Refer to Figures 18-2a and 18-2b. Locate once again the following:

1. *Parotid gland* and *parotid duct,* which open on the inside of the cheek, opposite the last premolar tooth
2. *Submandibular gland* and *submandibular duct,* which open into the mouth cavity in front of the parotid duct
3. *Sublingual gland* and *sublingual duct,* which open into the mouth cavity in almost the same place but slightly anterior to the submandibular duct

In addition to the glands just noted, the cat has two additional pairs of salivary glands:

1. *Molar glands*—located under the mucosa of the lower lip and extending from the *masseter* muscle to the canine tooth. Several small ducts open into the mouth cavity.
2. *Infraorbital glands*—lie in the ventrolateral floor of the orbits. A single duct from each gland opens into the mouth cavity near the molar tooth.

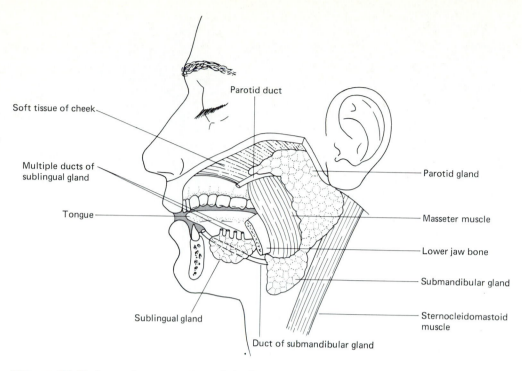

**Figure 44-2** General organization of the human mouth.

## THE HUMAN MOUTH

The floor of the human mouth (see Figure 44-2) lies in a horseshoe around the tongue and is continuous laterally with the gum, or *gingiva,* and *centrally* with the tongue (Color Plate 48). Near the front end are the openings of the submandibular and sublingual salivary glands (Figure 44-2 and Color Plate 49). These glands empty their secretions into the mouth. Here, the fluids aid in swallowing, chewing, and digestion.

The tongue is made up of skeletal muscle. In addition to its role in chewing and swallowing, the tongue is important to humans because it assists in speech and contains receptors for taste. The rough surface of the tongue is due to several types of *papillae* (elevations) that are heavily supplied with taste buds. Taste buds are accumulations of taste cells that carry stimuli to the central nervous system (see Color Plate 33). The sense of taste provided by the taste buds appears to be limited to four basic types—bitter, salty, sour, and sweet. These tastes are localized in various areas of the tongue. (Refer to Exercise 32.)

The human is supplied with two sets of teeth: 20 deciduous teeth (baby or milk teeth) and 32 permanent teeth. Teeth are categorized into four groups (as shown in Figure 44-3): *Incisors* are used for cutting food, *canines* for tearing food, and *premolars* (*bicuspids*) and *molars* for grinding. The parts of each tooth are the *crown,* the portion above the gum; the *root,* the structure embedded in the jaw; and the *neck,* the constricted region between the crown and the root (see Figure 44-3).

### Materials

Identify the human mouth parts using the following materials:

1. Models of human adult skull with teeth
2. Hand mirrors
3. Wooden tongue depressors

### Procedure

This procedure is to be performed by students in pairs.

1. Examine a human skull model and with the aid of Figure 44-3, identify the major categories of adult teeth.
2. Use a hand mirror to examine your own mouth. Identify the major categories of teeth present. Do you have a full set of 32 teeth?
3. Examine your laboratory partner's mouth, using a tongue depressor, and identify the base of the tongue, gum tissue, visible parts of teeth, and major categories of teeth. Use Figures 44-2 and 44-3 for reference. Does your laboratory partner have a full set of teeth?

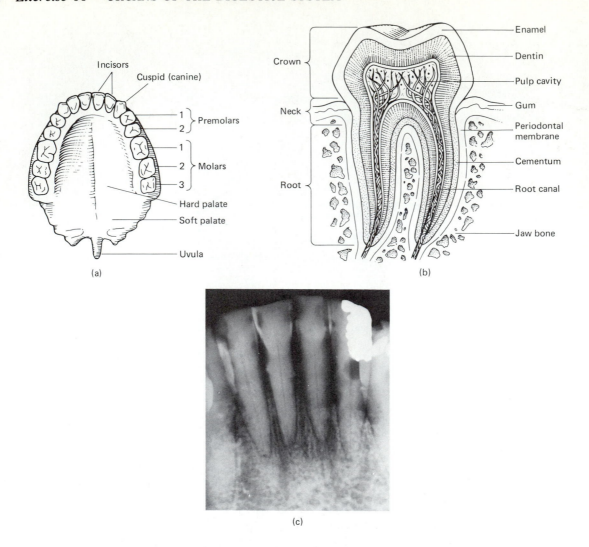

**Figure 44-3** Types and structure of human teeth: (a) major types of permanent teeth; (b) structure of a tooth; (c) x-ray of a tooth—note the appearance of the filling.

## THE DIGESTIVE SYSTEM OF THE CAT

Trace the esophagus of the cat from the thoracic cavity, where it can be seen lying dorsal to the trachea. Follow it to and through the diaphragm. Below the diaphragm the digestive tract enlarges to form the stomach (Figure 44-4).

Study Figures 44-4 and 44-5, and identify the different parts of the stomach. The pouch that rises above the entrance of the esophagus (above the dotted lines) is the *fundus.* The central area is the body of the stomach. Posterior to the body is the *pyloric region,* which ends at the *pylorus* (a sphincter muscle). The left surface, called the *greater curvature,* is convex and is the

embryonic dorsal surface. The right surface, the *lesser curvature,* is concave and is the embryonic ventral surface. The stomach is actually lying on its right side. At this time, note the *gastrosplenic ligament* (a mesentery connecting the stomach and the spleen), which is an extension of the *greater omentum.*

Next, begin to label Figure 44-5 in the spaces provided. If you have any problems, ask your instructor for help. Complete it as you read on.

Make a longitudinal slit in the stomach, in order to observe the inner wall of the organ. If it contains food, remove it with a spatula or a spoon. Observe that the inner stomach wall is thrown into prominent longitudinal folds. The folds (*rugae*) are characteristic of the stomach lining and will be seen in all stomachs

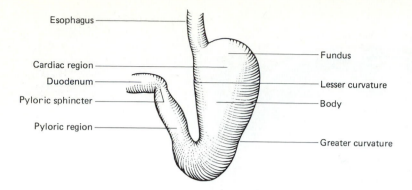

**Figure 44-4** Stomach of the cat.

except those extremely distended with food. The pyloric region of the stomach opens into the first part of the small intestine, the *duodenum*. The duodenum courses posteriorly for about 3 inches (the *descending portion of the duodenum*), then doubles back for another 2 inches (the *ascending portion of the duodenum*), at which point it terminates in a sharp backward bend and becomes known as the *jejunum*. The rest of the small intestine consists of jejunum and *ileum*. It is not possible to distinguish between them on gross examination, but they differ microscopically. The small intestine terminates at the end of the ileum.

Cut longitudinally into the small intestine, and feel the inner lining, which is velvety to the touch. The lining of the mammalian small intestine is thrown into millions of tiny folds that can be felt but not seen. The folds, called *villi*, serve to increase the surface area of the small intestine, to facilitate the absorption of digested foods.

The *mesentery proper* or *mesointestine*, which supports the small intestine, is attached to the dorsal body wall over a small area in the region of the duodenum but is extensive within the intestinal area. Within it is a large lymph node termed the *pancreas of Aselli*. Locate it.

The *large intestine*, or *colon*, begins at the ileocecal valve and is considerably larger in diameter than the small intestine. A blind sac, the *cecum*, marks the beginning of the colon. It is short and vestigial and does not give rise to an *appendix*, as it does in the human. The *ascending colon* and *transverse colon* are short. Trace them to the *descending colon*, which merges imperceptibly with the *rectum*, which ends at the *anus*. The mesentery attaching the colon to the dorsal body wall is the *mesocolon*.

## Associated Digestive Organs of the Cat

The *salivary glands* (parotid, submaxillary, and sublingual) have already been observed during the dissection of the muscles of the neck and jaw (see Figure 18-2). Review their anatomy and their ducts. Saliva acts on food as it passes from the *mouth* to the *pharynx, esophagus,* and *stomach.*

The *pancreas* in the cat consists of two regions. (See Figure 44-6.) Lying beside the *descending duodenum* is the *head of the pancreas*. Continuous with it, passing to the left side of the body, dorsal to the stomach, is the *tail of the pancreas*. The two *pancreatic ducts* are the *duct of Wirsung* within the head of the pancreas and the *duct of Santorini*, within the tail. The duct of Wirsung joins the *common bile duct* from the *liver* and opens into the duodenum near the *pylorus* at an internal swelling termed the *ampulla of Vater*. Expose the pancreatic duct by gently removing parts of the pancreatic tissue from the dorsal surface with a dissecting needle at the place where the common bile duct joins the small intestine on the dorsal surface. Trace this pancreatic duct posteriad in the pancreas tissue. The accessory pancreatic duct of Santorini usually joins the duct of Wirsung before entering the duodenum.

The *liver* is divided into five lobes (described in Exercise 34). Locate again the *right medial lobe* (the gallbladder lies within it), the *right lateral lobe,* and the *right caudal lobe*. On the left side of the cat are the *left medial lobe* and the *left lateral lobe*. The mesentery lying between the liver and the duodenum is the *lesser omentum*. In it is the long greenish *common bile duct,* along with several blood vessels. The *common bile duct* is formed by the confluence of *hepatic ducts* from each liver lobe, and the *cystic duct* from the gallbladder. Find as many of these ducts as possible.

Bile from the liver and gallbladder and pancreatic juice from the pancreas flow together and enter the duodenum at the ampulla of Vater.

Be able to demonstrate all organs, mesenteries, and ducts mentioned in this exercise.

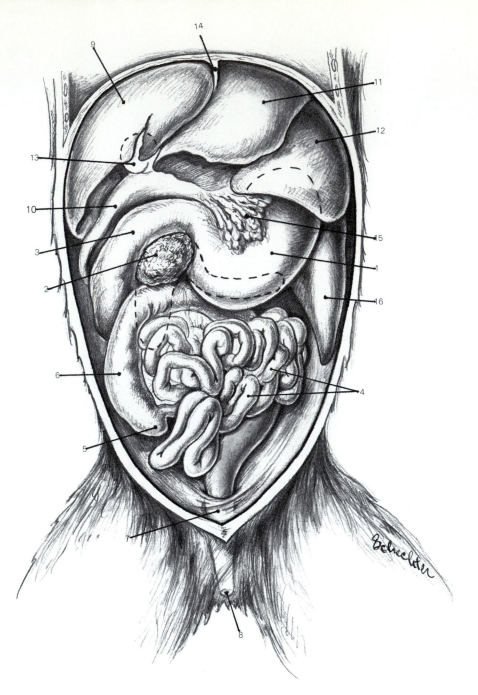

**Figure 44-5** Digestive system of the cat.

1. __Stomach__

2. __Pancreas (head)__

3. __Duodenum__

4. __Small intestine__

5. __Cecum__

6. __Ascending colon__

7. __Rectum__

8. __Anus__

9. __Liver—Rt. medial lobe__

10. __Liver—Rt. lateral lobe__

11. __Liver—Left medial lobe__

12. __Liver—Left lateral lobe__

13. __Gall bladder__

14. __Falciform ligament__

15. __Greater omentum__

16. __Spleen__

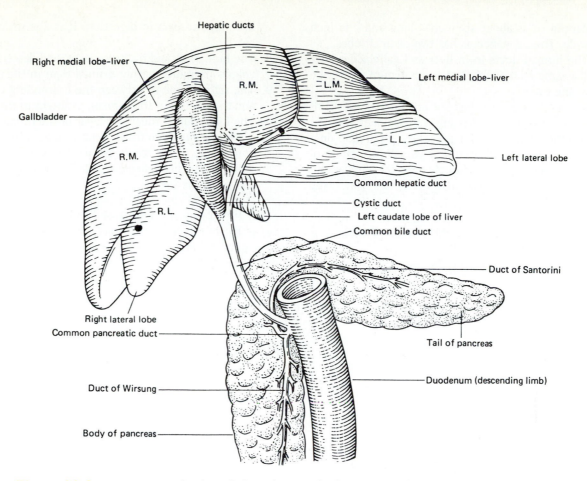

**Figure 44-6** Cat digestive glands and their ducts—the liver is raised to reveal the gallbladder and ducts.

## The Greater Omentum and Other Mesenteries of the Cat

The *greater omentum,* in which large quantities of fat are deposited, was noted in Exercise 34 and has been reflected back over the abdominal cavity to the left side of the cat. It is actually a large membranous sac containing the *lesser peritoneal cavity.* (The *abdominal cavity* is known as the *greater peritoneal cavity.*) It has become greatly displaced, due to many changes in position of the stomach during its embryonic development. The opening into this sac, the *epiploic foramen,* or *foramen of Winslow,* can be located by inserting a probe into an oval entrance located dorsal to the stomach and medial to the right lateral lobe of the liver. Place the probe or your little finger into this opening, and it will be completely enclosed by the greater omentum.

All mesenteries discussed thus far have been dorsal mesenteries. Most ventral mesenteries disappear during embryonic development, but remnants of two of them remain. Observe the small ventral mesentery between the liver below the diaphragm and the midventral body wall. This remnant of one of the many embryonic ventral mesenteries is the *falciform ligament.* Another ventral mesentery connects the *urinary bladder* and the ventral body wall. Locate each of these mesenteries on the cat.

## THE HUMAN DIGESTIVE SYSTEM

The human digestive system is much like that of the cat. There are some similarities and differences:

1. There are three pairs of salivary glands in the human, versus five pairs in the cat.
2. There are four human liver lobes, versus five in the cat.
3. The human pancreas is smaller and restricted to the region between the stomach and the duo-

denum. It is about six inches long and one inch wide. The cat pancreas has two distinct lobes.

4. The human ducts from liver and gallbladder have the same relationship to the duodenum as they do in the cat.

5. The single pancreas has but one duct in most humans, the duct of Wirsung, which opens into the common bile duct. An accessory pancreatic duct (duct of Santorini) occurs in a small percentage of people and opens separately into the duodenum. Cats have two ducts that drain two pancreatic masses.

6. The jejunum and the ileum together are approximately 20 feet long in the human. This area is shorter in the cat.

7. The appendix arises from the cecum in the human but plays no part in digestion. It is absent in the cat.

8. The human ascending, transverse, and descending colons are more sharply defined than in the cat.

9. The sigmoid colon, between the descending colon and the rectum in the human, is absent in the cat.

10. The longitudinal muscular layer of the human colon is arranged into three bands (teniae coli); when contracted, these bands gather the colon into a series of pouches called *haustra*. These pouches are absent in the cat.

Study the drawings (Figures 44-7, 44-8, 44-9, and 44-10) of the human digestive tract, and become familiar with all of its parts. Compare its length and shape with that of the cat.

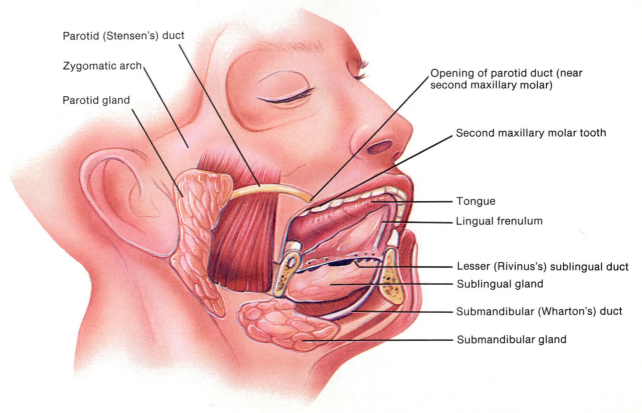

**Figure 44-7** Location of human salivary glands. (Courtesy of Douglas Merrill.)

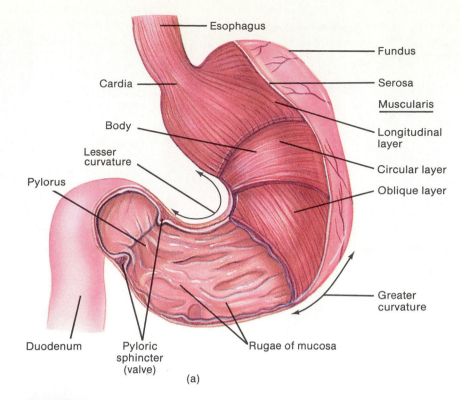

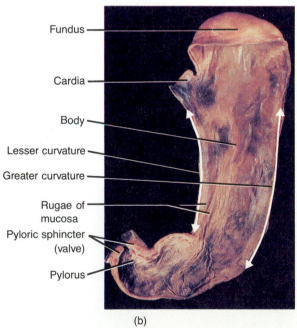

**Figure 44-8** External and internal anatomy of the stomach. (a) Diagram. (From G. J. Tortora, *Principles of Human Anatomy,* 6th ed., New York: HarperCollins, 1992.) (b) Photograph of the internal surface showing rugae. (From J. A. Gosling, P. F. Harris et. al., Atlas of Human Anatomy, Gower Medical Publishing Ltd., 1985.)

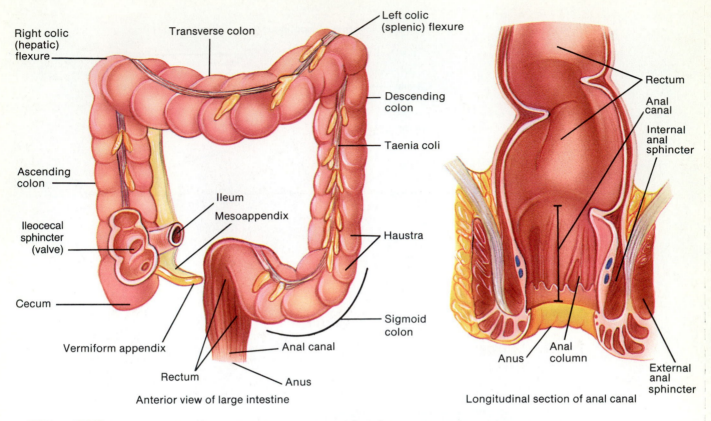

Right colic (hepatic) flexure

Transverse colon

Left colic (splenic) flexure

Descending colon

Taenia coli

Ascending colon

Ileum

Mesoappendix

Ileocecal sphincter (valve)

Haustra

Cecum

Vermiform appendix

Sigmoid colon

Rectum

Anal canal

Anus

Anterior view of large intestine

Rectum

Anal canal

Internal anal sphincter

Anus

Anal column

External anal sphincter

Longitudinal section of anal canal

**Figure 44-9** Large intestine. (From G. J. Tortora, *Principles of Human Anatomy,* 6th ed., New York: HarperCollins, 1992.)

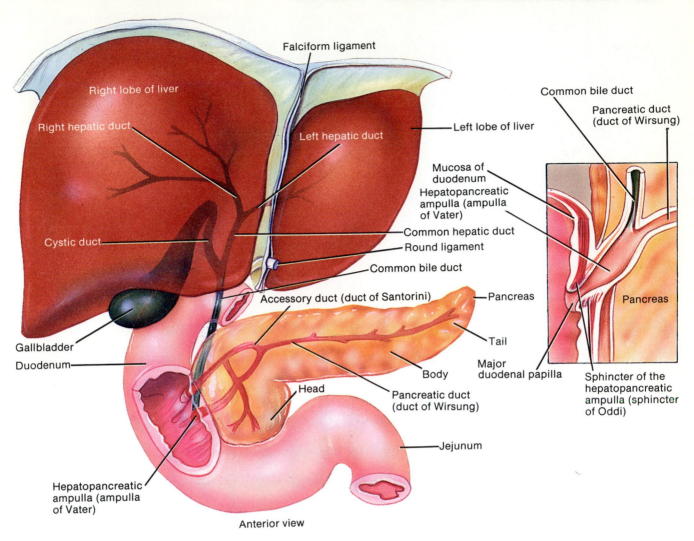

Falciform ligament

Right lobe of liver

Right hepatic duct

Left hepatic duct

Left lobe of liver

Cystic duct

Mucosa of duodenum

Hepatopancreatic ampulla (ampulla of Vater)

Common hepatic duct

Round ligament

Common bile duct

Accessory duct (duct of Santorini)

Pancreas

Gallbladder

Tail

Duodenum

Body

Head

Major duodenal papilla

Pancreatic duct (duct of Wirsung)

Jejunum

Hepatopancreatic ampulla (ampulla of Vater)

Anterior view

Common bile duct

Pancreatic duct (duct of Wirsung)

Pancreas

Sphincter of the hepatopancreatic ampulla (sphincter of Oddi)

**Figure 44-10** Human liver and pancreas. (From G. J. Tortora, *Principles of Human Anatomy,* 6th ed., New York: HarperCollins, 1992.)

# LABORATORY REVIEW *44*

NAME _____

LAB SECTION _____ DATE _____

## *ORGANS OF THE DIGESTIVE TRACT*

### FIGURE LABELING AND MATCHING

1. Label the Figure Q44-1, and compare all parts in the human digestive system to those of the cat, as to relative size, shape, and length.

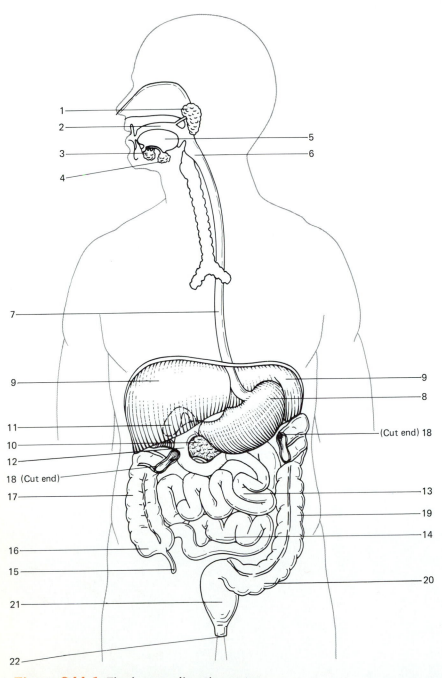

**Figure Q44-1** The human digestive system.

## ANSWERS

1. _____
2. _____
3. _____
4. _____
5. _____
6. _____
7. _____
8. _____
9. _____
10. _____
11. _____
12. _____
13. _____
14. _____
15. _____
16. _____
17. _____
18. _____
19. _____
20. _____
21. _____
22. _____

## ANSWERS

a. _____

b. _____

c. _____

d. _____

e. _____

f. _____

g. _____

h. _____

i. _____

j. _____

k. _____

1. Pancreatic duct

2. Falciform ligament

3. Esophagus

4. Haustra

5. Parotid

6. Rugae

7. Ileum

8. Gallbladder

9. Pancreas

10. Cystic duct

11. Appendix

12. Hepatic duct

13. Gastrosplenic ligament

14. Liver

15. Pancreas of Aselli

16. Jejunum

17. Pharynx

18. Teniae coli

19. Mesocolon

20. Common bile duct

2. Identify the types and/or categories of teeth shown in Figure Q44-2.

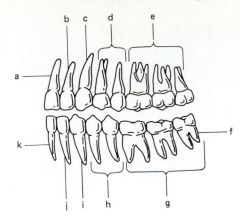

**Figure Q44-2** The types and/or categories of teeth.

3. Match the definitions on the left with the parts of the human digestive system listed on the right.

a. _____ A salivary gland

b. _____ Organ that stores bile

c. _____ Duct from liver that opens into duodenum

d. _____ Duct leading from gallbladder

e. _____ Ventral mesentery supporting liver

f. _____ Mesentery supporting large intestine

g. _____ Tube that transports food to stomach

h. _____ Part of digestive tract that opens into large intestine

i. _____ Pouches within large intestine

j. _____ Part of digestive tract through which food passes from mouth

k. _____ Lymph gland located in mesentery of small intestine

l. _____ Middle part of small intestine

m. _____ Vestigial structure attached to cecum

n. _____ Four-lobed organ ventral to stomach

o. _____ Striated lining of stomach

# Metabolism: Enzyme Action

## After completing this exercise, you should be able to

1. Define or explain *optimum, enzyme, substrate, hydrolysis,* and *end product,* as they relate to metabolism
2. List three specific enzymes and the substrates and end products associated with each one
3. Perform experiments to determine the effect of pH, temperature, and substrate concentration on enzymatic activity
4. Summarize the effects and importance of pH and temperature on enzymatic activity

The carbohydrates, fats, and proteins ingested as food must undergo specific changes before they can be absorbed from the digestive tract. Other substances—such as water, vitamins, and minerals—are ready for immediate absorption. Complex foods are prepared for absorption by *mechanical* and *chemical* activities. Mechanical activities include chewing, swallowing, and several different types of peristaltic contractions that break down large particles, mix the food with mucus and enzymes, and propel it through the digestive tract. Chemical activities break down the large molecules (substrates) that we eat into molecules that are small enough so that they can be absorbed through the walls of the digestive tract, transported through the body, and eventually used by body cells.

This exercise is concerned with selected tissues and chemical actions associated with metabolic enzymes (Figure 45-1). Several different enzymes and the factors that affect their activities are also considered. The procedures include specific tests used in the detection and identification of carbohydrates, lipids, and proteins.

## HISTOLOGY OF THE ORGANS AND STRUCTURES ASSOCIATED WITH DIGESTION

### Materials

The following materials should be provided for class use:
1. Prepared slides of
   a. Salivary glands
   b. Stomach, showing mucosa and chief parietal and mucous cells
   c. Small intestine, showing epithelium and goblet cells
   d. Large intestine, showing mucosa
   e. Pancreas
2. Anatomical charts of digestive system components

### Procedure

This procedure should be performed by students individually.

1. Examine the prepared histological slides.
2. Sketch and label a representative field showing the characteristic cells of each organ or related structure in the results and observations section. Use Color Plates 31 and 49–52 as guides.
3. Answer the questions in the results and observations section.

## DEMONSTRATION OF EXTRACELLULAR DIGESTION

Microorganisms are used here to show the extracellular nature of digestive processes.

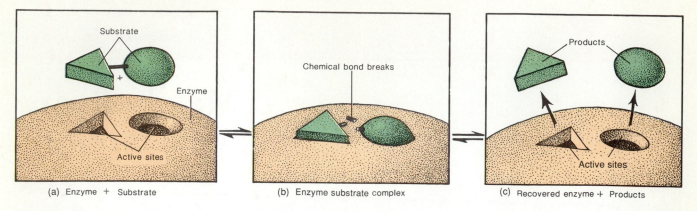

(a) Enzyme + Substrate     (b) Enzyme substrate complex     (c) Recovered enzyme + Products

**Figure 45-1** Enzyme action in a decomposition reaction: The enzyme and substrate molecules combine to form an enzyme–substrate (ES) complex. During combination, the substrate is changed into products. Once the products are formed, the enzyme is recovered and may be used again to catalyze a similar reaction. (G. J. Tortora, *Introduction to the Human Body*, 2nd ed., New York: HarperCollins, 1991.)

## Materials

The following materials should be provided for class demonstration:

1. One uninoculated starch agar plate
2. One starch agar plate centrally inoculated with the bacterium *Bacillus subtilis,* but not incubated
3. One starch agar plate inoculated with *B. subtilis* and incubated for 24 hours
4. Lugol's iodine solution in a container with a dropper
5. Container with disinfectant for disposal of cultures

## Procedure

This procedure is to be demonstrated by the instructor.

1. Examine the starch agar plates. Sketch the appearance, and indicate the color of each on one side of the appropriate circle in the results and observations section.
2. Your instructor will flood the plates with an iodine solution.
3. After 15 seconds, examine the plates. Sketch their appearance, and indicate the colors formed on the other side of the appropriate circles in the results and observations section.
4. Answer the questions in the results and observations section.

## SALIVARY AMYLASE ACTIVITY

Starch is one of the major carbohydrates in the human diet. To use starch, the body must first break down the large starch molecule into smaller glucose molecules.

The process by which starch is hydrolyzed into smaller units begins in the mouth, by the action of an enzyme *amylase*. In the next set of experiments, the progress of this enzymatically catalyzed reaction and the various factors that can effect it are followed by observing the disappearance of the reacting substrate (starch), using the iodine test.

## Materials

The following materials should be provided for class use:

1. Three stock starch solutions—5 percent, 1 percent, and 0.5 percent
2. Three buffer solutions, with pH of 4, 7, and 9
3. Lugol's iodine or other similar iodine reagent in a dropper bottle
4. Pieces of paraffin or sugarless gum
5. Pyrex test tubes
6. Test-tube racks
7. 12-ml Erlenmeyer flasks
8. Porcelain spot plates
9. Gauze
10. Glass or plastic funnels
11. 50-ml graduated cylinders
12. Water baths—4°C, 37°C, and 70°C
13. Distilled water
14. Container for the disposal of saliva and related materials

## Procedure 1: Salivary Amylase Stock Preparation

This procedure is to be performed by students in groups of four.

1. Two students should produce and collect 20 ml of saliva each. The sample should be measured and pooled in a clean Erlenmeyer flask. (Note that saliva excretion can be stimulated by drinking a glass of water and then chewing on a small piece of paraffin or sugarless gum.)
2. Add an equal amount of room temperature distilled water, and mix the contents of the flask. Pass the mixture through a gauze filter system, to remove most of the mucus. The resulting preparation is your amylase stock solution for the remaining procedures in this exercise. (If additional enzyme material is needed, repeat this procedure.)

## Procedure 2: Effects of Temperature on Amylase Activity

This procedure is to be performed by students in pairs.

1. Label four test tubes with numerals 1–4, respectively, and place them in a test-tube rack.
2. Add 5 ml of amylase stock solution to Tubes 1, 2, and 3.
3. Place the tubes in the temperature environment specified next. Keep the tubes at these temperatures for at least 10 minutes (to allow all the solution to reach the desired temperature).

    Tube 1: 4°C

    Tube 2: 37°C

    Tube 3: 70°C

    Tube 4 (control): room temperature (25°C). This is the control tube for the starch–iodine reaction.

4. Add 5 ml of the 1 percent starch solution to each tube, shake gently, and return the mixtures to their water baths to incubate for 5 minutes.
5. Remove a small sample from each tube, place each into a separate depression on a spot plate, and test immediately for starch hydrolysis by adding 2 drops of the iodine reagent.
6. Enter your findings in the results and observations section.
7. Wash and rinse the tubes used.

## Procedure 3: Effects of pH on Amylase Activity

This procedure is to be performed by students in pairs.

1. Number the four test tubes, and place them in a rack.
2. Add 3 ml of the amylase stock solution to each tube.
3. Add the following substances to the tubes, as indicated:

    Tube 1: 3 ml buffer solution, pH 4

    Tube 2: 3 ml buffer solution, pH 7

    Tube 3: 3 ml buffer solution, pH 9

    Tube 4: 3 ml distilled water, pH 7

4. Shake and mix the contents, and place them in a 37°C water bath for 10 minutes.
5. Add 3 ml of 1 percent starch solution to each tube, mix, and continue to incubate at 37°C.
6. Almost immediately, test the contents of each tube for starch hydrolysis, as described in Procedure 2, Step 5.
7. Continue to incubate the preparations at 37°C, and take additional samples at 15 and 30 minutes. Test for starch hydrolysis, as in Step 6.
8. Enter your findings in the results and observations section.
9. Wash and rinse the tubes used.

## Procedure 4: Effect of Substrate Concentration on Amylase Activity

This procedure is to be performed by students in pairs.

1. Number 12 test tubes, and place them in a rack in two rows of 6 tubes each.
2. Add 1 ml of the amylase stock solution to each tube.
3. Add the following substances to the tubes, as specified:

**Row I**

Tube 1: 1 ml 1 percent starch solution
Tube 2: 2 ml 1 percent starch solution
Tube 3: 3 ml 1 percent starch solution
Tube 4: 5 ml 1 percent starch solution
Tube 5: 10 ml 1 percent starch solution
Tube 6: 5 ml distilled water

**Row II**

Tube 1: 1 ml 5 percent starch solution
Tube 2: 2 ml 5 percent starch solution
Tube 3: 3 ml 5 percent starch solution
Tube 4: 5 ml 5 percent starch solution
Tube 5: 10 ml 5 percent starch solution
Tube 6: 5 ml distilled water

4. Shake the tubes gently to mix the contents, and place them in a 37°C water bath for 10 minutes.
5. Take samples and test for starch hydrolysis at 1 minute, 15 minutes, 30 minutes, and 60 minutes, as described in Procedure 2, Step 5.
6. Enter your findings in the results and observations section.
7. Wash and rinse the tubes used.

## DIGESTION OF FATS

### Materials

The following materials should be provided for class use:

1. The following substances in appropriate containers
   a. Safflower oil
   b. Distilled water
   c. Ethyl alcohol
   d. Xylene
   e. Bile salts (5 percent solution)
2. 5-ml pipettes with bulbs
3. Test tubes
4. Test-tube racks

### Procedure

This procedure is to be performed by students in pairs.

1. Number five test tubes and place them in a rack.
2. Add 2 ml of safflower oil to each tube.
3. Carefully add the following substances to the tubes, as specified:

   Tube 1: 5 ml distilled water

   Tube 2: 5 ml ethyl alcohol

   Tube 3: 3 ml xylene

   Tube 4: 3 ml bile salts

   Tube 5: no additions

4. Shake the tubes gently to mix the contents, and examine all tubes immediately. Look for a complete mixing of the contents of the tubes.
5. Repeat your examinations at 15 and 30 minutes.
6. Enter your findings in the results and observations section.

## DIGESTION OF PROTEIN

### Materials

The following materials should be provided for class use:

1. Small pieces of coagulated egg albumin
2. The following solutions in appropriate containers
   a. 1 percent trypsin
   b. 1 percent pepsin
   c. 0.5 percent sodium carbonate
   d. 0.8 percent hydrochloric acid
3. Distilled water
4. Biuret test reagents
5. Test tubes
6. Test-tube racks
7. 5-ml pipettes with bulbs
8. 37°C water bath

### Procedure: The Actions of Pepsin and Trypsin on Egg Albumin

This procedure is to be performed by students in pairs.

1. Prepare a test-tube rack containing two rows of six tubes each. Number the tubes accordingly (1–6 in each row).
2. Add the following substances to the tubes, as specified:

**Row I**

Tube 1: 5 ml pepsin + 5 ml distilled water
Tube 2: 5 ml pepsin + 5 ml sodium carbonate
Tube 3: 5 ml pepsin + 5 ml hydrochloric acid
Tube 4: 5 ml distilled water + 5 ml sodium carbonate
Tube 5: 5 ml distilled water + 5 ml hydrochloric acid
Tube 6: 5 ml distilled water

**Row II**

Tube 1: 5 ml trypsin + 5 ml distilled water
Tube 2: 5 ml trypsin + 5 ml sodium carbonate
Tube 3: 5 ml trypsin + 5 ml hydrochloric acid
Tube 4: 5 ml distilled water + 5 ml sodium carbonate
Tube 5: 5 ml distilled water + 5 ml hydrochloric acid
Tube 6: 5 ml distilled water

3. Add several pieces of coagulated egg albumin to each tube, shake gently to mix, and place all tubes in a 37°C water bath.
4. Examine each tube at 15-minute intervals during a 90-minute period, for evidence of the digestion of the egg albumin. This type of reaction is indicated by a disintegration of the protein and a general clearing of the protein material.
5. At the end of 90 minutes, add 4 drops of the Biuret solution to each tube, to test for the presence of the end products of protein digestion. The formation of a pink to violet color indicates the presence of polypeptides. (Refer to Color Plate 53.)
6. Record your findings in the results and observations section.
7. Wash and rinse the tubes used.

## RESULTS AND OBSERVATIONS

### *Histology of the Organs and Tissues Associated with Digestion*

1. Sketch and label representative portions of the tissues indicated.

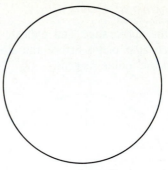

Large intestine

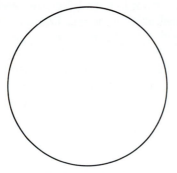

Salivary glands

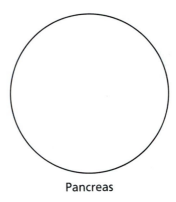

Pancreas

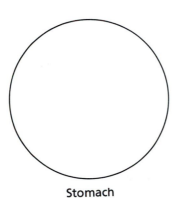

Stomach

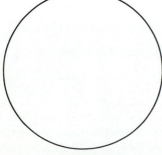

Small intestine

2. Indicate the specific metabolically associated product(s) of the structures or cells listed.

a. Salivary glands _____

b. Chief cells _____

c. Parietal cells _____

d. Mucous cells _____

e. Brunner's glands _____

f. Goblet cells (small intestine) _____

g. Pancreas _____

_____

## Demonstration of Extracellular Digestion

1. Sketch the appearance, and indicate the color of the starch agar plates before and after the application of the iodine reagent.

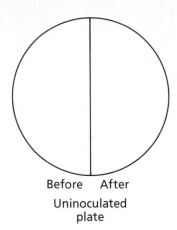

Before     After

Uninoculated
plate

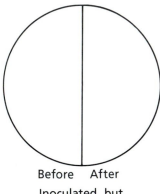

Before     After

Inoculated, but
not incubated plate

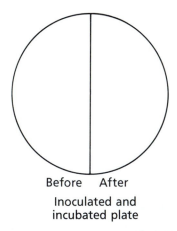

Before     After

Inoculated and
incubated plate

2. In which of the plates was starch hydrolysis observed? _____

3. Why must an inoculated plate be incubated to produce a positive reaction? _____
_____

4. What color develops after the addition of iodine in the cases of

   a. Starch hydrolysis _____

   b. Presence of starch _____

## Salivary Amylase Activity

1. Enter in Table 45-1 the reaction to iodine at different temperatures, and indicate whether or not starch hydrolysis occurred. Indicate a positive reaction with a "+" and a negative reaction with a "−."

2. Enter in Table 45-2 the results of the tests for starch hydrolysis on substances at different pHs. Indicate a positive reaction with a "+" and a negative reaction with a "−."

3. Enter in Table 45-3 the results of the tests for starch hydrolysis on substances of different substrate concentration. Indicate a positive reaction with a "+" and a negative reaction with a "−."

4. Summarize your conclusions regarding the influence of temperature, pH, and substrate concentration on enzyme activity. Use an additional page, if needed. _____
_____
_____

5. What would have happened in the tests of amylase activity if alcohol had been used as the diluting fluid, instead of distilled water? _____
_____
_____

## Digestion of Fats

1. Enter in Table 45-4 your observations of the degree to which the contents of the tubes mixed.

2. If there were any observable differences in any of the tubes at 15 or 30 minutes, explain why. _____
_____
_____

3. Where is bile produced in the body? _____

4. What is the function of bile? _____
_____

## Digestion of Protein

1. Enter your observations and test results for protein digestion in Table 45-5. Indicate evidence of protein and a positive test reaction with a "+" and no evidence and a negative test reaction with a "−."

**Table 45-1**

| Tube | Temperature | Color After Addition of Iodine | Starch Hydrolysis |
|------|-------------|-------------------------------|-------------------|
| 1 | 4°C | | |
| 2 | 37°C | | |
| 3 | 70°C | | |
| 4 | 25°C | | |

**Table 45-2**

| Tube | Substance | Immediate | 15 Minutes | 30 Minutes |
|------|-----------|-----------|------------|------------|
| 1 | Buffer, pH 4 | | | |
| 2 | Buffer, pH 7 | | | |
| 3 | Buffer, pH 9 | | | |
| 4 | Distilled water, pH 7 | | | |

**Table 45-3**

| Row | Tube | Concentration Amount (ml) | Concentration Solution | 1 Minute | 15 Minutes | 30 Minutes | 60 Minutes |
|-----|------|---------------------------|------------------------|----------|------------|------------|------------|
| I | 1 | 1 | 1% starch | | | | |
| I | 2 | 2 | 1% starch | | | | |
| I | 3 | 3 | 1% starch | | | | |
| I | 4 | 5 | 1% starch | | | | |
| I | 5 | 10 | 1% starch | | | | |
| I | 6 | 5 | distiled water | | | | |
| II | 1 | 1 | 5% starch | | | | |
| II | 2 | 2 | 5% starch | | | | |
| II | 3 | 3 | 5% starch | | | | |
| II | 4 | 5 | 5% starch | | | | |
| II | 5 | 10 | 5% starch | | | | |
| II | 6 | 5 | distiled water | | | | |

2. Which of the preparations produced a positive Biuret reaction? _____

_____

3. What is digestion? _____

_____

**Table 45-4**

| Tube | Substance Mixed with Oil | Immediate | 15 Minutes | 30 Minutes |
|------|--------------------------|-----------|------------|------------|
| 1 | Distilled water | | | |
| 2 | Ethyl alcohol | | | |
| 3 | Xylene | | | |
| 4 | Bile salts | | | |
| 5 | None | | | |

**Table 45-5**

| Row | Tube | Solution | 15 Minutes P | T | 30 Minutes P | T | 45 Minutes P | T | 60 Minutes P | T | 75 Minutes P | T | 90 Minutes P | T |
|-----|------|----------|---|---|---|---|---|---|---|---|---|---|---|---|
| I | 1 | Distilled water | | | | | | | | | | | | |
| I | 2 | Sodium carbonate | | | | | | | | | | | | |
| I | 3 | HCl | | | | | | | | | | | | |
| I | 4 | Sodium carbonate | | | | | | | | | | | | |
| I | 5 | HCl | | | | | | | | | | | | |
| I | 6 | — | | | | | | | | | | | | |
| II | 1 | Distilled water | | | | | | | | | | | | |
| II | 2 | Sodium carbonate | | | | | | | | | | | | |
| II | 3 | HCl | | | | | | | | | | | | |
| II | 4 | Sodium carbonate | | | | | | | | | | | | |
| II | 5 | HCl | | | | | | | | | | | | |
| II | 6 | — | | | | | | | | | | | | |

# LABORATORY REVIEW 45

NAME _____

LAB SECTION _____ DATE _____

## METABOLISM: ENZYME ACTION

### COMPLETION, DEFINITION, AND INTERPRETATION

1. Are the pH optimums the same for most enzymes? Explain. _____

   _____

   _____

2. What is an enzyme–substrate complex? Use labeled diagrams to explain your answer.

3. List three major factors that affect enzyme action. Briefly explain how each affects enzymatic activity.

   a. _____

   _____

   b. _____

   _____

   _____

   c. _____

   _____

4. How are enzymes named? _____

   _____

   _____

5. Define or explain the following:

   a. Enzyme _____

   b. Substrate _____

   c. Optimum _____

   d. Active site _____

## ANSWERS

1. _____

2. _____

3. _____

4. _____

5. _____

## MULTIPLE CHOICE

Enter the correct answers in the spaces provided.

1. What color formation indicates a positive starch hydrolysis reaction after the addition of an iodine solution?
   a. purple
   b. yellow
   c. green
   d. black
   e. blue

2. Which of the following can affect enzymatic activity?
   a. temperature
   b. pH
   c. substrate concentration
   d. enzyme concentration
   e. choices *a* through *d*

3. In which of the following would fats be dissolvable?
   a. distilled water
   b. bile salts
   c. xylene
   d. ethyl alcohol
   e. choices *b, c,* and *d* only

4. If a pink color formed when Biuret solution was added to an unknown specimen, which of the following would this color be indicating to be present?
   a. distilled water
   b. starch
   c. fat
   d. polypeptides
   e. sodium carbonate

5. If iodine has been added to an unknown solution, which rapidly turns purple, which of the following does the solution probably contain?
   a. amylase
   b. albumin
   c. starch
   d. glucose
   e. bile

# Digestion and Chemistry

## After completing this exercise, you should be able to

1. Perform general procedures and chemical tests to demonstrate the presence of carbohydrates, lipids, and proteins
2. Apply these procedures to identify the contents of an unknown specimen

Several tests and procedures, which are used in the identification of carbohydrates, lipids, proteins, and other substances of dietary importance, are demonstrated in this exercise. Be sure to handle all chemicals carefully. Observe safety rules.

## CARBOHYDRATE IDENTIFICATION

### Materials

The following materials should be provided for class use:

1. 1 percent known solutions in dropper bottles
   a. Glucose
   b. Lactose
   c. Maltose
   d. Albumin
   e. Starch
   f. Gelatin
2. Test materials in dropper bottles
   a. Honey
   b. Sugar-free drink
   c. Nondiet cola drink
   d. Distilled water
   e. Syrup
   f. Milk
   g. Saccharin
   h. Orange juice
3. Test reagents
   a. Benedict's reagent
   b. Lugol's iodine
   c. Molisch test reagents—5 percent alpha-naphthol alcoholic solution and concentrated sulfuric acid
4. Pyrex test tubes, test-tube racks, and test-tube holders
5. 1-ml and 10-ml pipettes with bulbs
6. Boiling water bath
7. Wax pencils or other marking devices

## Procedure 1: Molisch Reaction

The Molisch reaction is a general test for carbohydrates. The formation of a purple or pink color, upon the addition of the Molisch reagents indicates the presence of a carbohydrate.

This procedure is to be performed by students in pairs.

1. Label a test tube for each of the 1 percent known solutions. Place the tubes in a rack.
2. Introduce 10 drops of each known solution into its labeled tube. Use a separate pipette for each. Examine each preparation for any distinctive color, smell, or other noticeable property. Enter your findings in the results and observations section.
3. Add 2 drops of the 5 percent alpha-naphthol solution to each tube. Mix the contents by gently tapping the tube.
4. Tilt each tube, and carefully add 5 ml of the concentrated sulfuric acid ($H_2SO_4$), so that it runs slowly down the inside surface. The acid is heavier and should form a layer beneath the test solutions.

5. Examine each tube for the formation of a pink or purple color at the boundary between the test solution and the sulfuric acid. Record your findings in the results and observations section.
6. Dispose of the chemical solutions as directed by your instructor.
7. Repeat Steps 1–6 with the test materials.

## Procedure 2: Reducing Sugars Test

Benedict's reagent is used to detect the presence of a reducing sugar. When the cupric ion ($Cu^{2+}$) in the reagent reacts with a reducing sugar, a red precipitate of cuprous oxide forms.

This procedure is to be performed by students in pairs.

1. Label a test tube for each of the 1 percent known solutions. Place the tubes in a rack.
2. Introduce 10 drops of each known solution into its labeled tube. Use a separate pipette for each.
3. Using a 10-ml pipette, add 5 ml of Benedict's reagent to each test tube. Mix the contents by gently tapping the tubes.
4. Place all the tubes in a boiling water bath for 5 minutes.
5. Examine each tube for evidence of the presence of a reducing sugar (red precipitate) (see Color Plate 55). Enter your findings in the results and observations section.
6. Dispose of the chemical solutions as directed by your instructor.
7. Repeat Steps 1–6 with the test materials.

## Procedure 3: Starch Test

The presence of starch generally is indicated by the formation of a blue-black color, upon the addition of an iodine reagent.

This procedure is to be performed by students in pairs.

1. Label a test tube for each of the 1 percent known solutions. Place the tubes in a rack.
2. Introduce 10 drops of each known solution into its labeled tube. Use a separate pipette for each.
3. Add 3 drops of the iodine reagent to each tube, and mix.
4. Examine each tube for the formation of a blue-black color. Enter your findings in the results and observations section.
5. Dispose of the chemical solutions as directed by your instructor.
6. Repeat Steps 1–5 with the test materials.

## PROTEIN IDENTIFICATION

### Materials

The following materials should be provided for class use:

1. 1 percent known solutions
   a. Albumin
   b. Cysteine
   c. Glucose
   d. Glycine
   e. Phenylalanine
   f. Starch
   g. Tryptophan
2. Test materials
   a. Casein
   b. Hair
   c. Gelatin
3. Test reagents
   a. Biuret reagents—6 molar (M) sodium hydroxide (NaOH) and 0.5 percent copper sulfate ($CuSO_4$)
   b. 1 molar (M) lead acetate ($PbAc_2$) solution
   c. Concentrated nitric acid ($HNO_3$)
   d. Glacial acetic acid
   e. Litmus paper
4. 1-ml and 5-ml pipettes with bulbs
5. Pyrex test tubes, test-tube racks, and test-tube holders
6. Boiling water bath
7. Wax pencils or other marking devices

### Procedure 1: Biuret Test

The Biuret test is named for the purple compound that forms when a dilute alkaline solution of copper sulfate reacts with substances containing two or more peptide bonds.

This procedure is to be performed by students in pairs.

1. Label a test tube for each of the 1 percent known solutions and test materials. Place the tubes in a rack.
2. Introduce 1 ml of each solution into its labeled tube. Use a separate pipette for each.
3. Add 2 ml of the sodium hydroxide solution and 10 drops of the copper sulfate to each tube. Shake the tube gently to mix the contents.
4. Look for the presence of a pink to purple color (Color Plate 53). Enter your findings in the results and observations section.
5. Dispose of the chemical solutions as directed by your instructor.

## Procedure 2: Sulfur Test

The sulfur test is used to detect the presence of amino acids in proteins that contain sulfur. When lead acetate is added to compounds containing sulfur, lead sulfide is formed. Lead sulfide is black.

This procedure is to be performed by students in pairs.

1. Label a test tube for each of the 1 percent known solutions. Put the tubes in a rack.
2. Introduce 1 ml of each known solution into its labeled tube. Use a separate pipette for each.
3. Add 2 ml of sodium hydroxide to each tube, and place the tubes in a boiling water bath for 60 minutes.
4. Remove the tubes.
5. Neutralize the contents of each tube by carefully adding glacial acetic acid until blue litmus paper (pH indicator) turns red.
6. Add 10 drops of lead acetate.
7. Look for a black color. Enter your findings in the results and observations section.
8. Dispose of the chemical solutions as directed by your instructor.

## Procedure 3: Xanthoproteic Test

This test detects the presence of benzene rings in amino acids. Such compounds produce yellow to red colors when exposed to nitric acid.

This procedure is to be performed by students in pairs.

1. Label a test tube for each of the 1 percent known solutions and the test materials. Place the tubes in a rack.
2. Introduce 1 ml of each known solution into its labeled tube. Use a separate pipette for each.
3. Carefully add 1 ml of concentrated nitric acid by letting the acid run down the side of each tube.
4. Place the tubes in a boiling water bath for 10 seconds.
5. Remove the tubes, and observe the reactions. Enter your findings in the results and observations section.
6. Dispose of the chemical solutions as directed by your instructor.

## Procedure 4: Denaturation by Heat

Some proteins become *denatured* (permanently altered) and precipitate out of solution when exposed to boiling temperatures.

This procedure is to be performed by students in pairs.

1. Label a test tube for each of the 1 percent known solutions. Place the tubes in a rack.
2. Introduce 5 ml of each known solution into its labeled tube. Use a separate pipette for each.
3. Place the tubes in a boiling water bath for 3 minutes.
4. Examine each tube for the presence of a solid mass or precipitate. Enter your findings in the results and observations section.
5. Dispose of the chemical solutions as directed by your instructor.

## PROPERTIES OF FATS

### Materials

The following materials should be provided for class use:

1. Solvents
   a. Distilled water
   b. Ethyl alcohol
   c. 0.55 percent sodium chloride
   d. Xylene
2. Test solutions in dropper bottles
   a. 1 percent albumin
   b. Cottonseed or immersion oil
   c. Gelatin
   d. 2 percent glucose
   e. 2 percent starch
3. Test tubes and test-tube racks
4. 5-ml pipettes with bulbs
5. Brown paper strips (7.5 × 5 cm)
6. Wooden applicator sticks
7. Wax pencils or other marking devices

### Procedure 1: Solubility Testing

This procedure is to be performed by students in pairs.

> **CAUTION**
>
> (*Note:* Make certain that there are no open flames in the area where the following tests are being performed.)

1. Label a test tube for each of the solvents. Place the tubes in a rack.
2. Introduce 5 ml of each solvent into its labeled tube. Use a separate pipette for each.

3. Add 6 drops of the test-material albumin to each tube and shake.
4. Examine each tube immediately and after 15 minutes, to see whether the test material dissolved. Enter your findings in the results and observations section.
5. Dispose of the chemical solutions as directed by your instructor.
6. Repeat Steps 1–5 with each of the other test materials.

### Procedure 2: Grease Spot Test

This procedure is to be performed by students in pairs.

1. On a brown paper strip, label a small area for each of the test materials.
2. Using a separate applicator stick for each, apply a small amount of each test material to its labeled area.
3. Examine each immediately and after 30 minutes for the presence of a grease spot. Enter your findings in the results and observations section.

## IDENTIFICATION OF CONTENTS OF UNKNOWN SAMPLE

### Materials

The following materials should be provided for class use:

1. Test reagents provided earlier for carbohydrate, lipid, and protein tests
2. Pyrex test tubes, test-tube racks, and test-tube holders
3. 1-ml and 10-ml pipettes with bulbs
4. Boiling water bath
5. One unknown sample (10 ml) per student

### Procedure

This procedure is to be performed by students individually.

1. Obtain an unknown sample from your instructor.
2. Examine the unknown specimen for color, smell, and other distinctive properties.
3. Divide your specimen into several samples, and identify its contents by performing whichever of the preceding procedures you think are necessary.
4. Enter your findings in the results and observations section.
5. Dispose of all chemical solutions as directed by your instructor.

## RESULTS AND OBSERVATIONS

### Carbohydrate Identification

1. In Table 46-1, enter your observations of physical properties of the known solutions and test materials.
2. In Table 46-2, enter the chemical test results.
3. Which of the materials tested showed the presence of a reducing sugar? _____

_____

### Protein Identification

1. Enter your observations of the reactions to the chemical tests in Table 46-3.
2. Which of the materials tested contain peptide bonds?

_____

3. Which of the amino acids tested contain a benzene ring? _____

_____

4. How many of the materials tested showed the presence of sulfur? Which ones? _____

_____

5. Which of the materials tested solidifed when heated? _____

### Properties of Fats

1. Indicate in Table 46-4 the results of the fat-solubility test.
2. Which of the substances tested produced an obvious grease spot on the brown paper strip?_____

_____

3. What other types of reactions besides those of solubility did you find? _____

_____

### Identification of Contents of Unknown Sample

1. List the tests used, and enter your findings in Table 46-5.
2. On the basis of the tests performed and the results obtained, the unknown sample _____

_____ contains _____

_____

**Table 46-1**

| Materials | Color | Smell | Other Properties |
|---|---|---|---|
| Glucose | | | |
| Lactose | | | |
| Maltose | | | |
| Albumin | | | |
| Starch | | | |
| Gelatin | | | |
| Honey | | | |
| Sugar-free drink | | | |
| Cola drink | | | |
| Distilled water | | | |
| Syrup | | | |
| Milk | | | |
| Saccharin | | | |
| Orange juice | | | |

**Table 46-2**

| Materials | Carbohydrate Test | Reducing Sugars Test | | Starch Test | |
|---|---|---|---|---|---|
| | Molisch Reaction | Reaction with Benedict's Reagent | Reducing Sugar Present | Reaction with Iodine | Starch Present |
| Glucose | | | | | |
| Lactose | | | | | |
| Maltose | | | | | |
| Albumin | | | | | |
| Starch | | | | | |
| Gelatin | | | | | |
| Honey | | | | | |
| Sugar-free drink | | | | | |
| Cola drink | | | | | |
| Distilled water | | | | | |
| Syrup | | | | | |
| Milk | | | | | |
| Saccharin | | | | | |
| Orange juice | | | | | |

**Table 46-3**

| Material | Biuret Reaction | Sulfur Reaction | Xanthoproteic Reaction |
|---|---|---|---|
| Albumin | | | |
| Cysteine | | | |
| Glucose | | | |
| Glycine | | | |
| Phenylalanine | | | |
| Starch | | | |
| Tryptophan | | | |
| Casein | | | |
| Hair | | | |
| Gelatin | | | |

**Table 46-4**

| Solvent | Albumin | | Cottonseed Oil | | Gelatin | | Glucose | | Starch | |
|---|---|---|---|---|---|---|---|---|---|---|
| | 0 Time* | After 15 Minutes | 0 Time* | After 15 Minutes | 0 Time* | After 15 Minutes | 0 Time* | After 15 Minutes | 0 Time* | After 15 Minutes |
| Distilled water | | | | | | | | | | |
| Ethyl alcohol | | | | | | | | | | |
| Sodium chloride | | | | | | | | | | |
| Xylene | | | | | | | | | | |

*Immediate reaction.

**Table 46-5**

| Test or Procedure | Description of Results |
|---|---|
| General appearance | |
| | |
| | |
| | |
| | |
| | |
| | |
| | |
| | |
| | |

# LABORATORY REVIEW *46*

## *DIGESTION AND CHEMISTRY*

## COMPLETION, DEFINITION, AND INTERPRETATION

1. Complete Table Q46-1.

**Table Q46-1**

| Chemical Test Reagent | Test Substance | Reaction Produced |
|---|---|---|
| Iodine | Glucose | a. |
| Iodine | Starch | b. |
| Benedict's solution | c. | Red precipitate |
| Benedict's solution | Galactose | d. |
| Molisch reagents (alphanaphthol and concentrated sulfuric acid) | Albumin | e. |
| Molisch reagents | Sucrose | f. |
| Ninhydrin | g. | Purple color |
| Ninhydrin | Starch | h. |
| Biuret test reagent | Starch | i. |
| Biuret test reagent | Albumin | j. |
| Iodine | Orange juice and starch solution | k. |

2. Define or explain the following:

   a. Reducing sugar _____

   _____

   b. Carbohydrate _____

   _____

   c. Vitamin _____

   _____

   d. Hydrolysis _____

   _____

   e. Lipase _____

3. Differentiate between the following:

    a. Monosaccharide and polysaccharide _____

    _____

    b. Amino acid and peptide _____

    _____

    c. Exocrine and endocrine _____

    _____

4. What chemical elements are present in the compounds listed?

    a. All carbohydrates _____ ;

    b. All proteins _____ ;

    c. All lipids _____ ;

    d. All nucleic acids _____ .

# Cholesterol Testing

**After completing this exercise, you should be able to**

1. Differentiate between *atherosclerosis* and *arteriosclerosis*
2. Describe several properties of cholesterol and its role in atherosclerosis
3. Distinguish among *triglycerides, low-density lipoproteins,* and *high-density lipoproteins*
4. List conditions that can cause elevated total serum cholesterol levels
5. List conditions that can cause decreased levels of total serum cholesterol
6. Perform a laboratory test for the determination of total serum cholesterol levels

*Atherosclerosis* is a vascular disease associated with lesions, known as *atherosclerotic plaques* (Figure 47-1), on and in the walls of high-pressure arteries. These include the vessels in the heart, brain, kidneys, arms, and legs. The disease is a form of *arteriosclerosis,* which includes several conditions characterized by abnormal thickening or hardening of the arteries. The two conditions differ, in that atherosclerosis involves the depositing of fatty substances, especially *cholesterol* and *triglycerides* (ingested fats), while arteriosclerosis is associated with individual or combination deposits of calcium, lipid, and connective tissue.

*Atherosclerotic vascular disease* and its complications, which include cerebral infarction, coronary artery disease, coronary or myocardial infarction (heart attack), and pulmonary obstruction, are the leading causes of death in Western society. Most epidemiological analyses show a strong correlation of serum cholesterol levels with coronary disease.

Cholesterol is a fatlike substance derived from either the breakdown of ingested fats or formed by the liver. It is found throughout the body, especially in blood, bile, and brain tissue. Most of the body's cholesterol is formed in the liver and flows with bile into the small intestine, where it mixes with its dietary sources. Cholesterol and triglycerides cannot dissolve in water and thus cannot travel in the blood in their unaltered forms. They are made water-soluble by combining with proteins produced by the liver and intestine *(apoproteins),* and the complexes thus formed are termed *lipoproteins,* which vary in size, weight, and density.

Two of the major classes are *low-density lipopro-*

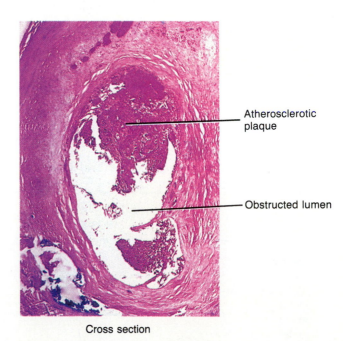

Cross section

**Figure 47-1** Photomicrograph of the cross-section of an artery partially obstructed by an atherosclerotic plaque. (Courtesy of Sklar/Photo Researchers.)

**655**

*teins (LDLs)* and *high-density lipoproteins (HDLs).* There is an important difference between the two: *LDL* contains 60 to 70 percent of the *total serum cholesterol.* It also appears to pick up cholesterol and deposit it in body cells, including—under abnormal conditions—smooth-muscle fibers in arteries. Thus, as blood levels of LDL increase, the risk of coronary artery disease and related conditions *increases.* In contrast, as blood levels of HDL increase, the risk of coronary artery disease and other related conditions *decreases.* A third type of lipoprotein, known as a *very low-density lipoprotein (VLDL),* contains about 10 to 15 percent of total serum cholesterol and a large amount of triglyceride (fat). Its role in the development of atherosclerotic plaques is not clear. The amount of *total cholesterol (HDL, LDL, and VLDL,* combined) in blood is one commonly used indicator of risk for coronary artery disease.

Total cholesterol concentrations are reliable indicators of an individual's tendency toward atherosclerosis. Elevated cholesterol levels *(hypercholesterolemia)* are considered to be one of the factors accompanying vascular disorders. Although cholesterol itself does not cause atherosclerosis, excess amounts of lipoproteins (hyperlipoproteinemia) containing cholesterol do contribute to the formation of plaque deposits that cause artery blockage (Figure 47-1). Thus, cholesterol levels can be used to evaluate the risk of future heart attack (myocardial infarction). Determinations of total cholesterol, such as the one considered in this exercise, are used mainly in the study of individuals subject to cardiovascular disease and may occasionally be included with tests dealing with liver function and disease diagnosis.

The range of normal cholesterol values varies in adults and appears to be influenced by several factors, including age, diet, sex, race, physical activity, stress, and seasonal changes. Table 47-1 shows the total serum cholesterol values according to age. Elevated total cholesterol levels may occur in conditions such as ather-

osclerosis; infections involving the kidney, liver, and pancreas; diabetes mellitus; hypothyroidism; leukemia; pregnancy; early starvation; possible hereditary disorders associated with large amounts of cholesterol in the blood *(hypercholesterolemia);* and increased levels of lipoproteins in the blood *(hyperlipoproteinemia).* In general, as the total cholesterol level increases above 150 mg/dl (a *deciliter* [dl] is $\frac{1}{10}$ of a liter), the risk of coronary and related diseases slowly begins to rise. Above 200 mg/dl, the risk increases even more rapidly. The chance of a heart attack has been found to double with every 50 mg/dl increase in total cholesterol once the level goes over 200 mg/dl.

Decreased total cholesterol levels may develop with conditions such as extensive liver disease, malnutrition, some forms of anemia, severe infections, terminal cancer, tuberculosis, intestinal abstruction, and hepatitis.

This exercise uses the AccuMeter® Cholesterol Test. The procedure incorporates a handheld cassette that consists of three plastic parts (Figure 47-2): *a cover, a movable tab, and a base.* The cover holds the *measurement scale,* the *blood application well,* and *two windows* through which the functioning of the test reagents and the completion of the assay can be observed. The movable tab and base operate together to facilitate a sequence of steps in the procedure and the associated chemical reactions. The assay is performed in four simple steps (see Figure 47-3):

1. The blood sample is obtained from the finger.
2. The blood is applied to the well and allowed to filter.
3. The tab on the cassette is pulled, to start the reactions.
4. The results are then read directly from the cassette.

The reactions involved can be summarized as follows. The blood sample first comes in contact with a buffered protein solution containing horseradish peroxide (HRP). Cholesterol within the sample is converted to hydrogen peroxide by two enzymes (cholesterol esterase, and cholesterol oxidase) on the chromatography paper. The hydrogen peroxide makes contact with a uniformly coated dye in a distal region of the cassette (the measurement scale), and the hydrogen peroxide and the dye react with the HRP to form a purple peak.

Once the chemical reactions begin, a color change is visible at the proximal end of the paper strip (the first window), to show that the device is functioning properly. When the entire solution has reached the top of the measurement scale, an indicator produces a green color (seen through the second window), to signal that the assay is complete. The height of the color peak is proportional to the amount of hydrogen peroxide and thus the concentration of cholesterol within the sample. The height (in millimeters) is converted to cholesterol (in mg/dl) by means of a conversion table (Table 47-2).

**Table 47-1    Total Normal Cholesterol Values, According to Age**

| Age | Normal Cholesterol Values (in mg/dl) |
|---|---|
| Newborn | 50–100 |
| 1 year | 70–175 |
| Adolescence years | 135–240 |
| 20–29 | 144–275 |
| 30–39 | 165–295 |
| 40–49 | 170–315 |
| 50–69 | 175–340 |
| Over 70 | 130–245 |

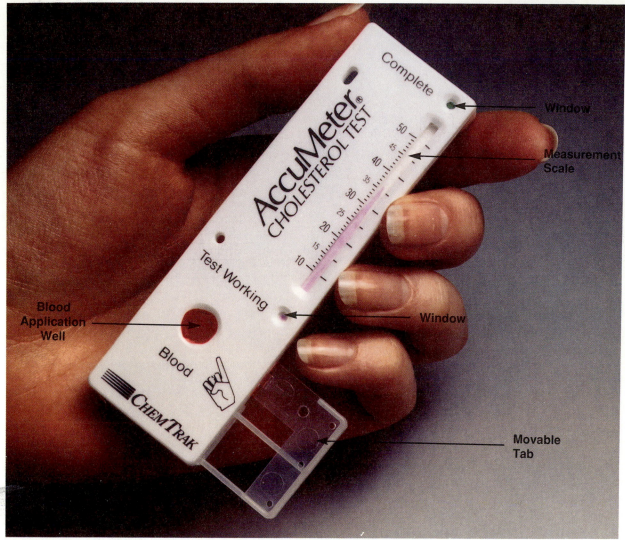

**Figure 47-2** The AccuMeter® is a handheld cassette consisting of three plastic parts: a *cover,* a *movable tab,* and a *base*. (Courtesy of CHEM TRAK, Sunnyvale, California.)

## Materials

The following materials should be provided for class use:

1. Sterile, disposable blood lancets
2. Commercial alcohol prep-pads
3. Sterile cotton squares
4. AccuMeter® Cholesterol Test cassettes (1 per student)
5. Disinfectant solution for disposal of blood-containing materials
6. Disposable surgical gloves

## Procedure

This procedure should be performed by students in pairs.

### CAUTION

Working with blood or other body fluids will not pose any danger if the following precautions are observed:

1. Place any and all blood or materials containing blood into the disinfectant containers provided.
2. Dispose of any blood-containing materials as indicated by the instructor.
3. In cases of blood spillage, wipe the area with a disinfectant-soaked paper towel, and dispose of all materials as indicated by the instructor.
6. Always wash your hands after handling blood and associated materials.
5. The use of disposable surgical gloves.

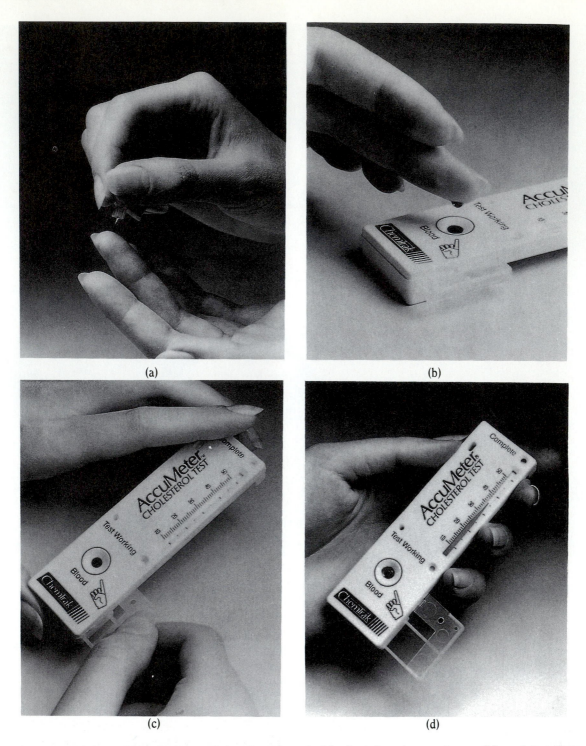

(a)

(b)

(c)

(d)

**Figure 47-3** Steps in the use of the AccuMeter®: (a) obtaining the specimen; (b) applying the blood specimen to the reaction well; (c) starting the reaction by pulling the "tab"; (d) reading the result. (Courtesy of CHEM TRAK, Sunnyvale, California.)

**Table 47-2   AccuMeter® Cholesterol Result Chart[a]**

| Cassette Reading | Cholesterol mg/dl[b] | Cassette Reading | Cholesterol mg/dl |
|---|---|---|---|
| 16.0 | 123 | 31.0 | 248 |
| 19.0 | 128 | 31.5 | 254 |
| 19.5 | 132 | 32.0 | 259 |
| 19.5 | 136 | 32.5 | 265 |
| 20.0 | 141 | 33.0 | 270 |
| 20.5 | 145 | 33.5 | 276 |
| 21.0 | 150 | 34.0 | 281 |
| 21.5 | 155 | 34.5 | 287 |
| 22.0 | 159 | 35.0 | 293 |
| 22.5 | 164 | 35.5 | 298 |
| 23.0 | 169 | 36.0 | 304 |
| 23.5 | 173 | 36.5 | 310 |
| 24.0 | 178 | 37.0 | 316 |
| 24.5 | 183 | 37.5 | 322 |
| 25.0 | 188 | 38.0 | 328 |
| 25.5 | 192 | 38.5 | 334 |
| 26.0 | 197 | 39.0 | 341 |
| 26.5 | 202 | 39.5 | 347 |
| 27.0 | 207 | 40.0 | 353 |
| 27.5 | 212 | 40.5 | 360 |
| 28.0 | 217 | 41.0 | 367 |
| 28.5 | 222 | 41.5 | 373 |
| 29.0 | 227 | 42.0 | 380 |
| 29.5 | 233 | 42.5 | 387 |
| 30.0 | 238 | 43.0 | 394 |
| 30.5 | 243 | 43.5 | 401 |

[a]National Cholesterol Education Program Guidelines:

| **Total Cholesterol** | **Classification** |
|---|---|
| Less than 200 mg/dl | Desirable |
| 200 to 239 mg/dl | Borderline high |
| 240 mg/dl or greater | High |

[b]mg/dl = milligrams per ceciliter

1. Examine the AccuMeter®, and locate its various parts (Figure 47-2).
2. Select the finger that will be used, and wipe it with a cotton square that has been moistened, but not saturated, with alcohol. Do not use a finger that will be involved later in holding a pencil or other such activity.
3. Unwrap the blood lancet. Note that the lancet is to be used only once and then discarded as indicated by the instructor.
4. Hold the finger in a downward position, and gently squeeze it.
5. Quickly puncture the ball of the finger with the lancet (Figure 47-3a).
6. Keeping the finger in a downward position, drip 3 drops of blood into the cassette well (Figure 47-3b).
7. Pull the tab on the right side of the AccuMeter® to start the reaction (Figure 47-3c).
8. Wait 12 minutes until the "completion-indicator window" turns green.
9. Read the result on the scale, and refer to Table 47-2 for your cholesterol reading.
10. Enter your scale reading and cholesterol value in the results and observations section.
11. Enter the readings and values of three other students in the class, for comparison purposes.
12. Dispose of all blood-containing materials as indicated by your instructor.
13. Wash your hands.
14. Answer the questions in the results and observations section.

## RESULTS AND OBSERVATIONS

1. Enter the AccuMeter® readings and cholesterol values in Table 47-3.
2. Compare the cholesterol values to those in Table 47-2

a. Were all values within the normal range? ____
b. If not, what factors could account for the values observed? _____

**Table 47-3**

| Student | AccuMeter® Reading | Cholesterol Value (mg/dl) | Approximate Age (in years) |
|---|---|---|---|
|  |  |  |  |
|  |  |  |  |
|  |  |  |  |
|  |  |  |  |
|  |  |  |  |

# LABORATORY REVIEW *47*

## *CHOLESTEROL TESTING*

### LISTING, COMPLETION, AND DEFINITION

1. List seven factors that may influence adult serum-cholesterol levels.

   a. _____

   b. _____

   c. _____

   d. _____

   d. _____

   f. _____

   g. _____

2. List eight conditions in which elevated serum-cholesterol levels may occur.

   a. _____

   b. _____

   c. _____

   d. _____

   e. _____

   f. _____

   g. _____

   h. _____

3. List eight conditions in which decreased serum-cholesterol levels may occur.

   a. _____

   b. _____

   c. _____

   d. _____

   e. _____

   f. _____

   g. _____

   h. _____

4. Define or explain the following:

   a. Cholesterol _____

   _____

   b. Hypercholesterolemia _____

   _____

**ANSWERS**

c. Hyperlipoproteinemia _____

_____

d. Atherosclerotic plaque _____

_____

## MULTIPLE CHOICE

Insert the correct answers in the spaces provided.

1. _____

2. _____

3. _____

4. _____

5. _____

6. _____

7. _____

1. In atherosclerosis, which of the following substances are deposited in arterial walls?
   a. protein
   b. polysaccharides
   c. nucleic acids
   d. triglycerides
   e. amino acids

2. The lesion in atherosclerosis is called an "atherosclerotic [*what?*]"
   a. LDL
   b. plaque
   c. HDL
   d. VLDL
   e. choices *a* and *c*

3. What are two major classes of lipoproteins that form the total cholesterol of an individual?
   a. VLDL
   b. HDL
   c. LDL
   d. choices *a* and *b*
   e. choices *b* and *c*

4. As blood levels of LDL increase, what happens to the risk of coronary artery disease?
   a. remains the same
   b. increases
   c. decreases

5. When the risk of coronary artery disease decreases, what are likely to be the associated blood levels of HDL?
   a. remains the same
   b. increases
   c. decreases

6. What is the factor considered in ratio to total cholesterol that is used for predicting the risk of developing coronary artery disease and related conditions?
   a. HDL
   b. LDL
   c. plaque
   d. VLDL
   e. choices *a* and *d* only

7. Which of the following terms describe a condition of elevated cholesterol levels?
   a. hypocholesterolemia
   b. hyperlipoproteinemia
   c. atherosclerosis
   d. arteriosclerosis
   e. hypercholesterolemia

# The Urinary and Reproductive Systems

### THE URINARY SYSTEM

The major function of the urinary system is to regulate the volume and composition of the extracellular fluid. The accumulation of unused or toxic by-products of metabolism and changes in the individual's environment, food ingestion, or metabolic activities may upset the balance or constancy of the body's internal environment. The kidneys—major components of the urinary system—are essential homeostatic organs because of their ability to remove wastes and to alter the volume and composition of blood through the production and excretion of urine.

The composition of urine can vary greatly. Some variation in the amount of substances normally found in the urine may reflect altered body activity or food and fluid intake. Elevated levels of these common components or the presence of substances normally not found in urine may indicate disturbed metabolic functions or abnormal renal activity.

Disorders and diseases of the kidneys may result in severe impairment of function. If the kidneys cannot adequately regulate the composition of blood, an artificial device must be used in order to maintain life. The so-called artificial kidney uses a dialysis mechanism involving a membrane through which substances from the blood may diffuse.

Because of the importance of the kidneys and related structures, the functioning of the urinary system is routinely tested. Evaluation of kidney function usually involves determination of the composition and volume of urine. Such tests also may reveal the presence of microbial infection.

This section considers some specific physiological principles of renal function, the normal and abnormal components of urine, and the principle of hemodialysis. It examines the procedures involved in routine tests and detection of bacterial infection. It also demonstrates the pH value and buffering capacity of several substances involved in the acid–base balance of body fluids.

## THE REPRODUCTIVE SYSTEM

Exercises in this section describe the structures and functions of organs associated with sex-cell production and the stages of pregnancy and birth.

All eukaryotic cells exhibit the complex mechanism of *mitosis,* which ensures the even distribution of hereditary material. This process is quite efficient, and errors rarely occur. Mitosis results in the formation of two new cells, each containing the same number and types of chromosomes as in the original cell. The chromosome number normally remains constant from one generation of body, or somatic, cells to the next. In *meiosis,* or reduction division, the chromosome number is reduced by one-half. The process is associated with the sexual process of fertilization. In *fertilization,* the nuclei of two different cells, each containing one-half the normal number of chromosomes (haploid), fuse and establish the normal number of chromosomes (diploid). Errors during the stages of chromosome separation can result in genetic abnormalities. *Karyotyping,* a laboratory procedure, can be used to detect a number of chromosomal abnormalities. Exercises in this section also deal with meiosis, stages of early development, ovulation detection, pregnancy testing, and karoytyping.

## GENERAL REFERENCES

Greep, R. O. (Ed.). *Reproductive Physiology IV.* Baltimore: University Park Press, 1983.

Lenin, A. *The Cycling Female.* San Francisco: Freeman, 1979.

Stamey, T. A., and R. W. Kindrachuk. *Urinary Sediment and Urinalysis.* Philadelphia: Saunders, 1984.

Sullivan, L. D., and J. J. Grantham. *Physiology of the Kidney,* 2nd ed. Philadelphia: Lea & Febiger, 1982.

Tortora, G. J., and S. R. Grabowski. *Principles of Anatomy and Physiology,* 7th ed. New York: HarperCollins Publishers, 1993.

## SPECIFIC REFERENCES

Asman, J. (Ed.). *Infertility: Diagnosis and Management.* New York: Springer-Verlag, 1984.

Beaconsfield, P., G. Birdwood, and R. Beaconsfield. "The Placenta." *Scientific American,* 243: 94–102, 1980.

Fuchs, F., and A. Klopper (Eds.). *Endocrinology of Pregnancy,* 3rd ed. Philadelphia: Harper & Row, 1983.

Hirschhorn, N., and W. B. Greenough III. "Progress in Oral Rehydration Therapy." *Scientific American,* 264: 50–56, 1991.

Iannucci, L. "The Perplexities of Pregnancy." *FDA Consumer,* 24: 16–19, 1990.

Lewis, R. "Genetic Screening—Fetal Signposts on a Journey of Discovery." *FDA Consumer* 24: 17–23, 1990.

Nightingale, E. L., and M. Goodman. *Before Birth: Prenatal Testing for Genetic Disease.* Cambridge, MA: Harvard University Press, 1990.

Wassarman, P. M. "Fertilization in Mammals." *Scientific American,* 259: 78–84, 1988.

# Anatomy of the Urogenital System

## After completing this exercise, you should be able to

1. Name the parts of the kidney
2. Note the retroperitoneal position of the kidney
3. Point out the major regions seen in a coronal section of the kidney
4. Trace the pathway of wastes from the blood through the microscopic nephrons and gross structures of the urinary system
5. Name the parts of the reproductive systems of both sexes
6. Recognize the similarity of the location in the body of the reproductive organs of each sex

## Materials

The following materials should be provided for class use:

1. Cat dissection specimens
2. Dissecting tools
3. Model of human kidney
4. Charts of human kidney

The organs of the *urogenital system* (the urinary system plus the genital system) are closely related during embryonic development. In the lower chordates, the urinary, reproductive, and digestive systems have a common outlet from the body, the *cloaca*. During the course of its development, the human embryo passes through the cloacal stage until a subdivision of this region takes place. The result is a separate opening for fecal wastes (the anus), and for genital and urinary products (the urogenital sinus). In the human female, there has been a further division of the urogenital region, resulting in separate openings of the urinary tract and the genital tract. There are three separate openings in the human female. In the male (both cat and human), reproductive products and urine share the same duct and open together.

## DISSECTION OF THE CAT URINARY SYSTEM

If you have not already done so, remove the *parietal peritoneum,* which covers the kidney. Also, carefully remove any fat around the kidney. Because the kidneys are not actually inside the abdominal cavity, they are said to be located *retroperitoneally.* The right kidney in the cat is more anterior (superior) than the left. This situation is reversed in the human where the left kidney is superior to the right.

Identify once again the *renal artery* and *renal vein* associated with the concave medial surface of the kidney, which is called the *hilum.* Emerging from this same region is the broad yellow *ureter.* Follow this to the *urinary bladder.*

The urinary bladder collects urine produced by each kidney and stores it. When the urine is expelled, it leaves the urinary bladder through the *urethra,* which opens into the urogenital sinus and then to the outside of the body. The urethra is exposed in the subsequent dissection of the reproductive organs.

Just above and medial to the kidney is the hard, somewhat oval *adrenal gland.* In the cat, the adrenal gland is not located directly above the kidney, as it is in the human, where it is called the *suprarenal gland.* Crossing the adrenal gland is the prominent *adreno-lumbar vein.*

Without removing the kidney from the body or severing its attachment to the ureter, cut through it longitudinally with a sharp knife. Open it to see the *cortex, medulla, pelvis,* and its single *pyramid* (Figure 48-1).

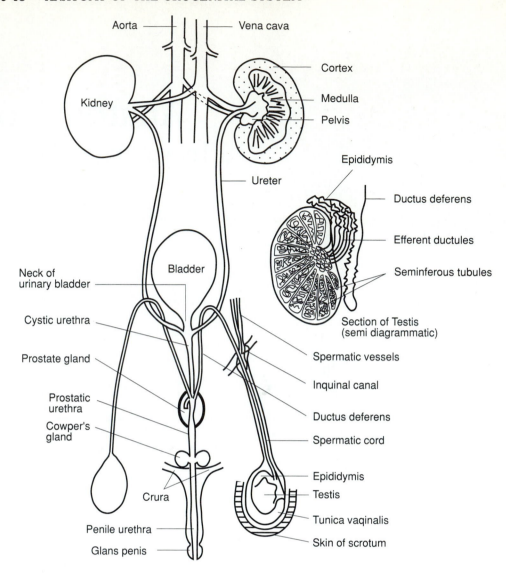

**Figure 48-1** Urogenital tract of the cat.

## THE HUMAN URINARY SYSTEM

The same organs you located in the cat are found in the human (Figure 48-2). There are some differences to note:

1. In the cat, the right kidney is anterior to the left. In the human, this is reversed, with the left kidney superior to the right (Figure 48-2).
2. The cat has only 1 pyramid emptying into the renal pelvis, whereas the human has 9–12 (Figure 48-3). Between pyramids are the renal columns.
3. The urinary bladder of the cat has a short neck between it and the urethra. The human bladder has no neck (Figures 48-1 and 48-5).
4. The human female has separate openings for the genital and urinary tracts (Figure 48-5). In the cat,

the urine opens into the small urogenital sinus, and together urine and reproductive products exit through the urogenital aperture (Figure 48-8).

Urine is produced by millions of microscopic units called *nephrons* (Figure 48-4). Most of the nephron is located within the *cortex* area of the kidney and has an intimate association with the vascular system. Wastes are taken from the blood at the *glomerulus* into *Bowman's capsule* of the nephron. They pass through the *proximal tubule, loop of Henle, distal tubule,* and *collecting tubule.* The collecting tubules empty into *papillary ducts* at the *renal papillae.* There, the papillary ducts empty into the *minor calyces,* which in turn empty into *major calyces.* The major calyces empty into the *renal pelvis.* Each kidney has a large duct, the *ureter,* which transports urine from the renal pelvis to the *urinary bladder,* where it is temporarily stored. Pe-

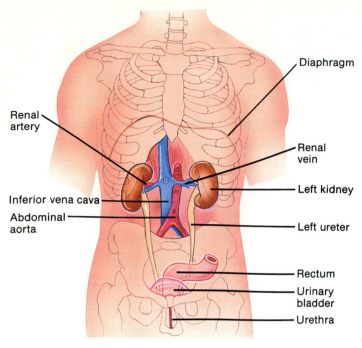

Anterior view

**Figure 48-2** Organs of the male human urinary system in relation to surrounding structures. (From G. J. Tortora and S. R. Grabowski, *Principles of Anatomy and Physiology,* 7th ed., New York: HarperCollins, 1993.)

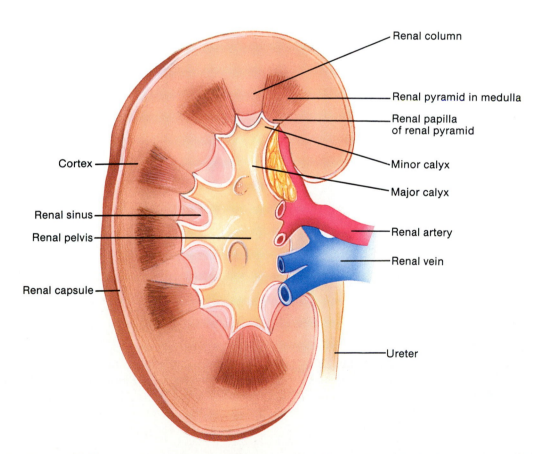

**Figure 48-3** Coronal section through the human kidney. (From G. J. Tortora and S. R. Grabowski, *Principles of Anatomy and Physiology,* 7th ed., New York: HarperCollins, 1993.)

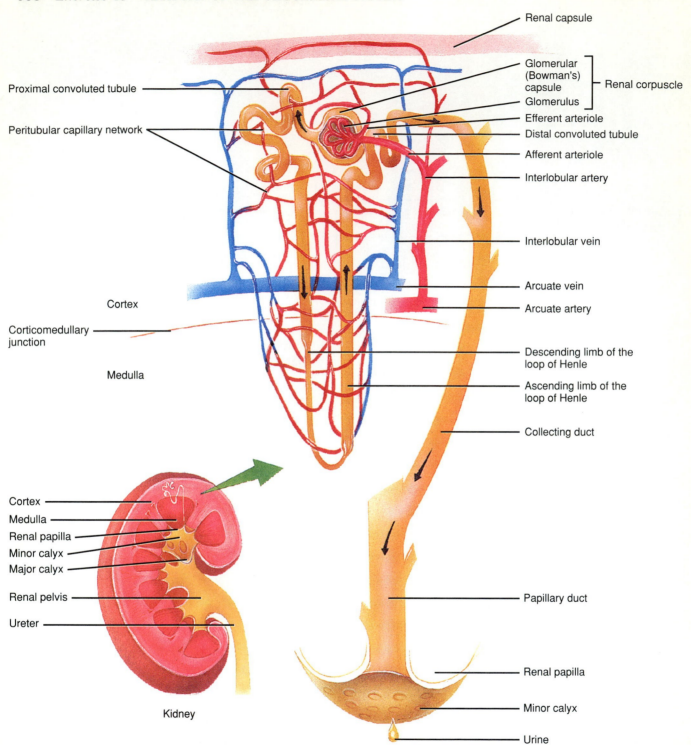

**Figure 48-4** Parts of a human nephron and its blood supply: The smaller arrows indicate the path of fluid and solutes. (From G. J. Tortora and S. R. Grabowski, *Principles of Anatomy and Physiology,* 7th ed., New York: HarperCollins, 1993.)

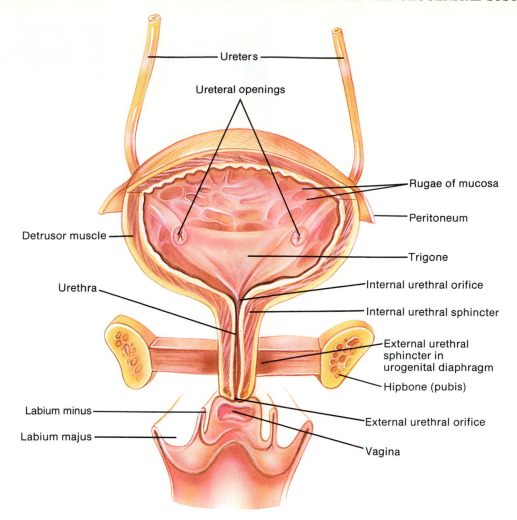

- Ureters
- Ureteral openings
- Rugae of mucosa
- Peritoneum
- Detrusor muscle
- Trigone
- Internal urethral orifice
- Urethra
- Internal urethral sphincter
- External urethral sphincter in urogenital diaphragm
- Hipbone (pubis)
- Labium minus
- External urethral orifice
- Labium majus
- Vagina

**Figure 48-5** Human urinary bladder and female urethra. (From G. J. Tortora and S. R. Grabowski, *Principles of Anatomy and Physiology,* 7th ed., New York: HarperCollins, 1993.)

riodically, urine is voided through the *urethra* (Figures 48-5 and 48-12). Note the striations in the kidney medulla. With your probe, locate the collapsed pelvis. The kidney is covered by a connective tissue capsule, the *renal capsule.* Find all parts on the appropriate figures (Figures 48-3 and 48-4).

## THE REPRODUCTIVE SYSTEM OF THE MALE CAT

### Materials

The following materials should be provided for class use:

1. Cat dissection specimens
2. Dissecting tools
3. Bone forceps
4. Models of male and female human reproductive organs
5. Charts of reproductive organs

Return to the testis. To see it more clearly remove the tough loose outer layer of tissue, termed the *tunica vaginalis,* that surrounds it. Lying on the side of the testis (seen through another layer, the *tunica albuginea*) and running the length of the organ, is a small, highly convoluted tubule, the *epididymis.* Other ducts for the transport of spermatozoa cannot be seen without further dissection and the use of the microscope.

Remove one of the testes, and cut it in half longitudinally. Many septa create chambers that contain the seminiferous tubules within which the spermatozoa are formed. Spermatozoa leave the testis through the rete testes and vasa efferentia, which join the epididymis noted earlier.

The vas deferens can be seen in its full length if the *pelvic girdle* is split at the *pubic symphysis.* Using

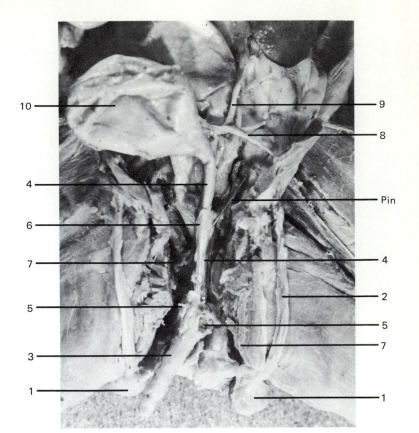

1. Testis
2. Spermatic cord
3. Penis
4. Urethra
5. Bulbourethral gland
6. Prostate gland
7. Split pubic bone
8. Vas deferens
9. Ureter
10. Urinary bladder

**Figure 48-6** Male cat with a split pelvis.

a nonrazor scalpel, cut through the muscle in the midline between the legs. If you are exactly at the midpoint, your scalpel blade will sink into the cartilage of the symphysis. Cut through the symphysis enough to weaken it, then place your hands on both legs, and force them apart, breaking the joint and revealing the area posterior to the urinary bladder (Figures 48-6 and 48-7). Clear away all connective tissue, and trace the entire length of the urethra. Posteriorly, it enters the penis through which it opens to the outside of the body.

The urethra is subdivided into a *cystic urethra,* between the urinary bladder and the prostate gland; the *prostatic urethra,* from the prostate gland to the penis; and a *penile urethra,* which traverses the penis and opens to the outside (Figure 48-1).

At the point of entry of the vas deferens into the urethra, there is a large, hard mass of tissue surrounding the urethra. This is the *prostate gland.* About an inch farther down the urethra there are paired hard masses, the *bulbourethral glands* or *Cowper's glands.* The prostate and bulbourethral glands secrete *semen* directly into the urethra. The cat does not have *seminal vesicles,* which are a third set of associated reproductive glands present in the human.

## THE REPRODUCTIVE SYSTEM OF THE FEMALE CAT

The uterus is the *bicornuate type,* meaning that it has two large lateral extensions (horns) from the central body portion. The body of the uterus is located dorsal to the urinary bladder (Figure 48-8).

From the body of the uterus, trace the paired *uterine horns* anteriorly (Figure 48-9). The uterine horns are very small in an immature female, large in a mature cat, and enormous in a pregnant cat. In the case of the pregnant cat (Figures 48-9 to 48-11), the horns will contain *embryos.*

At the anterior end of each uterine horn is a small oval body, the *ovary,* surrounded by the expanded end of the *oviduct.* The coiled oviducts may be seen lateral to the ovary, connecting with the uterine horn. The opening of the *oviduct,* which only partially surrounds the ovary, is the *ostium.*

To observe the posterior end of the reproductive and urinary systems, follow the instructions given for the male reproductive tract, and split the pubic symphysis (Figure 48-6). Trace the body of the uterus pos-

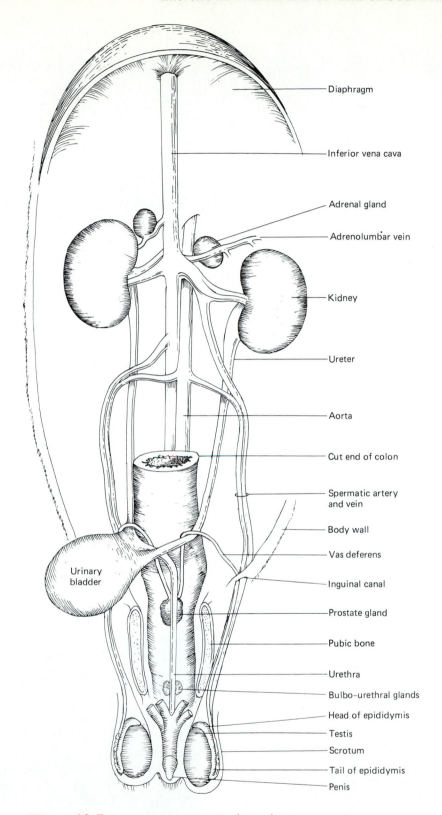

**Figure 48-7** Reproductive system of a male cat.

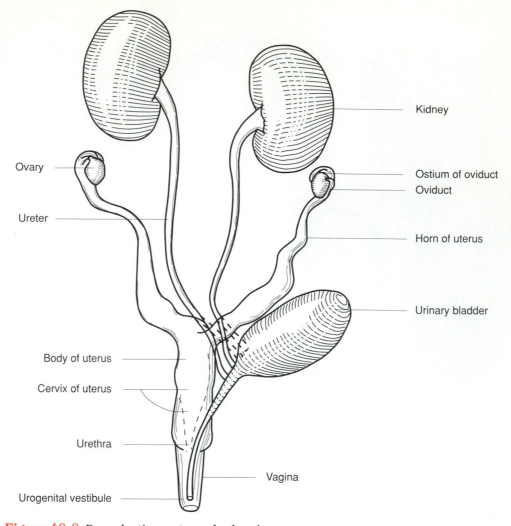

**Figure 48-8** Reproductive system of a female cat.

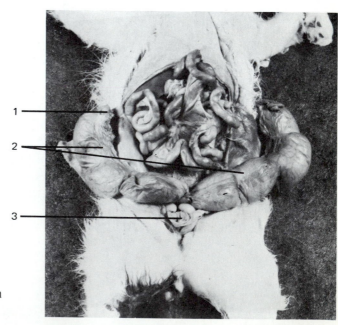

1. Ovary
2. Uterine horns
3. Urinary bladder

**Figure 48-9** Pregnant cat showing uterine horns in a late stage of pregnancy.

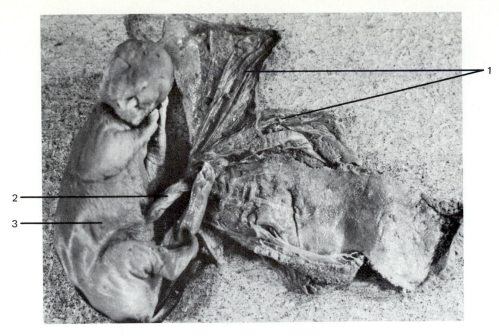

1. Placenta
2. Umbilical cord
3. Embryo

**Figure 48-10** Cat embryo and placenta.

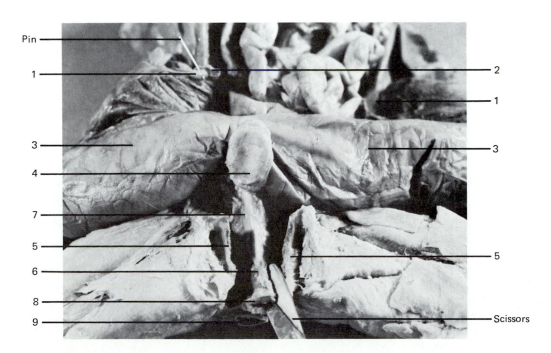

1. Ovary
2. Oviduct
3. Uterine horn
4. Urinary bladder
5. Pubic symphysis (split)

6. Vagina
7. Body of uterus
8. Urogenital sinus opening
9. Anal opening

**Figure 48-11** Pregnant female cat with a split pelvis. Scissors are shown inserted in the vagina.

teriorly as it passes dorsal to the *urethra* from the *urinary bladder*.

The uterus terminates inside the vagina at the urogenital aperture. If the cat has large uterine horns or is pregnant, cut into the vagina from the urogenital sinus (aperture) craniad until you reach the *cervix* (i.e., neck region) of the uterus. Make your cut to one side so as not to damage the urethra.

## THE HUMAN REPRODUCTIVE SYSTEMS—MALE AND FEMALE

The anatomy of the cat reproductive system is almost identical to that of the human. See Figure 48-12 for a diagram of the human male reproductive tract (shown in lateral view from a midsagittal section). The presence of the *seminal vesicles* in the human male (a third pair of reproductive glands in addition to the prostate gland and the bulbourethral glands) is the only major difference between the cat and the human. Find all parts of the male cat reproductive tract, and compare them with Figure 48-12.

The anatomy of the female reproductive tracts of the cat and the human are also similar (Figure 48-13). The main difference is in the shape of the uterus. The human has a *simplex* uterus, which usually accommodates one fetus at a time. Note its shape in Figure 48-13. The cat has a *bicornuate* uterus, with horns capable of accommodating several fetuses of the typical multiple birth. The oviduct (the *fallopian tube* or the *uterine tube*) is more prominent in the human than it is in the cat. Study Figure 48-13, and compare the size and shape of the various organs in the two species.

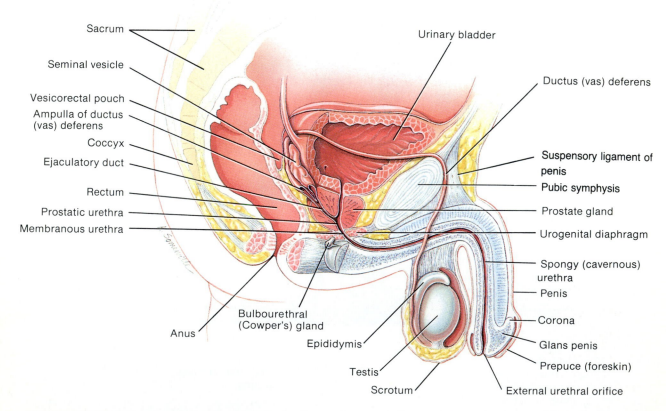

**Figure 48-12** Human male organs of reproduction and surrounding structures (seen in sagittal section). (From G. J. Tortora and S. R. Grabowski, *Principles of Anatomy and Physiology,* 7th ed., New York: HarperCollins, 1993.)

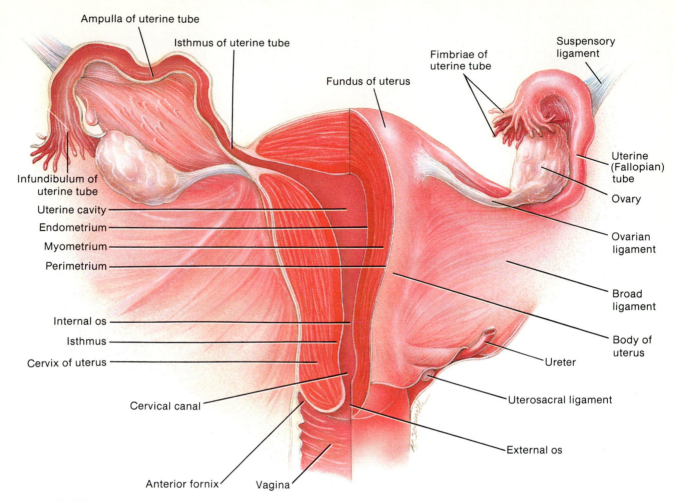

**Figure 48-13** The human uterus and associated structures (seen in posterior view): The left side of the figure has been sectioned to show internal structures. (From G. J. Tortora and S. R. Grabowski, *Principles of Anatomy and Physiology,* 7th ed., New York: HarperCollins, 1993.)

# LABORATORY REVIEW *48*

## *ANATOMY OF THE UROGENITAL SYSTEM*

### TRACING, COMPLETEION, AND FIGURE IDENTIFICATION

1. Trace the pathway of a waste product from the bloodstream through the kidney to its elimination from the body. Some of the steps are provided. You fill in the rest.

   a. Afferent arteriole _____

   b. _____

   c. _____

   d. Proximal convoluted tubule _____

   e. _____

   f. _____

   g. Collecting duct _____

   h. _____

   i. _____

   j. Renal pelvis _____

   k. _____

   l. _____

   m. Urethra _____

2. What anatomical feature of the human female reproductive tract makes it difficult for her to have multiple births? Explain. _____

   _____

   _____

   _____

3. The anatomy of the male reproductive tract in both the cat and the human is still tied to the urinary system, while in the female, the two systems have separated. Explain the relationship of the reproductive system and the urinary system in both male and female. _____

   _____

   _____

   _____

   _____

   _____

   _____

   _____

4. Identify the parts of the human kidney in Figure Q48-1.

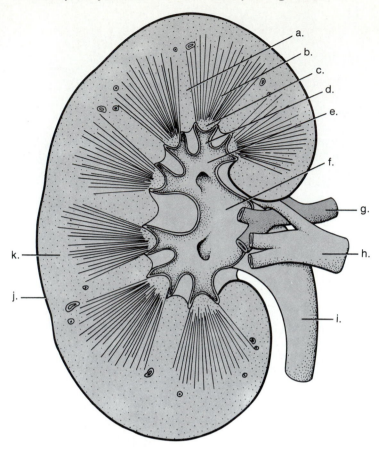

**Figure Q48-1** The human kidney. (From G. J. Tortora, *Principles of Human Anatomy,* 3rd, 4th, 5th eds., New York: HarperCollins, 1983, 1986, 1989.)

a. _____     g. _____

b. _____     h. _____

c. _____     i. _____

d. _____     j. _____

e. _____     k. _____

f. _____

5. Trace the pathway of a human spermatozoan from its place of origin to the site of fertilization of an ovum. Some of the answers are provided. You fill in the rest.

a. <u>Seminiferous tubules</u>

b. _____

c. _____

d. <u>Urethra</u>

e. <u>Vagina</u>

f. _____

g. _____

h. <u>Fallopian tube or oviduct</u>

# Fluid and Electrolyte Balance and a Demonstration of Dialysis

1. Describe the process of urine formation
2. List the general parts of a nephron
3. Measure the rate of urine formation after ingesting a known quantity of a test solution
4. Determine selected physical and chemical characteristics of urine samples after ingesting a known quantity of a test solution
5. Describe the dialysis process
6. Perform a procedure to demonstrate dialysis
7. List three conditions in which dialysis can be used

## URINE FORMATION, FILTRATION, AND PHYSIOLOGICAL REGULATION OF FLUID AND ELECTROLYTE BALANCE

*Urine* is an extremely complex water solution of various inorganic and organic substances that are the waste products of metabolism or are derived from ingested foods and fluids. The most important *inorganic* constituents of urine include ammonia, chlorides, phosphates, and sulfates. The chief *organic* components are creatinine, urea, and uric acid. Urine is a valuable indicator of various normal and abnormal situations because most pathological conditions originating in the kidneys and various diseases involving other body organs alter its composition. Certain chemical substances and large numbers of microorganisms appear in urine only when pathological conditions are present. (Exercise 50 describes representative disease conditions and the substances that are associated with them.)

Urine formation is a process essential to the body's capability to rid itself of metabolic waste products. The kidneys are among the organs that assume a major responsibility for the regulation of the volume, pH, and chemistry of body fluids. Functional kidney performance depends on its basic units, the *nephrons* (Figure 49-1). These delicate, microscopic structures vary the concentration, volume, and composition of urine, according to the needs of the body. Each nephron consists of a *glomerulus* (a tuft of capillaries enclosed by a double-walled cup termed the *renal* or *malpighian corpuscle*), the *proximal convoluted tubule,* the *loop of Henle,* the *distal convoluted tubule,* and the *collecting duct,* which leads to the *renal pelvis.* Each kidney contains approximately 1 million nephrons.

Urine formation begins in the glomerulus, where water and small molecules are filtered from the blood. The blood supply to this filtering unit is provided by the afferent arteriole, which in turn is supplied by a renal artery. The resulting filtrate passes through long tubules. Physiologically important substances (such as glucose, water, electrolytes, including sodium, chloride, and phosphates, and some ions) are selectively reabsorbed, while other substances (such as ammonia, potassium, hydrogen ions, and foreign materials such as penicillin) are secreted into the urine. The final tubule fluid, urine, that enters the bladder is quite different in composition from the glomerular filtrate. The amount of filtrate that forms in both kidneys every minute is known as the *glomerular filtration rate (GFR).* An average, normal adult secretes about 180 liters (48 gallons) each day. This rate amounts to approximately 125 ml/minute.

The GFR is regulated by two major mechanisms: *renal autoregulation* and *neural regulation. Renal autoregulation* refers to the ability of the kidneys to maintain a constant blood pressure and GFR, in spite of

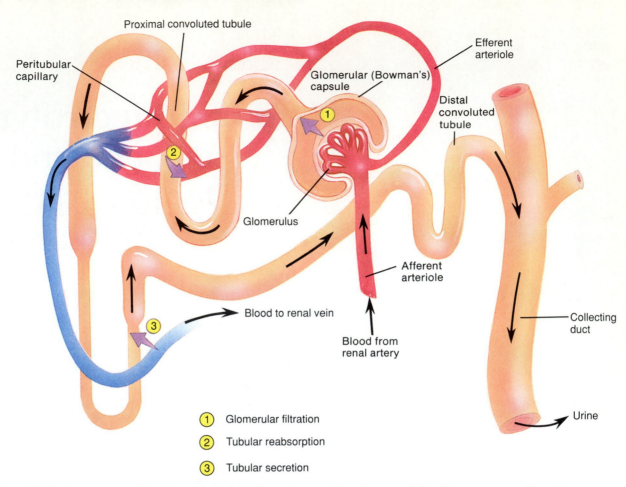

Proximal convoluted tubule

Peritubular capillary

Efferent arteriole

Glomerular (Bowman's) capsule

Distal convoluted tubule

① Glomerulus

② Glomerulus

Afferent arteriole

Blood to renal vein

③ Blood from renal artery

Collecting duct

Urine

① Glomerular filtration

② Tubular reabsorption

③ Tubular secretion

**Figure 49-1** A summary of nephron function. Note the numbered parts of the diagram, as well as the arrows. ① As blood flows from the glomerulus into the glomerular (Bowman's) capsule, it is filtered by the endothelial–capsular membrane. As the filtrate passes through the nephron ②, some components are selectively reabsorbed into blood, and ③ other substances are secreted from the blood into the filtrate for elimination in urine. (From G. J. Tortora and S. R. Grabowski, *Principles of Anatomy and Physiology*, 7th ed., New York: HarperCollins, 1993.)

changes in the body's arterial pressure. *Neural regulation* involves the sympathetic division of the autonomic nervous system. Under strong sympathetic stimulation, the adrenal medulla secretes epinephrine, which causes a decrease in the GFR.

The continuous adjustment of urine volume and electrolyte content within the tubules is controlled by several substances and conditions, including (a) *antidiuretic hormone (ADH)* from the neurohypophysis; (b) the sodium-retaining hormone, *aldosterone,* from the adrenal cortex blood flow; and (c) the acid–base balance.

*ADH,* also known as *vasopressin,* is regulated by osmoreceptors in the hypothalamus. When the body is dehydrated, the concentration of water in the blood falls below normal levels and consequently stimulates these receptors to bring about the release of ADH into the bloodstream and its transportation to the kidneys. The kidneys respond by decreasing the urine output, thereby conserving water. If, on the other hand, the blood contains higher-than-normal water concentrations, the os-

moreceptors stop ADH secretion. The kidneys then release large quantities of urine, and the volume of water in body fluids is brought back to normal. Angiotensin II is an active hormone associated with the release of ADH and water retention by the kidneys. Secretion of ADH also can be influenced by various chemical factors, such as alcohol, nicotine, and other drugs, and by physical and mental factors, such as pain, trauma, blood loss, and stress.

*Aldosterone* acts on some specific cells in the kidneys, to increase their reabsorption of sodium ($Na^+$) ions from the urine, return them to the blood, and thus prevent rapid depletion of sodium from the body. At the same time, aldosterone stimulates excretion of potassium ($K^+$) ions, so that large amounts of potassium are lost in the urine.

The control of aldosterone secretion involves several mechanisms. One of these is the *renin–angiotensin pathway*. A decrease in blood pressure—caused by a drop in blood volume due to dehydration, sodium ($Na^+$)

ion deficiency, or severe blood loss—stimulates some specific kidney cells to secrete the enzyme *renin* into the bloodstream. This enzyme converts *angiotensinogen*, a protein produced by the liver, into *angiotensin I*, which is then converted into *angiotensin II* by an enzyme in the lungs. Angiotensin II stimulates the adrenal cortex to produce more aldosterone. Aldosterone brings about increased Na$^+$ reabsorption, and water is also reabsorbed, thus increasing fluid volume and restoring blood pressure to normal.

Another mechanism for the control of aldosterone is potassium (K$^+$) ion concentration. An increased K$^+$ concentration in extracellular fluid directly stimulates aldosterone secretion and causes the kidneys to eliminate excess K$^+$. A decreased K$^+$ concentration in extracellular fluid reverses the process.

In the first part of this exercise, the effects of salt, sugar, grapefruit juice (a source of potassium), and coffee or tea (sources of caffeine) on the volume, pH, specific gravity, and chloride content of urine are considered. Students who are diabetic or are under medical treatment for a kidney or related problem should not participate in the first portion of this exercise. The principle of dialysis and factors that can affect the process are considered in the second portion of this exercise.

## HEMODIALYSIS

*Dialysis* is a process by which certain substances in solution can be separated from one another by virtue of the difference in their rates of diffusion through a semipermeable membrane. *Hemodialysis* represents one type of dialysis that is used to correct a variety of conditions in the body that affect renal excretion. In situations such as drug overdose, temporary stoppage of urine formation, and extensive kidney destruction, hemodialysis may mean the difference between life and death.

Two types of commercial hemodialysis systems, frequently referred to as *artificial kidneys,* are in general use. Both operate on the same principle. Blood from the patient is circulated by a pump past one side of a semipermeable membrane, termed a *dialyzing membrane* and usually made of cellophane. Circulating on the other side of the membrane is a *dialyzing solution,* containing electrolytes in the same concentration as in normal plasma. As the blood and solution pass on either side of the membrane, excess electrolytes in the blood diffuse through the membrane into the solution because of the concentration gradient.

In one type of hemodialysis system—the *coil type*— the dialyzing membranes are single or parallel flat cellophane tubes wrapped around a central core, supported by plastic screening. The dialyzing or bathing solution

is pumped through the concentric layers and over the flattened tubing of the system. The coil assembly is presterilized and is generally discarded after use.

The second type of system, the *plate* or *layer dialyzer,* contains cellophane sheets instead of tubes (Figure 49-2). The sheets are held apart within plastic frames to allow passage of the fluid to be dialyzed between them. The dialyzing solution flows on the outside of the cellophane sheets. All parts of this system are sterilized before use.

The dialyzing solution is temperature controlled and has a composition that chemically approximates normal extracellular fluid. Specific needs of individuals on hemodialysis may require alterations in the extracellular fluid. Dialysis begins when blood from one of the patient's arteries is pumped through the system. After removal of the unwanted substances, the blood passing through the dialysis system is returned to the patient's circulation by a vein.

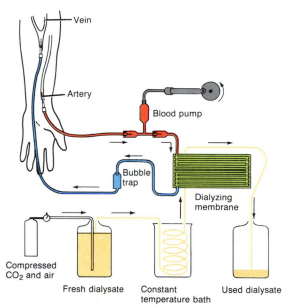

**Figure 49-2** The operation of an artificial kidney. The blood route is indicated in red and blue. The route of the dialysis is indicated in gold. (From G. J. Tortora and S. R. Grabowski, *Principles of Anatomy and Physiology,* 7th ed., New York: HarperCollins, 1993.)

## EFFECTS OF SODIUM, POTASSIUM, CAFFEINE, AND WATER ON URINE COMPOSITION

### Materials

1. The following materials should be provided for class use:
   a. Urinometers (cylinder and hydrometer float device)

b. pH indicator paper strips (pHydrion or nitrazine test papers)

c. Thermometers

d. 250-ml beakers

e. Filter paper

f. Test tubes

g. Eyedroppers

h. 20 percent potassium chromate solution in dropper bottle

i. 2.9 percent silver nitrate solution

j. Disposable drinking cups

k. 1-ml pipettes, graduated in 0.1 ml, with bulbs

l. Test-tube racks

m. Liquid soap and brushes for washing containers

n. Containers of disinfectant

2. The following materials should be provided for students in groups of five:

a. Five urine-specimen containers

b. Five 500-ml graduated cylinders

c. Five 250-ml beakers

d. Test solutions in the volumes indicated:
Cola drink (nondietetic), 750 ml
Distilled water, 1250 ml (500 ml is to be used with a salt tablet)
Unsweetened, strong coffee or tea, 750 ml
Unsweetened grapefruit juice, 750 ml

e. One salt tablet (4.5g)

3. Disposable surgical gloves

---

### CAUTION

Working with urine or other body fluids will not pose any danger if the following precautions are observed:

1. Place any and all urine or materials containing urine into the disinfectant containers provided.

2. Dispose of any urine-containing materials as indicated by the instructor.

3. In cases of urine spillage, wipe the area with a disinfectant-soaked paper towel, and dispose of all materials as indicated by the instructor.

4. Always wash your hands after handling urine and associated materials.

5. The use of disposable surgical gloves.

---

## Procedure 1: Volume, Color, Turbidity, and Odor

This procedure is to be performed by students individually, but they are to work in groups of five. The instructor will indicate the facilities to use in order to obtain urine samples.

1. Obtain a urine specimen container, and empty your bladder completely. Note the time, and enter it in the appropriate space of the results and observations section. This sample is your normal urine specimen.

2. Use one of the 500-ml graduated cylinders to determine the volume of the normal specimen. Record the information in the results and observations section. The normal volume per 24 hours is 1 to 2 liters.

3. Note the color, the presence of turbidity (cloudiness), and any odor. Record your findings in Table 49-3 of the results and observations section. Normal, freshly voided urine is generally yellow to amber, is clear, and does not have a disagreeable odor.

## Procedure 2: Specific Gravity and pH

This procedure is to be performed by students individually, but they are to work in groups of five.

1. Pour about 30 ml of the urine sample into a small beaker. Carefully pour the urine from the beaker into a clean urinometer cylinder until it is three-fourths full. Obtain more of the specimen if necessary. If foam forms on the surface, remove it with a piece of filter paper. Discard the paper as indicated by your instructor.

2. Obtain a thermometer, and check the reading on it. If the reading is not 37°C, (carefully) shake it down to that level. Insert the thermometer in the urinometer, and read the temperature after 3 minutes. The temperature of the sample is used in the determination of specific gravity in the next step. Clean the thermometer as indicated by your instructor.

3. Float the hydrometer in the cylinder. Make certain that it is not touching the bottom or sides of the cylinder. Adjust the value of the hydrometer reading according to the temperature, to obtain the specific gravity. For each 3° above 25°C, add 0.001 to the hydrometer reading; for each 3° below 25°C, subtract 0.001. Record your findings in the results and observations section. The specific gravity of normal urine is 1.002 to 1.030. Remove the hydrometer, and clean it as indicated by your instructor.

4. Dip a pH indicator test strip paper in the urinometer cylinder three consecutive times. Remove excess urine by touching the paper to the inside surface of the cylinder. After 1 minute, compare the color of the strip with the color guide on the strip container or other available guide. Normal urine has an average pH of 6 but ranges from 4.8 to 7.5. Record your findings in the results and observations section.

5. Clean the urinometer as indicated by your instructor.

## Procedure 3: Chloride Concentration Determination

This procedure is to be performed by students individually, but they are to work in groups of five.

1. Place two test tubes in a rack.
2. Using a 1-ml pipette, introduce 0.5 ml of the urine sample into each tube.
3. Add 1 drop of 20 percent potassium chromate to each tube.
4. Gently shake the tubes. The solution in the tubes should be yellow.
5. With an eyedropper, add 1 drop at a time of the silver nitrate solution provided to the first tube. Gently shake the tube after each drop. Count the number of drops required to change the color of the urine solution from yellow to brown. *Note:* One drop of the silver nitrate solution added is equivalent to 61 mg of chloride ion ($Cl^-$) per 100 ml of urine.
6. Calculate the chloride concentration of the urine sample by multiplying the number of silver nitrate drops added by 61. The value should be reported in milligrams/100 ml of urine.
7. Repeat Steps 5 and 6 with the second tube. Average your findings, and enter them in Table 49-3 of the results and observations section.
8. Empty the tubes as indicated by your instructor, and clean all glassware and other materials.

## Procedure 4: Effects of Various Test Solutions on Urine Composition

This procedure is to be performed by students in groups of five.

1. Determine within your group the individuals who will drink a particular test solution or material. The test solutions include distilled water, cola drink, unsweetened grapefruit juice, and unsweetened strong coffee or tea. A salt tablet with water also will be used.

2. All members of the group should obtain the necessary volume of their test solution; 750 ml should be consumed. The individual selecting the salt tablet should swallow it with 500 ml of distilled water.
3. Drink the test solution as quickly as possible. If an individual is unable to consume the entire volume designated, the quantity consumed should be recorded in Table 49-2. Your instructor will indicate the pH, specific gravity, and chloride concentration of each test solution. Enter this information in Table 49-2.
4. Begin timing as soon as you have consumed the test solution entirely. Be prepared to obtain a urine sample at four 30-minute intervals. A total of four urine specimens will be collected.
5. Perform Procedures 1, 2, and 3 with each urine specimen collected. Enter the findings obtained with your test solution in Table 49-3 of the results and observations section.
6. Compare your findings with those obtained by other members of your group, and answer the questions in the results and observations section.

## RESULTS AND OBSERVATIONS

1. In Table 49-1, list the food and drink items you consumed approximately 2 hours before starting the experiment. Indicate the approximate quantity of each item.
2. a. Approximate time of urinary bladder emptying prior to the laboratory exercise: _____
   b. Time at which normal urine specimen was obtained at the beginning of the laboratory exercise: _____
3. a. Enter the pH, odor, specific gravity, and relative chloride content of each test solution provided for your group in Table 49-2.
   b. Which test solution did you use? _____

## Table 49-1

| Food or Drink Item | Quantity | Food or Drink Item | Quantity |
|---|---|---|---|
|  |  |  |  |
|  |  |  |  |
|  |  |  |  |
|  |  |  |  |

**Table 49-2**

| Test Solution | pH | Odor | Specific Gravity | Relative Chloride Content | Volume Consumed |
|---|---|---|---|---|---|
| Distilled water | | | | | |
| Unsweetened cola drink | | | | | |
| Unsweetened coffee or tea | | | | | |
| Unsweetened grapefruit juice | | | | | |
| Salt tablet and water | | | | | |

4. Enter the findings obtained with your urine samples in Table 49-3.
   a. To calculate your rate of urine production for the normal sample, find the difference in time between emptying the bladder before the exercise and taking the normal urine sample. Then calculate the number of milliliters produced per minute. Enter the result in Table 49-3.
   b. Calculate the rate of urine formed for each interval after drinking the test solution, in ml/minute. Enter the results in Table 49-3.

5. a. Were any of the properties observed with the normal urine sample changed by the test solution consumed? If so, which ones? _____
   _____

   b. Offer an explanation for any changes observed.
   _____
   _____

6. a. Were any of the properties observed with the

**Table 49-3**

| Property | Experimental Samples | | | | |
|---|---|---|---|---|---|
| | Normal Sample | 30 Minutes | 60 Minutes | 90 Minutes | 120 Minutes |
| Volume (in ml) | | | | | |
| Color | | | | | |
| Odor | | | | | |
| Turbidity (clouding) | | | | | |
| pH | | | | | |
| Specific gravity | | | | | |
| Relative chloride content (in mg/ml) | | | | | |
| Rate of urine formation (in ml/minute) | | | | | |

normal urine samples of your group changed by the test solutions they consumed? If so, which ones? _____

_____

b. Offer an explanation for any changes observed.

_____

_____

## DEMONSTRATION OF DIALYSIS

### Materials

1. The following materials should be provided per four students:
   a. Two pieces of dialysis tubing, each 20 cm (about 8 inches) long
   b. One 50-ml beaker and two 250-ml beakers
   c. One small funnel
   d. Four Pyrex test tubes (10–15 ml capacity)
   e. One 100-ml graduated cylinder
2. The following materials should be provided for class use:
   a. 2 percent glucose solution
   b. 5 percent starch solution
   c. 10 percent sodium chloride (NaCl) solution
   d. 10 percent egg albumin solution
   e. Benedict's reagent (for glucose determinations)
   f. Silver nitrate reagent (for $Cl^-$ determinations)
   g. Iodine reagent (for starch determinations)
   h. Sodium hydroxide (NaOH) and copper sulfate ($CuSO_4$) reagents (for protein determinations)
   i. Distilled water, kept at 4°C, room temperature (25°C), and 37°C
   j. String for dialysis bags
   k. Boiling water bath

### Procedure 1: Preparation of Dialysis Bags

This procedure is to be performed by students in groups of four.

1. Soak the dialysis tubing in distilled water (25°C) until soft.
2. While the dialysis tubing is softening, mix in a 50-ml beaker with 10 ml each of the albumin, glucose, sodium chloride, and starch solutions.
3. When the tubing material is soft, prepare a dialysis bag by tying one end of the tube with a small piece of string. Make certain that the knot is tight enough so that the system will not leak when filled.
4. Have a laboratory partner hold a small funnel in the open end of the dialysis bag. Fill the bag with the prepared chemical mixture. Leave only enough

space at the top of the bag so that the tube can be tied again to seal the bag.
5. Close the open end of the dialysis bag by tying a knot with another piece of string.
6. Carefully rinse the filled dialysis bag with distilled water (25°C).
7. Place the bag in a 250-ml beaker containing distilled water at 4°C. The bag should be completely submerged.
8. Prepare another dialysis bag by repeating Steps 1–6. Submerge the second bag in a beaker containing distilled water at 37°C.

### Procedure 2: Testing the Dialysate

This procedure is to be performed by students in groups of four. The instructor will demonstrate the testing procedures for the *dialysate* (the chemicals passing out of the dialysis bag) before the students perform the procedure.

1. Immediately after submerging the dialysis bags in the beakers, carefully pour about 20 ml of the water from each beaker into separate graduated cylinders.
2. Divide the sample of the 4°C water into four separate test tubes (about 5 ml each). To determine which of the substances added to the dialysis mixture are present in the dialysate, perform the following tests, using one test tube for each test.
   a. The *Biuret* or *protein (egg albumin) test*—Add approximately 5 ml of the sodium hydroxide and 2 ml of the copper sulfate reagents. The formation of a pink-to-violet color is a positive reaction. Refer to Color Plate 53.
   b. *Chloride ion ($Cl^-$) test*—Add 8 drops of the silver nitrate solution. The formation of a white precipitate is a positive result.
   c. *Glucose test*—Add 10 ml of the Benedict's reagent, and place tube in a boiling water bath for about 5 minutes. The formation of an orange color is a positive result (see Color Plate 55).
   d. *Starch test*—Add 3 drops of the iodine reagent. The formation of a purple-to-black color is a positive result.
3. Remove 20-ml samples at 15 minutes, 30 minutes, 1 hour, 2 hours, and 48 hours (or during the next laboratory period).
4. With each sample, perform the tests listed in Step 2.
5. Enter your findings in the results and observations section.
6. Repeat Steps 2–5 with the second dialysis system (37°C water).
7. Answer the questions in the results and observations section.

## RESULTS AND OBSERVATIONS

1. Describe the test results that would indicate the presence or absence of the substances listed in Table 49-4.
2. Enter in Table 49-5 the results of the tests performed. Indicate the presence of a substance with a "+" or its absence with a "−."
3. What other methods could be used to test for the following substances?

   a. Glucose _____

   b. Albumin _____

4. Briefly explain the relationship between dialysis and kidney function. _____

   _____

   _____

5. Briefly describe the appearance of the dialysis bags at the beginning and at the end of the experiment.

   _____

   _____

**Table 49-4**

| Substance | Positive Result | Negative Result |
|-----------|-----------------|-----------------|
| Chloride ion | | |
| Egg albumin | | |
| Glucose | | |
| Starch | | |

6. What general conclusions can you draw from the results of this experiment? _____

   _____

   _____

**Table 49-5**

| Time of Sampling | Chloride Ion | | Egg Albumin | | Glucose | | Starch | |
|------------------|------|------|------|------|------|------|------|------|
| | 4°C | 37°C | 4°C | 37°C | 4°C | 37°C | 4°C | 37°C |
| Immediate | | | | | | | | |
| 15 minutes | | | | | | | | |
| 30 minutes | | | | | | | | |
| 1 hour | | | | | | | | |
| 2 hours | | | | | | | | |
| 48 hours | | | | | | | | |

# LABORATORY REVIEW *49*

## FLUID AND ELECTROLYTE BALANCE AND A DEMONSTRATION OF DIALYSIS

### COMPLETION AND DEFINITION

1. Briefly describe the general events involved with urine formation. _____

_____

_____

_____

2. List the three distinct processes involved with urine formation.

    a. _____

    b. _____

    c. _____

3. Define or explain the following:

    a. Glomerulus _____

    _____

    b. Nephron _____

    _____

    c. Antidiuretic hormone _____

    _____

    d. Aldosterone _____

    _____

4. Give the function of the following structures in urine formation:

    a. Glomerulus _____

    _____

    b. Proximal tubule _____

    _____

### DEMONSTRATION OF DIALYSIS

5. Define *dialysis*. _____

_____

_____

6. What is hemodialysis? _____

_____

**ANSWERS**

7. List and briefly describe two types of systems used for hemodialysis.

   a. _____

   b. _____

8. List three conditions in which hemodialysis is necessary.

   a. _____

   b. _____

   c. _____

9. Define or explain the following:

   a. Electrolyte _____

   b. Osmotic pressure _____

   c. Artificial kidney _____

   d. Reagent _____

**MULTIPLE CHOICE**

Insert the correct answers in the spaces provided.

1. _____
2. _____
3. _____
4. _____
5. _____
6. _____
7. _____
8. _____
9. _____
10. _____

1. Which of the following are not among the chief organic components of urine?
   a. creatinine
   b. chloride
   c. uric acid
   d. urea
2. Which of the following are not among the most important inorganic constituents of urine?
   a. chlorides
   b. ammonia
   c. uric acid
   d. phosphates
   e. sulfates
3. Which of the following is not a component of a nephron?
   a. glomerulus
   b. loop of Henle
   c. glomerular capsule
   d. collecting duct
   e. proximal convoluted tubule
4. Which of the following structures leads to the renal pelvis?
   a. glomerulus
   b. loop of Henle
   c. glomerular capsule
   d. collecting duct
   e. proximal convoluted tubule

5. In which of the following structures does urine formation begin?
   a. glomerulus
   b. loop of Henle
   c. capsule
   d. collecting duct
   e. proximal convoluted tubule
6. Which of the following substances is or are not selectively reabsorbed during urine formation?
   a. glucose
   b. water
   c. any of the electrolytes
   d. ammonia
   e. choices *c* and *d* only
7. Which of the following is or are secreted in urine?
   a. glucose
   b. water
   c. any of the electrolytes
   d. ammonia
   e. choices *c* and *d* only
8. Which of the following is or are known to stimulate the release of ADH into the bloodstream?
   a. body dehydration
   b. higher-than-normal water concentration in the blood
   c. normal water concentration in the blood
9. Which of the following substances increases aldosterone release?
   a. increased $K^+$
   b. decrease in $Na^+$
   c. increased ammonia levels
   c. choices *a* and *c* only
   e. choices *a* and *b* only
10. Which of the following specifically stimulates the adrenal cortex to produce aldosterone?
    a. renin
    b. angiotensin I
    c. angiotensin II
    d. angiotensinogen
    e. $Na^+$

# Routine Urinalysis

1. List eight substances normally found in urine
2. List five abnormal constituents of urine and their associated causes
3. Describe, perform, and interpret the basic tests included in a routine urinalysis
4. Perform and interpret the results of a microbiological examination of a urine specimen
5. Describe the appearance of urine in cases of diabetes insipidus and diabetes mellitus
6. Define or explain *albuminuria, glycosuria,* and *glomerulonephritis*

*Urine* is a watery solution containing nitrogenous waste and inorganic salts that are removed from blood plasma and eliminated by the kidneys. Table 50-1 lists the general properties of normal urine. Table 50-2 lists the average amounts of solid constituents of normal urine.

Urine and blood are analyzed more frequently than any other body fluid. As with blood, the composition of urine is a functional indicator of the state of body health at any given period of time. Some abnormal constituents and the types of conditions responsible for their appearance in urine are listed in Table 50-3. The detection of several such abnormal constituents can be achieved by tests that are based on color changes caused by enzymes or chemical reagents. These tests may be qualitative (showing the presence of an abnormal constituent) or quantitative (showing the concentration of the constituent). For most of these tests (Table 50-4), the color-reacting substance has been impregnated on plastic strips so that a urine sample can be easily applied.

An example of a diagnostic urinary test strip demonstrating the features described here is Phenistix. It is used for the detection of the hereditary metabolic defect, phenylketonuria (PKU). A baby born with this defect is unable to synthesize adequate amounts of the

**Table 50-1  Physical Properties of Urine**

| Property | Normal Feature | Abnormal Features |
|---|---|---|
| Color | Generally light yellow to amber | Dark red or brown |
| Odor | Aromatic | Fishy, ammonialike |
| Turbidity (cloudiness) | Clear | Hazy, may contain obvious shreds of mucus |
| Specific gravity (at 25°C) | 1.002–1.030 | Any major change from normal range |
| pH (hydrogen ion concentration) | 6.0 average, but ranges from 4.8 to 7.5 | Any major variation from normal range |
| Volume | 1–2 liters per 24 hours | Any major variation from normal range |

**Table 50-2  Solid Constituents of Normal Urine**

| Constituent | Daily Normal Values |
|---|---|
| Bilirubin | 0.02 mg |
| Calcium | 30–75 mg |
| Creatinine | 1 g |
| Glucose | 130 mg |
| Sodium | 125–260 mEq (milliequivalents) |
| Urea nitrogen | 6–17 g |
| Uric acid | 250–750 mg |
| Average amount of solids | 60 g |

**Table 50-3  Abnormal Constituents of Urine and Associated Causes**

| Abnormal Constituent and Term Used to Indicate Its Presence | Associated Causes |
|---|---|
| Acetone bodies or ketone bodies (acetonuria or ketosis) | Diabetes mellitus<br>Starvation<br>Diarrhea in children<br>Pernicious vomiting of pregnancy |
| Albumin (albuminuria) | Damaged kidney tubules<br>Chemical poisoning<br>Prolonged exercise<br>Cold baths<br>Altered glomeruli |
| Bile | Obstructive jaundice (blockage of bile duct resulting in the yellowing of tissues) |
| Bilirubin | Liver disorders<br>Obstructive jaundice |
| Glucose (glycosuria) | Diabetes mellitus |
| Hemoglobin | Conditions involving marked blood destruction, such as<br>Chemical poisoning<br>Poisonous snake bites<br>Malaria<br>Transfusion with incompatible blood |
| Leukocytes and other components of pus (pyuria) | Cystitis (infection of the urinary bladder)<br>Infection of the kidney<br>Urethritis (infection of the urethra)<br>Tuberculosis of the kidney |
| Red blood cells (hematuria) | Infection of glomeruli<br>Tumors of the urinary tract<br>Kidney and bladder stones<br>Tuberculosis<br>Circulatory system disorders |
| Sugars, galactose, and fructose | Severe liver disease<br>Inherited metabolic defect |

**Table 50-4    Examples of Commercially Available Urinary Test Strips**

| Product Name and Manufacturer | Tests for the Presence of | Color Reaction for Definite Positive Reaction | Significance of Results | Limitations |
|---|---|---|---|---|
| Chemstrip (BioDynamics) | Leukocytes<br><br>Nitrates | Purple<br><br>Red-violet | Both tests used in combination detect possible urinary tract infection | Certain antibiotic drugs interfere<br><br>Large amounts of ascorbic acid (orange juice, vitamin C tablets) decrease sensitivity |
| Clinistix (Ames) | Glucose | Purple-blue | Results can be used to adjust diet and/or medication for diabetics and others | Vitamin C, aspirin, and some drugs may produce a false-negative result |
| Combistix[a] (Ames) | pH<br><br><br>Glucose<br><br>Protein | Orange-yellow and green-blue<br><br>Green-brown<br><br>Green-blue | Test results may provide information concerning kidney function, acid–base balance, and carbohydrate metabolism | False-positive reactions may occur with alkaline, highly buffered specimens |
| Diastix (Ames) | Glucose | Green-brown | Highly specific for glucose | Generally none |
| Ketostix (Ames) | Urinary ketones (acetoacetates and acetone) | Lavender-purple | Detects ketoacidosis | Phthalein medications may produce false-positive reactions |
| Multistix (Ames) | pH | Broad range of colors, depending on pH | Provides information on acid–base balance | Contamination (runoff) from other reaction patches causes color changes |
| | Glucose | Green-brown | Provides information on carbohydrate metabolism | Vitamin C and aspirin may cause false reactions |
| | Protein | Green-blue | Provides information on kidney function | Highly alkaline urine and some drugs interfere with test results |
| | Bilirubin | Various shades of tan | Provides information on liver disorders and diseases | Various drugs and vitamin C may cause false reactions |
| | Ketone | Maroon | Detects ketoacidosis, indication of carbohydrate metabolism disorders | Certain drugs interfere with test results |
| | Nitrite | Pink | Detection of urinary tract infection | Drugs and vitamin C may cause false positive results |
| | Urobilinogen | Brown-orange | Sensitive to elevated urobilinogen secretion | Certain drugs may produce false positive reactions |
| | Blood | Green spots | Highly sensitive to the presence of hemoglobin and indicates possible infection | Vitamin C, microorganisms, and elevated protein may produce false reactions |
| Phenistix (Ames) | Urinary phenylketones | Blue-gray | Test used for PKU screening | Aspirin, excess bilirubin, and phenothiazines interfere with the test |

[a]A variation of this product, Hema-Combistix, incorporates a test for the detection of occult blood in urine.

enzyme phenylalanine hydroxylase. As a consequence, the amino acid phenylalanine is not converted to the compound tyrosine but instead is converted to phenylpyruvate. Excessive amounts of phenylpyruvate in the baby's body can damage its central nervous system. If PKU is detected early, foods that contain phenylalanine can be eliminated from the diet and brain damage can be avoided. The Phenistix is based on a reaction between a ferric ion reagent and phenylpyruvic acid, to produce a gray-green color. The diagnostic reagent strip with a test patch containing the reagent is immersed in the urine specimen, tapped to remove excess urine, and in 30 seconds compared to a color chart on the strip container. As little as 8 mg of phenylpyruvate per deciliter can be detected. A blood test generally is recommended in association with positive Phenistix results. Table 50-4 describes several commercially available urine diagnostic test strips. Tests using reagent strips are performed in this exercise.

Microscopic examination of a drop of urine that has been centrifuged may show the abnormal presence of blood and other cells, casts, and renal calculi (Figure 50-1). *Casts* are small masses of material that have hardened in the lumens of tubules and have assumed the shape of the lumen. Renal calculi are tiny stones formed within the kidney.

Normal urine is generally sterile or contains 1000 or fewer bacteria per milliliter, these few being the microorganisms making up the normal urethral flora. The presence of a microbial infection in any portion of the urinary system usually produces significant changes in urine specimens. The presence of large numbers and kinds of bacteria in urine, or *bacteriuria,* can be determined by specific bacteriological techniques and the use of commercially produced dipsticks or related devices (see Figure 50-2).

To be of value, a urine analysis must be done on a representative specimen, which should be examined within 30 minutes after it is passed or be stored under refrigeration. Because an early morning specimen is concentrated, it is more likely to reveal abnormalities than urine collected at some other time. In some laboratories, standard determinations are semiautomated.

Routine analysis of a urine specimen involves a description of physical properties (color, turbidity, odor, specific gravity, and pH; see Table 50-1) and determination of the presence or absence of abnormal constituents. A microscopic examination of urinary sediment and tests for the detection and identification of microorganisms such as bacteria also may be required. Several of these aspects of urinalysis are considered in this exercise.

## DESCRIPTION OF PHYSICAL PROPERTIES

### Materials

The following materials should be provided for class use:

1. Urinometer (cylinder and hydrometer float device)
2. pH indicator paper strips (pHydrion or nitrazine test papers)
3. Two freshly voided urine samples
4. Thermometers
5. 250-ml beakers
6. Filter paper
7. Liquid soap and brushes for washing of containers
8. Containers with disinfectant
9. Disposable surgical gloves

---

### *CAUTION*

Working with urine or other body fluids will not pose any danger if the following precautions are observed:

1. Place any and all urine or materials containing urine into the disinfectant containers provided.
2. Dispose of any urine-containing materials as indicated by the instructor.
3. In cases of urine spillage, wipe the area with a disinfectant-soaked paper towel, and dispose of all materials as indicated by the instructor.
4. Always wash your hands after handling urine and associated materials.
5. The use of disposable surgical gloves.

---

### Procedure

This procedure is to be performed by students individually.

1. Pour about 30 ml of one of the samples into a beaker. Carefully pour the urine from the beaker into a clean urinometer cylinder until it is three-fourths full. Obtain more of the specimen if necessary. If foam forms on the surface, remove it with a piece of filter paper. Discard the paper as directed by your instructor.
2. Note the color, odor, and turbidity of the sample. (Refer to Table 50-1.) Record your findings in the results and observations section.

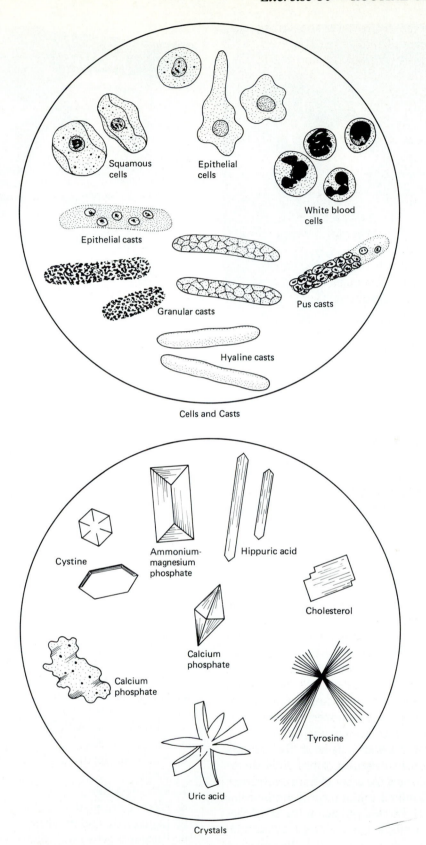

**Figure 50-1** Elements found in urine sediments.

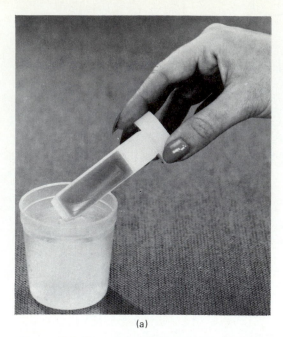

(a)                                                   (b)

**Figure 50-2** One approach to testing urine for the presence of an infectious disease caused by bacteria: (a) This dip paddle device contains nutrients on which bacteria can grow; it is dipped into a urine specimen to collect the microorganisms. (b) After an incubation period, evidence of an infection is clear; each of the small clumps on the dipstick is a bacterial colony containing 1000 organisms. (Courtesy of Medical Technology Corporation.)

3. Check the reading on a thermometer. If it is not at 37°C, shake it down to that level. Insert the thermometer in the urinometer, and read the temperature after 3 minutes. The temperature of the sample is used in the determination of specific gravity in the next step. Clean the thermometer as directed by your instructor.

4. Float the hydrometer in the cylinder. Make certain that it is not touching the bottom or the sides of the cylinder. Adjust the value of the hydrometer reading according to temperature to obtain the specific gravity. For each 3° above 25°C, add 0.001 to the hydrometer reading; for each 3° below 25°C, subtract 0.001. Record your findings in the results and observations section. Remove the hydrometer, and clean it as directed by your instructor.

5. Dip a pH indicator test strip paper in the urinometer cylinder three consecutive times. Remove excess urine by touching the paper to the inside surface of the cylinder. After 1 minute, compare the color of the strip with the color guide on the strip container or other available guide. Compare your findings with the values in Table 50-1. Record your findings in the results and observations section.

6. Clean the urinometer as directed by your instructor.

7. Repeat Steps 1–6 with the second urine sample.

## DETERMINATION OF PRESENCE OF ABNORMAL CONSTITUENTS

### Materials

The following materials should be provided for class use:

1. N-Multistix (Ames) and Phenistix (Ames) reagent strips with their respective containers showing test results or individual or other combined test reagent strips for urinalysis, such as Albustix, Clinistix, Ketostix, Hemastix, and Bili-Labstix.
2. Two freshly voided urine samples
3. 250-ml Pyrex beakers
4. 10-ml pipettes with bulbs
5. Benedict's reagent in dropper bottles
6. Beta-phenylpyruvic acid test solution in a dropper bottle
7. Pyrex test tubes
8. Electric hot plate or other heating system
9. Liquid soap and brushes
10. Pasteur pipettes with bulbs

### Procedure 1: N-Multistix Reagent Strips

This procedure is to be performed by students individually.

1. Pour about 125 ml of one of the urine samples into a beaker.
2. Remove one N-Multistix reagent strip from its container. Be careful not to handle the specific color squares. Examine the strip and the identification or test-interpretation portion of the label on the container. Note that this strip contains material for determinations of pH, protein, glucose, ketones, bilirubin, blood, nitrite, and urobilinogen.
3. Dip the test strip into the urine sample, and tap the excess liquid off on the inside top surface of the beaker. Wait 15–30 seconds, and compare the color reactions with the test interpretation label. (Refer to Color Plate 56.) Record your findings in the results and observations section. Discard the test strip as directed by your instructor.
4. Repeat Steps 1–3 with the second urine sample.

## Procedure 2: Phenistix Reagent Strips

This procedure is to be performed by students individually.

1. Remove one Phenistix reagent strip from its container. Be careful not to handle the specific color squares. Examine the strip and the identification or test-interpretation portion of the label on the container. Note that this strip contains material for the determination of phenylpyruvate.
2. Place 2 drops of the beta-phenylpyruvic acid test solution on the strip reaction side. Wait 30 seconds, and compare the results with the color chart on the container. Dispose of the strip as indicated by your instructor.
3. Pour about 25 ml of one of the urine samples into a clean beaker.
4. Obtain a fresh Phenistix strip, and dip it into the urine sample. Tap the excess liquid off on the inside top surface of the beaker. Wait 30 seconds, and compare the color reactions with the test interpretation label. Record your findings in the results and observations section. Discard the test strip as directed by your instructor.
5. Repeat Steps 3 and 4 with the second urine sample.

## Procedure 3: Quantitative Test for Glucose

This procedure is to be performed by students individually.

1. Pipette 5 ml of Benedict's reagent into two separate Pyrex test tubes.
2. Add 10 drops of the urine sample to each tube and mix by tapping the tubes. Place the tubes in a beaker of water on a hot plate or other heating system and boil for 5 minutes. Allow the tubes to cool, and determine the quantity of glucose according to Table 50-5 (see Color Plate 55). Record your findings in the results and observations section.
3. Clean the equipment used as directed by your instructor.
4. Repeat Steps 1–3 with the second urine sample.

## URINARY SEDIMENT EXAMINATION

### Materials

The following materials should be provided for class use:

1. Sedi-stain (concentrated stain preparation for urinary sediment examination) with dropper
2. Centrifuge and centrifuge tubes

### Table 50-5 Benedict's Reagent Test Results

| Color | Glucose | Result Reported |
|---|---|---|
| Clear, white turbidity, or green turbidity | Zero | Negative |
| Yellow-green | 0.5–1.5 | Weak positive |
| Yellow-orange | 1.5–4.0 | Moderately positive |
| Red | 4.10 and over | Strongly positive |

3. Urine disposal container
4. Two freshly voided urine samples
5. Glass slides and cover slips
6. Microscopes
7. 100-ml capacity graduated cylinders
8. Disposable Pasteur pipettes with bulbs
9. Containers with disinfectant

## Procedure

This procedure is to be performed by students individually.

1. Use a graduated cylinder to transfer a urine sample to a centrifuge tube.
2. Centrifuge your sample as directed by your instructor. Usually, a speed of 1500 revolutions per minute (rpm) for 5 minutes is sufficient.
3. Without disturbing the sediment in the tube, pour off the fluid into the urine-disposal container or as directed by your instructor.
4. Add 2 drops of the Sedi-stain preparation to the urine sediment, and mix by tapping the tube with your index finger.
5. Transfer 1 drop of the stained material to a glass slide. Cover the specimen with a cover slip.
6. Examine the preparation under low and high power. With the aid of Figure 50-1, identify the various items in the sediment. Sketch a representative field in the results and observations section.
7. Repeat Steps 1–6 with the second urine sample.

## DETECTION OF BACTERIURIA

### Materials

The following materials should be provided for class use:

1. Bacturcult tubes with counting strip (one per student)
2. Urine specimen with low bacterial count
3. Urine specimen with high bacterial count
4. Sterile small beakers
5. Urine disposal container with disinfectant

### Procedure

This procedure is to be performed by students in pairs.

1. Pour a small amount of one of the urine specimens into a beaker.
2. Open one Bacturcult tube, and pour the urine specimen into it.

3. *Immediately* pour the urine back into the beaker, and allow all the fluid to drain completely from the Bacturcult tube for a few seconds.
4. Replace the screw cap snugly on the tube, loosen the cap about a half turn, and incubate the tube in your laboratory drawer for at least 48 hours.
5. Dispose of the urine specimen as directed by your instructor.
6. Have your laboratory partner repeat Steps 1–5 with the other urine specimen.
7. After incubation, place the counting strip around an area of the tube where there appears to be an even distribution of bacterial colonies. Such colonies appear as variable-sized specks on the sides of the tube. (Refer to Color Plate 57.)
8. Count the number of colonies within the circle, and enter your findings in the results and observations section. If the number of colonies is fewer than 25, no urinary infection exists. A count of greater than 50 is a definite indication of a urinary tract infection.

## UNKNOWN SPECIMENS

### Materials

The following materials should be provided for class use:

1. All materials listed for the other procedures
2. Containers for urine specimen collection (one per student)
3. Unknown urine specimen (one per student)

### Procedure

This procedure is to be performed by students individually.

1. Collect a sample of your own urine in the container provided by your instructor.
2. Obtain an unknown specimen from your instructor.
3. Perform the following procedures and tests on both specimens, and time yourself while carrying out the entire analysis. Enter your findings, and indicate your time in the results and observations section.
   a. Description of physical properties (color, odor, turbidity, pH, and specific gravity)
   b. Chemical determinations for bilirubin, blood, glucose (qualitative and quantitative), ketones, nitrite, and urobilinogen
   c. Bacteriuria determination
   d. Sediment examination
4. Discard the specimens and materials used as directed by your instructor. Clean all glassware.

## RESULTS AND OBSERVATIONS

### *Description of Physical Properties*

1. Enter your findings in Table 50-6.
2. Were there any unusual properties associated with the urine samples? If so list them. _____

_____

_____

### *Determination of Presence of Abnormal Constituents*

1. Enter your findings in Table 50-7. Indicate the following items: the color (C) of the reagent test strip; the units (U) represented by the color reaction if appropriate; the color (NC) of a normal result; and the units (NU) that correspond to the normal result if appropriate.
2. Were any abnormal results observed with the urine specimens? Explain. _____

_____

_____

**Table 50-6**

| Property | Urine Sample 1 | Urine Sample 2 |
|---|---|---|
| Color | | |
| Odor | | |
| Turbidity | | |
| Specific gravity | | |
| pH | | |

**Table 50-7**

| Test | Urine Sample 1 | | | | Urine Sample 2 | | | |
|---|---|---|---|---|---|---|---|---|
| | C | U | NC | NU | C | U | NC | NU |
| Protein | | | | | | | | |
| Glucose | | | | | | | | |
| Ketone | | | | | | | | |
| Bilirubin | | | | | | | | |
| Blood | | | | | | | | |
| Nitrite | | | | | | | | |
| Phenylpyruvic acid | | | | | | | | |
| Urobilinogen | | | | | | | | |
| Benedict's | | | | | | | | |

3. Why should urine specimens be refrigerated if tests cannot be performed soon after collection?

_____

_____

## *Urinary Sediment Examination*

1. Sketch the results of your urinary sediment examination.

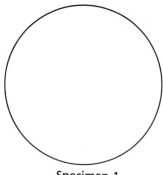

Specimen 1

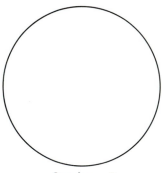

Specimen 2

2. What did your urine specimens contain? _____

_____

_____

## *Detection of Bacteriuria*

1. Indicate the bacterial counts for each of the urine specimens tested.

   a. Specimen 1 _____

   b. Specimen 2 _____

2. Which of the specimens showed signs of an infection? _____

3. Was this procedure simple to perform? _____

_____

## *Unknown Specimens*

1. Enter your findings in Table 50-8.
2. Were any of the results obtained with the unknown specimen abnormally high? If so, indicate which ones. _____

_____

_____

3. How long did it take you to perform the urinalysis?

_____

4. How did your time compare with others in your laboratory? _____

_____

**Table 50-8**

| Procedure or Test | Results for Student Specimen | Results for Unknown Specimen #_____ |
|---|---|---|
| Color | | |
| Turbidity | | |
| pH | | |
| Bilirubin[a] | | |
| Blood[a] | | |
| Glucose (qualitative) | | |
| Glucose (quantitative) | | |
| Ketone[a] | | |
| Nitrite[a] | | |
| Phenylpyruvic acid[a] | | |
| Protein[a] | | |
| Urobilinogen[a] | | |
| Bacterial count[b] | | |
| Urinary sediment examination | | |

[a]These determinations are to be made with reagent strips.
[b]This result can be determined only after 24 hours.

# LABORATORY REVIEW *50*

## *ROUTINE URINALYSIS*

### COMPLETION, INTERPRETATION, AND DEFINITION

1. In Table Q50-1, distinguish between diabetes mellitus and diabetes insipidus as to the properties of urine.

**Table Q50-1**

| Property of Urine | Diabetes Mellitus | Diabetes Insipidus |
|---|---|---|
| Volume | a. | b. |
| Specific gravity (low, normal, high) | c. | d. |
| Presence of sugar | e. | f. |
| Color | g. | h. |

2. What would a count of 200,000 bacterial colonies on a dipstick device indicate?

_____

3. List five abnormal constituents of urine, and indicate the possible cause for their appearance.

| ABNORMAL CONSTITUENT | POSSIBLE CAUSE |
|---|---|
| a. _____ | _____ |
| b. _____ | _____ |
| c. _____ | _____ |
| d. _____ | _____ |
| e. _____ | _____ |

4. Define or explain the following:

a. Albuminuria _____

b. Casts _____

_____

c. Urea _____

d. Uric acid _____

e. Creatinine _____

f. Glycosuria _____

g. Jaundice _____

h. Phenylketonuria _____

**ANSWERS**

5. List three microbial diseases of the urinary system.

a. _____

b. _____

c. _____

## MULTIPLE CHOICE

Insert the correct answers in the spaces provided.

1. _____

2. _____

3. _____

4. _____

5. _____

6. _____

7. _____

8. _____

9. _____

10. _____

11. _____

1. In the hereditary metabolic disorder known as PKU, foods containing which of the following amino acids cannot be metabolized?
   a. phenylalanine hydroxylase
   b. tyrosine
   c. phenylpyruvate
   d. glycine
   e. phenylalanine
2. Which of the following is an individual with PKU unable to synthesize?
   a. tyrosine
   b. phenylalanine
   c. phenylpyruvate
   d. phenylalanine hydroxylase
   e. Phenistix
3. What would be the normal volume of urine per 24 hours in liters for an average adult?
   a. 1–2
   b. 2–3
   c. under 1
   d. 3–4
   e. 5
4. What would a brown-colored urine specimen be considered?
   a. normal
   b. abnormal
   c. a variation of a normal value
5. What should be the specific gravity value of normal urine?
   a. 6
   b. 4.8–7.5
   c. 1.002–1.030
   d. 24
   e. 25
6. The pH of urine normally can range from a low of 4.8 to a high of what?
   a. 6.0
   b. 7.5
   c. 8.0
   d. 6.5
   e. 10
7. Which of the following is not a normal constituent of urine?
   a. sodium
   b. calcium
   c. acetone bodies
   d. uric acid
   e. glucose

8. Hematuria indicates the presence of which of the following in urine?
   a. fructose
   b. red blood cells
   c. bile
   d. leukocytes
   e. urea

9. Which of the following conditions may result in the presence of large amounts of albumin in urine?
   a. damaged kidney tubules
   b. diabetes mellitus
   c. chemical poisoning
   d. malaria
   e. choices *a* and *c* only

10. What is the term for the small masses of body secretions and other materials that harden in tubules and that assume various shapes and appear in urine?
    a. bacteria
    b. casts
    c. renal calculi
    d. urethral flora
    e. choices *a* and *b* only

11. In each milliliter of normal urine, up to what number of bacteria may be found?
    a. 1000
    b. 5000
    c. 10,000
    d. 100,000
    e. 1,000,000

# Gamete Formation, Stages of Early Development, and Karyotyping

## After completing this exercise, you should be able to

1. Distinguish between *mitosis* and *meiosis*
2. Describe and identify the stages of meiosis
3. Recognize characteristic structures of ovarian tissue and the general appearance of spermatozoa
4. Identify early stages of division of the zygote
5. Prepare a karyotype from a metaphase spread of chromosomes
6. List and describe four structural changes of chromosomes
7. Recognize normal and selected abnormal karyotypes

---

As indicated in Exercise 3, chromosomes carry the hereditary chemical material, deoxyribonucleic acid (DNA). All cells of the body, except *gametes* (the sex cells), contain two complete chromosome sets. For humans, the total cellular number is 46, or 23 pairs. The 46 is referred to as the *diploid number* or *2n*. Gametes contain one set of 23 chromosomes; this *haploid number* is *n*.

Reproduction includes the process by which genetic material is passed from generation to generation. One chromosome from each pair comes from the mother, and the other comes from the father. *Homologous chromosomes,* the two chromosomes of a pair, contain genes that control the same genetic traits. If a chromosome holds a gene for eye color, its homologous chromosome will have a gene for eye color.

A gene that dominates or controls is termed the *dominant gene,* and the associated trait expressed is said to be a *dominant trait.* The gene that is inhibited is referred to as the *recessive gene,* and the trait it controls is the *recessive trait.* Genes that control the same inherited trait, such as height, eye color, or hair color, and that occupy the same position on homologous (paired) chromosomes are known as *alleles.*

Twenty-two (44) of the 23 pairs are referred to as the *autosomes* and the remaining pair as the sex chromosomes. In the female, one chromosome—the *X chromosome*—has a homologous X chromosome, while in the male, one *X chromosome* is replaced by a smaller one, the *Y chromosome.* Thus, under normal conditions, human body cells contain 46 chromosomes, including either two X chromosomes for a female, or one X and one Y for a male.

In preparation for the sexual reproduction (*fertilization*) process, diploid cells divide into haploid cells through the special nuclear division process, termed *meiosis. Meiosis,* or reduction division, is a process by which the chromosome number is reduced by half. Whereas mitosis involves one division of the nucleus following a single duplication of the normal or diploid number of chromosomes (described in Exercise 3), meiosis consists of two distinct nuclear divisions, resulting in four new nuclei. Meiosis is associated with the production of sex cells.

In the testes of the male, meiosis results in the production of gametes termed *spermatozoa* through the process of *spermatogenesis* (Color Plate 59). In the ovaries of the female, meiosis results in the production of gametes termed ova (singular is *ovum*), through the process of *oogenesis* (Color Plate 58). Figure 51-1 diagrammatically shows meiosis in the formation of female sex cells (oogenesis) and male sex cells (spermatogenesis). Only three chromosome pairs are used in this figure, to show how they are distributed during the process.

During spermatogenesis, *spermatogonia* (diploid cells) undergo two stages of nuclear division. During the first division, each pair of homologous chromo-

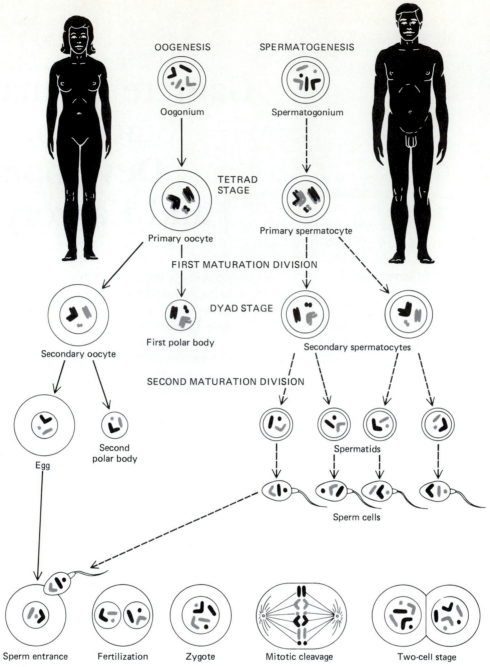

**Figure 51-1** Meiosis in female sex-cell production *(oogenesis)* and male sex-cell production *(spermatogenesis)*. Each diploid spermatogonium produces four haploid sperm cells. Meiosis is essentially the same in the female, except that the division of the cytoplasm is unequal, and only one functional haploid ootid is produced from each diploid oogonium.

somes (each consisting of two halves, termed *chromatids*) form a *tetrad* (4). During this time, the cells are known as *primary spermatocytes*. The pairs of chromosomes next migrate to opposite ends of the cell, thereby completing the first nuclear (maturation) division and resulting in the formation of two new cells

termed *secondary spermatocytes*. These cells are haploid. It is important to note that at this stage, while the chromosomes are haploid, each chromosome consists of two halves (chromatids). In the second nuclear (maturation) division, the chromatids separate, forming *spermatids*. These cells also are haploid but contain half the number of chromosomes as found in the spermatogonia. When the spermatids mature, they form *spermatozoa.* Figure 51-1 shows that at the completion of spermatogenesis, four spermatozoa have developed from one primary spermatocyte.

During the first stage of oogenesis, the *oogonium* forms tetrads and develops into a *primary oocyte.* Once the primary oocyte is released by the ovary (ovulation), and is fertilized by a spermatozoan, it will undergo division resulting in the formation of two cells of unequal size. The larger of the two received most of the cytoplasm. This cell is termed the *secondary oocyte.* The smaller one, which is known as the *first polar body* eventually disintegrates. During ovulation, the secondary oocyte is released into the uterine tube. If fertilization occurs, the secondary oocyte undergoes the second nuclear (maturation) division, forming two haploid cells, again each of unequal size. The larger of the two cells is known as an *ootid,* and the smaller one is the *second polar body.* The ootid develops into one ovum. The process of oogenesis is similar to spermatogenesis, however, it results in only one gamete, the *ovum.*

In organisms that reproduce sexually, the union of the haploid sperm produced by a male parent and the haploid ovum produced by the female parent forms the diploid fertilized egg or *zygote.* The zygote undergoes a type of mitotic division known as *cleavage.* The cell division is so rapid that there is insufficient time for a cell to enlarge between divisions. The number of cells increases, but the cells become smaller and smaller, so

that the mass of cells that develops is about the same size as the original zygote and known as the *blastula* (Figure 51-2). The blastula generally consists of cells organized inside a hollow sphere that surrounds a fluid-filled cavity, the *blastocoel* (Figure 51-2b). Processes collectively known as *gastrulation* convert the single-layered blastula into a multilayered structure with functionally specialized areas. These changes in the developing cellularly specialized *embryo* depend on a series of cellular movements and the segregation of specific groups of cells. Demonstration slides are used in this exercise to show these early stages of development.

Since the discovery that human somatic (body) cells contain 46 chromosomes, a number of diseases have been found to be related to abnormalities in chromosome structure and number. Detailed cytogenetic procedures have been established for studying chromosomes. *Karyotyping* is one method commonly used to demonstrate many chromosomal *aberrations* (deviations from the normal). The method (Figure 51-3) involves the growing of cells, such as leukocytes or skin cells, in tissue culture for a defined period, stopping cell division during metaphase by the use of a drug such as colchicine, staining the chromosomes, and photographing the metaphase spread of chromosomes (Figure 51-4a). Once the photograph is made, the individual chromosomes are cut out and placed in order, according to size, shape, and staining pattern. The longest chromosome is 7 to 8 μm, while the shortest one is about 1 to 2 μm. In a routine analysis, a chromosomal count is done. There may be more than the normal number

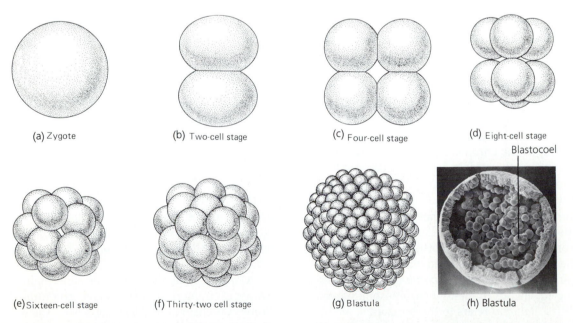

(a) Zygote     (b) Two-cell stage     (c) Four-cell stage     (d) Eight-cell stage

Blastocoel

(e) Sixteen-cell stage     (f) Thirty-two cell stage     (g) Blastula     (h) Blastula

**Figure 51-2** Early stages of development (a) through (g): the repeated mitotic divisions (known as *cleavage*) until the formation of the blastula. (h) A scanning photomicrograph of the internal features of a blastula (from S. Komazaki, *Cell Differentiation,* 24: 25–32, 1988).

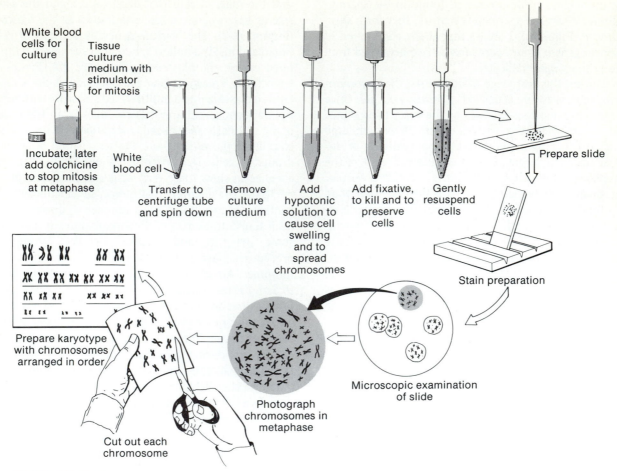

**Figure 51-3** The steps in a karyotyping procedure.

or a rearrangement of those present. Differences from the normal number or chromosome arrangement are usually indicative of clinically important genetic defects.

When paired, numbered, and grouped in a standard way, the pattern formed by the chromosomes is termed a *karyotype* (Figure 51-4b and Table 51-1). The grouping system makes use of the location of the centromere associated with the chromosome. A chromosome with a centromere in the middle, which divides it into two equal arms, is termed *metacentric*. If the centromere is somewhat nearer one end, creating a distinct long and short arm, the chromosome is *submetacentric*. When the centromere is very near to one end, so that the chromosome has a very short arm and longer long arm, the chromosome is termed *acrocentric*.

Staining methods termed *banding techniques* are used in cytogenetics to identify chromosomes and abnormalities. *Banding* refers to the horizontal, alternating light and dark crossbands that result from the staining. Different staining techniques are used for different purposes. Some identify and differentiate the chromosomal pairs. Such techniques routinely are used

for detecting and defining chromosomal aberrations. Other procedures stain specific chromosome areas to identify special problem situations. The most popular banding technique is G (Giemsa stain) banding. Other techniques are Q (quinacrine dye) bands and C (centromere staining) bands. Figure 51-5 is a diagrammatic representation of banding patterns revealed by staining and of the system used to specify location on the chromosome. The chromosome pairs are identified by number. The short and long arms are labeled *p* and *q*, respectively. The bands are specified by a two-digit numbering system. Thus, if the chromosome shown in Figure 51-5 were chromosome 1, the light band at the tip of the short arm would be designated 1 *p*36, and the light band at the bottom end of the long arm would be 1 *q*44.

Chromosomal aberrations or abnormalities are of two types: numerical changes and structural changes. Numerical changes are deviations from the total complement number of 46. Such changes may result from a misdivision during metaphase called *nondisjunction*. In this situation the sister chromatids fail to separate, resulting in one daughter cell with an extra chromo-

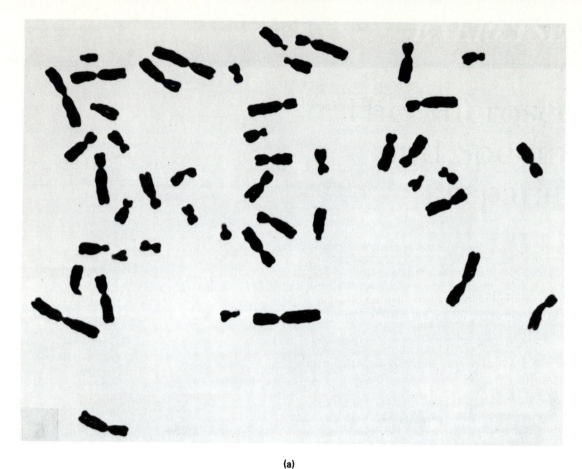

(a)

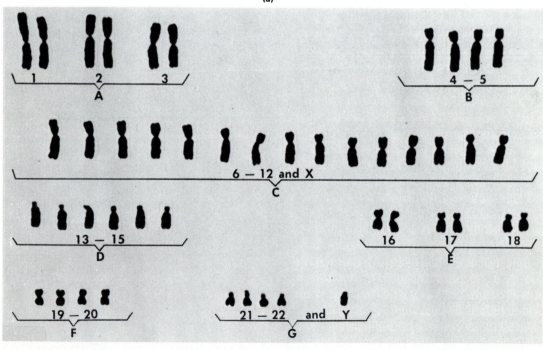

(b)

**Figure 51-4** Karyotyping: (a) a metaphase spread from a normal male cell; (b) a karyotype of a normal male. In determining the karyotype, the chromosomes are grouped and arranged according to their shapes and sizes. The X chromosome is located at the right of group C. The Y chromosome is located at the right of group G. (From L. Larson, *American Journal of Medical Technology,* 49: 687–697, 1983).

**Table 51-1  The Grouping System for Chromosomes in a Karyotyping Procedure**

| Group | Description |
|---|---|
| Group 1–3 or A | The three largest pairs of chromosomes, all of which are metacentric |
| Group 4–5 or B | Two slightly smaller submetacentric pairs |
| Group 6–12 and X or C | Different-sized autosomes, some metacentric and some submetacentric, and X chromosome (sex chromosomes are sometimes placed separately at the lower right of a karyotype as in Figure 51-6) |
| Group 13–15 or D | Three pairs, nearly as large as those in Group 6–12, which are submetacentric and have satellites (tiny chromatic masses attached to the short arm) |
| Group 16–18 or E | Three smaller pairs, which are metacentric or submetacentric and lack satellites |
| Group 19–20 or F | The two smallest metacentric pairs |
| Group 21–22 and Y or G | The two smallest acrocentric autosomes, which have satellites, and the Y chromosome, which is also acrocentric and typically larger than the others, but the size varies |

some and one with a missing chromosome. One extra chromosome at a pair location is called *trisomy;* one missing is referred to as *monosomy.* The consequences depend on the chromosomes involved. An example of a trisomy involving an autosome is Down's syndrome or trisomy 21, that is, an extra chromosome at pair 21 (Figure 51-6). Abnormalities involving sex chromosomes include Klinefelter's syndrome, an XXY combination in a male, and Turner's syndrome, a single X chromosome in a female (Table 51-2).

In a karyotype description numerical changes are indicated by a " + " or a " − " placed before a chromosome number to indicate an extra or missing whole chromosome (see the caption for Figure 51-5).

There are several types of structural changes in chromosomes. A *deletion* (del) occurs when a chromosome is broken and a particular fragment is lost. This form of aberration usually results in severe malformations in the individual.

A *translocation* (t) involves a break in a chromosome and the transfer of part or all of one chromosome to another. In a karyotype description, translocations are indicated by a "t" placed before the numbers of chromosomes involved, given in parenthesis. The break points are also indicated. For example, t(2:10) (p24;q26) specifies a translocation involving chromosomes 2 and 10, breaks occurring on the short arm (*p*) at band 24 and on the long arm (*q*) at band 26.

An *inversion* (inv) results when a section breaks off a chromosome and reattaches in an inverted position. The section may be one end of a chromosome or a fragment created by two breaks within the chromosome.

An *isochromosome* (i) is caused by an unequal division of the centromere.

A *ring chromosome* is formed when the ends of both the long and short arms of a chromosome break off and then reattach to each other to form a ringlike structure.

This exercise includes the preparation of karyotypes and identification of chromosomal parts and abnormalities. Table 51-2 lists several human chromosomal aberrations.

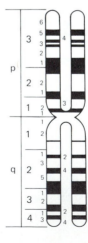

**Figure 51-5** A diagrammatic representation of the human chromosome's banding patterns revealed by the different kinds of staining techniques. The letter *p* represents the short arm, *q* the long arm. Two numbers specify the bands.

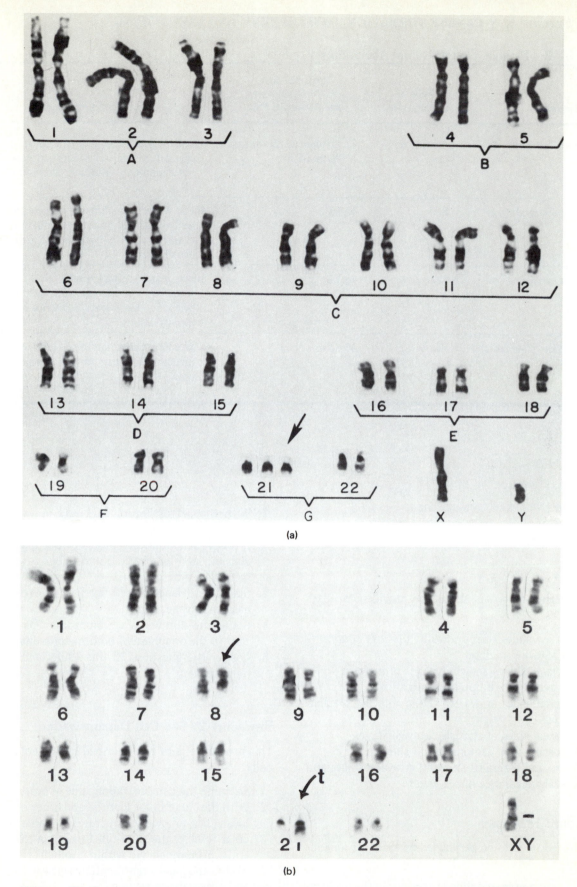

**Figure 51-6** Karyotypes: (a) a karyotype of a patient with trisomy 21 *(arrow)*— G *(Giemsa)* bands can be seen on the individual chromosomes; (b) a G-stained karyotype of a patient, showing a translocation (t) involving pairs 8 and 21. Note also that the Y chromosome is missing (−Y) and that therefore the individual has only 45 chromosomes. This condition is found in individuals with a form of acute leukemia. The karyotype is expressed as 45,XY, t(8:21). (From A. Cork, *American Journal of Medical Technology*, 49: 703–714, 1983.)

## Table 51-2 Human Chromosomal Aberrations

| Abnormality | Type of Chromosome | Total Number of Chromosomes and Abnormal Pair | Resulting Conditions |
|---|---|---|---|
| Down's syndrome | Autosome | 47, trisomy 21 (shown in Figure 51-6) | Growth stunted: characteristic facial changes; mental retardation; increased susceptibility to infection; affects both sexes |
| Klinefelter's syndrome | Sex | 47,XXY | Affects males; *atrophic* (abnormally small) testes; enlarged breasts; secondary sex characteristics underdeveloped or lacking; usually sterile |
| Turner's syndrome | Sex | 45,X– | Affects females; small, webbed neck; secondary sex characteristics lacking; ovaries underdeveloped; absence of menstruation |
| XXX female (superfemale) | Sex | 47,XXX | Generally appears normal; slightly depressed mental ability; can produce normal children |

## DEMONSTRATIONS OF MEIOSIS— SEX CELLS AND CLEAVAGE STAGES

### Materials

The following materials should be provided for class use:

1. Prepared slides of roundworm (*Ascaris equorum*) eggs showing meiosis
2. Models or charts showing stages of meiosis
3. Prepared slides of human spermatozoa
4. Prepared slides of ovarian tissue showing graafian follicles
5. Prepared slides of starfish cleavage stages
6. Microscopes for demonstration slides
7. Charts and diagrams showing cleavage stages and general appearance of sex cells

### Procedure 1: Meiosis

This procedure is to be performed by students individually.

1. Examine a prepared slide of portions of the oviducts of an *Ascaris*. The preparation should show eggs undergoing meiosis.

2. With the aid of Figure 51-1 and other material showing the stages of meiosis, identify metaphase I, anaphase I, and telophase I of the first meiotic division, and anaphase II and telophase II of the second meiotic division.
3. Sketch and label these stages in the results and observations section.
4. Count the number of chromosomes (as best you can) in the anaphase of both meiotic divisions.
5. Answer the questions in the results and observations section.

### Procedure 2: Sex-Cell Demonstration

This procedure is to be performed by students individually.

1. Examine the demonstration slide of ovarian tissue.
2. With the aid of Color Plate 58 and other references, locate the following structures: graafian or vesicular follicle, the fluid-filled cavity of the follicle, and a primary oocyte within a follicle.
3. Sketch and label a representative view in the results and observations section.
4. Examine the slide containing the spermatozoa.
5. With the aid of Color Plate 59 and other references,

identify the head and thinner tail portions. Look for abnormalities such as two heads or no tail. Sketch and label a normal and an abnormal spermatozoon in the results and observations section.

6. Answer the questions in the results and observations section.

### Procedure 3: Cleavage Stages Demonstration

This procedure is to be performed by students individually.

1. Examine the demonstration slide showing cleavage stages in the starfish.
2. Identify the following stages: zygote, 2-cell, 4-cell, 8-cell, 16-cell, and blastula. (See Figure 51-2.)
3. Sketch and label the stages as indicated in the results and observations section.
4. Answer the questions in the results and observations section.

## KARYOTYPING

### Materials

The following materials should be provided per each two students:

1. Two copies of Figure 51-7 (both removed from students' manuals) or similar reproductions of photographs of metaphase chromosome spreads

2. One pair of scissors
3. A millimeter ruler
4. Rubber cement or similar material with which to paste

### Procedure

This procedure is to be performed by students individually.

1. Examine the chromosome spread in Figure 51-7a.
2. Carefully cut out the chromosomes, and arrange them in pairs according to approximate length and general appearance. Refer to Figure 51-4b.
3. Paste the chromosomes on one of the karyotype analysis sheets provided in the results and observations section. Use Figure 51-4b as a guide.
4. Compare your karyotype with Figure 51-4b, and determine the number of chromosomes; whether the karyotype is for a female or male, and whether the karyotype is an abnormal one for a human.
5. Record your findings in the results and observations section.
6. Repeat this entire procedure with the chromosome spread in Figure 51-7b.

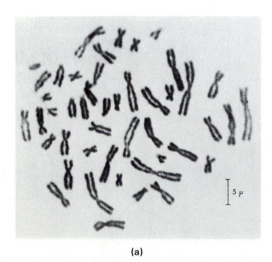

(a)

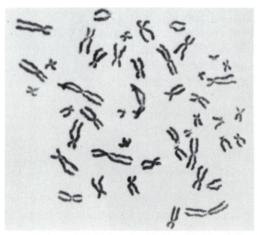

(b)

**Figure 51-7** Metaphase chromosome spreads.

## RESULTS AND OBSERVATIONS

### *Meiosis*

1. Sketch and label stages of meiotic division.

Meiosis I

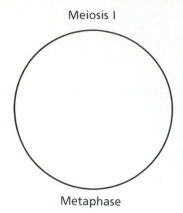

Metaphase

Meiosis I

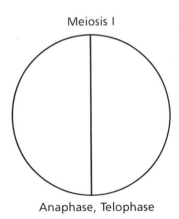

Anaphase, Telophase

Meiosis II

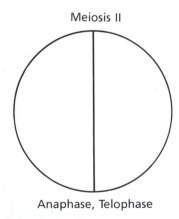

Anaphase, Telophase

2. Were the same numbers of chromosomes in anaphase also in meiosis I and meiosis II? If not, explain. _____

_____

_____

### *Sex Cells*

1. Sketch representative views of a graafian follicle and representative spermatozoa.

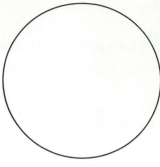

Graafian follicle

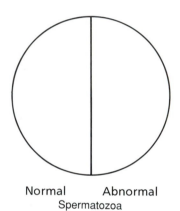

Normal          Abnormal
Spermatozoa

2. Were abnormal spermatozoa present? If so, what percentage of the cells observed were abnormal?

_____

3. What factors are responsible for abnormal spermatozoa formation? _____

_____

4. Complete the following diagram of possible meiotic divisions.

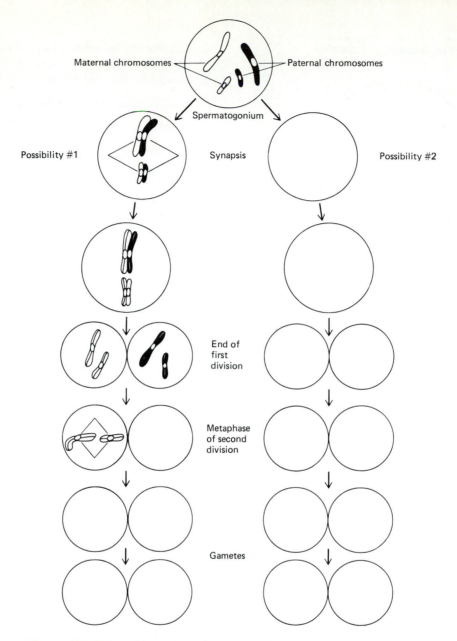

**Figure 51-8** Possible meiotic divisions in spermatogenesis. Only two pairs of chromosomes are shown.

## Cleavage Stages

1. Sketch stages of cleavage as indicated.

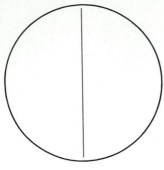

Zygote    Two-cell stage

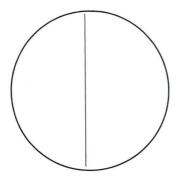

Eight-cell stage    Blastula

2. Compare the zygote and blastula stages. Which one is larger? _____

3. Are each of the cells in the eight-cell and blastula stages the same size? _____

## Karyotyping

1. Examine the karyotypes you prepared from Figures 51-7a and 51-7b.

2. Which is the karyotype of a human female? _____ Which is the karyotype of a human male? _____

3. Are the karyotypes normal? If not, list the aberrations. _____

_____

_____

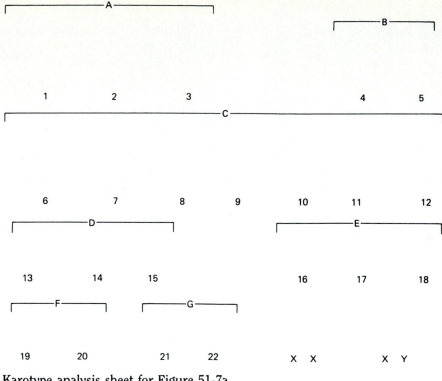

Karotype analysis sheet for Figure 51-7a.

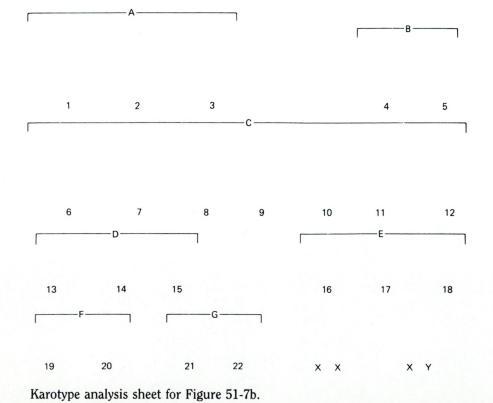

Karotype analysis sheet for Figure 51-7b.

# LABORATORY REVIEW *51*

## GAMETE FORMATION, STAGES OF EARLY DEVELOPMENT, AND KARYOTYPING

### COMPLETION, DEFINITION, AND INTERPRETATION

1. List three features of meiosis that distinguish it from mitosis.

    a. _____

    b. _____

    c. _____

    d. _____

2. Define or explain the following:

    a. Chromosome _____

    b. Chromatid _____

    c. Haploid _____

    d. Diploid _____

    e. Meiosis _____

    f. Down's syndrome _____

    g. Nondisjunction _____

3. List and describe four specific structural chromosomal aberrations.

    a. _____

    b. _____

    c. _____

    d. _____

4. Examine the karyotype in Figure Q51-1.

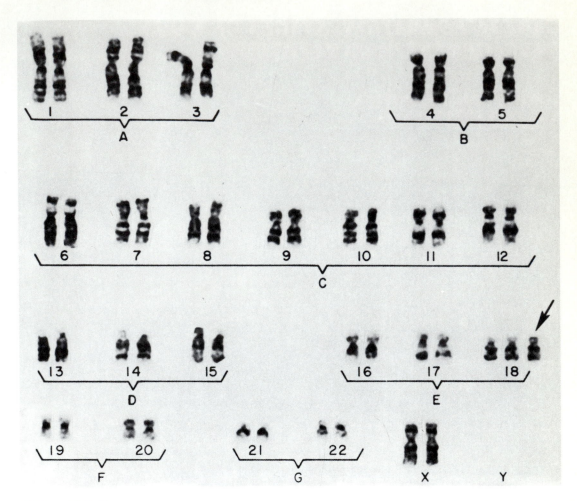

**Figure Q51-1** (From L. Larson, *American Journal of Medical Technology,* 49: 687–697, 1983.)

a. Is the arrangement of chromosomes correct for a normal human? _____

b. Is the karyotype typical of a normal human? _____

## MULTIPLE CHOICE

Insert the correct answers in the spaces provided.

1. What is the total number of chromosomes for all cells of the body except gametes?
   a. 23
   b. 46
   c. 92
   d. 184
   e. 48

2. What is the total number of chromosomes in each human gamete?
   a. 23
   b. 46
   c. 92
   d. 184
   e. 48

3. Which of the following is the term for X and Y chromosomes?
   a. dominant chromosomes
   b. sex chromosomes
   c. autosomes
   d. gametes
   e. body chromosomes
4. Which of the following chromosome arrangements do the body cells of a normal female contain?
   a. 1X
   b. 2Xs
   c. XY
   d. Y
   e. either choice *a* or *b*
5. Which of the following chromosome arrangements do the body cells of a normal male contain?
   a. 1X
   b. 2Xs
   c. XY
   d. Y
   e. either choice *a* or *b*
6. What are the genes that control the same inherited trait and occupy the same site on homologous chromosomes?
   a. recessive
   b. dominant
   c. homologous
   d. chromatids
   e. alleles
7. What are the halves of chromosomes that develop during spermatogenesis?
   a. recessive
   b. dominant
   c. homologous
   d. chromatids
   e. alleles
8. In the meiotic process, how many spermatozoa does one primary spermatocyte normally give rise to?
   a. 2
   b. 3
   c. 4
   d. 5
   e. 6
9. In the meiotic process, how many sex cells does one primary oocyte normally give rise to?
   a. 1
   b. 2
   c. 3
   d. 4
   e. 6
10. What is the term for when one extra chromosome is added to a chromosome pair?
    a. monosomy
    b. trisomy
    c. translocation
    d. karyotype
    e. deletion

**ANSWERS**

1. _____
2. _____
3. _____
4. _____
5. _____
6. _____
7. _____
8. _____
9. _____
10. _____

# Ovulation Prediction, Human Chorionic Gonadotropin, and Pregnancy Testing

**After completing this exercise, you should be able to**

1. Perform and interpret the results of a luteinizing hormone detection test
2. Perform and interpret the results of a slide pregnancy test
3. List three conditions that could produce false-positive pregnancy test results
4. Explain the immunological basis of a slide test for pregnancy

Ovulation is the process during which a secondary oocyte (refer to Exercise 51) is released from an ovary. The few days around ovulation (two days before, the day of ovulation, and one day after) mark the most likely time for fertilization to take place. A woman's body normally produces a small quantity of *luteinizing hormone (LH)*. However, the LH level greatly increases for a short time in the middle of the menstrual and ovarian cycles. This LH increase or *surge* causes ovulation. If ovulation is to take place, it will occur within 40 hours after the LH surge begins.

In this exercise, a test kit that can be used as an aid in predicting the time of ovulation is examined. The *OvuGen test* uses an enzyme-linked immunoassay (ELISA) and is designed to produce two visually detectable purple dots: one for a urine sample and one for a reference standard control (Figure 52-1). The control provides a standard for a visual comparison each time the test for the LH surge is performed. If the urine sample develops a purple dot that is darker than or as dark as that of the reference standard, this is interpreted as an LH surge associated with ovulation.

Another hormone, human chorionic gonadotropin (HCG), appears in the urine and serum of pregnant women in large concentrations. The hormone is a glycoprotein produced by placental trophoblastic cells and plays a major role in maintaining the developing corpus luteum. The corpus luteum maintains the state of pregnancy by the production of the hormones progesterone and estradiol until the placenta and fetus begin the secretion of steroid and protein hormones. Following implantation of a fertilized ovum, HCG levels increase rapidly in urine and may reach detectable levels shortly after conception. Laboratory test demonstration of the hormone's presence is generally indicative of pregnancy. However, false-positive results may be obtained if the urine and serum contain high levels of LH, if the individual is menopausal, or if some specific cancerous states exist, such as choriocarcinoma and chorioepithelioma, or in situations of a spontaneous abortion. Following spontaneous abortion, the results remain positive for approximately one week.

The immunoenzymatic test for pregnancy considered in this exercise is an accurate, rapid, and simple procedure. It clearly replaces the various tests that require specialized equipment, radioactive reagents, or the use of animals such as frogs, rats, mice, or rabbits.

The Pregnospia test is an immunoenzymatic assay. It uses a revolutionary type of diagnostic tool, the monoclonal immunoglobulin (antibody). The production of pure or monoclonal (single-type) antibodies is achieved by an in vitro hybridization process in which an antibody-producing cell is fused with a myeloma (particular cancer) cell under special experimental conditions. In this process, the genetic materials of both cells are combined to create a hybrid cell that contains desirable characteristics from both of the parent cells.

The Prognosia test is an agglutination immunoassay. The monoclonal antibodies mentioned earlier are coated onto special gold particles, which have an intense

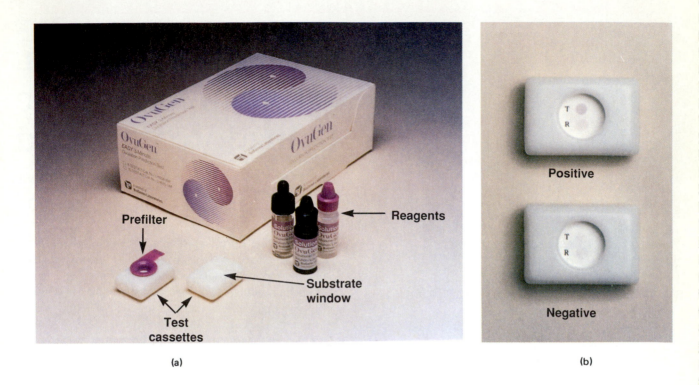

**Figure 52-1** Luteinizing hormone (LH) detection materials, including the test cassette. (Courtesy of BioGenex Laboratories, San Ramon, California).

reddish-purple color in solution. These antibodies are noted for their specificity for special sites on the intact HCG molecule and one of its polypeptide subunits. When the HCG antibody preparation reacts with HCG present in a urine specimen, connecting molecular bridges are formed, binding the gold particles together and thereby causing them to clump or agglutinate. The resulting agglutination is evidenced by a distinct color change from the original reddish-purple color of the HCG antibody preparation. If there is no HCG present in the urine sample, the agglutination reaction will not change color. This result is interpreted as a negative reaction.

> ### *CAUTION*
> Working with urine or other body fluids will not pose any danger if the following precautions are observed:
>
> 1. Place any and all urine or materials containing urine into the disinfectant containers provided.
> 2. Dispose of any urine-containing materials as indicated by the instructor.
> 3. In cases of urine spillage, wipe the area with a disinfectant-soaked paper towel, and dispose of all materials as indicated by the instructor.
> 4. Always wash your hands after handling urine and associated materials.
> 5. The use of disposable surgical gloves.

## LUTEINIZING HORMONE DETECTION (OVUGEN OVULATION PREDICTION TEST)

### Materials

The following materials should be provided per two students. All items should be available in the commercial kit.

1. Four test cassettes in individual pouches
2. Four disposable droppers (with bulbs if appropriate)
3. OvuGen solution A (Vial A)
4. Conjugate solution (Vial B)
5. Substrate solution (Vial C)
6. Stop solution (Vial D)
7. One known positive urine control (labeled " + ")
8. One known negative urine control (labeled " − ")
9. Two unknown specimens (one per student) (labeled "1" and "2")
10. One marking pen or similar device
11. Container for disposing of droppers and prefilters
12. Container of disinfectant
13. Disposable surgical gloves

### Procedure

The procedure is to be performed by students in pairs.

#### *General*

1. Obtain four test cassettes and other materials needed to perform the procedure (Figure 52-1).
2. Examine the cassettes, and note the (removable) purple prefilter.
3. Label each cassette for each specimen (positive control, negative control, and each unknown).

#### *Solution Preparation*

1. Add OvuGen solution (Vial A) to the conjugate solution.
2. Close the vial tightly, and mix the contents by turning it upside down at least five times.
3. Allow the mixed solution to stand for at least 10 minutes before use.

#### *Test*

1. Using a disposable dropper, place 7 drops of one of the unknown urine samples onto the prefilter area of the labeled test cassettes (Figure 52-2a). Allow the system to drain.

2. Add 7 drops of conjugate solution by means of the vial's dropper into the prefilter (Figure 52-2b). Allow the system to drain.
3. Remove the prefilter, and place it in the container for disposal.
4. Add 7 drops of the substrate solution into the cassette (test window) (Figure 52-2c). Wait 3 minutes.
5. Add 7 drops of the stop solution (Figure 52-2d).
6. Examine the cassettes for a color change. A test dot that is darker purple than or that matches the reference dot indicates an LH surge. A test dot that is lighter than the reference dot or that does not appear is a negative result (Figure 52-2e).
7. Repeat Steps 1 through 6 with each of the controls and with the second unknown urine specimen.
8. Interpret your findings, and enter them in the results and observations section.
9. Check your results with the instructor.
10. Answer any questions pertaining to this exercise in the laboratory review section.

## RESULTS AND OBSERVATIONS

Enter your findings in Table 52-1.

## HUMAN CHORIONIC GONADOTROPIN DETECTION

### Materials

The following materials should be provided per each two students: (All reagents are available in the commercial kit.)

1. Four tubes of Organon Accusphere reagent (gold solid particles bonded with monoclonal antibodies against HCG)
2. Buffer (2 ml)
3. One 5-ml pipette, calibrated in 0.1-ml units, and a rubber bulb
4. Three 1-ml pipettes, calibrated in 0.1-ml units, and rubber bulbs
5. One known positive urine control
6. One known negative urine control
7. Two unknown specimens (one per student)
8. One marking pen or similar device
9. Container for disposing of pipettes

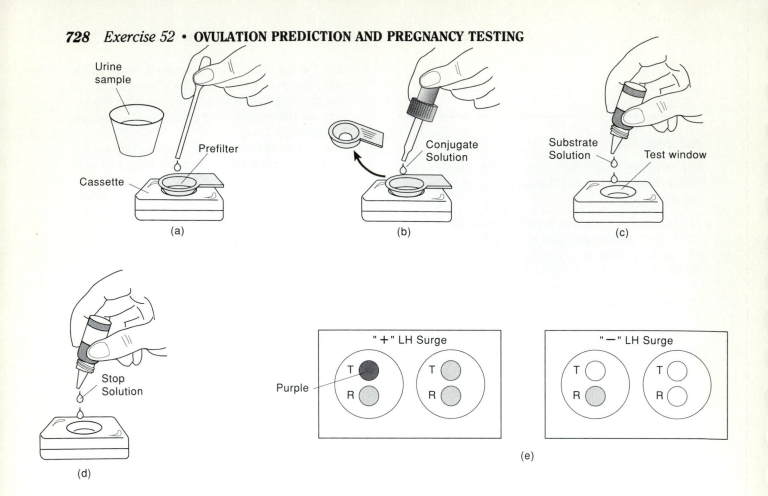

**Figure 52-2** The luteinizing hormone detection procedure.

**Table 52-1**

| Specimen | Color Formed | | LH Present | Indicate of LH Surge |
| | Reference | Test | | |
| --- | --- | --- | --- | --- |
| Known positive | | | | |
| Known negative | | | | |
| Unknown 1 | | | | |
| Unknown 2 | | | | |

## Procedure

This procedure is to be performed by students in pairs.

1. Obtain four tubes containing the Accusphere reagent.
2. Number the tubes accordingly: 1 (known positive), 2 (known negative), 3 (unknown), and 4 (unknown).
3. Remove the rubber stoppers from the four tubes.
4. With the 5-ml pipette, introduce 0.4 ml of the buffer provided to each tube.
5. Replace the stoppers.
6. Gently shake each tube for approximately 15–30 seconds until the Accusphere reagent is completely resuspended.
7. Using separate 1-ml pipettes, add 0.1 ml of the positive, negative, and each of the unknown samples to their respectively marked tubes.
8. Gently shake the tubes for about 30 seconds to achieve thorough mixing.
9. Place the tubes in a rack or on a flat surface, and wait 5 minutes.
10. Examine the tubes for a color change. If a change does not occur within this time period, continue to observe the tubes for another 25 minutes.
11. Compare the unknown tubes with the results shown in the positive and negative samples (Tubes 1 and 2).
12. Interpret your findings, and enter them in the results and observations section.
13. Check your results with the instructor.
14. Answer any questions pertaining to this exercise in the laboratory review section.

## RESULTS AND OBSERVATIONS

Enter your findings in Table 52-2.

**Table 52-2**

| Specimen | Agglutination | HCG Present | Indicative of Pregnancy (positive/negative) |
|----------|---------------|-------------|---------------------------------------------|
| Known positive | | | |
| Known negative | | | |
| Unknown 1 | | | |
| Unknown 2 | | | |

# LABORATORY REVIEW 52

NAME _____

LAB SECTION _____ DATE _____

## OVULATION PREDICTION, HUMAN CHORIONIC GONADOTROPIN, AND PREGNANCY TESTING

### COMPLETION, DEFINITION, AND INTERPRETATION

1. What does an LH surge indicate? _____
_____

2. a. Is the ovulation prediction test simple to perform? _____

   b. Should LH-surge detection testing be performed only once or more often during a several-day period? Explain your answer. _____
_____

3. What is HCG? _____
_____

4. Briefly explain the principle involved in the urinary chorionic gonadotropin (UCG) in vitro test. _____
_____
_____
_____

5. List three advantages of the in vitro immunoenzymatic test for the detection of pregnancy, when compared to the use of laboratory animals for the same purpose.

   a. _____

   b. _____

   c. _____

6. What is the appearance of the results of the UCG test or a comparable test?

   a. Negative result _____

   b. Positive result _____

7. What conditions produce a false-positive result? _____
_____
_____

8. List two limitations of the in vitro test.

   a. _____

   b. _____

## ANSWERS

1. _____
2. _____
3. _____
4. _____

## MULTIPLE CHOICE

Insert the correct answers in the spaces provided.

1. To which of the following does ovulation refer?
   a. fertilization
   b. increased production of LH
   c. secondary oocyte release
   d. HCG production
   e. choices *b* and *c* only
2. From what does ovulation result?
   a. fertilization
   b. increased production of LH
   c. secondary oocyte release
   d. HCG production
   e. choices *b* and *c* only
3. Which of the following appears in the urine and serum of a pregnant woman?
   a. LH
   b. secondary oocyte
   c. primary oocyte
   d. HCG
   e. corpus luteum
4. Which of the following increases rapidly in urine, following the implantation of a fertilized ovum?
   a. LH
   b. secondary oocyte
   c. primary oocyte
   d. HCG
   e. corpus luteum

# Instructions for Physiograph Applications

A variety of devices are currently in use to demonstrate and record physiological activities and data accurately. Among these instruments are multichannel recording devices, known as physiographs, and the classic old-style kymograph.

The *physiograph* is a modern device designed to be an expandable recording system. The basic components of a physiograph system generally include (1) the input transducer (converts the energy of the physiological event into a proportional electrical signal), (2) the amplifier (provides the required increase in signal strength to permit visualization), (3) the coupler (renders the input energy from the transducer or the signal from the subject compatible with the channel amplifier), and (4) the output transducer (provides visual or audio representation of information). In studying physiological events such as heart action, muscle contraction, blood pressure, or cerebral activity, a detection device first is used to pick up the electrical signal from the tissue preparation or the experimental subject. Examples of such devices include an electrode for electrical signals, as in cases of neural activity, and an input transducer for nonelectrical signals, as in situations of temperature measurement.

A transducer converts the physiological event into an electrical signal. Because physiological events often produce weak signals, an amplifier is used to enlarge the magnitude of the event. The amplified signals are converted by an output transducer into some form of visual or audio display that can be observed and measured. The display can be in the form of an electrical meter, an oscilloscope, or a chart recorder.

Various arrangements of physiograph components are possible. Figure A-1 shows an arrangement example. The components of a physiograph are contained in a main frame assembly, which provides power, devices for the control and monitoring of circuitry and instrument calibration, and locations for mounting accessory plug-in units (Figure A-2). Table A-1 lists and describes both the basic components and the accessories (Figure A-3) used with physiographs.

The *kymograph* is a moving-drum recording instrument. It is the classical device used for the measurement of movements produced by muscle contractions. The kymograph is not considered here.

In this appendix, the features of physiological recording devices are demonstrated. This material may serve as a general frame of reference for procedures

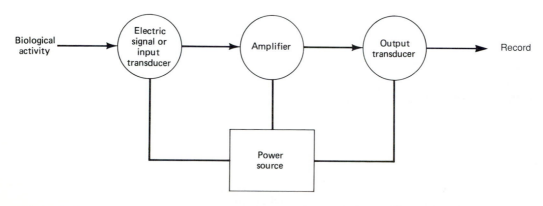

**Figure A-1** The general arrangement of physiograph instrumentation components.

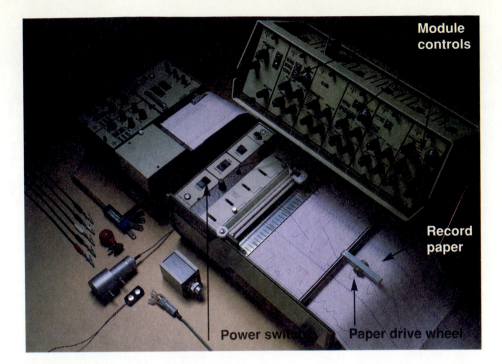

(a)

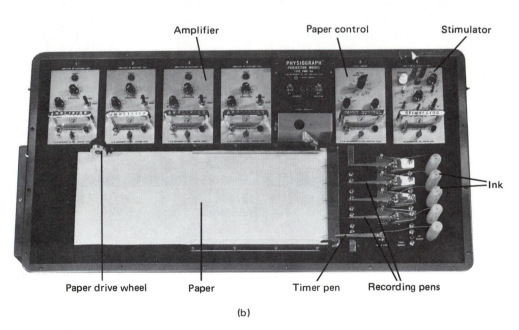

(b)

**Figure A-2** Examples of physiographs: (a) a Gilson Biophysical and Industrial Recorder and its components (courtesy of Gilson Medical Electronics, Inc.); (b) physiograph (courtesy of Nacro Bio-Systems).

**Table A-1   Physiograph Components and Their Functions**

| Physiograph Component | Functions |
|---|---|
| Amplifier | Increases the strength of the electrical signal while keeping the proportional relationship of the electrical signal obtained from the transduced physiological activity. |
| | Amplifiers contain power supply units to convert alternating current (AC) to direct current (DC) and to either increase or decrease voltage requirements. |
| Amplifier controls | The types of control are determined by the function of an amplifier. Controls include<br>1. *Gain control:* adjusts sensitivity of amplifier to the strength of the incoming signal<br>2. *Mode control:* enables instrument to accept either AC or DC input signals<br>3. *Centering (zero) control:* moves the zero line on a recording system in either a positive or a negative position<br>4. *Calibration control:* used for checking or establishing the sensitivity of the amplifier |
| Bellows pneumograph | Photoelectric transducer used for recording respiration |
| Blood pressure transducer | Photoelectric transducer for the measurement of direct blood pressure |
| Coupler | Renders the input energy from the transducer (or the signal) from the subject compatible with the channel amplifier |
| Electrode | Device used to carry an electrical current into or away from a biological specimen |
| Myograph | Transducer used primarily for the quantitative measurement of muscular contractions |
| Oscilloscope (cathode ray tube) | Output transducer that provides a visual display of voltage signals on a fluorescent screen |
| Power supply (for amplifier) | Converts alternating current (AC) to direct current (DC); either increases or decreases the voltage needed by the amplifier |
| Recording device | Converts electrical energy into rotary motion; receives electrical signals from the amplifier and drives the pen across a moving chart, thereby providing a record of the event being measured |
| Stimulator | Device used to produce electrical shocks of desired voltages in experimental situations |
| Stylus | Writing instrument used to indicate time, beginning and ending of an event, etc. |
| Thermistor | Transducer that monitors temperature changes |
| Transducer | Converts a physiological event into a proportional electrical signal; the most sensitive components in a transducer are a light bulb and photoelectric cell |

dealing with muscle measurements of laboratory animals.

## Identification of Physiograph Components

Although physiographs from different manufacturers vary in appearance or design, they generally consist of the same basic components. The parts of a physiograph and their functions are described in Table A-1. Use this information as you examine and locate the parts of the physiograph.

## Materials

The following materials should be provided for class examination and demonstration:

1. Physiographs
2. Accessory devices and attachments, including those used for ECG, pulse detection, respiratory measurements, and electromyograms
3. Reference diagrams or instruction manuals

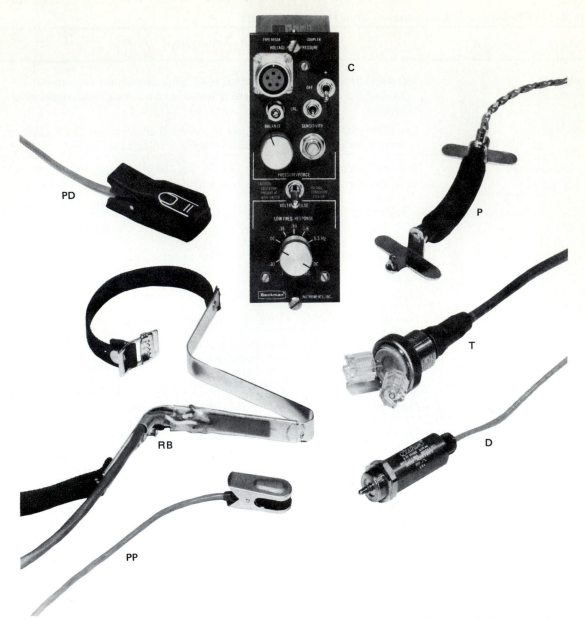

**Figure A-3** Various accessories used in physiological measurement. The display includes a coupler component (C), bellows pneumograph (P), general-purpose transducer (T), ECG accessory pulse-rate detector (D), ear/nose transducer for pulse pickup (PP), respiration belt using a strain gauge bridge (RB), and photoelectric pulse detector (PD). (Courtesy of Sensor Medics Corporation.)

## Procedure

This procedure is to be performed by students individually.

1. Examine the physiograph and accessories provided. With the aid of Figures A-1 and A-2, locate and learn the functions of each component listed in Table A-1, and others indicated by your instructor.

2. Your instructor will indicate the following control and other appropriate switches associated with the physiological functions:

a. Turning on the instrument
b. Changing the paper speed
c. Increasing the intensity of electrical signal or recording
d. Recording the time
e. Recording the event
f. Balancing the transducer
g. Increasing the amplifier sensitivity
h. Stopping or starting the chart movement

# Laboratory Animal Preparation

## Procedure: Pithing

This procedure is to be performed by students in pairs.

1. Hold the frog in your left hand, with its hind limbs extended and its dorsal surface facing toward you (Figure B-1a).
2. Place your forefinger over the frog's snout (nose), and bend the frog's head downward until it forms a sharp angle with the trunk of its body. Use your thumb and remaining fingers to hold the frog's limbs securely.

3. Take the dissecting needle in your right hand and run the tip down the midline of the frog's skull, to locate a depression or groove between the skull and the first vertebra. This depression is about 2–3 mm behind the posterior border of the animal's eardrums (Figure B-1a). It marks the location of the *foramen magnum,* which is a large opening of the skull, through which the spinal cord emerges.
4. Push the needle in a forward and downward direction with a sharp movement through the skin, the depression, the underlying tissues, and the foramen

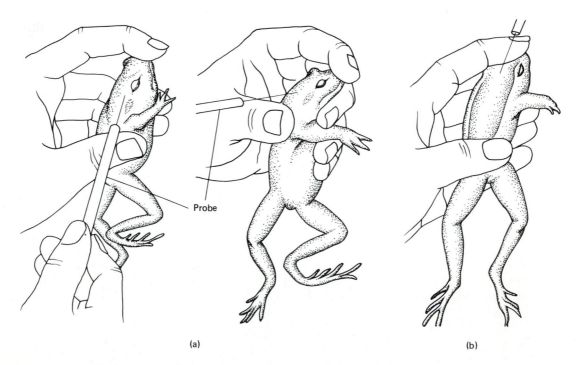

(a)

(b)

**Figure B-1** Pithing: (a) the single-pithing procedure—shows the holding position for the process, and the location of the foramen magnum of the skull (note the bent appearance of the legs); (b) the double-pithing procedure—shows the insertion of the dissection needle into the spinal cord, which causes the frog's legs to extend and stiffen initially.

magnum into the brain. Twist and turn the needle, to destroy the brain. Remove the needle, and stop any bleeding by pressing a gauze pad over the depression. This completes the *single-pithing process.*

5. To double-pith the frog, insert the needle again between the front vertebra and the skull, but in a downward or caudal direction into the spinal cord.

Twist and turn the needle to destroy the spinal cord (Figure B-1b).

The legs of the frog should extend during Step 5, indicating entry of the needle into the spinal cord. Successful double pithing results in the animal's limbs becoming limp soon thereafter. The frog is now ready for procedures requiring pithed animals.

# Laboratory Animal Blood-Specimen Preparation Procedures

When living animals are to be used in obtaining blood specimens, care always must be taken to avoid causing unnecessary irritation or injury.

## Materials

The following items should be provided for the bleeding procedures:

1. One live healthy laboratory mouse or rat
2. One pair of animal handling gloves (optional)
3. One mouse or rat restrainer (The size of the restrainer is determined by the weight of the laboratory animal to be used; the restrainer should be made of a transparent plastic material, such as acrylic, to allow for maximum visibility.) (Acrylic restrainers are obtainable from Harvard Bioscience, An Ealing Division, Pleasant Street, South Natick, MA 01760.)
4. One pair of sterile scissors (for mice)
5. Sterile, individually wrapped blood stylets (for larger animals, such as rats)
6. Sterile cotton or gauze pads
7. Alcohol prep-pads or similar sterilizing gauze pads (obtainable from MED PAK Corporation, Poco, WY 25159)
8. Slides or test tubes with sterile saline, in which to collect blood specimens
9. Container with disinfectant for the disposal of blood-containing items.

## Procedure

This procedure is to be performed by the instructor or by students under the direct supervision of an instructor.

1. Carefully remove the laboratory animal from its cage or container by grasping it firmly behind the neck. If the animal is a laboratory rat, a pair of animal gloves should be worn.
2. Insert the animal into the restrainer (see Figure C-1).
3. Position the adjustable tail plate in the restrainer so that the animal's tail is exposed.
4. Unwrap an alcohol-containing prep-pad package, remove the alcohol pad, and wipe the end of the laboratory animal's tail. Allow the tail to dry.
5. If the laboratory animal to be used is a rat, proceed to Step 8; if a mouse, take a sterile pair of scissors, and—while squeezing the tail—snip off approximately ⅛ inch. Continue to squeeze the tail until drops of blood appear.
6. The blood may be used to make blood smears directly or may be allowed to drip into a test tube containing sterile saline. In the event that blood drops do not appear with the initial cut of the mouse's tail, an additional ⅛ inch of tail material can be removed.
7. After enough blood has been obtained, pressure should be applied with a sterile gauze pad, to stop any unnecessary bleeding. Once the bleeding has

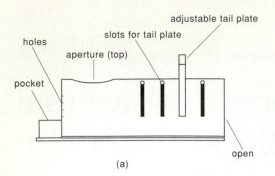
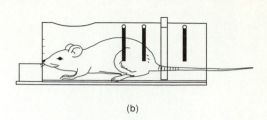

**Figure C-1** Obtaining blood specimens from laboratory animals (mice or rats): (a) an example of an animal restrainer; (b) a mouse positioned in a restrainer, in order to obtain a blood specimen from its tail.

been stopped, the animal should be returned to its cage. This procedure should provide enough blood for 4–6 blood smears. Additional mice should be used if more smears are needed. This is the end of the procedure when using mice.

8. If the laboratory animal to be used is a rat, use a sterile stylet to puncture the animal's tail near the tip (about 1 inch from the end). Rubbing the area with a fresh gauze pad will help to bring blood to the surface of the tail. Squeeze the tail until drops of blood appear. The blood may be used for the same types of specimens as mentioned for the mouse (see Step 6).

9. After enough blood has been obtained, pressure should be applied with a sterile gauze pad, to stop any unnecessary bleeding.

10. With the aid of laboratory gloves, carefully remove the rat from its restrainer, and return the animal to its cage.

# Color Atlas

Plate 1  The letter "e" under low power.

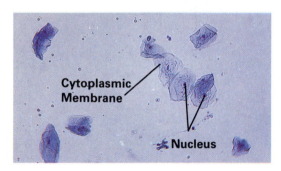

Cytoplasmic Membrane

Nucleus

Plate 2  Cheek cells: Note the presence of a nucleus and a cytoplasmic membrane in these cells.

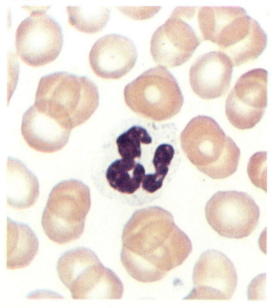

Plate 3  A human blood smear showing erythrocytes (red blood cells) and a neutrophil.

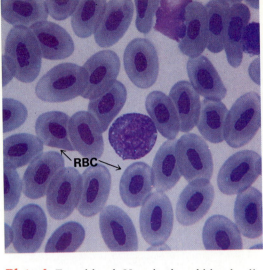

RBC

Plate 4  Frog blood: Here both red blood cells (RBCs) and white blood cells are nucleated. Compare this photo with that of human blood cells.

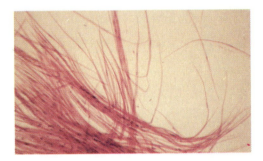

Plate 5  A preparation of smooth muscle cells. (Courtesy Carolina Biological Supply Company.)

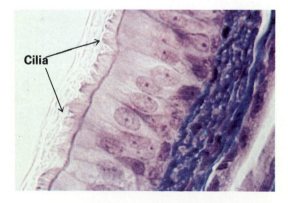

Cilia

Plate 6  A view of tracheal epithelium: Pseudostratified epithelium bearing cilia and underlying connective tissue are shown in this micrograph. (Courtesy Carolina Biological Supply Company.)

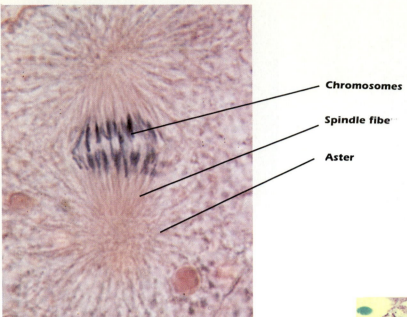

Chromosomes

Spindle fibe

Aster

**Plate 7** The appearance of an aster: Note the strand composition of the cellular component.

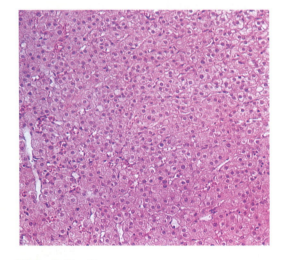

**Plate 8** The liver.

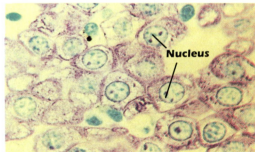

Nucleus

**Plate 9** DNA and RNA in animal cells, stained with Azure B: The DNA in the nucleus appears bluish-green and the RNA in the smaller, nucleolus appears purple. (Courtesy of Ward's Natural Science Establishment, Inc.)

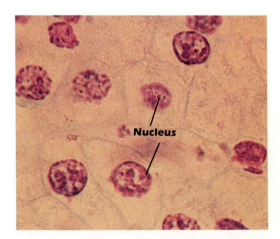

Nucleus

**Plate 10** Feulgen reaction showing the purple appearance of DNA in cell nuclei.

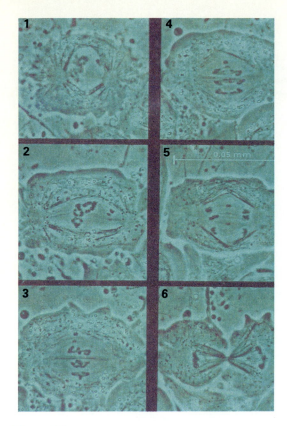

**Plate 11** A living cell undergoing mitosis as seen under phase contrast microscopy.

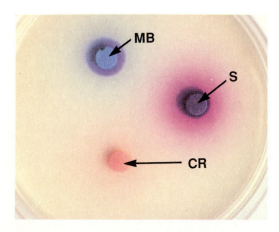

**Plate 12** Dye diffusion zones: crystal violet (CR), methylene blue (MB), and safranin (S).

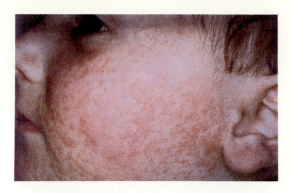

**Plate 13** The appearance of a facial rash.

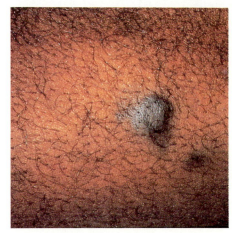

**Plate 14** Kaposi's sarcoma, a tumor associated at times with the disease acquired immune deficiency syndrome (AIDS).

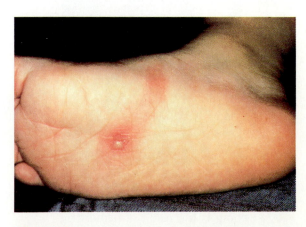

**Plate 15** The features of a local infection.

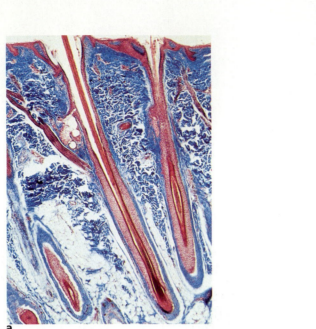

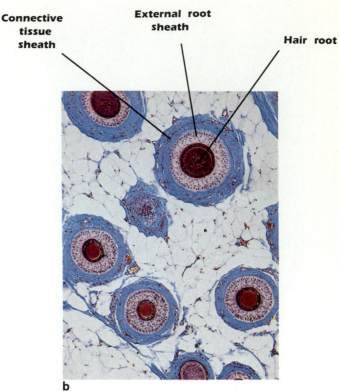

Connective tissue sheath

External root sheath

Hair root

**Plate 16** Human scalp, showing the features of hair follicles: (a) longitudinal section—particular structures included here are the hair follicle, sebaceous glands, and arrector pili; (b) this transverse section shows the central cortex, the surrounding cuticle layer, internal root sheath, external root sheath, and the connective tissue sheath. (Courtesy of Fisher Scientific Company.)

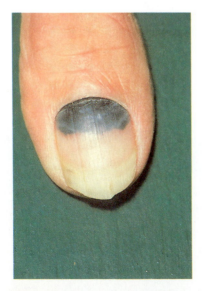

**Plate 17** The blackening of a nail due to a physical injury. The blackening is the result of hemorrhaging.

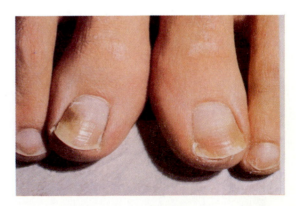

**Plate 18** The discoloration of the nails shown is a side effect of certain medications.

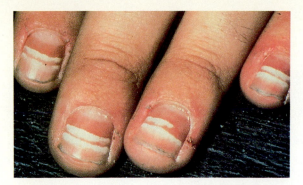

**Plate 19** The white lines are associated with a genetic factor.

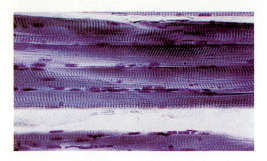

**Plate 20** Skeletal muscle. A longitudinal section. (Courtesy Carolina Biological Supply Company.)

**Plate 21** A cross-section of skeletal muscle. (Courtesy Carolina Biological Supply Company.)

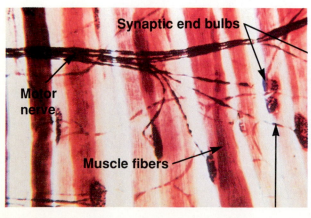

Synaptic end bulbs

Motor nerve

Muscle fibers

**Plate 22** The parts of a motor end plate.

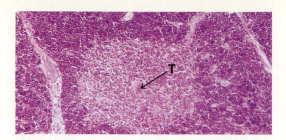

T

**Plate 23** The thymus gland: Both the lighter inner medulla region and the outer cortex region of a lobule are shown. A centrally located thymic (T) corpuscle also is present.

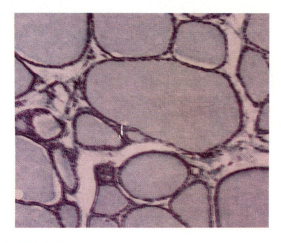

**Plate 24** Appearance of cells in a normal thyroid gland.

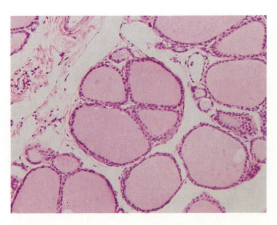

**Plate 25** Appearance of cells in a thyroid gland specimen from a goiter.

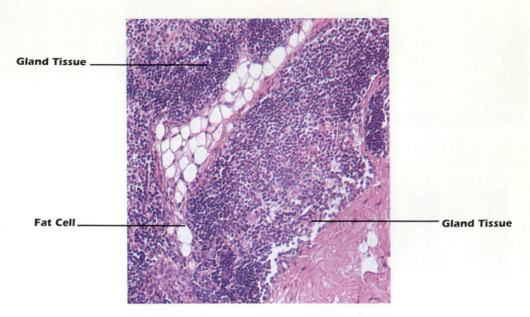

Plate 26 Cellular structure of a portion of the parathyroid.

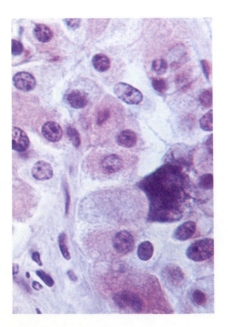

Plate 27 A higher magnification of the cells in the parathyroid gland.

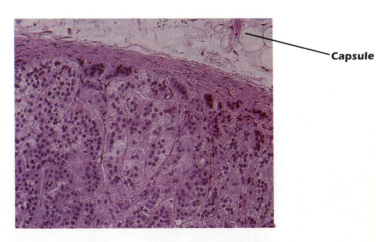

Plate 28 The adrenal gland: Capsule and the zone glomerulosa. This zone contains secretory cells, arranged in irregular ovoid clumps, separated by connective tissue strands.

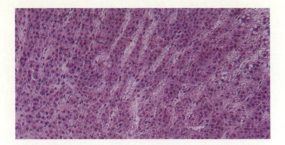

**Plate 29** The zona fasciculata, the intermediate and broadest of the three zones of the adrenal cortex. This zone consists of narrow cords of large secretory cells, with lightly staining cytoplasm.

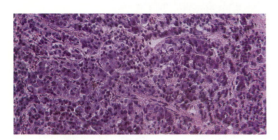

**Plate 30** The adrenal medulla consists of closely packed, irregular clumps of secretory cells. These cells have large granular nuclei and dark-blue- to purple-staining cytoplasm.

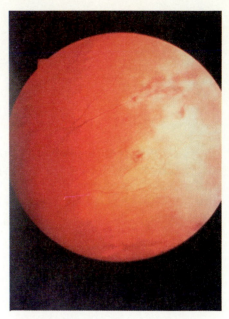

**Plate 32** Actual view of the retina.

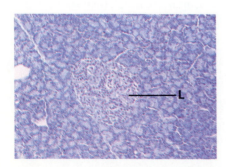

**Plate 31** The pancreas: The centrally located islets of Langerhans (L) and the surrounding acini are shown.

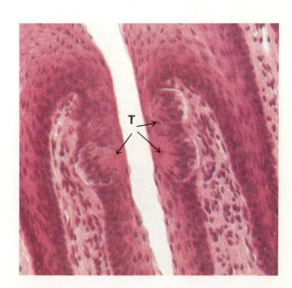

**Plate 33** Microscopic view of taste buds (T).

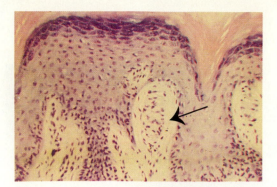

**Plate 34** Meissner's corpuscles are small, oval, encapsulated sensory organs located in the dermal papillae immediately beneath the skin epithelium. (Courtesy of Ward's Natural Science Establishment, Inc.)

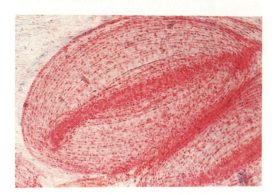

**Plate 35** The pacinian corpuscle, an encapsulated nerve ending associated with the sensations of touch and pressure. Dendritic endings are enclosed within many onionlike layers of connective tissue. Such layers absorb some of the pressure when the stimulus is applied and maintained.

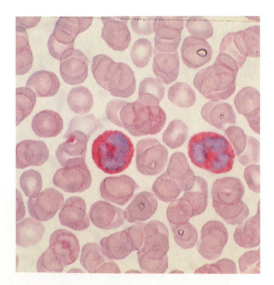

**Plate 36** A blood smear, showing eosinophils.

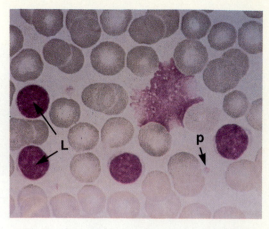

**Plate 37** A number of lymphocytes (L) and platelets (P) in a blood smear.

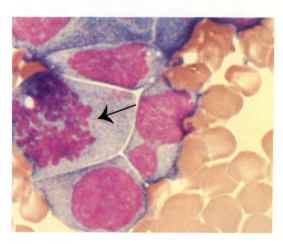

**Plate 38** A bone-marrow preparation, showing the features of an acute monocytic leukemia. A cell in mitosis (arrow) can be seen in this smear. Bone marrow is usually obtained by aspiration, which involves inserting a needle into the marrow cavity and withdrawing a sample. The sternum, ilium, iliac crest, and vertebral spinous process are the most common sites used for aspiration. (Courtesy of Dr. I. Mirchandani, Wayne State University.)

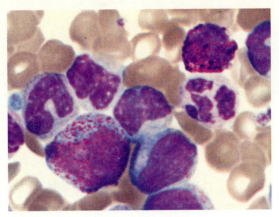

**Plate 39** Chronic granulocytic leukemia. (Courtesy of Dr. I. Mirchandani, Wayne State University.)

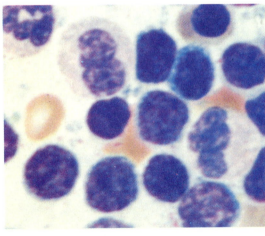

**Plate 40** Chronic lymphocytic leukemia. (Courtesy of Dr. I. Mirchandani, Wayne State University.)

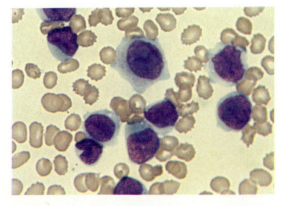

**Plate 41** Acute myelomonocytic leukemia. (Courtesy of Dr. I. Mirchandani, Wayne State University.)

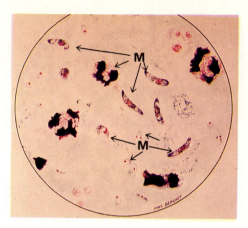

**Plate 42** Stages of the malarial (M) parasite *Plasmodium falciparum*.

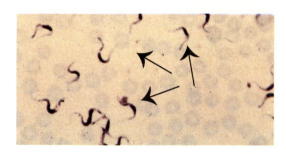

**Plate 43** The causative agents (arrows) of African sleeping sickness.

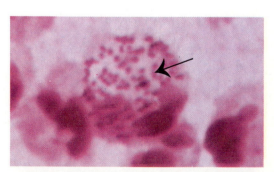

**Plate 44** The intracellular location of *Neisseria gonorrhoeae* (arrow), the causative agent of the sexually transmitted disease gonorrhea.

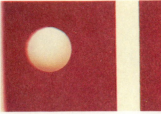

**Plate 45** A Tallquist reaction showing a normal hemoglobulin content.

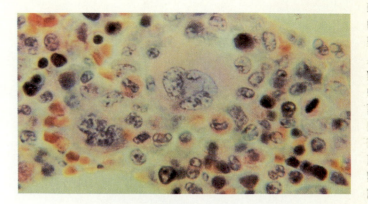

**Plate 46** Bone-marrow smear.

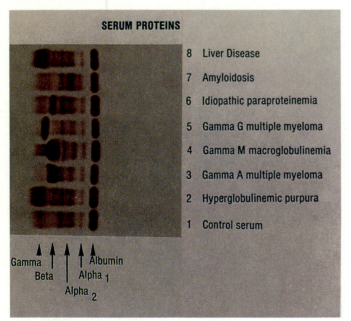

SERUM PROTEINS

8 Liver Disease

7 Amyloidosis

6 Idiopathic paraproteinemia

5 Gamma G multiple myeloma

4 Gamma M macroglobulinemia

3 Gamma A multiple myeloma

2 Hyperglobulinemic purpura

1 Control serum

Gamma Beta Alpha₂ Alpha₁ Albumin

**Plate 47** An electrophoretic pattern, showing the appearance of serum proteins of a normal specimen, cases of liver disease, and other abnormal conditions.

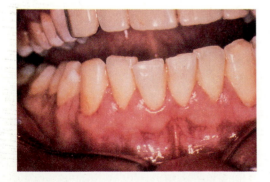

**Plate 48** A severe form of the gum disease, gingivitis.

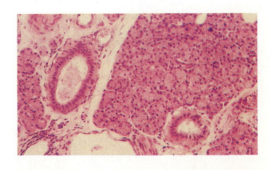

**Plate 49** The submandibular salivary gland.

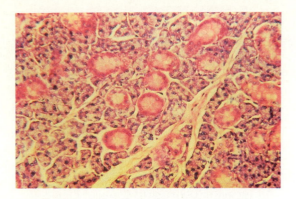

Plate 50 The submaxillary salivary gland.

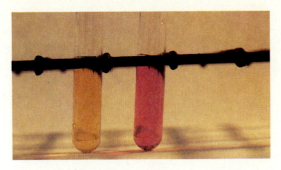

Plate 53 The appearance of a positive (purple) Biuret test.

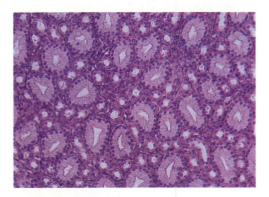

Plate 51 The parotid gland.

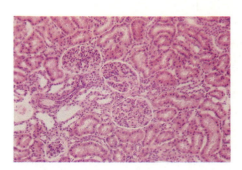

Plate 54 A microscopic view of the kidney.

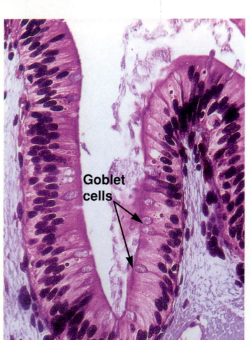

Goblet cells

Plate 52 A portion of intestine showing secretory cells (goblet cells) and columnar cells.

Plate 55 The results of a Benedict's test. Left tube shows no glucose, while the one in the center and the one on the right clearly contain glucose.

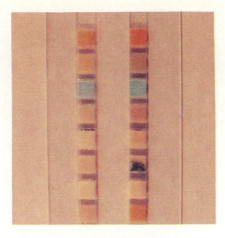

Plate 56 Multistix™ reactions: The strip on the left is normal.

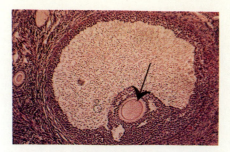

Plate 58 Human ovary showing a follicle (arrow) in the first stages of growth. (Courtesy of Carolina Biological Supply Company.)

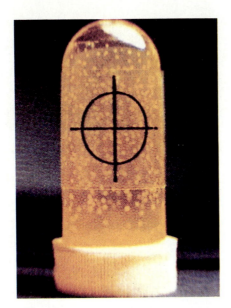

Plate 57 Bacturcult results showing a bacterial count under 100,000/ml. Note the appearance of the counting strip.

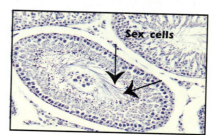

Plate 59 Microscopic view of human testes.

# Index